THE STEPHEN BECHTEL FUND

IMPRINT IN ECOLOGY AND THE ENVIRONMENT

The Stephen Bechtel Fund has

established this imprint to promote

understanding and conservation of

our natural environment.

The publisher gratefully acknowledges the generous contribution to this book provided by the Stephen Bechtel Fund.

BIRDS OF THE SIERRA NEVADA

BIRDS *of the* SIERRA NEVADA

Their Natural History, Status, and Distribution

Edward C. Beedy and Edward R. Pandolfino

Illustrated by Keith Hansen

Foreword by Rich Stallcup

UNIVERSITY OF CALIFORNIA PRESS
Berkeley   Los Angeles   London

University of California Press, one of the most distinguished university presses in the United States, enriches lives around the world by advancing scholarship in the humanities, social sciences, and natural sciences. Its activities are supported by the UC Press Foundation and by philanthropic contributions from individuals and institutions. For more information, visit www.ucpress.edu.

University of California Press
Berkeley and Los Angeles, California

University of California Press, Ltd.
London, England

Library of Congress Cataloging-in-Publication Data

Beedy, Edward C.
  Birds of the Sierra Nevada : their natural history, status, and distribution / Edward C. Beedy and Edward R. Pandolfino ; illustrated by Keith Hansen.
     pages cm
  Includes bibliographical references and index.
  ISBN 978-0-520-27493-8 (cloth : alk. paper) — ISBN 978-0-520-27494-5 (pbk. : alk. paper)
  1. Birds—Sierra Nevada (Calif. and Nev.) 2. Birds—Sierra Nevada (Calif. and Nev.)—Identification. I. Pandolfino, Edward R. II. Title.
  QL683.S54B444 2013
  598.097944—dc23

                              2012043369

Manufactured in China
19 18 17 16 15 14 13
10 9 8 7 6 5 4 3 2

The paper used in this publication meets the minimum requirements of ANSI/NISO Z39.48-1992 (R 2002) (*Permanence of Paper*). ♾

Cover illustration: Flock of "Sierra Nevada" Gray-crowned Rosy-Finches (*Leucosticte tephrocotis dawsoni*) in an Alpine setting. Painting by Keith Hansen.

TO STEVEN P. MEDLEY (1949–2006),
A FRIEND OF SIERRA BIRDS,
WHO INSPIRED AND SUPPORTED
OUR EFFORTS TO CREATE THIS BOOK

CONTENTS

FOREWORD

Defining avifaunal boundaries for the Sierra Nevada is not an easy task. The long, rolling hills west of the crest and the steep eastern escarpment are obvious features in the center of the range, but as the mountainous highlands dwindle in southern California and vaguely become the Cascades in the north, drawing appropriate lines becomes more difficult. The authors of this book have done a masterful job in this regard by including nearby Great Basin habitats at the base of the Sierra on the East Side.

The species accounts are thorough and scholarly. Each contains etymology (the origin of the bird's common and scientific names), and expansive, very informative sections on natural history, status, and distribution. Population dynamics and changes in distribution are clearly presented. This was made possible by the authors' careful scrutiny of Christmas Bird Count, Breeding Bird Survey, and eBird data. I have birded frequently in all the Sierra special habitats for more than forty years and still learn something new from each account. Although this entirely new book is not a field guide that rehashes identification clues, the splendid, color paintings of each species (presented on-page with each species account) by Keith Hansen should leave no one wondering what kind of bird he or she has seen. The abundance of illustrations and precision of every detail is simply amazing. How does he do that?

John Muir, following an eloquent but spirited discussion of the Water Ouzel (American Dipper), said: "And so I might go on, writing words, words, words; but to what purpose? Go see him [the ouzel] and love him, and through him as through a window look into Nature's warm heart." So get out there and up there, but go with this book in your pack or car. If you are already a birder, you will see in new light, but if you are new to the magic and wonder of birds, *beware*—your life may change in amazing ways. Sharper long-distance vision, unthinkable intellectual challenges, and the urge to leave work and go bird the mountains are but some of the more common maladies.

RICH STALLCUP (1944–2012),
PRBO CONSERVATION SCIENCE

PREFACE

A day spent in the Sierra paying close attention to every aspect of landscape, weather, plants, and wildlife can be timeless, a healthy bit of immortality captured in a single day. Or, as John Muir put it: "Another glorious Sierra day in which one seems to be dissolved and absorbed and sent pulsing onward we know not where. Life seems neither long nor short, and we take no more heed to save time or make haste than do the trees and stars. This is true freedom, a good practical sort of immortality."

The Sierra, Muir's "Range of Light," is rightly noted for its spectacular landscapes, but for the visitor who takes time to watch and listen, even the most unassuming corner is filled with wonders, delights, and surprises. The mountains hum with activity of all sorts of animals, but none are as readily observed and enjoyed as the birds. Nearly 300 species are regular visitors, and each has its own unique story to tell about where it came from, where it is going, and how it uses these mountains. And these stories are not static. In the past several decades, new species of birds have colonized the Sierra, some species have nearly vanished, populations of others have grown and spread, while some have dwindled.

We wanted to create a book that would update our knowledge of all the birds of the Sierra and would enhance and deepen the experience of a day in this range for the serious ornithologist as well as the casual hiker. We wanted to take full advantage of the latest research that takes us deeper into the natural history of these birds. We also wanted to reap the benefit of the much finer scale of status and distribution information now available. The number of people who have the skills to find and identify these birds has increased greatly and with that, our knowledge of behaviors and ranges of Sierra birds has likewise expanded.

Words alone could never do justice to the wonder and diversity of these birds. Therefore, while not making any attempt to create a "field guide," we chose to prepare a richly illustrated book showing all the regularly occurring birds of the Sierra. We went well beyond our own personal experience and tapped into the priceless local knowledge of birders and ornithologists who live, work, and play in the Sierra. Our hope is that readers will find that time spent with this book alters the way they experience a day in the Sierra, helping them find, in that day, "true freedom, a good practical sort of immortality."

EDWARD C. BEEDY
EDWARD R. PANDOLFINO
KEITH HANSEN

ACKNOWLEDGMENTS

In a region as vast and dynamic as the Sierra Nevada, no one could ever acquire or maintain detailed, up-to-date knowledge of the status, distribution, and natural history of every bird. To make this book as current and accurate as possible, we relied heavily on the experience and boundless generosity of the following local and regional experts and avid Sierra birders: Dan Airola, Elizabeth Ammon, Bob Barnes, Matt Brady, Ryan Burnett, Walt Carnahan, Mark Chichester, Rudy Darling, Katy Delaney, David DeSante, Bruce Deuel, Colin Dillingham, Jon Dunn, Todd Easterla, Mary Jo Elpers, Jim Estep, Tom Hahn, Aaron Haiman, Rob Hansen, Matt Heindel, Tom and Jo Heindel, Josh Hull, Rodd Kelsey, Barney Kroeger, Jack Laws, Kelli Levinson, John Lockhart, Jim Lomax, Tim Manolis, Guy McCaskie, Chris McCreedy, Mac McCormick, Joe Medley, Bob Meese, Martin Meyers, Bartshé Miller, Joe Morlan, Kristie Nelson, Will Richardson, Deren Ross, Susan Sanders, Dave Shuford, Rich Stallcup, Susan Steele, John Sterling, Brad Stovall, Jerry Tecklin, Phil Unitt, Bruce Webb, Bud and Margaret Widdowson, John Wilson, and Bill Yeates. We also thank all those who participated in any of the 44 Breeding Bird Surveys or 25 Christmas Bird Counts we used to detect and validate population trends of Sierra birds. Any mistakes that remain are entirely the responsibility of the authors.

We are indebted to Peter Pyle, Steve Howell, David Sibley, and David DeSante for their careful reviews of Keith Hansen's art. We thank Steve Beckwitt for preparing the maps, which deftly balance clarity and detail. Elliot Minner performed the digital magic needed to put the paintings into the format needed for this book. Katrina Beedy spent many hours helping us to prepare the index. Tim Messick's botanical expertise was essential for defining the various ecological zones and bird habitats of the Sierra.

Steve Granholm, coauthor of *Discovering Sierra Birds*, offered his generous and enthusiastic support for using ideas and text from that original book. We thank Shirley Beedy, Bill and Lorraine Dicke, Burr Heneman, and Janet Visick for reviewing portions of the manuscript and improving the quality of our prose.

Dan Airola, Bob Barnes, Chris Conard, Kathleen Lynch, Martin Meyers, Kristie Nelson, Dave Quady, Phil Robertson, Susan Sanders, Rich Stallcup, and John Sterling spent many hours patiently reviewing earlier versions of the manuscript and making suggestions, which improved our book immeasurably. We also benefited from thoughtful peer reviews by David

DeSante, John Marzluff, Will Richardson, and David Shuford.

Barbara Moulton expertly guided us through the process of finding the best publisher for this book, and Phyllis Faber and Karl Olson offered excellent guidance about publication. We are deeply grateful to Blake Edgar, Kate Hoffman, Lynn Meinhardt, and Chuck Crumly of the University of California Press and to David Peattie at BookMatters and Amy Smith Bell for their help, patience, and steady support throughout.

We are indebted to the David and Lucile Packard Foundation for its generous grant in support of Keith's artwork, and to Cole Wilbur and Mark Valentine for their help and guidance in securing this grant. We also appreciate the efforts of the Yosemite Association, and later the Yosemite Conservancy, for managing these funds. Thanks to Keith's friends at the Bolinas Museum who nourished him with support, on many levels, for more than 20 years. Generous grants from Sierra Foothills Audubon Society and Sacramento Audubon Society were also much appreciated.

Finally, we acknowledge the essential and loving support of our wives, Susan Sanders, Kathleen Lynch, and Patricia Briceño, who tolerated the many hours we spent hunched in front of computers, or over drawing boards, and understood our need to take all those trips into the Sierra to conduct our required "research."

Introduction

Our goal in creating *Birds of the Sierra Nevada* is to offer a beautifully illustrated and user-friendly book to everyone who is interested in Sierra birds. We hope that the book will enrich the experiences of visitors to the Sierra who want to know about the birds they see, inform Sierra residents who want a deeper understanding of the birds they observe daily, and engage the interest of serious ornithologists who want detailed, up-to-date, and well-researched information about Sierra birds.

The origins of this book date back to 1985, when *Discovering Sierra Birds* was published jointly by the Yosemite Natural History Association (YA) and the Sequoia Natural History Association. That book was out of print by the mid-1990s, and in 1998 the YA president, Steve Medley (to whom this book is dedicated), asked us to revise and expand on that book with broader geographical and species coverage; a stronger focus on status, distribution, and conservation of Sierra birds; and new artwork. What started as a revision of that book has evolved into an entirely new volume. This compilation and distillation of our own storehouses of bird knowledge, and those of scores of other birders and naturalists, has contributed to the species accounts and illustrations. Our hope is that this book stimulates greater interest in the birds and natural history of the Sierra and inspires some readers to work for broader protection of its remaining wild areas.

BOUNDARIES AND SUBREGIONS

Defining exact boundaries for the Sierra ultimately requires making some arbitrary choices. One could use characteristics like soil types or plant communities to define borders, but such features are not always helpful for observers on the ground. Instead, we wanted to use boundaries that were easy to understand and identify for birders and other natural historians. We also wanted to define the Sierra in a broad sense, including areas east of the crest that, although not necessarily in the heart of the range, are

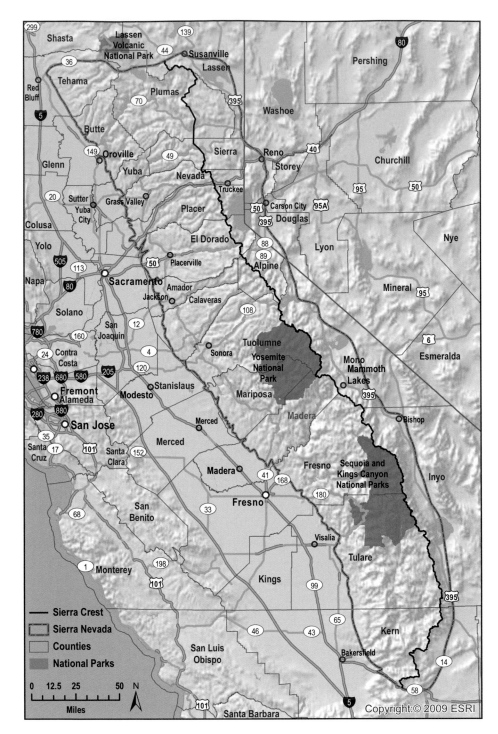

MAP 1 Political and road map

strongly influenced climatically and biologically by this great range.

For the purposes of this book, we define the Sierra as extending from Highway 36 (near Lake Almanor in Plumas County) in the north to Highway 58 (Kern County) in the south. The western border follows the 500-foot elevation contour except for a small portion south of Porterville, where it follows U.S. Forest Service ecological zone boundaries, rising to approximately 1,200 feet at Highway 58 (see Map 1). The eastern border is roughly defined by Highways 395 and 14 but also includes large East Side lakes, reservoirs, and wetlands that attract huge numbers of Sierra birds and that are immediately adjacent to the region (e.g., Mono Lake, Bridgeport Reservoir, Topaz Lake, Carson Valley, and Honey Lake) and the Owens Valley floor. We define the West Side as the region west of the Sierra crest down to the 500-foot contour, and the East Side as the region east of the crest roughly to Highway 395. While this definition of West Side and East Side is mostly consistent with watershed boundaries (i.e., West Side waters flow west and East Side waters flow east), a few watersheds do not follow this general rule. For example, Sierra Valley (Plumas and Sierra Counties) is within the west-flowing Feather River watershed, is east of the crest, and includes biotic communities associated with the East Side. Therefore, from a bird habitat perspective, Sierra Valley really belongs to the East Side.

The Sierra's accessibility makes observing birds especially easy and rewarding. Public lands are plentiful, with huge pristine areas preserved in Yosemite, Sequoia and Kings Canyon National Parks, and numerous National Forest Wilderness Areas. Major all-weather highways, such as Interstate 80 and U.S. 50, transect the range, as do many state routes; other highways, such as Highway 49, run mainly north and south through the foothills. In a morning's drive, one can traverse the West Side from California's Central Valley nearly up to tree line at Sonora Pass on Highway 108 or Tioga Pass on Highway 120 through Yosemite National Park. A long day's drive down Highway 395 takes one through most of the habitats characteristic of the East Side.

NOMENCLATURE, TAXONOMY, AND SUBSPECIES

We group species accounts into their respective families. The order of families and species and all common and scientific names follow the American Ornithologists' Union's *Check-list of North American Birds* (1998 [7th edition], plus all changes up to the 52nd supplement published in the *Auk* [2011, volume 128]). Howell et al. (2009) proposed using a standardized species order for field guides that would be more useful for identification purposes and would not change with each revision of the official *Check-list*. While we agree that a standardized order is appropriate for field guides, where identification is the main purpose, we think it important to use the most current taxonomic order for publications that focus on natural history. Embedded in this taxonomic sequence is our best current knowledge about the interrelationships and evolutionary history of each species. Thus taxonomic order is inherently an element of natural history and should be used in this context.

We discuss subspecies occurring in the Sierra when we have significant information about their status and distribution (such as differing winter versus summer populations), when the subspecies are identifiable in the field, or when current research suggests that a species may be split in the future. Because naming and recognition of subspecies is dynamic and sometimes controversial, we only cite subspecies that are widely accepted. We also provide a common name to identify a subspecies when that name is frequently used and widely recognized.

In the "origins of names" sections of the species accounts, we have used the following abbreviations for the derivations of common and scientific names: Anglo-Saxon (AS.), French (Fr.), Greek (Gr.), Italian (It.), Latin (L.), Old English (OE.), Old French (OF.), Old German (OG.), Old Icelandic (OI.), Old Latin (OL.), Spanish (Sp.), and Swedish (Sw.).

SPECIES INCLUDED AND ABUNDANCE CATEGORIES

One of our most challenging decisions was where to draw the line between rare species and the more common ones that required full species accounts. As we write this, 442 species have been observed at least once in the Sierra as we define the region (see Appendix 1 for the complete list). We decided to include full accounts of the most regularly occurring species and identified 276 species in 54 families that met the threshold of being abundant to uncommon as defined below. For all these species, we provide illustrations and family and species accounts. Another 166 species have been seen in the Sierra region but are considered rare, casual, or accidental visitors. Their status and distribution are described briefly in Appendix 2. In all cases, our assessment of relative abundance and seasonal status is based on a combination of our own personal experience, the experience of the many experts we consulted, and data gleaned from Christmas Bird Counts, Breeding Bird Surveys, and eBird. In recent years, both the quantity and quality of eBird data have dramatically improved, allowing us to verify anecdotal experience with quantitative data.

The approximate abundance of each species is described using the categories below. Each category is based on the relative frequency that an experienced birder might expect to see or hear a given species in its favored habitat and in the appropriate season during peak birding hours. These categories reflect the likelihood of detecting a species in a given habitat and season; it may be more or less numerous at any particular site; rare, casual, and accidental species are discussed only in Appendices 1 and 2.

ABUNDANT. Encountered on every day afield, usually many individuals.

COMMON. Encountered on most days afield, sometimes many individuals.

FAIRLY COMMON. One or a few individuals encountered on most days afield.

UNCOMMON. Encountered on relatively few days afield, never in large numbers; often missed unless a special search is made.

RARE. Seldom encountered and often highly localized; at least a few individuals occur in the region in all or most years.

CASUAL. Not encountered in the region in most years, but a pattern of occurrence may exist over many years or decades.

ACCIDENTAL. Encountered in the region on one or a few occasions (<5) and the species is far out of its normal range.

BIRD SEASONS

The lives of birds are tied intimately and inextricably to the passage of seasons, and major changes take place in Sierra bird life as the year progresses. Birds migrate, stake out territories, court, nest, molt, and shift their habitats in response to seasonal cues such as changing day length and weather patterns. Climate varies dramatically from the foothills, with their mild winters and hot, dry summers, through the wetter, cooler middle elevations, up to the vast Subalpine and Alpine zones with frigid, long winters and short summers. Birds living at different altitudes follow radically different yearly schedules, as do birds of different species. Taking such variability into account, the "bird seasons" we have used throughout the book do not follow strict calendar dates, but rather capture seasonal changes from a bird's perspective using the standard definitions from *North American Birds*:

WINTER. December–February

SPRING. March–May

SUMMER. June–July

FALL. August–November

ILLUSTRATIONS

Because this book is not intended to be used as a field guide, we did not attempt to include illustrations of all plumages of every bird. Most illustrations are of adult birds, except where indicated otherwise.

BIRD FINDING THE SIERRA

Although this book is not intended as a bird-finding guide, the species accounts include many examples of when and where to find particular species in the Sierra. The first section of the bibliography includes two excellent sources for more detailed information: Kemper 1999 and Schram 2007. Joe Morlan's "California Birding Pages" (http://fog.ccsf.cc.ca.us/jmorlan/) also includes links to information on hundreds of locations, with the very latest tips on where to find birds in California, including the Sierra. Bruce Webb moderates an electronic discussion list (http://groups.yahoo.com/group/sierra-nevadabirds/) that provides almost daily updates on Sierra bird observations, including rarities.

CITATIONS OF PUBLISHED SOURCES

Because we wanted to create a book that would appeal to a broad spectrum of readers, we decided not to include full citations within the text in every case where we include information from a published source. Instead, we provide a full bibliography of all published references we consulted while writing this book, organized by topics and family groups. In most cases, it will be clear which facts are based on which source. For natural history we relied heavily on our own experience and information published in *Birds of North America*, available from Cornell Laboratory of Ornithology, both in print and online through subscription (http://bna.birds.cornell.edu/bna/). Frequently consulted publications (e.g., Gaines 1992; Grinnell and Miller 1944; and Shuford and Gardali 2008) are noted in the first section of the bibliography.

Ecological Zones and Bird Habitats

The Sierra offers an extraordinary variety of bird habitats, from the rolling foothill grasslands, through oak studded savannas and giant conifer forests, up to alpine meadows and chilly, windswept peaks, and over the crest to the lakes, forests, and sagebrush flats of the East Side as well as Joshua tree woodlands of the southern desert regions. No wonder Sierra bird life is so varied! The West Side boasts an elevation gradient unequaled in the 48 contiguous states, spanning nearly 14,000 feet from the lowest foothills to the highest peaks (see Map 2). Most of the Sierra lies west of the divide, and the East Side drops off sharply to the Great Basin. On this steep eastern escarpment, altitudinal vegetation zones overlap extensively, making them less apparent than on the West Side. North of Lake Tahoe the main crest is flanked on both sides by other ridges and the elevation of the crest itself is lower, making the distinction between the western and eastern

Sierra less obvious. The Kern Plateau in southeastern Tulare County is an area that does not fit neatly into the zones described below. It includes ecological elements from both sides and is an area where species normally associated with the East Side (e.g., Pinyon Jay) occur on the West Side, and where other species are found at much higher altitudes than elsewhere in the Sierra (e.g., Lawrence's Goldfinch).

In this book we have recognized seven major ecological zones: Foothill, Lower Conifer, Upper Conifer, Subalpine, Alpine, East Side, and Desert (see Table 1; Map 3). Note that the elevation ranges are approximate and overlap considerably. Local differences in slope, soils, rainfall, and other factors alter the exact range of any ecological zone. As discussed in this chapter, most of these zones include several distinct bird habitats. Sierra watersheds and key locations are shown in Map 4. Common and scientific names of all plant species are provided in Appendix 4.

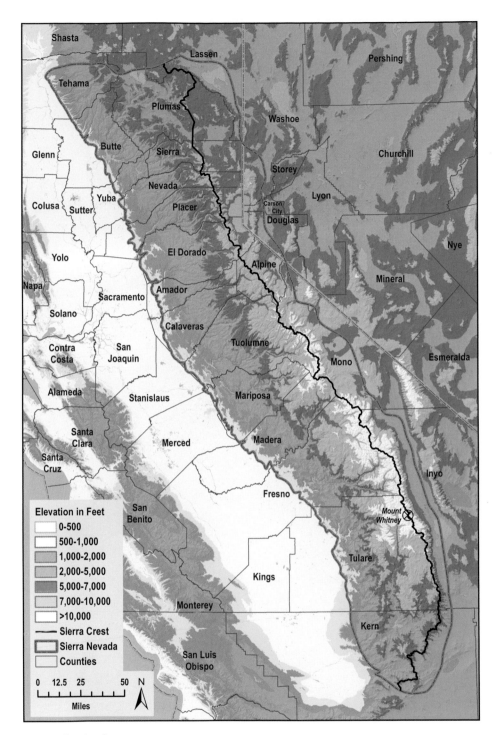

MAP 2 Elevational zones

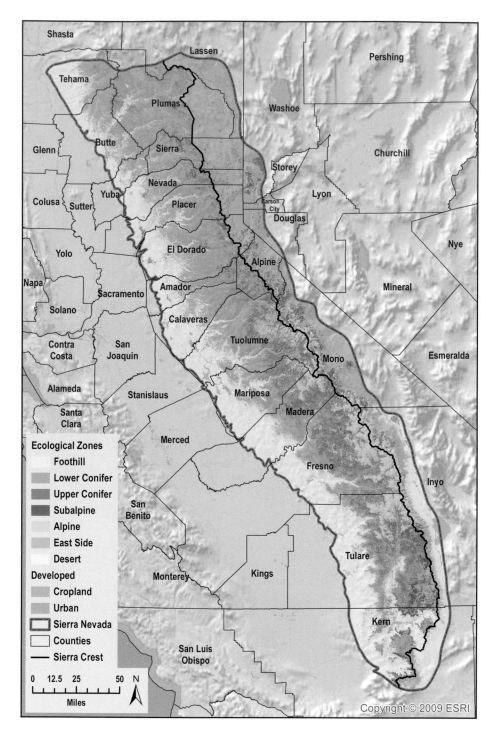

MAP 3 Ecological zones

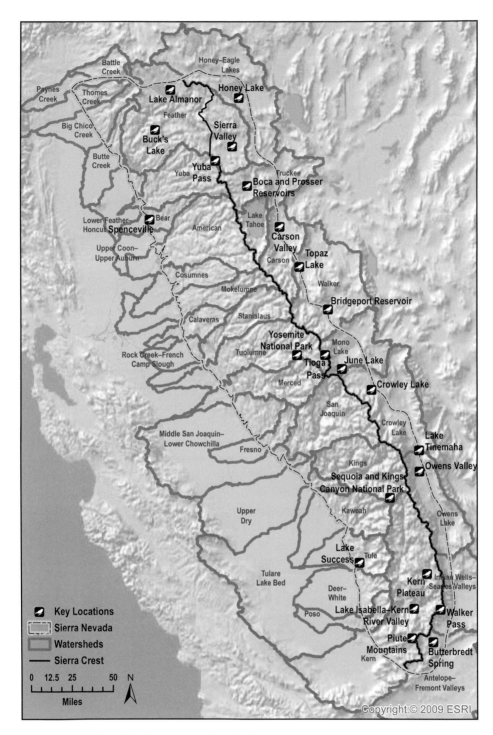

Battle
Creek

Honey–Eagle
Lakes

Paynes
Creek

Thomes
Creek

Honey Lake

Lake Almanor

Feather

Sierra
Valley

Big Chico
Creek

Buck's
Lake

Butte
Creek

Yuba
Pass

Yuba

Boca and Prosser
Reservoirs

Truckee

Lower Feather–
Honcut

Bear

Spenceville

American

Lake
Tahoe

Carson
Valley

Topaz
Lake

Upper Coon–
Upper Auburn

Carson

Cosumnes

Walker

Mokelumne

Bridgeport Reservoir

Calaveras

Stanislaus

Yosemite
National Park

Mono
Lake

June Lake

Rock Creek–French
Camp Slough

Tuolumne

Tioga
Pass

Merced

Crowley Lake

San
Joaquin

Crowley
Lake

Lake
Tinemaha

Middle San Joaquin–
Lower Chowchilla

Fresno

Owens Valley

Kings

Sequoia and Kings
Canyon National Park

Owens
Lake

Upper
Dry

Kaweah

Lake
Success

Tule

Indian Wells–
Searles Valleys

Tulare
Lake Bed

Deer–
White

Kern
Plateau

Walker
Pass

Poso

Lake Isabella–Kern
River Valley

Piute
Mountains

Butterbredt
Spring

Kern

Antelope–
Fremont Valleys

Copyright:© 2009 ESRI

Key Locations

Sierra Nevada

Watersheds

Sierra Crest

0   12.5   25   50   N

Miles

MAP 4 Watersheds and key locations mentioned in the text

TABLE 1

Approximate elevations and total area of Sierra ecological zones

Ecological Zone	Northern Sierra	Southern Sierra	Area (square miles)
Foothill[a]	<500–4,500	<500–5,500	8,337
Lower Conifer[a]	2,000–6,500	2,500–7,000	7,994
Upper Conifer[a]	4,500–8,500	5,000–9,500	1,706
Subalpine[b]	7,500–10,000	8,000–11,500	1,580
Alpine[b]	8,500–10,800	9,500–14,500	2,224
East Side[c]	3,500–8,500	3,000–9,000	4,397
Desert[b]	–	2,000–7,000	1,515
Total Area			27,753

[a] West Side only
[b] Both sides
[c] East Side only

FOOTHILL ZONE

Annual Grasslands

North <500 to 2,000 feet; South <500 to 2,500 feet

Many travelers pass through annual grasslands without registering them as "habitat." This open (less than 10 percent tree cover), gently rolling terrain is parched to a golden brown in summer but transforms to vivid green in late fall through early spring. Patches of this habitat in the Sierra are found in eastern Tehama and Butte Counties and westernmost El Dorado County, but the largest expanses are found from western Tuolumne County south into Kern County. Because no naturalists were present in pre-European times to document the conditions, we do not know what plant species dominated this landscape, and there is considerable controversy about whether perennial grasses or forbs were the most abundant plants. Even before European settlement, Native Americans had been managing these areas with fire for thousands of years. In any case, they are now dominated by introduced grasses brought by European settlers. A high diversity of indigenous plants still survive in vernal pools or intermixed with the non-native species. In many areas woodland and chaparral were cleared to create grazing land and are now annual grasslands. In spring, wildflowers still flourish in the foothills, and specialized blooms form rings around the receding waters of vernal pools. Recent research has shown that cattle grazing—by far the dominant land use in this habitat—is actually beneficial for most native vernal pool plants and most grassland birds, at least partly because grazing keeps aggressively invasive plants like star thistle and medusa head in check and prevents thatch buildup that inhibits growth of many native plants.

Wintering Horned Larks and American Pipits flock together in grazed pastures and plowed furrows, and Savannah Sparrows forage in the deeper grasses, each species constantly wary of the risk of a Prairie Falcon attack. Annual grasslands provide abundant food and cover for high numbers of rodents and other small mammals and therefore support an impressive variety of raptors with winter migrants augmenting resident Red-tailed Hawks and American Kestrels and species like Ferruginous and Rough-legged Hawks visiting only from fall through early spring. In spring, American Kestrels, Western Kingbirds, and Loggerhead Shrikes stake out territories on fence lines or on "tombstones" of ancient metamorphic rock that rise abruptly

FIGURE 1 Foothill zone

here and there. The lovely songs of Western Meadowlarks can be heard almost any month of the year. On hot afternoons Turkey Vultures float lazily above low ridges and hillsides.

Oak Savannas
North < 500 to 3,000 feet; South <500 to 2,500 feet

Savannas dominated by blue oak occupy more than a million acres of the Sierra foothills. These habitats have 10 to 30 percent tree cover and vary from grasslands with a few widely spaced trees to denser stands that may also support interior live oaks, California buckeyes, and occasional gray pines. The extent of oak savanna in the foothills has been reduced through conversion to developed and agricultural land uses, woodcutting, and historical efforts to improve grazing land by clearing trees. Range managers have now learned that having oaks on the land actually improves the forage for cattle by allowing the shaded grasses to retain water.

This quintessentially California landscape is threatened by a widespread lack of natural regeneration of oaks. Although the impacts of grazing have certainly affected the ability of these trees to reproduce, it has been shown that simply removing grazers does not aid regenera-

tion. Other factors, like competition with non-native plants for scarce water (mainly at the seedling or sapling stages), are also critical to oak survival. In fact, removing grazers entirely can inhibit oak regeneration because of the deep thatch and explosion of invasive weeds that usually follows. Encouraging results have been seen with a combination of protection of seedlings and saplings from grazing and management of competing vegetation, followed by a well-managed grazing regimen. Western Scrub-Jays have historically been the main agents of oak regeneration because acorns they bury but fail to recover have a good chance of germinating; since acorns do not roll uphill, jays and squirrels must move them there. Oaks have evolved a boom-and-bust cycle of acorn production to guarantee that in boom years there will be far too many acorns for the jays to consume.

Oak savannas provide perching and nesting sites for several species of raptors as well as for a stunning variety of songbirds like Lark Sparrows, Western Kingbirds, Bullock's Orioles, and Western Meadowlarks. As the oaks age or die, they provide essential nesting habitat for a variety of cavity nesters such as Acorn and Nuttall's Woodpeckers, Oak Titmice, Ash-throated Flycatchers, White-breasted Nuthatches, Bewick's

Wrens, and both Violet-green and Tree Swallows. Some winters, large numbers of richly colored Lewis's Woodpeckers visit oak savannas.

Oak Woodlands
North 1,500 to 4,500 feet; South 2,000 to 5,500 feet

Above the grasslands and oak savannas, denser oak groves of more than 30 percent tree cover crowd the hillsides of the western Sierra. In addition to blue and interior live oaks, three other species dominate: valley oaks, California black oaks, and canyon live oaks. All are long-lived and sprout from their stem bases when cut or top-killed by fire. Otherwise, there are considerable differences in their ecology. Valley oaks are large deciduous trees that dominate some riparian forests and open woodlands on fertile, deep soils. Similar to blue oaks, valley oaks suffer from limited recruitment of saplings. Blue oaks tolerate thinner soils and drier conditions than valley oaks and are the most likely oak to be found in uplands well away from drainages. Interior and canyon live oaks are evergreens and vary in growth from low shrubs to large trees. In steep ravines and river canyons these oaks cling to rocky slopes, while toyon, California bay laurel, and redbud grow in cooler glades below. Gray pines, with their wispy, gray green needles, grow along with the oaks in the northern and central Sierra and in the extreme south.

In spring, oak woodlands come alive with the songs of Orange-crowned and Black-throated Gray Warblers, and the roughly whistled notes of Ash-throated Flycatchers. The Hutton's Vireo's monotonously repetitive song is a frequent sound in stands dominated by live oaks. White-breasted Nuthatches search the deeply furrowed bark of these trees and sound their nasal, hornlike calls. Northern Pygmy-Owls hide in these canyons by day and emerge at dusk to hunt for songbirds. In the Mother Lode country of the central Sierra, Highway 49 cuts through broad expanses of foothill woodland and chaparral, where one might still spot a Greater Roadrunner. On nearly any turnout,

the strident calls of Western Scrub-Jays or Oak Titmice can be heard. California Quail work through tangles of poison oak, and Acorn Woodpeckers flash back and forth between trees, flycatching, chattering, and tending their acorn caches.

Foothill Chaparral
North <500 to 4,500 feet; South 500 to 5,500 feet

Impenetrable seas of brush cover hot, dry slopes the length of the Sierra foothills. Interspersed with foothill woodlands, chaparral vegetation generally occupies the steeper, more arid exposures, and the most extensive stands occur south of the San Joaquin River. Turnouts along the new Priest Grade (Highway 120), near Ash Mountain in Sequoia National Park (Highway 198), and the near vertical slopes above the South Fork Kern River (Highway 178) provide easy places to view these habitats. Visitors to foothill chaparral will notice pungent odors of chamise, whiteleaf manzanita, buckbrush, coffeeberry, and shrubby oaks filling the air. These shrubs grow together in thickets forbidding to people but offering shade and shelter to birds. Chaparral birds usually sing, defend territories, and forage in the cool, early morning hours.

Presunrise visits to these arid shrublands are often rewarded by a chorus of Common Poorwills, Wrentits, California Thrashers, Lazuli Buntings, and Spotted and California Towhees. As early as January, Anna's Hummingbirds defend patches of shrubs with squeaky calls, and in spring Blue-gray Gnatcatchers and "Bell's" Sage Sparrows (in chamise chaparral) raise their families within this protective cover. Birds can be particularly abundant in foothill chaparral habitats because they exist below the snow zone and because many native shrubs, such as toyon and poison oak, produce fruits that attract such species as American Robins, Cedar Waxwings, and Hermit Thrushes.

Foothill chaparral is a fire-prone system and the health and diversity of this habitat depends on fire. Many of the shrubs can survive fire and sprout from their burned stumps. Many others produce seeds than can only germinate after a fire. This habitat goes through a postfire succession, analogous to that of conifer forests, but at a much-accelerated pace. The open ground following a fire soon fills with a dazzling array of wild flowers. In the first postfire years birds like Rufous-crowned Sparrows, Lazuli Buntings, and Lawrence's Goldfinches find conditions perfect. As the chaparral grows denser over time, new species appear and others depart. When the habitat becomes heavily overgrown and the shrubs reach their maximum height and density, bird diversity tends to decline, awaiting the next fire to begin the cycle again.

LOWER CONIFER ZONE

Ponderosa Pine Forests
North 2,000 to 6,000 feet; South 2,500 to 7,000 feet

Rising above the heat and haze of the Foothill zone, the Lower Conifer zone is where many people first feel they have reached the mountains. Breezes rustle the trees and, though hot in summer, these forests are distinctly cooler than the lowlands. They also receive more rainfall and snow, enabling them to survive the summer drought. Historically, ponderosa pines were the most common and widespread trees in the Lower Conifer zone because they tolerate hotter and drier climates than most other West Side conifers. Also called "mid-montane conifer forests" by some authors, a variety of other conifers including incense cedars, white firs, Douglas-firs, and sugar pines may now outnumber the ponderosas in mixed stands depending on fire history, elevation, and local conditions.

Before European settlement, these forests experienced frequent, low- to mid-intensity wildfires (primarily surface fires) that were a major factor influencing stand density, structure, and species composition. A policy of fire exclusion, or suppression, during the 20th century, along with the selective harvest of many large pines, has significantly changed fire behavior and led to an increase in fire severity and the number of infrequent but high-intensity, stand-destroying fires. In areas where fire has been prevented for many years, shade-tolerant white firs and incense cedars often outnumber the pines and oaks. In many ponderosa pine forests, kit-kit-dizze (a member of the rose family) covers the forest floor, and its pungent odor permeates the forest and clings to boots and clothing, earning it another name: "mountain misery."

Large snags (i.e., greater than 24 inches diameter-at-breast-height) and decaying portions of living trees offer nesting cavities for Pileated Woodpeckers, Northern Flickers, and Western Screech-Owls. A variety of woodpeckers, Red-breasted Nuthatches, and Brown Creepers patrol the bark of conifers, while Warbling Vireos, Yellow-rumped Warblers, Black-headed Grosbeaks, and Western Tanagers make music from above. Near campgrounds and other developed areas, Steller's Jays squawk and patrol their picnic tables, and Brewer's Blackbirds strut across the pavement.

Pine-Oak Woodlands
North 2,000 to 6,500 feet; South 3,000 to 7,000 feet

Within the Lower Conifer zone, hardwood species like California black oaks, Pacific madrones, and bigleaf maples often intermingle with pines and other conifers. Black oaks, with dark trunks and bright green leaves, grow in patches mixed in with conifers—especially on open, rocky ridges and in forest clearings. These deciduous oaks turn gold in fall like the aspens, willows, and cottonwoods. The madrones and maples favor cool, wet drainages. Black oaks harbor hordes of caterpillars and flying insects that attract Nashville Warblers, Black-throated Gray Warblers, and Cassin's Vireos to forage and sing. The high diversity of birds in these habitats is driven by an abundance of insects and nutrient-rich acorns. Accordingly, such acorn-consuming species as Western Scrub-Jays, Steller's Jays, Acorn Woodpeckers,

FIGURE 2 Lower Conifer zone

Mountain Quail, and Band-tailed Pigeons are common residents.

Oaks also provide nutritious sap that exudes from wounds in the bark caused by insects, tree falls, fire, and Red-breasted Sapsuckers. The sap provides feeding opportunities for the sapsuckers as well as for hummingbirds and warblers like Yellow-rumps and Orange-crowns. The high-protein seeds of many broadleaved trees are eaten by a long list of birds, including White-breasted and Red-breasted Nuthatches, Chestnut-backed and Mountain Chickadees, Dark-eyed Juncos, and Spotted Towhees.

UPPER CONIFER ZONE

Mixed Conifer Forests
North 5,500 to 7,500 feet; South 6,000 to 8,000 feet

As one proceeds up the West Side, ponderosa pines of the Lower Conifer zone blend into the cooler, moister, mixed conifer forests of the Upper Conifer zone. Mixtures of four or five species of conifers are typical in these forests, as the name of this habitat suggests. At places such as Crane Flat in Yosemite (Highway 120) and near the Giant Forest in Sequoia National Park, Jeffrey pines, with large cones and an aroma like vanilla rising from their bark, outnumber ponderosas. Shade-tolerant white firs

are often the most abundant trees, but usually incense cedars, Douglas-firs, sugar pines, and even a few red firs grow there, too. Black oaks reach this high but are fewer than farther down. Where the soil is rocky or wet, especially near meadows, lodgepole pines may grow in scattered stands. Giant sequoias, the world's largest known living things, occur naturally only in the western Sierra, primarily in the Upper Conifer zone. A few grow as far north as Placer County, but most of the 75 groves are south of the Kings River and the largest, most majestic stands are in Sequoia National Park and Giant Sequoia National Monument.

Historically, many mixed conifer forests were comprised of large, thick-barked, fire-resistant trees, which were widely spaced with open understories. Fire suppression throughout the Sierra in the past century has greatly reduced the number and frequency of beneficial, low- to moderate-intensity fires that may have burned for weeks, or even months, cleansing the forest of the excess buildup of "ladder fuels" such as dead branches, small trees, and brush that fuel severe crown fires. There has also been a widespread increase in shade-tolerant tree species such as incense cedars and white firs within formerly pine-dominated stands.

In unburned and/or unlogged mixed conifer

FIGURE 3 Upper Conifer zone

forests, huge conifers provide nesting habitat for Northern Goshawks, Spotted and Great Horned Owls, and foraging habitat and singing perches for Western Tanagers, Black-headed Grosbeaks, and Warbling Vireos. Hermit Warblers sound their buzzy songs from the deeper, shaded woods while White-headed and Pileated Woodpeckers call loudly from ancient snags. Golden-crowned Kinglets thrive in these forests, whispering high-pitched notes throughout the day.

Red Fir Forests
North 6,500 to 8,500 feet; South 7,000 to 9,000 feet

Red fir forests receive the heaviest Sierra snows. In these shady groves deep drifts often last long into summer, gradually releasing moisture into the soil. Small numbers of red firs grow on north slopes in the mixed conifer zone, but higher up their dark, silent columns reign supreme. Entire forests consist of only these trees, with rings of chartreuse lichen circling their trunks above the snow line. Often lodgepole and western white pines are mixed in, as are white firs at the lower edge of these forests. Particularly common in the northern and central Sierra, red firs grow only in limited areas farther south. At Porcupine Flat and Badger

Pass in Yosemite and near Soda Springs on Interstate 80, impressive stands of red firs and lodgepole pines grow together. Lodgepole pines may form extensive stands in this zone, particularly in cold sites, around meadows, and on dry, rocky slopes. Heavy snows and deep forest litter prevent much growth of shrubs, grasses, or flowers; most birds in red fir forests search bark surfaces and sprays of conifer needles for food.

Red fir forests often seem lacking in birds, but if one listens carefully, rhythmic tapping of Williamson's Sapsuckers may be heard. The characteristic sounds of Mountain Chickadees, Hermit Thrushes, Townsend's Solitaires, Yellow-rumped Warblers, and Golden-crowned Kinglets echo through towering trees, and the complex, hyperspeed song of Pacific Wrens rises from the deep drainages. Flocks of Pine Siskins, Red Crossbills, or Evening Grosbeaks may fly overhead. Open areas with shrubs, wildflowers, and grasses draw Dark-eyed Juncos, Chipping Sparrows, and Cassin's Finches that search the ground for insects and seeds. Some "edge" species such as Mountain Quail and "Mountain" White-crowned Sparrows use trees and shrubs for cover but feed in open areas nearby. At dusk, listen for the *peents* and *booms* of Common Nighthawks overhead and, as darkness comes, the incessant tooting of Northern Saw-whet Owls.

Mountain Chaparral
North 4,500 to 8,500 feet; South 5,000 to 9,500 feet

Brushfields of huckleberry oak, greenleaf manzanita, snowbrush, or chinquapin occupy steep, rocky slopes or forest clearings created by treefalls, logging, or fire. Especially prevalent on south-facing exposures, montane chaparral habitats may grow in impenetrable thickets but can be quite open on steep slopes or recently established sites. Some soil types are capable of only supporting chaparral. Although usually dry, these shrubfields are cooler than their foothill counterparts. Widespread above the foothills, montane chaparral habitats can be easily seen along Interstate 80 near Dutch Flat, Highway 120 west of Crane Flat in Yosemite, and on the slopes below Giant Forest in Sequoia National Park. Here the whistled melodies of Fox Sparrows may mingle confusingly with the similar songs of Green-tailed Towhees. Although they are often found in moist streamside habitats, Yellow Warblers and MacGillivray's Warblers also nest in montane chaparral. Mountain Quail sneak through the underbrush and nest under this protective cover. At the edges where chaparral meets forest, you may find species such as Dusky Flycatchers and Western Wood-Pewees.

In some places, montane chaparral is an early stage of succession following a stand-replacing fire, destined to be gradually replaced by forest. In other spots, where the soils cannot support forest, the chaparral may represent a relatively permanent condition.

SUBALPINE ZONE

Lodgepole Pine Forests
North 7,500 to 9,000 feet; South 8,000 to 9,500 feet

Mountain lakes and springs spawn tiny streams that meander through meadow grasses and open stands of lodgepole pine. These pines also grow in dense groves that dominate the Subal-pine zone, especially in the northern and central Sierra on both slopes, where they are sometimes joined by mountain hemlocks. Here and there, majestic specimens of western juniper occupy isolated, prominent cliffside locations. These trees, second only to the giant sequoia in longevity among Sierra trees, often bear the scars of numerous lightning strikes. Echo Summit (U.S. 50), Donner Summit (Interstate 80), and Tuolumne Meadows in Yosemite National Park have good examples of lodgepole pine forests. Hardy Mountain Chickadees reside in these forests year-round, moving along sprays of conifer needles and swinging down to inspect them from below. Dusky Flycatchers and Ruby-crowned Kinglets sound their familiar calls from forest edges and open ridge lines, while flocks of Red Crossbills and Pine Grosbeaks rove the treetops, even in winter. Hairy and Black-backed woodpeckers along with Williamson's Sapsucker and Northern Flickers attack dead and dying lodgepole pines.

Lodgepoles growing at the edges of wet meadows and streams provide important nesting strata and cover for species closely associated with riparian habitats (this includes many migrating Neotropical songbirds such as flycatchers, vireos, warblers, and so on). The abundance of nesting and migratory songbirds draws predatory birds such as Cooper's and Sharp-shinned Hawks, and Northern Goshawks. These hawks may nest in lodgepole pine forests but more frequently use them as hunting sites for ambushing prey, primarily birds but also rodents.

Subalpine Pine Forests
North 9,000 to 10,000 feet; South 9,500 to 11,500 feet

Backpackers head for these high forests where serrated ridges, perpendicular cliffs, and massive granite domes form spectacular vistas. On both sides of the Sierra, wind-sculpted white-bark pines grow with stunted lodgepoles on exposed ridges and north-facing slopes up to treeline. In the southern Sierra, erect stands of

FIGURE 4 Subalpine zone

foxtail pine largely replace the gnarled white-barks. The higher slopes of the Great Western Divide in Sequoia and Kings Canyon National Parks and the areas near Tioga Pass in Yosem-ite National Park have good examples of subal-pine pine forests. Clark's Nutcrackers and Pine Grosbeaks bound from tree to tree searching for pine nuts but dive for cover if the shadow of a Prairie Falcon skims by. Mountain Bluebirds, Dark-eyed Juncos, and "Mountain" White-crowned Sparrows forage in open, rocky areas. Here they are also joined by Common Ravens that prey on small birds and their eggs and nestlings and search for road-killed animals along high mountain roads.

ALPINE ZONE

Alpine Fell-Fields

North 8,500 to 10,800 feet; South 9,500 to >14,500 feet

The line of demarcation between the upper subalpine pine forests and treeless Alpine meadows is known as "timberline," the point where trees cease to exist because of extremes in climate, shallow rocky soils, and high eleva-tion. Here, freezing temperatures are possible any day of the year, and even the intense high-elevation sun does little to warm the cool, clear air. Life in these habitats is severe, and most plants are stunted by high winds, bitter cold, and shallow soils. Herbaceous plants dominate Alpine fell-fields, but stunted willows and other dwarf shrubs occur here as well. North of the Tahoe region, few peaks exceed 9,000 feet and alpine areas are limited to the summits of a few mountains, such as the Sierra Buttes in Sierra County, Castle Peak in Nevada County, and Granite Chief in Placer County. Farther south, many peaks exceed 13,000 feet, includ-ing Mount Whitney (14,496 feet), the highest mountain in the contiguous 48 states. Alpine environments are particularly extensive in Yosemite and Kings Canyon National Parks. Most Sierra Alpine fell-fields occur above the reach of highways, but a short hike from Car-son Pass (Highway 88), Sonora Pass (Highway 108), or Tioga Pass (Highway 120) puts one amid the rugged cliffs, glacial cirques, and rock gardens of this zone.

In summer, wildflowers bloom profusely in Alpine fell-fields, and Dark-eyed Juncos and "Mountain" White-crowned Sparrows forage near shrubby willows. Gray-crowned Rosy-Finches feed on low turf or snow banks, and American Kestrels hover above. Mountain Bluebirds perch on boulders and dart out to snatch flying insects, the stunning blue of their

FIGURE 5 Alpine zone

plumage rivaling the alpine sky. American Pipits, Horned Larks, and Dark-eyed Juncos, as well as small mammals, attract raptors (especially during migration) such as Sharp-shinned and Cooper's Hawks, Prairie Falcons, and Red-tailed Hawks. The summer flowering season coincides with an influx of hummingbirds that are either in migration or seeking relief from the summer drought and heat of the foothills. Rufous (in migration) and Calliope Hummingbirds can be abundant around particularly lush flower patches. Rock faces, boulder fields, and talus dominate much of the landscape in the Alpine zone, providing abundant habitat for Rock Wrens.

EAST SIDE ZONE

Pine Forests

North 3,500 to 7,500 feet; South 3,000 to 8,500 feet

Growing in the Sierra rain shadow, pine forests of the East Side are generally more open and have smaller trees than the similar forests of the West Side. Easily seen along Highway 395, especially to the north of Mono Basin, they are dominated overwhelmingly by Jeffrey pines, but ponderosa pines are sometimes found as scattered individuals, or isolated pockets within

larger Jeffrey pine stands, from Tahoe Basin north. Farther south, ponderosas are almost entirely replaced by Jeffrey pines, where higher elevations and associated colder temperatures combine to make soils drier. Some of the larger East Side basins, such as those surrounding Lake Tahoe and Mammoth Lakes, have extensive red fir and mixed conifer forests much like those on the West Side. Other tree species found on the East Side include western juniper, white fir, and lodgepole pine. East Side pine forests often have a dominant understory of Great Basin shrubs such as big sagebrush, bitterbrush, and rabbitbrush, along with numerous grasses and herbs. Some characteristic birds of these forests include Gray and Dusky Flycatchers, Pygmy Nuthatches, Clark's Nutcrackers, and Red Crossbills.

Pinyon-Juniper Forests

North 3,500 to 8,500 feet; South 3,000 to 9,000 feet

At lower elevations of the East Side is an even drier habitat, pinyon-juniper forest, which is almost nonexistent on the West Side except at the southern end, where extensive stands occur in the drainages of the South Fork Kings River, the Kern River, and the South Fork Kern River.

FIGURE 6 East Side zone

Highway 395 crosses through miles of this habitat where shrubby single-leaf pinyon pines grow in open stands, sometimes mingling with Utah junipers and Jeffrey pines interspersed between Great Basin shrubs like sagebrush, curl-leaf mountain mahogany, bitterbrush, and rabbitbrush. Bird species that tend to be more common in these forests than in pine forests are Pinyon Jays and Townsend's Solitaires; Juniper Titmice occur in these habitats in the Mono Basin.

Great Basin Scrub
North 3,500 to 6,500 feet; South 3,000 to 7,000 feet

What may appear as a monotony of silvery-gray to olive-green shrubs in the lowlands of the East Side is often a floristically diverse assemblage of aromatic plants. Bitterbrush and sagebrush are the dominant shrubs, but a wide variety of other shrubs, perennial and annual forbs, and bunchgrasses may be found. Scattered emergent pines and large stands of curl-leaf mountain mahogany are also common. In large areas only the shrubs remain, forming the dominant habitat of the Great Basin. Sagebrush scrub also extends up to high elevations on dry hillsides of the East Side and locally west of the crest, as near Donner Summit. Driving the back roads through these scrubby habitats at dawn or dusk, one might flush a Greater Sage-Grouse, Common Nighthawk, or Common Poorwill, while birds that may be spotted any time of day are Say's Phoebes, Gray Flycatchers, Sage Thrashers, and Vesper and Brewer's Sparrows.

DESERT ZONE

Joshua Tree Woodlands
North none; South 2,000 to 7,000 feet

Joshua trees usually grow in widely scattered stands interspersed with a variety of evergreen and deciduous shrubs typical of Mojave desert scrub habitats. While Joshua trees sometimes grow in pure stands, more often there are a few singleleaf pinyon pines, Utah junipers, or Mojave yuccas somewhere in the vicinity. Joshua trees are generally found at moderate elevations in broad valleys with deep soils situated between mountains and mesas. Bird life in Joshua tree woodlands is similar to desert scrub habitats, but Joshua trees offer lookout posts, song perches for a few species such as Ladder-backed Woodpeckers, Cactus Wrens, and Scott's Orioles that are not found elsewhere

FIGURE 7 Desert zone

in the Sierra region. Joshua tree woodlands are easily observed along Highway 178 on both sides of Walker Pass in Kern County.

Desert Scrub and Washes
North none; South 2,000 to 7,000 feet

In the Sierra, desert scrub habitats are usually dominated by creosote bushes standing up to six feet high with considerable open ground in between. Creosote bushes are often surrounded by other desert plants like catclaw acacia, desert agave, burrowbrush, rabbitbrush, teddybear chollas, or beavertail pricklypear cactus. These habitats receive little rainfall with cold, dry winters and hot, dry summers. In the Sierra region, these habitats can only be found in Inyo and Kern Counties, where they can be accessed easily from Highways 14, 58, 178, and 395. Small mammals, birds, and reptiles provide prey for Red-tailed Hawks, Prairie Falcons, and Loggerhead Shrikes. Plants in desert wash habitats tend to be taller than in surrounding desert scrub habitats, and some typical species include blue paloverde, desert ironwood, smoketree, catclaw acacia, mesquite, and the invasive tamarisk. A stroll down a desert wash might offer views of a Greater Roadrunner, Phainopepla, or a flock of Black-throated Sparrows.

SPECIAL HABITATS

The habitats described in this section are of limited extent in the Sierra but provide key resources for birds. Because of their importance and because each occurs across a wide range of ecological zones, they are described separately here.

Riparian Forests

Only small fragments remain of the shady jungles of sycamores, cottonwoods, and willows that once flourished along rivers of the Sierra foothills. Most were cleared long ago for lumber, firewood, and agriculture or inundated by large reservoirs. By far the largest and most impressive remaining lowland riparian forest in the Sierra is along the South Fork Kern River, upstream from Lake Isabella. This riparian oasis is more than a mile wide, includes more than 3,000 acres, and extends for miles—it is now protected by the California Department of Fish and Game, the U.S. Forest Service, and Audubon California. The South Fork Kern River riparian forests support breeding populations of a number of special status bird species that breed nowhere else in the Sierra, such as Yellow-billed Cuckoos, "Southwestern"

FIGURE 8 Riparian forest (photograph by Phil Robertson)

Willow Flycatchers, Brown-crested Flycatchers, Vermilion Flycatchers, and Summer Tanagers.

Overall, remaining lowland riparian forests along West Side rivers and streams support a higher density and diversity of breeding and migratory birds than any other Sierra habitat. Numbers of migratory birds in riparian areas can be more than 10 times greater than found in the surrounding uplands. Draped in wild grapes, berry vines, or poison oak, these lush groves offer moisture and shade during the hot summers. Breeding birds like Yellow Warblers, Yellow-breasted Chats, Warbling Vireos, House Wrens, and Black-headed Grosbeaks sing above the constant din of humming insects, while Downy and Nuttall's Woodpeckers drill on branches and flake off bark. Other typical Sierra riparian birds include Black Phoebes, Pacific-slope Flycatchers, Ash-throated Flycatchers, House Wrens, Orange-crowned Warblers, Spotted Towhees, Song Sparrows, and House Finches.

In the Lower and Upper Conifer zones, riparian forests mostly become narrow, discontinuous corridors of dogwood, black cottonwoods, and aspens or, more typically, tall willows or alders. In the Subalpine zone, shrubby willows and alders form a narrow and patchy border along most streams. Narrow stands of mois-ture-loving deciduous trees also border some ponds and lakes throughout the high Sierra. In all these ecological zones, deciduous trees often replace conifers along streams, especially in steep terrain. Song Sparrows and Yellow and Wilson's Warblers breed in these broad-leaved trees, and migrant vireos, warblers, sparrows, finches, and tanagers flock to them in late summer and fall. Turbulent mountain streams are home to Common Mergansers, American Dippers, and rarely Harlequin Ducks. Nearby sandbars may harbor nesting Spotted Sandpipers or perhaps a family of Killdeer.

Aspens commonly form large broad-leaved woodlands in the high mountain regions and dominate riparian areas along the entire East Side, where they are joined by cottonwoods and willows. The pale, light green leaves or yellow autumn hues and white bark provide a dramatic contrast to the various coniferous forest communities that span their range in the Sierra. Aspen woodlands provide important breeding, resting, and foraging habitat for a diverse array of birds that eat the buds, flowers, seeds, and catkins of quaking aspens and supply a yearlong food source for Sooty Grouse. Typical nesting birds include Red-breasted Sapsuckers, Hairy Woodpeckers, Northern Flickers, Warbling Vireos, Tree and Violet-green Swallows,

FIGURE 9 Mountain meadows

MacGillivray's and Yellow Warblers, and Song Sparrows. Breeding raptors may include Great Horned Owls, Cooper's Hawks, and Northern Goshawks.

Mountain Meadows

Mountain meadows usually begin as lakes that are gradually filled with sediment from the surrounding uplands and may someday become forests. Grasses, sedges, and rushes share the moist central portions with wild onions, corn lilies, shooting stars, and countless other wildflowers, while young trees gradually colonize the edges. These meadows are most numerous and extensive in the glaciated terrain of the Subalpine zone but are also scattered throughout the Lower and Upper Conifer zones of the Sierra. They range in size from small forest clearings to wide expanses such as Tuolumne Meadows in Yosemite National Park, the largest example of these high-country flower gardens in the Sierra. Perazzo Meadow in Sierra County is another spectacular, and recently protected, mountain meadow paradise. One can walk for miles in any direction through these vast expanses, encountering Calliope Hummingbirds, Golden Eagles, Prairie Falcons, American Kestrels, Willow Flycatchers, Dark-eyed Juncos,

Chipping Sparrows, and Mountain Bluebirds along the way. Unfortunately, many of the Sierra's largest and most spectacular meadows, such as Hetch Hetchy Valley in Yosemite National Park, are now buried beneath the waters of human-made reservoirs.

Growths of shrubby willows tracing the main stream courses in wet mountain meadows harbor birds such as Willow Flycatchers, Lincoln's Sparrows, and MacGillivray's Warblers. Abundant mosquitoes, dragonflies, and butterflies dance in the air while Western Wood-Pewees and Olive-sided Flycatchers perched at meadow's edge scan for these and other insects. Wet meadows stay green all summer, while dry grassy sites atop domes and ridges usually turn brown and go to seed by the end of August. The margins of many Sierra meadows are ringed by aspens or cottonwoods that display a brilliant gold when framed against a blue October sky. Whether rimmed with broad-leaved trees or ranks of conifers, forest-meadow edges attract birds like Western Bluebirds, Yellow-rumped Warblers, Chipping Sparrows, Dark-eyed Juncos, and Purple Finches that forage among meadow plants for insects and seeds and take to the trees for cover.

The edges of forests and mountain meadows are among the most rewarding for birdwatch-

FIGURE 10 Freshwater marshes

ers. During the night, cold air flows down into meadow depressions and by dawn these areas become significantly colder than nearby forests. Like humans, birds seek warmth at dawn and usually flock to the sunny sides of mountain meadows. Sandhill Cranes and Wilson's Phalaropes sometimes nest in wet meadows on the East Side. Dense forests near meadow edges provide nesting habitat for raptors such as Northern Goshawks and Spotted Owls. Great Gray Owls primarily search for prey in these mountain meadows.

Freshwater Marshes

Freshwater marshes are distinguished from deep water aquatic habitats and wet meadows or grassland habitats by the presence of rushes, sedges, or tall, erect, grasslike plants such as tules and cattails that are rooted in soils that are permanently or seasonally flooded or inundated. Marshes can occur in basins or depressions at all elevations, aspects, and exposures, but they are most common on level to gently rolling topography below about 4,000 feet but can be found above 8,000 feet in the southern Sierra. On the West Side, freshwater marshes are mostly confined to the shallow edges of reservoirs and ponds; the largest ones are at

Lake Almanor in the north and Lake Isabella in the south. On the East Side, sizable marshes can be found at Honey Lake, Sierra Valley, and Carson Valley, at such large lakes as Bridgeport Reservoir, Lake Tahoe, and surrounding Mono Lake. Freshwater marshes also occur as fringes around reservoirs such as Lake Crowley where the slopes are gentle enough to create a rim of shallow water. On slow-moving meadow streams, freshwater marshes can occur as narrow bands over long distances.

Freshwater marshes teem with bird life. Where shorelines are clogged by tules, cattails, or rushes, these aquatic habitats harbor Canada Geese, Mallards, Cinnamon Teal, Virginia Rails, Soras, American Coots, Common Gallinules, Wilson's Snipe, Marsh Wrens, Common Yellowthroats, and throngs of Red-winged Blackbirds, while Great Blue Herons stand like statues in the shallows.

Reservoirs, Lakes, and Ponds

Most Sierra "lakes" are actually artificial reservoirs. More than 150 reservoirs exist on West Side rivers and creeks, and the largest of these (more than 5,000 surface acres) are Lake Almanor and Lake Oroville (Feather River), New Melones Lake (Stanislaus River), Don Pedro Reservoir

FIGURE 11 Reservoirs, lakes, and ponds (Mono Lake)

(Tuolumne River), Lake McClure (Merced River), Millerton Lake (San Joaquin River), Pine Flat Lake (Kings River), Lake Success (Tule River), and Lake Isabella (Kern River). On the East Side, examples of large reservoirs include Lake Davis (Plumas County), Stampede (Sierra County), Prosser Creek and Boca (Nevada County), Topaz Lake (Douglas County, Nevada), Bridgeport Reservoir, and Crowley Lake (Mono County).

Fish-eating birds such as Ospreys and Bald Eagles patrol the surfaces of these reservoirs, while Double-crested Cormorants, American White Pelicans, Western and Clark's Grebes, Common Loons, and Common Mergansers dot the surface and dive for similar prey. A wide diversity of waterfowl, shorebirds, and other aquatic birds can be found in natural East Side lakes and wetlands such as Honey Lake and Lake Tahoe. Mono Lake hosts almost a million Eared Grebes and thousands of Wilson's and Red-necked Phalaropes in fall migration as well as the largest California Gull breeding colony in California.

Smaller lakes, ponds, and sewage treatment plants attract ducks like American Wigeon, Northern Shovelers, Ring-necked Ducks, Buffleheads, Hooded Mergansers, and Ruddy Ducks as well as California Gulls and flocks of shorebirds where mudflats are exposed. However, because most Sierra rivers flow through steep-sided canyons, extensive mudflats are scarce, but they do occur around reservoirs, lakes, and ponds with gentle slopes where ducks like Northern Shovelers, Gadwalls, American Wigeon, Northern Pintails, and Green-winged and Cinnamon Teal dabble in the shallows.

Although unequaled in beauty, clear lakes in the Alpine and Subalpine zones offer little food for birds. Probably the most common visitors to these waters are California Gulls traveling to and from their Great Basin breeding grounds. Spotted Sandpipers and American Dippers may patrol their margins for aquatic insects but never in large numbers.

Old Growth Forests

If a forest can avoid severe fire and the chainsaw for a couple of centuries, it may attain a state commonly known as "old growth." This represents a "climax" stage and can, under the right conditions, last for many more centuries with little change in structure or species composition. Small, localized stands of old growth forests occur in the Lower Conifer zone, but most of the remaining old growth in the Sierra exists in the Upper Conifer zone; the most extensive stands are in Yosemite and Sequoia National Parks. The best estimate is that at least 67 percent of Sierra mixed conifer forests were in this condition at the time of European settlement of California. Today less than 12 percent remains, much of it in isolated fragments. Old growth forests include huge trees, a wide variety of tree sizes and species, very large snags, and a relatively sparse understory with the fallen boles of ancient giants scattered about the forest floor. They tend to be cool, dark, and quiet. Although these forests harbor relatively few birds compared to some other Sierra habitats, they are by far the most productive habitats for some species, such as Spotted Owls and Northern Goshawks. While no bird species uses old growth exclusively, some species, such as Brown Creepers, Pileated Woodpeckers, and Great Gray Owls, are strongly associated with these ancient forests.

One could think of old growth as a final stage. However, from nature's perspective, it is just part of a cycle of succession. Imagine an old growth forest that finally meets a combination of conditions that allows a severe fire to reach the crowns of the trees and kills all or nearly all of them. Within hours of the fire's end, bark beetles from as far away as 100 miles have sensed the conditions needed to reproduce. They arrive and lay the eggs of their next generation in the snags. Once they hatch and larvae begin to grow and consume the dead wood, a woodpecker feast of magnificent proportions begins. By the following spring, the ground is ablaze in a new fire of wildflowers of stunning variety and abundance. Within a few years a healthy growth of shrubs appears and along with them, the suite of bird species that needs this mountain chaparral. At this point the snags are becoming ideal homes for cavity-nesting birds. Gradually, sun-loving trees begin to grow among the shrubs. Once those trees attain the size to provide shade, the shrubs begin to disappear and shade-adapted trees begin to take hold, destined to overtop and shade out the earlier tree species. At each stage a different suite of birds finds their ideal conditions. Given enough time, this forest may once again achieve old growth status.

Recent Burns

Terms like "tragic" and "devastating" are often linked to large forest fires in media reports. However, this judgment misses the point that fire, even intense stand-replacing fire, is a crucially important part of the Sierra ecosystem and is necessary to maintain the biodiversity of the region. Birders know well that time spent at the site of a fairly recent fire can be the most exciting and productive part of a day in the field. At lower elevations on the West Side, Rufous-crowned Sparrows are among the first birds to take advantage of these newly open areas. Lazuli Buntings will appear in large numbers within the first few years. Blue-gray Gnatcatchers, California Thrashers, Bewick's Wrens, and an amazing variety of sparrows all favor these areas. Where large fire-killed snags remain, Black-backed Woodpeckers may appear quickly to take advantage of the insects that infest these trees. A host of other cavity-nesting birds—like Hairy Woodpeckers, Northern Flickers, Mountain Chickadees, and Mountain and Western Bluebirds—soon occupy these areas.

Olive-sided Flycatchers sally out from the tops of highest snags to catch insects on the wing. These burned areas will generally support a higher diversity of birds and a different suite of species than the surrounding forest. Thus the mosaic of habitats created by fires of varying intensity over varying periods is a key force driving the Sierra's remarkable diversity of birdlife.

Rocks and Cliffs

In the northern Sierra, ancient volcanoes spewed molten lava across the landscape, but the central and southern Sierra have a different geologic history and display vast expanses

FIGURE 12 Developed habitats (cropland)

of glacier-polished granitic cliffs, domes, and scattered boulders. Some of the earth's finest rock work graces the canyons of the Merced and Tuolumne Rivers in Yosemite and the high country of Sequoia and Kings Canyon National Parks. Rock crevices and ledges high on steep canyon walls provide nesting sites for swifts, Golden Eagles and Prairie and (more and more frequently) Peregrine Falcons. Rock Wrens, Rosy-Finches, and the occasional introduced White-tailed Ptarmigan scuttle confidently across alpine talus slopes, and Canyon Wrens inhabit jumbles of boulders in river gorges.

DEVELOPED HABITATS

Forage Crops, Irrigated Pastures, and Croplands

Forage crops, such as grass hay and alfalfa, nearly always consist of a single species that may be annual or perennial. Most forage crops are planted in the spring and harvested in summer or fall and in the Sierra mostly include hay and alfalfa. For the most part, forage crops and irrigated pastures are planted in fertile soils in alluvial valley bottoms or gently rolling terrain in the low to mid-elevations of the Sierra, and most are found on the East Side. They are often adjacent to annual grasslands, ephemeral or perennial streams, or irrigation canals, which increases their attractiveness to such birds as Northern Harriers, Red-tailed Hawks, American Kestrels, Mourning Doves, Western Kingbirds, American Crows, Black-billed Magpies (East Side only), Western Meadowlarks, Brewer's Blackbirds, and Red-winged Blackbirds. By far the most impressive examples of these habitats are in the Sierra Valley (Plumas and Sierra Counties), where, in conjunction with marshlands and wet meadows, these habitats attract impressive numbers of waterfowl and wintering raptors, nesting Sandhill Cranes, shorebirds, and waterfowl.

Orchards and Vineyards

Compared to all-natural habitats, orchards and vineyards are relatively barren of breeding birds. In some parts of the Foothill zone, large-scale conversion of annual grasslands, oak savannas, and oak woodlands to orchards and vineyards has resulted in direct losses of bird habitat. Typical birds that forage in orchards include Mourning Doves, American Crows, Yellow-billed Magpies, American Robins, and House Finches. Compared with orchards, vineyards are usually grown in rolling hills with deeper, well-drained soils. They are managed intensively and the soil under the vines is generally sprayed and barren to prevent the growth of grasses and other herbs, which reduces their value to bird life. Huge flocks of introduced European Starlings visit vineyards, especially in late summer and fall when they may consume entire crops of ripening grapes. A few native birds that forage in vineyards opportunistically include Mourning Doves, Western Scrub-Jays, American Crows, American Robins, Western Bluebirds, Cedar Waxwings, Yellow-rumped Warblers, Dark-eyed Juncos, and House Finches. The robins, bluebirds, and warblers mainly take advantage of fruits left over after fall harvest.

Urban/Suburban

In terms of bird habitats, urban and suburban areas usually offer a patchy mosaic of ornamental plantings, vacant lots, and remnant native habitats that occur between structures. Ornamental plantings in older neighborhoods are often mature, introduced evergreen and winter-deciduous trees that may be as much as 100 years old. These ornamental species range in height from approximately 20 to 50 feet high at maturity and are typically much smaller and younger than the occasional remnant oaks, pines, or incense cedars in these neighborhoods. Small lawns and mature hedges are also characteristic and include many introduced fruiting species that attract a variety of birds. Riparian or stream habitats occurring within urban and suburban landscapes usually accommodate the greatest number of species, such as Anna's and Rufous Hummingbirds (at houses with feeders), American Crows, Steller's Jays, Western Scrub-Jays, American Robins, Northern Mockingbirds, House Finches, Cedar Waxwings, Brown-headed Cowbirds, and Brewer's Blackbirds, native denizens of parking lots. Introduced non-natives like European Starlings and House Sparrows are found almost exclusively in such areas. Bird feeders, an ever-increasing feature of these areas, attract and help sustain many of these birds.

Recent Trends in Sierra Bird Populations and Ranges

In this chapter we review changes in populations and ranges of birds of the Sierra over almost four decades. We used data from Breeding Bird Surveys (BBS) and Christmas Bird Counts (CBC) and supplemented those data with observations cited in *North American Birds*, numerous publications (see the bibliography), and the personal experience of ourselves and many regional experts. We also compare our analyses of BBS data with those of Sauer et al. (2011). Those authors used a different definition of Sierra boundaries, a different time frame, and more sophisticated statistical methodology (see Appendix 3). We have also, where possible, attempted to put these trends into larger state- or continent-wide perspective.

There are 44 BBS routes and 25 CBC circles within the Sierra region (see Map 5). Most of these have been run throughout much of the past 35 years, providing a relatively robust, long-term dataset to explore. BBS routes are run during the spring and summer breeding season, and CBCs are run from late December into early January. Since many of the species counted during these surveys are migratory, the breeding season and wintering season birds

may often be of different subspecies or from separate populations. Therefore, one should not necessarily expect that a given species should reflect similar trends from BBS and CBC data. BBS routes are limited to roads, so much of the Sierra (in particular the higher elevations) is not surveyed by these routes. CBC circles are almost completely limited to lower elevations that are accessible in the winter. Although both these sources have their limitations, when the data show significant and consistent trends over a number of circles or routes, one can be fairly confident they reflect real trends. We limited our analyses to species that are relatively common and widespread throughout the Sierra. Appendix 3 describes the methods we used to detect trends for each species.

POSITIVE TRENDS

Species that showed statistically significant positive population trends from either BBS or CBC data are summarized in Table 2. Of the 117 species we analyzed, more showed positive trends than negative ones. Population increases for several of the species can be attributed to

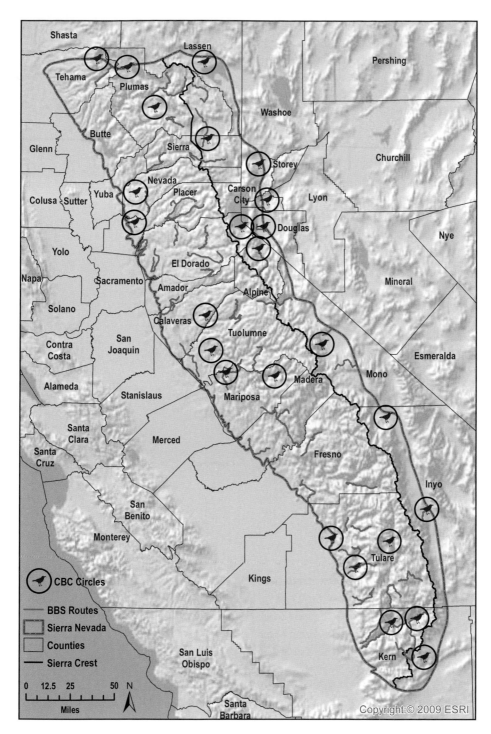

MAP 5 Locations of Breeding Bird Survey (BBS) routes and Christmas Bird Count (CBC) circles used for trend analyses

direct or indirect effects of human activities. The California Department of Fish and Game first introduced Wild Turkeys into the state in 1908. However, those and all subsequent introductions (mainly using captive-bred birds) over the next 50 years proved unsuccessful. It was only when wild-captured birds from Texas began to be released from 1959 to 1988 that California populations began to become self-sustaining. Data from BBS and CBC surveys reflect the success of those efforts, with huge increases starting in the mid-1980s and continuing up to the present.

Large and growing, resident flocks of Canada Geese may be the result of a combination of factors including historical California Department of Fish and Game introductions (i.e., prior to the early 1970s), releases from private waterfowl breeders, and native wintering birds failing to migrate and producing nonmigratory offspring. Dramatic recoveries in populations of Bald Eagle and Osprey are likely due in large part to banning of DDT in North America in the 1970s. Eagles also benefitted from protection under the state and federal Endangered Species Acts. Increasing human population in the Sierra and the bird feeders which usually accompany that increase could be contributing to increases in Anna's Hummingbirds and Lesser Goldfinches. These same factors have allowed House Finches to expand their year-round range into higher-elevation portions of the Sierra. Nest box programs are likely benefiting species like Wood Ducks, Western Bluebirds, and Tree Swallows. Planting of exotic, winter-fruiting trees in newly developing areas of the Sierra could help explain the increasing winter populations of American Robins, Cedar Waxwings, and Western Bluebirds, although increasing average winter temperatures could also be a factor.

We found increases in seven species associated with chaparral/early successional habitats (Bewick's Wren, Blue-gray Gnatcatcher, MacGillivray's Warbler, Spotted and California Towhees, Rufous-crowned Sparrow, and Lazuli

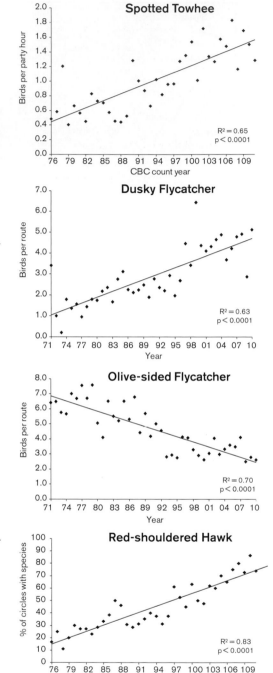

FIGURE 13 Charts showing trends for selected species based on Christmas Bird Count (CBC) data (Spotted Towhee and Red-shouldered Hawk) or Breeding Bird Survey data (Dusky Flycatcher and Olive-sided Flycatcher). All trend lines are based on linear regression. Note that for CBC data, Count Year 76 is winter 1975–76 and so on.

Bunting) or forest-chaparral edges (Dusky Flycatcher) (Figure 13). The increase in large-scale, stand-replacing fires in the Sierra in recent years may have expanded these types of habitats to the benefit of these species.

The trends for Brown-headed Cowbirds have been complex. Consistent with their historical spread throughout California in the past century, cowbirds followed human expansion and cattle grazing up into the Sierra, and our BBS data show a steady increase in numbers up to 1990. However, since then, the trend has been flat to negative. Analyses of BBS data by Sauer et al. (2011) showed a significant negative trend. CBC data show a slightly positive trend that is not statistically significant.

NEGATIVE TRENDS

Species that showed statistically significant declines based on either CBC or BBS data are summarized in Table 3. For most of the species on this list, a variety of different sources have documented long-term, widespread population declines throughout their North American ranges. Degradation of winter habitat in Central and South America has been implicated in the long-term declines seen in Olive-sided Flycatchers and Western Wood-Pewees. In addition, some forestry practices like postfire salvage logging may have contributed to the alarming negative population trend for Olive-sided Flycatchers (see Figure 13) in the West. Reduced quality of montane riparian habitats is a likely contributor to declines of Western Wood-Pewees, Yellow Warblers, and Wilson's Warblers. Brood parasitism by Brown-headed Cowbirds is also implicated in Yellow Warbler declines. Killdeer, Horned Larks, and Lark Sparrows are among the 70 percent of grassland-associated birds that have shown highly significant population declines throughout North America.

The steady decline in American Kestrel numbers across the continent has been studied intensively but remains largely unexplained.

Extensive development in the lower elevations of the West Side of the Sierra Nevada may be partly to blame for the negative population trends for Lark and Chipping Sparrows, as both species are particularly sensitive to habitat fragmentation. Habitat fragmentation in its low-elevation breeding range may also be responsible for the declines in Bullock's Orioles. Reasons for significant declines in year-round Sierra residents such as Mountain Chickadees, Purple Finches, and Cassin's Finches are uncertain as most of their habitats are relatively undisturbed.

The negative trend for Rough-legged Hawks in the Sierra is reflected throughout the western, eastern, and southern parts of this species' winter range in North America. However, this decline is offset by much higher numbers wintering in the northern Great Plains, possibly as a result of reduced snowpack in that region in recent decades. Negative trends for Brewer's Blackbird and Violet-green Swallow are perplexing but consistent with state-wide trends in California. Indeed, of all Sierra species generally thought to benefit from human development, Brewer's Blackbird was the only one to show a significant decline. Also surprising was our finding that American Robins and Mourning Doves both show significant negative trends from BBS data but significant positive trends from CBC data, the only species to show this discrepancy.

As pointed out in the introduction to this section, the migratory nature of these species means that the winter and summer populations may be from different sources. In this case, milder winter conditions and/or the presence of winter-fruiting trees might explain the winter increases. American Robins were the only species to show significant contradictory results between our BBS analyses and those of Sauer et al. (2011), who reported a positive trend for robins. The difference in definitions of the Sierra region may explain this as the region they used included areas to the north and excluded areas to the south as well as lower elevations on

TABLE 2
Species showing significant positive trends.

Species	CBC	BBS	BBS[a]
Canada Goose	+, ns	+	+
Wood Duck	+	+	n/a
Mallard	−, ns	+	+
Hooded Merganser	+	n/a	n/a
Wild Turkey	+	+	n/a
California Quail	+	+, ns	+
Great Blue Heron	−, ns	n/a	+
Osprey	n/a	+	n/a
Bald Eagle	+, ns	+	n/a
Red-shouldered Hawk	+	+	n/a
Eurasian Collared-Dove	+	+	n/a
Mourning Dove	+	−	−, ns
Anna's Hummingbird	+	+	+
Downy Woodpecker	+	−, ns	n/a
Northern Flicker	+	+, ns	−, ns
Dusky Flycatcher	n/a	+	+, ns
Black Phoebe	+, ns	+, ns	+
Cassin's Vireo	n/a	+	+
Hutton's Vireo	+, ns	+	+
Warbling Vireo	n/a	+	−, ns
Western Scrub-Jay	+	−, ns	−, ns
American Crow	+	+, ns	+, ns

(continued)

both the East and West Sides. A possible shift in breeding range uphill and/or to the north could be responsible for this discrepancy.

RANGE EXPANSIONS

A number of species have either expanded their range into or within the Sierra in recent decades. Since the first Great-tailed Grackles crossed the Colorado River in the 1960s, they have spread rapidly through California. While still largely confined to lower-elevation areas on both slopes, numbers and range within the Sierra increase each year. Many experts thought the grackle expansion was as rapid as any could be, but the spread of the Eurasian Collared-Dove has proceeded with truly breathtaking speed. Just since 2005, this species has gone from a rarity to a common year-round resident on both the East and West Sides, with breeding records now in every Sierra county.

As noted earlier, the recovery of populations of Osprey and Bald Eagle have allowed those species to spread their breeding range into the Sierra. In the first 25 years of the BBS, only 3 Ospreys and no Bald Eagles were recorded on any Sierra routes. Since the early 1990s, nearly 100 Ospreys and 15 Bald Eagles have been

TABLE 2 *(continued)*

Species	CBC	BBS	BBS[a]
Common Raven	+	+	+
Tree Swallow	n/a	+	+
Oak Titmouse	+	+, ns	−, ns
White-breasted Nuthatch	+	+, ns	−, ns
Bewick's Wren	+, ns	+	−, ns
Blue-gray Gnatcatcher	n/a	+	+
Western Bluebird	+	+, ns	+
American Robin	+	−	+
Cedar Waxwing	+	n/a	n/a
Yellow-rumped Warbler	+, ns	+	+, ns
MacGillivray's Warbler	n/a	+	+, ns
Western Tanager	n/a	+, ns	+
Spotted Towhee	+	+	+, ns
California Towhee	+	+, ns	+, ns
Rufous-crowned Sparrow	+	+	n/a
Song Sparrow	+, ns	+, ns	+
Golden-crowned Sparrow	+	n/a	n/a
Dark-eyed Junco	+	+, ns	−, ns
Lazuli Bunting	n/a	+	+
Great-tailed Grackle	+	+	n/a
Brown-headed Cowbird	+, ns	+, ns	−
Lesser Goldfinch	+	+	+

[a]BBS data from Sauer et al. 2011.
ns = Trend not statistically significant. + = Positive population trend. − = Negative population trend. n/a = Insufficient data to analyze.

recorded, and these species are now consistently found on nearly a half-dozen BBS routes. It is less clear why Red-shouldered Hawks have increased so dramatically in California in recent decades, but that population growth has been accompanied by a spread into the Sierra with the species now much more numerous on the West Side and regular in areas on the East Side, where they were never observed before the 1990s (see Figure 13).

Warmer temperatures in the Sierra in the past several decades may have encouraged some species to move upslope. Hutton's Vireos are now recorded on twice as many BBS routes than in the 1970s and 1980s, due mainly to the species now showing up on higher-elevation West Side routes. Anna's Hummingbirds have also shown a breeding season upslope trend on both the East and West Sides, probably in response to widespread use of feeders. Mourning Doves, recorded on fewer than half of the Sierra CBCs prior to the 1990s, are now found on nearly 90 percent of the counts. We are also beginning to see Hooded Orioles creeping up into western foothills following development (and planted palms).

Common Ravens are now found regularly on lower-elevation West Side BBS routes and

TABLE 3
Species showing significant negative trends.

Species	CBC	BBS	BBS[a]
Rough-legged Hawk	–	n/a	n/a
American Kestrel	–	–	–, ns
Killdeer	–, ns	–, ns	–
Mourning Dove	+	–	n/a
Downy Woodpecker	+, ns	–, ns	–
Olive-sided Flycatcher	n/a	–	–
Western Wood-Pewee	n/a	–	–
Black-billed Magpie	–	–, ns	n/a
Horned Lark	–	n/a	n/a
Violet-green Swallow	n/a	–	–
Barn Swallow	n/a	–, ns	–
Mountain Chickadee	–, ns	–, ns	–
Bushtit	+, ns	–, ns	–
Ruby-crowned Kinglet	+, ns	–, ns	–
Wrentit	n/a	+, ns	–
American Robin	+	–	–
Orange-crowned Warbler	n/a	+, ns	–
Yellow Warbler	n/a	–	–
Wilson's Warbler	n/a	–	–
Yellow-breasted Chat	n/a	–	n/a
Chipping Sparrow	n/a	–	–
Lark Sparrow	–, ns	–	–, ns
Brewer's Blackbird	–, ns	–	–
Bullock's Oriole	n/a	–, ns	–
Purple Finch	–, ns	–	–
Cassin's Finch	–, ns	–, ns	–

[a]BBS data from Sauer et al. 2011.
ns = Trend not statistically significant. + = Positive population trend.
– = Negative population trend. n/a = Insufficient data to analyze.

CBC circles where they were historically rare; they have also expanded their range into the high Sierra in dramatic fashion (see Common Ravens in the "Family and Species Accounts" chapter). Somewhat counterintuitive has been the apparent southward expansion of the breeding ranges of Buffleheads and Ring-necked Ducks in the Sierra. Hooded Mergansers, whose breeding range has extended southward into the Central Valley and whose wintering populations in California have increased in the past 20 years, are now found much more frequently on Sierra CBCs.

Barred Owls have been expanding their range westward and southward throughout the Pacific Northwest and have become relatively common in northwestern California. This expansion into the range of the Spotted Owl, a species already in decline due to habitat loss, has added another threat to that species. Barred Owls compete with, hybridize with, and even prey upon Spotted Owls. This species has only recently begun to spread into the Sierra, but records as far south as the Grant Grove in Tulare County suggest that it is just a matter of time before we begin to see similar negative impacts on the Sierra Nevada subspecies of Spotted Owl.

Many readers may be surprised to learn that Chestnut-backed Chickadees are a relatively recent addition to the Sierra avifauna. In Joseph Grinnell's (1904) assessment of this species' range, he considered it entirely absent from the range. Before 1951, the only record of this chickadee in the Sierra was a single specimen collected in Butte County (1939). In 1951 flocks were noted in El Dorado and Calaveras Counties and Chestnut-backed Chickadees have since spread through the lower-mid-elevations of the West Side of the range as far south as Madera County south of Yosemite National Park. The subspecies responsible for this colonization (*Parus rufescens rufescens*) ranges from Mendocino County northward. There is no indication that birds from the central or southern California coast (which have much paler sides than this subspecies) have made the leap across the Central Valley. It appears that this colonization has had little or no effect on Mountain Chickadee numbers, probably because of the Chestnut-backed Chickadee's preference for wetter, cooler areas with Douglas-fir, madrone, and other moisture-loving trees. As a result of this preference, the distribution of Chestnut-backed Chickadees in the Sierra is quite patchy.

RANGE CONTRACTIONS

While the news is mostly good, with many more Sierra bird species increasing in numbers and range than declining, we do have several species that have seen their ranges shrink substantially. For the most part, we have very poor understanding of the causes behind these changes. Two of the more perplexing examples of Sierra breeding range contractions are those of Ruby-crowned Kinglet and Swainson's Thrush. Based on their status in the early 20th century, Grinnell and Miller (1944) considered both species to be common on the West Side of the Sierra, from the northern edge of the range south to Tulare County. Breeding populations of both species are now highly localized and very sparsely distributed in the region. BBS routes rarely record either species, even in places where the species was considered to be historically common.

In the past few years the Grinnell Resurvey Project has systematically surveyed areas covered by Joseph Grinnell and his associates and found both species missing from nearly all the locations where they were previously present (Moritz 2007). No convincing hypothesis has surfaced to explain these range contractions. CBC data suggest that the wintering population of Ruby-crowned Kinglets in California is stable. BBS data show declines for the kinglet in the Sierra and Cascade Range but not in the Rockies. Swainson's Thrushes breeding in coastal California, while showing declines based on BBS data, still occupy most of their historical range in that region.

Greater Roadrunners, once relatively common and widespread in the western foothills of the Sierra, are now patchily distributed and rarely encountered in much of this part of their range. Historically, throughout most of California, conversion from rangeland to more intense forms of agriculture and increasing human development has coincided with the disappearance of this species. At least into the 1930s, Willow Flycatchers were still considered common throughout the Sierra in any area with suitable riparian/willow habitat. Today the species is the scarcest of all our Sierra flycatchers and persists in only few areas. Habitat loss from reservoir construction and water diversions and habitat degradation from overgrazing, combined with the incursion of Brown-headed Cowbirds into the Sierra, have all been blamed for their decline.

With Sierra temperatures predicted to continue rising, vegetation zones are likely to move upslope with changing climate. This could push species using the highest elevations, like American Pipits and Gray-crowned Rosy-Finches, toward extirpation from the Sierra. Because their breeding areas are difficult to reach, we have little data on breeding season trends for either species, but winter numbers of American Pipits in California have shown significant declines over the past 30 years.

Unanswered Questions about Sierra Birds

As much as we know about Sierra birds, there is even more we do not know. Unanswered questions abound, and many of these questions offer opportunities for amateur naturalists to make significant contributions. We urge all visitors to the Sierra to take careful notes and record all their observations using eBird (http://ebird.org/content/ebird/). Below we list just a few of these questions. Please refer to the "Family and Species Accounts" section of this book for more details.

What is the current breeding range of the Harlequin Duck in the Sierra?

Recent observations suggest that these rare ducks are beginning to recolonize the Sierra, but all known occurrences are in steep river canyons where most birders never venture. Careful breeding season observations by sturdy hikers who pay attention to birds will give us a much better understanding of their current distribution and conservation status.

What are the current breeding ranges of Ring-necked Ducks, Buffleheads, and Hooded Mergansers in the Sierra?

In recent years all three of these ducks appear to have extended their breeding ranges farther south into the Sierra than previously thought. Careful breeding season observations at mountain lakes could yield important information on distributional changes of these species.

Are rapidly increasing populations of Eurasian Collared-Doves and Great-tailed Grackles having an impact on other birds?

The rapid expansions of both these species in California seems likely to affect other birds. By making consistent obsevations of numbers of birds in a given area and by recording any changes in their breeding status and behavior, birders could help reveal impacts of these new arrivals.

How many Black Swifts breed in the Sierra?

Although we are aware of only a handful of breeding locations for this species in the Sierra, recent intensive surveys of likely sites (waterfalls with a specific set of characteristics) in the Rocky Mountains revealed many more birds than expected. Exploring similar sites in the Sierra could produce similar results.

Where do Williamson's Sapsuckers and Pine Grosbeaks spend the winter?

Neither species is commonly found at high elevations in winter. The sapsuckers are occasionally found at lower elevations (even down to the Central Valley) very rarely, but the grosbeaks seem to simply vanish in winter. Hardy birders venturing into the high Sierra in winter, and very observant birders anywhere, could add much to our knowledge of the movements and winter whereabouts of these species.

Do Black-backed Woodpeckers require recently burned forests?

The California Department of Fish and Game is currently evaluating whether this species warrants listing as either Threatened or Endangered. The species apparently prefers intensively burned forests over unburned forests and forests that have burned at lower intensities, but it also uses unburned forests. A better knowledge of their dependence on burned forests, and their occurrence in unburned forests, would be of great value to public agencies and private landowners.

What is the extent of the "return" of Bell's Vireos to the Sierra?

Previously extirpated from the region, this species appears to be staging a comeback. Birders would do well to learn their distinctive songs and be on alert for new locations, especially in riparian forests of the southern Sierra.

How are Purple Martins faring in the Sierra?

Recent increases in the extent and intensity of wildfires has created more areas with large fire-killed trees (snags) that may provide nesting habitat for martins. Visits to these landscapes in early summer could reveal new nesting locations as well as information about the presence of European Starlings and other competitors for nesting sites.

What are the exact range limits of the two Sierra subspecies of White-breasted Nuthatch?

These subspecies (which may someday be recognized as separate species) have distinct calls, as described in the "Family and Species Account" section of the book. Birders who learn the difference can help us understand the true ranges of these taxa in the Sierra.

What is the range of Pygmy Nuthatches on the West Side?

This species spills over onto the West Side here and there, especially in the northern Sierra. We do not have a precise knowledge of all the places where these nuthatches regularly occur, or of their seasonal status at these locations.

What happened to breeding Ruby-crowned Kinglets and Swainson's Thrushes in the Sierra?

Formerly fairly common and widespread as breeders, both species have disappeared from most of their historical breeding range in the Sierra. Other populations (kinglets in the Rocky Mountains and thrushes of the Pacific Coast) seem to be stable. Any breeding season observations of these species in the Sierra should be reported via electronic discussion lists and/or eBird.

What are the current breeding ranges of American Pipits and Gray-crowned Rosy Finches in the Sierra?

These species have been confirmed to breed at a handful of high-altitude locations. Birders visiting alpine regions of the Sierra in summer should be on the alert for them, as both species are likely to be affected by climate change.

Bird Conservation in the Sierra

HISTORY

Human impacts on the Sierra and its birdlife began thousands of years before European settlement. Native peoples used fire as a management tool to clear brush, maintain grasslands and meadows, make travel easier, and improve browse for game animals. Fire management, hunting, fishing, and gathering by Native Americans affected plant and animal communities in the Sierra and likely altered the relative abundance and distribution of some bird species long before the first Europeans arrived.

Changes caused by native peoples over a 10,000-year span prior to the 19th century paled in comparison to the impacts of European settlers and their large-scale, aggressive use and abuse of Sierra resources. Forests were logged extensively, with the largest trees targeted, and sheep and cattle swarmed over mountain meadows that had never experienced such intense grazing pressure. At lower elevations exotic annual grasses were introduced and dominated the grasslands and savannas so quickly that no naturalist was ever able to ob-

serve or describe the pre-European plant communities. Some speculate that native bunch grasses dominated, but the true nature of these original landscapes and most Sierra plant communities remains unknown.

The discovery of gold in the Sierra foothills in 1848 dramatically accelerated these impacts and added to them demolition of hillsides (typically with powerful water cannons) and diversions and excavations of nearly every stream on the West Side from Butte County south to Madera County, all in a frantic search for gold. The top predators, grizzly bears and wolves, were systematically exterminated. To encourage railroad companies to invest in the transcontinental railroad in the late 19th century, the United States gave large swaths of Sierra land to these companies by granting them every other section (a one-mile by one-mile square) of land along the proposed route. The result was a checkerboard of public-private ownership that has inhibited rational conservation planning ever since. One of the most direct effects on Sierra birds including ducks, geese, quail, and grouse came from "market hunting" to feed burgeoning gold rush populations as well as rampant shooting and persecution of raptors.

We can only speculate about the impacts on birds from all these changes because no comprehensive, large-scale studies of Sierra wildlife were conducted until the early decades of the 20th century. Logging of large trees almost certainly decreased the available habitat for species such as Northern Goshawks, Spotted Owls, and Pileated Woodpeckers. Grazing of mountain meadows degraded these habitats and must have reduced habitat for Willow Flycatchers, Yellow Warblers, and Lincoln's Sparrows. Gold-mining degraded streamside habitats for riparian-associated birds, including a diversity of migrating neotropical songbirds such as flycatchers, vireos, warblers.

Changes in public attitudes about nature and wildlife over the past few decades have eliminated or modified most of these historically damaging practices, and state and federal agencies are now tasked with managing public lands and wildlife resources using science-based approaches that help protect and enhance habitat for Sierra birds. Hunting is now highly regulated, and raptors are protected and mostly revered by the public. Logging, grazing, and mining practices are all regulated in an attempt to balance resource extraction with protection of natural resources, and riparian and wetland areas are being protected and enhanced as never before. The U.S. Forest Service and numerous nonprofit conservation organizations have actively worked to acquire land to eliminate the checkerboard pattern of ownership in the Sierra and strive to restore and enhance ecosystem processes and habitats where possible.

FIRE AND CURRENT FORESTRY PRACTICES

Fire regimes in the Sierra prior to European settlement likely consisted of frequent, low-intensity fires started by lightning or native people, maintaining an open-forest understory and supporting development of large areas of mature forest. Stand-clearing fires certainly also occurred and the resulting landscape was probably characterized by a mosaic of many different plant communities. Frequent chaparral fires maintained broad swathes of vegetation in early successional conditions and fire in grasslands and oak savanna created patches of open, treeless expanses. All these conditions would have produced a variety of habitats supporting a correspondingly high diversity of birds. Aggressive fire suppression in the Sierra began in the early 1900s and became more widespread and effective through the 1950s. Ironically, the major legacy of those years of fire suppression is a landscape that has higher densities of younger trees and denser understory, providing fuels that increase the frequency and size of large-scale, high-intensity fires.

More recently, forestry practices have recognized the important role of fire in the Sierra ecosystem. National Parks like Yosemite and Sequoia now have policies that allow most lightning-caused fires to burn, and also include controlled burns to reduce fuel levels. On publicly owned lands, fire-suppression policies have also been modified, and controlled burns are part of the management strategy. However, consideration given to commercial timber harvesting priorities, air quality, and nearby human residential areas all make implementation of these policies difficult. Experiments using selective logging to change fire behavior to reduce the risk of large stand-clearing fires are being conducted throughout the Sierra. It remains to be seen whether these practices are effective or economically practical. Their impact on wildlife is also difficult to predict. In any case, to the extent that birds have adapted to current conditions, any changes are likely to be detrimental to some species and beneficial to others.

In spite of the fact that old growth forests occupy less than one-fifth of their historical extent, these mature forests are still being logged both on private land and to a lesser extent on National Forests, further stressing birds that require this forest type. Even where protected from logging, the historical lack of fire has increased the density of shrub understory and smaller trees and increased the risk that this dwindling forest type could be lost to high-intensity fire. Clear-cutting and planting of even-aged, single-species tracks continues on private lands with a corresponding decrease in habitat diversity for birds and other wildlife.

One of the most problematic ongoing forestry issues is postfire salvage logging. When stand-clearing fires occur, the resulting landscape looks like a wasteland to the general public and represents an opportunity to harvest many large trees for commercial timber interests. However, a large and rapidly growing body of research (much of it conducted by the U.S. Forest Service) shows that removal of all or most of the largest standing dead trees (snags) is detrimental to a wide variety of cavity-nesting birds such as woodpeckers and bluebirds as well as other species that make extensive use of these landscapes for foraging (e.g., Olive-sided Flycatchers and Black-backed Woodpeckers). Much needs to be done to bring salvage logging policies in line with the best conservation science. In addition, forestry practices like herbicide application to suppress shrub growth and to accelerate regrowth of trees may not only be counterproductive, they may alter the succession of different habitat types that support diverse bird communities.

MOUNTAIN MEADOWS

Historical, unregulated overgrazing was so pervasive throughout the range that most experts in this field believe that not a single pristine Sierra meadow system remains. Many areas of these habitats have been altered right down to the basic hydrology that supports the entire meadow system. Eroded streams incise and cut deep channels, lowering water tables such that even when grazing is removed, the system cannot return to its prior wetland state without active intervention by restoration ecologists. However, grazing is now much reduced and more highly regulated, and efforts to restore

these meadows are producing encouraging results. The U.S. Forest Service has partnered with other federal and state agencies and non-profit conservation organizations to implement multiple restoration projects throughout the Sierra.

DAMS AND WATER DIVERSIONS

No major river system in the Sierra has been spared the impacts of dams or water diversions to provide water for human uses and to control flooding. While reservoirs have destroyed hundreds of miles of riparian habitat and drowned thousands of acres of meadows, likely contributing to the decline of the Harlequin Ducks and many meadow-dependent species, they have also created habitat that many species of water birds such as ducks, geese, and grebes have been quick to exploit. More recently, recovering populations of Bald Eagles and Ospreys are taking advantage of these human-made lakes to the point where they could be more numerous now in the Sierra than historically. Similarly, the massive diversion of foothill streams and the resulting canal systems that began with gold mining have led to the accidental creation of perennial wetlands in the north-central Sierra that have enabled the state-Endangered Black Rail to successfully colonize these areas. These changes are both a testament to the massive destructive power of humans and the remarkable adaptability of birds.

INTRODUCTIONS OF NON-NATIVE SPECIES

While European Starlings, Rock Pigeons, and House Sparrows may be the most visible of the non-native Sierra birds, no species has had a larger impact on native breeding birds than Brown-headed Cowbirds. Originally native to the Great Plains, where they followed herds of bison and pronghorns, cowbirds were unrecorded in California before 1870, although some may have been present on the East Side of the Sierra decades before that. By the 1930s, however, they were common and widespread in the Central Valley and Sierra foothills as they spread northward from Mexico and Arizona, taking advantage of the livestock grazing that accompanied human incursions into the Sierra.

Since then, they have moved steadily into higher and higher elevations. Sierra birds have not adapted to this brood parasite that lays its eggs in the nests of other birds. Effects remain relatively localized at higher elevations but are widespread in much of the low- and middle-elevation habitats that support more abundant cowbird populations. The populations and ranges of susceptible host species such as Willow Flycatchers and Yellow Warblers shrank dramatically because these species were unable to successfully fledge their own young while simultaneously raising a cowbird chick. The introduction of aggressive cavity nesters such as European Starlings and House Sparrows has likely impacted some native cavity-nesting birds. However, the fact that starlings and House Sparrows are mainly associated with urbanized areas has limited their impacts in the Sierra. It is hard to assess whether or not more recent intentional introductions (e.g., Wild Turkeys and White-tailed Ptarmigans) or rapid range expansions (e.g., Great-tailed Grackles and Eurasian Collared-Doves) will affect native Sierra birds in the future.

POLLUTION, PESTICIDES, AND OTHER ENVIRONMENTAL CONTAMINANTS

There may be no more compelling conservation success story than the one surrounding the banning of DDT in the United States in the late 1970s. This pesticide caused eggshell thinning, which had devastating impacts on populations of many bird species (see the "Family and Species Accounts" section of the book for the accounts of Bald Eagle and Peregrine Falcon as examples). Once the link was proven and the compound banned, the affected species recovered to the point that some have been removed from the Endangered Species List.

Continued diligence and careful research on pesticides is needed, however, as hundreds of new chemicals enter the environment every year and there is generally little testing for impacts on wildlife. Some have suggested links between environmental contaminants and recent widespread declines in some species such as Loggerhead Shrikes and American Kestrels. However, no solid evidence has emerged to confirm a link or identify a specific compound. Human development along the shores of Lake Tahoe has contaminated the once-pristine waters there, and air pollution from urban traffic has severely affected air quality in the foothills east of Sacramento, which has been shown to stress pine trees and other native vegetation and could be leading to subtle habitat changes. Direct impacts on birds have yet to be shown in either case, but degradation of water or air quality is likely to have widespread consequences across the entire natural community.

LAND USE CHANGES

Although many recent trends in human attitudes and practices are cause for optimism, land use changes in recent decades pose major threats to some Sierra habitats. The most serious threats are to the grassland, savanna, and chaparral habitats of the West Side foothills. Nearly all these lands are in private ownership and could be developed in the future. In recent decades residential and rural residential development has impacted the foothills more than any other part of the Sierra. In addition, thousands of acres have been converted from relatively wildlife-friendly cattle ranching to orchards and vineyards. As compared to the huge vineyards in the Central Valley that consume entire landscapes, much of the vineyard land in the Sierra occurs in smaller patches within a matrix of natural habitats. However, large contiguous areas of habitat are fragmented into smaller and smaller parcels, and efforts to protect homes and crops from the threat of fire (a natural, and critical, component of these ecosystems) impact all the land adjacent to these new land uses.

SIERRA BIRDS IN A CHANGING CLIMATE

Predicting the impacts of global climate change on the Sierra and its birds is particularly challenging. The nature of California's climate is inherently complex due to the effects of long-term and short-term weather pattern cycles in the Pacific Ocean and the highly varied topography of the state. While California has seen some warming during the past century, the changes are less dramatic than in many other regions of the United States. Most climate models predict a warmer Sierra climate, possibly including more precipitation, but with more of that precipitation falling as rain instead of snow. These models also predict that by the second half of the 21st century, temperatures in the Sierra foothills could increase by 3.5° to 7.5° F and the frequency of extremely hot days (greater than 100° F) could almost double. Data from recent studies from areas of the Sierra first surveyed a century ago suggest that birds are indeed gradually moving (generally upslope) to remain within relatively similar climate zones.

As the climate changes, bird species are expected to shift their distributions independently, in some cases resulting in combinations of co-occurring species that have not been seen before. Species using the highest altitudes for breeding (e.g., American Pipits, Gray-crowned Rosy-Finches) may be unable to find suitable habitat in the future. Changes in winter snowpack and spring temperatures could affect downstream riparian systems by reducing the amount of water and changing the seasonal timing of peak flows. These changes could alter the streamside vegetation and effect birds using those habitats.

HOPE FOR THE FUTURE

A historic 1994 decision by the State Water Resources Control Board reversed a long-term decline of the Mono Lake ecosystem caused by more than fifty years of water diversions from its principal tributary streams. This decision restored the stream flows and will eventually increase the surface elevation of Mono Lake to an average of 6,392 feet, which should ensure that this critical ecosystem will provide suitable habitat for myriad water birds in the future. Ongoing restoration efforts at Owens Lake and the upper Owens River, also degraded by historic water diversions, are also important steps toward restoring essential bird habitats in the eastern Sierra.

Two comprehensive reviews of the conservation status of the Sierra, the Sierra Nevada Ecosystem Project (SNEP) (California Resource Agencies 1996) and the Sierra Nevada Framework analysis (U.S. Forest Service 2001) came to the identical conclusions that this region faces urgent threats. In 2004 the SNEP report helped catalyze the formation of a new state agency, the Sierra Nevada Conservancy, which has the mission of initiating and supporting efforts that "improve the environmental, economic and social well-being of the Sierra Nevada Region, its communities and the citizens of California." This organization is tangible evidence of the high value that Californians place on protection of the Sierra and recognition that land use policies and management must focus on maintaining healthy ecosystems that provide high-quality water, spectacular scenery, and important wildlife habitat for all of California.

SUMMARY

Despite the challenges described here, we have good reason to be hopeful about bird conservation in the Sierra. Never before have so many people cared so deeply about nature in general, and birds in particular, and been willing to work to preserve natural habitats. Regulatory protections in place today would have been unthinkable even 50 years ago, and the new Sierra Nevada Conservancy provides a forum for collaborative, science-based approaches to managing Sierra bird and other wildlife populations. Most of the Sierra above the foothills is in public ownership, and many of the largest private landowners have strong commitments to good land stewardship.

Birds demonstrate to us again and again their astounding capacity to adapt. Indeed, the data summarized in the chapter "Recent Trends in Sierra Bird Populations and Ranges" suggest that more species in the Sierra are increasing than decreasing. More than almost any other organisms, birds are highly mobile, and many species can find and colonize new areas of habitat quickly. As human populations continue to increase and competition for critical resources like water and open space becomes more intense, we must combine our capacity to invent with our unique capacity to appreciate the inherent value of other species and create solutions that maintain viable, diverse populations of birds in the Sierra and elsewhere.

Family and Species Accounts

WATERFOWL · Family Anatidae

Of almost 30 waterfowl species observed regularly in the Sierra, most pass over during spring and fall migration and only about half of those breed in the region. Most members of this family frequent low-elevation marshes, ponds, and reservoirs, but a few visit Alpine lakes and turbulent streams of the high country.

Waterfowl are distinguished from other birds by their bills, which are blunt, somewhat flattened (except for mergansers), with a hard tip, or "nail." All species are excellent swimmers and well adapted to aquatic living, with webbed feet, long necks for underwater feeding, and thick coats of down. The region's ducks can be roughly divided into two groups based on foraging techniques and corresponding anatomy: dabbling ducks, which are generally agile on land and feed by tipping-up in shallow water; and diving ducks, which generally feed by diving and swimming under water. Diver's legs are positioned far back on their bodies to help them dive and swim but render them nearly helpless on land, requiring them to run across the water's surface before taking flight.

Swans and geese usually mate for life, and males and females wear identical plumage. Ducks are sexually dimorphic, with males displaying bright, well-marked breeding plumages and using elaborate courtship displays to attract the more subtly patterned females. Most duck species undergo a postbreeding molt into "eclipse" plumage that leaves the males looking very similar to females for about two months. The timing of this molt usually corresponds to a flightless period when flight feathers are replaced. This is followed by a molt of body feathers that produces the typical bright male plumage in time for the winter courtship season.

Ducks typically mate for a single season, with males taking no part in incubating eggs or caring for young and abandoning their mates soon after the last egg is laid. Waterfowl eggs are white, buff, or greenish in color and lack spots or other patterns. Geese and swans usually lay about five eggs, while most ducks lay five to more than ten. Females of most waterfowl incubate eggs for 20 to 30 days, and the precocial young can walk, swim, and feed immediately after hatching; they can fly after five to seven weeks—sooner for ducks than for swans and geese. Ducks are also sexually dimorphic in voice, due to an asymmetrical enlargement of the male's windpipe (the tracheal bulla). This causes them to emit low grunts and whistles rather than the louder, and perhaps more familiar, vocalizations of females. This diverse, cosmopolitan family includes about 145 living species and 4 recently extinct species. The family name was derived from L. *anas*, a duck.

Greater White-fronted Goose • *Anser albifrons*

ORIGIN OF NAMES "Greater" to distinguish it from the "Lesser" White-fronted Goose *(A. erythropus)*, a native of Eurasia; "white-fronted" for the margin of white feathers surrounding their bills; L. *anser,* goose; L. *albifrons,* white forehead.

NATURAL HISTORY Some hunters call these birds "speckle-bellies" or "specs," perhaps a more apt description than "White-fronted Goose." The bold, horizontal barring on the bellies of adult birds is visible in flight, unlike its namesake white patch ("front") at the base of the bill that can be seen only at close range. White-fronted Geese graze on grasses, fallen grain from harvested fields, and aquatic plants while on the wintering grounds. Winter flocks numbering in the tens of thousands roam rice fields and wetlands in search of abundant sources of food, where they mingle with white Snow and Ross's Geese. White-fronts travel in V-shaped wedges high above the landscape, frequently uttering loud, laughing calls—somewhat reminiscent of high-pitched barking dogs; these double or multisyllabic calls are unlike those of other geese.

STATUS AND DISTRIBUTION Greater White-fronted Geese breed in the high Arctic regions of Alaska and northwestern Canada, and they winter abundantly in California and other southern states. They are among the first of the northern breeders to arrive in California, and wintering flocks can be seen by mid-September. They are also the last geese to depart in spring, and migrating flocks fly over river canyons in the northern Sierra until at least mid-April.

West Side. Uncommon spring and fall migrants, a few individuals land on large, reservoirs; individuals rarely seen in flocks of Canada Geese on Sierra golf courses and wet meadows.

East Side. From late February through early April, thousands congregate in Sierra Valley and Honey Lake prior to departing for northern breeding grounds; uncommon spring and fall migrants and overwintering birds recorded at most large lakes and reservoirs.

TRENDS AND CONSERVATION STATUS White-fronted Geese have made a dramatic comeback across North America, including wintering populations in California. The lowest populations were recorded in the 1970s, but reduced harvest levels implemented in the 1980s allowed their populations to increase dramatically by the mid-1990s. Christmas Bird Count data showed a tenfold statewide increase from lows of fewer than 10,000 in the early 1970s to more than 110,000 in 2010. Creation of wintering habitat has increased winter survival, and earlier snowmelt at northern breeding areas has resulted in increased breeding habitat.

Snow Goose • *Chen caerulescens*

Juvenile

Adult

ORIGIN OF NAMES "Snow" for the white plumage of adults; Gr. *chen,* goose; L. *caerulescens,* bluish, for the blue color morph of this species.

NATURAL HISTORY Thought to mate for life, Snow Geese remain in family groups and have strong family ties maintained for many years. The young do not breed until at least three years old, so family units can be quite large. Flock formations are not as organized as for other geese, and Snow Geese fly in loose wedges and long diagonal lines that may extend for miles. Among the most social of all waterfowl, Snow Geese often concentrate in huge flocks sometimes numbering more than 100,000 birds milling about in productive feeding areas. They fly up to 30 miles between feeding and roosting sites, and large flocks descending into a field can resemble falling snow. Wintering and migratory birds roost on water but forage mostly on land. Entirely vegetarian, they are drawn to recently harvested rice fields and grain stubble but also forage on bulrushes and other marsh plants in winter and on freshly sprouted grasses in spring. Once classified as a separate species, the rare (in California) "Blue Goose" is a dark color morph of the far more abundant Snow Goose and represents a tiny fraction of the species' wintering population.

STATUS AND DISTRIBUTION Although they are abundant winter residents of the Central Valley, Snow Geese are only infrequent migrants over the Sierra in fall (early October to mid-November) and spring (mid-March to early April) that land occasionally on large lakes and reservoirs.

West Side. Uncommon spring, fall, and winter visitors to Lake Almanor but rare elsewhere; flocks sometimes numbering more than 500 birds cross the Sierra every year, primarily north of the Yosemite region.

East Side. Fairly common migrants through Sierra Valley and Honey Lake in spring and fall; rare or casual farther south.

TRENDS AND CONSERVATION STATUS Snow Goose populations in North America have quadrupled since the 1970s, possibly in response to increases in wintering carrying capacity created at refuges and in flooded rice fields. Their population explosion has impacted breeding habitat conditions throughout their range. Breeding birds are having major impacts on Arctic tundra environments by grubbing for roots and tubers disturbing the fragile soil and denuding large areas. Recently liberalized hunting seasons have been implemented to reduce populations to historic levels.

Ross's Goose • *Chen rossii*

ORIGIN OF NAMES "Ross's" for Bernard Rogan Ross (1827–1874), a trader for the Hudson's Bay Company in the mid-1800s who collected many bird specimens, including this goose.

NATURAL HISTORY Smallest of our wintering geese, Ross's Geese weigh only slightly more than Mallards but have much longer wingspans. They fly south from high Arctic breeding grounds in the company of Snow Geese, and the two species flock together in winter. In mixed flocks, Ross's can be identified by their smaller size relative to the larger and more abundant Snows. Ross's also have short, stubby bills and lack the "grin patches" of Snow Geese, and they seldom call in flight, unlike their larger cousins.

STATUS AND DISTRIBUTION Fall and spring migrants over the Sierra, Ross's Geese might be spotted among flocks of Snow Geese on large lakes and reservoirs. "Blue-morph" Ross's Geese can be found rarely in large flocks of "white" geese in the Central Valley but are accidental in the Sierra.

West Side. Rare spring, fall, and winter visitors at Lake Almanor and at other large reservoirs; casual south to Lake Isabella.

East Side. Uncommon in fall and winter in flocks of Snow Geese at Sierra Valley and Honey Lake; rare from Lake Tahoe south through the Owens Valley.

TRENDS AND CONSERVATION STATUS This species experienced a serious decline from the 1950s to 1970s, when its North American wintering population fell below 25,000 birds. However, full protection from hunting and the development of refuges and private wetlands have reversed this trend, and the wintering population in California now exceeds 100,000.

Cackling Goose • *Branta hutchinsii*

ORIGIN OF NAMES *Branta,* Modern English, related to "burnt" referring to the dark coloration of the head and neck; *hutchinsii* for Thomas Hutchins (1730–1789), an American frontiersman, surveyor, and geographer who collected the first specimen of a "small Canada goose" in northern Canada, where he was employed by the Hudson's Bay Company.

NATURAL HISTORY Based on recent studies, the "Canada Goose complex" has been divided into two groups, including seven types of large-bodied Canada Geese (see account below) that mostly breed at lower latitudes and a group of five (one extinct) smaller forms known as Cackling Geese that breed in the high Arctic. Three forms of Cackling Geese visit California in winter, including "Aleutian" Cackling Goose *(B. h. leucopareia),* "Taverner's" Cackling Goose *(B. h. taverneri),* and the form that once had exclusive use of the name, "Cackling" Cackling Goose *(B. h. minima).* The latter form is the one most likely observed with larger Canada Geese in the Sierra. Compared to Canada Geese, Cackling Geese are noticeably smaller, generally darker underneath, with shorter necks, smaller heads, and more delicate bills. The behavior and ecology of Cacklers and Canadas are similar and are covered in the following account.

STATUS AND DISTRIBUTION Cackling Geese are sometimes spotted in flocks of Canada Geese in fall, spring, and winter, but they leave for northern breeding grounds by late March.

West Side. Uncommon fall and winter visitors to the foothills below about 3,000 feet in the northern and central Sierra, and rare farther south; casual or accidental visitors to mountain lakes and meadows up to the Subalpine zone.

East Side. Rare; almost always seen in association with flocks of much more abundant Canada Geese.

TRENDS AND CONSERVATION STATUS "Aleutian" Cackling Goose populations dropped precipitously from the late 1960s to the mid-1980s, and this subspecies was listed as Endangered in 1967. The decline was due to a combination of predation from introduced Arctic foxes on their Aleutian Island breeding grounds, subsistence harvest in Alaska, and sport harvest on the wintering grounds, mainly in California. Careful management, including local control of foxes on the breeding islands and taking into account the need for fair and equitable bag limits among all hunters on the breeding and wintering grounds, promoted their rapid and encouraging recovery in recent decades. In 2001 the "Aleutian" Cackling Goose was removed from the Endangered Species list, and current populations are now well over 100,000 birds. The recent change in taxonomic status from a subspecies of Canada Goose to a full species has inspired birders to pay more attention to Cackling Geese, increasing reports and improving our knowledge of their status and distribution.

Canada Goose • *Branta canadensis*

ORIGIN OF NAMES L. *canadensis,* of Canada, where a large portion of their population lives.

NATURAL HISTORY The loud honking of Canada Geese is often heard before their V-shaped flocks appear overhead. After forming pair bonds in their second or third winter, mated birds usually stay together for life, like other geese and swans. Extremely clannish, Canada Geese stay in family groups throughout the year. These strong social bonds led to the isolation of small breeding populations and the evolution of new subspecies across North America. Seven forms of "large-bodied" Canada Geese are currently recognized, with the most common form in the Sierra being large

Great Basin "Honkers" *(B. c. moffittii)* that weigh up to 14 pounds. This is the only breeding sub-species in the Sierra, but other, smaller subspecies also mingle with Honkers in pastures, ponds, and golf courses in fall, winter, and spring. The smallest and darkest members of the "Canada Goose complex," Cackling Geese *(B. hutchinsii)*, are now recognized as a separate species (see account above).

Honkers nest commonly in the Great Basin and Sierra, and the breeding season usually extends from late February through mid-July. Monogamous pairs select nesting sites in a variety of natural and artificial sites, usually near water. Nesting pairs aggressively defend their territories, hissing menacingly if perceived predators (including humans) or neighboring geese venture too close to the nests. Artificial nesting structures are readily used, especially when located in farm ponds with irrigated pastures or golf courses nearby. They construct nests from local plant materials, usually cattails, tules, or small twigs. Nests are large, with outside diameters two feet or more, but the inner cup is only about nine inches and lined with soft down and plant fibers. Both parents care for the goslings and establish strong family ties that hold them together during migration and on the wintering grounds.

Primarily vegetarians, Canada Geese forage for grasses, clovers, and cultivated grains in pastures, wet meadows, and grain fields. More than most waterfowl, they have adapted their foraging behavior and migrations (or lack thereof) to exploit agricultural crops and other human-created food sources and habitats. Since Canada Geese are quick to use human-made habitats, they often multiply and become a nuisance in such settings because of their abundant droppings and aggressive behavior.

STATUS AND DISTRIBUTION Historically, most of California's Canada Goose population nested farther north in the Cascades and the Great Basin and visited the Sierra in transit between their breeding and wintering grounds. In recent decades, however, creation of reservoirs, artificial ponds, golf courses, and irrigated pastures has expanded the habitat available for these adaptable geese, and they are now year-round residents in many areas.

West Side. Common or locally abundant residents throughout the western foothills north of the Yosemite region; uncommon or rare farther south; at higher elevations nesting occurs at Lake Almanor, Lake Van Norden (in Placer and Nevada Counties), and other large reservoirs.

East Side. Common to locally abundant residents in suitable habitats; representative nesting areas include Sierra Valley, Honey Lake, Lake Tahoe, Bridgeport Reservoir, Mono Lake, and Owens Valley.

TRENDS AND CONSERVATION STATUS Prior to the early 1970s, resident Canada Goose populations outside of northeastern California were small and localized to a few areas in California, including Butte County (the Oroville area) and some breeding birds in the Bay Area. Some of these may have been the progeny of birds released by the California Department of Fish and Game decades ago or from private breeders or hunting clubs. Once a few pairs become established in urban or suburban settings, where they cannot be hunted, they reproduce at remarkable rates and have done so all across the state and country. Sierra populations reflect this overall trend. Before the mid-1980s, Canada Geese were very rarely recorded on any Sierra Breeding Bird Survey routes. They are now routinely found on nearly a third of those routes, and they are now common breeders in suitable habitats on both sides of the Sierra.

Tundra Swan • *Cygnus columbianus*

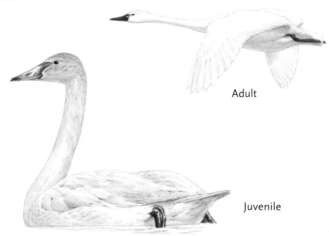

Adult

Juvenile

ORIGIN OF NAMES "Tundra" for the species' high Arctic breeding range; "swan" may be derived from an OE. word meaning "sound," or possibly from L. *sonus*, a sound; L. *cygnus*, swan; L. *columbianus*, of the Columbia River, where the type specimen of the species was collected.

NATURAL HISTORY When sitting on the water with necks erect, Tundra Swans tower above nearby geese and ducks; adults weigh up to 16 pounds and have 7-foot wingspans. Mated pairs stay together for life, and their young become sexually mature after four years. Strong family ties keep related groups together on 5,000-mile round-trip journeys to and from high Arctic breeding grounds. They usually arrive in California by mid-October and depart by mid-March.

Due to their large body mass, swans must run across the water's surface for a considerable distance before taking flight. Once aloft, their flight is rapid and direct, as V-shaped wedges cut through the sky. Like geese, swans have their legs positioned for good balance for walking and roosting in pastures and harvested grain fields. They are also powerful swimmers and forage in shallow water, wet fields, and irrigated pastures. Swans never dive but instead dip their heads and necks in shallow water and occasionally tip up. After a winter of frequent dipping into muddy or alkaline water, their heads may attain a rusty color. Preferred foods include leaves, stems, seeds, and tubers of aquatic plants as well as the seeds and young shoots of rice and other grain crops.

STATUS AND DISTRIBUTION Tundra Swans are mostly confined to rice-growing areas and wildlife refuges of the Central Valley and California's northeastern plateau in late fall and winter. Tens of thousands congregate in rice fields north of Marysville, Yuba County, and in the low foothills along Highway 20 between Grass Valley and Marysville, just west of the Sierra.

West Side. Common spring, fall, and winter visitors to Lake Almanor; rare at other foothill reservoirs north of the Yosemite region; casual migrants south to Lake Isabella.

East Side. Uncommon or rare winter visitors to most large lakes and reservoirs; most records in late December or January.

TRENDS AND CONSERVATION STATUS For more than 60 years Tundra Swans have been protected from hunting in California, allowing their populations to recover dramatically from the early 20th century, when they were shot indiscriminately for their plumage. Meat of older birds was considered too tough for human consumption.

Wood Duck • *Aix sponsa*

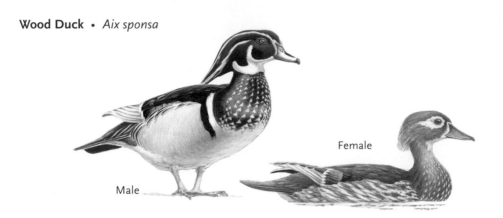

Female

Male

ORIGIN OF NAMES Wood refers to the species' preferred habitat; Gr. *aix,* waterfowl; L. *sponsa,* bride, a reference to the male's elegant plumage.

NATURAL HISTORY The exquisite breeding plumage of male Wood Ducks reflects metallic patterns of green and violet when bathed in sunlight. They are the only North American waterfowl with entirely iridescent wings and backs, and males display more different colors than any other Sierra bird. Females are similar in shape but wear more subdued tones. Their relatively small size (about half the size of Mallards), long tails, and broad wings enable graceful flight through dense woodlands. When disturbed, females make loud, nasal shrieks, and the males give low, squeaky whistles.

As their name suggests, Wood Ducks are partial to forested backwaters, where oaks, willows, cottonwoods, or dogwoods form dense tangles over water. They search for oaks with abundant acorns, an especially favored food. Wood Ducks usually hunt for acorns and other seeds in leaf litter of the forest floor but also land in trees and pluck them from the highest branches. They also forage in water for aquatic plants, insects, and other foods. Wood Ducks usually forage by bobbing at the water surface like puddle ducks but will also make shallow dives for submerged prey.

Courtship behavior begins in midwinter, when females fly to high perches to summon prospective mates; they are unique among Sierra ducks in their ability to perch in trees. Often, several males will compete for a single female until she selects her favorite. Mated pairs show great affection and often preen each other gently on their heads and backs. Females invite copulation by submerging their bodies in shallow water with outstretched heads and necks, as their chosen mate circles nearby.

Cavities in large trees are the Wood Duck's natural nesting sites. However, most pairs now use artificial nest boxes that have been installed widely in the Sierra foothills and throughout the state. Within a day of hatching, females coax their downy young from the nest. The tiny size and fluffy down of day-old hatchlings allow them to flutter uninjured to the ground from heights more than 50 feet and then walk away to the nearest wetland. Mothers alone guard their broods from predators—including raccoons, river otters, feral cats, opossums, large predatory fish (especially largemouth bass), and non-native bullfrogs—that may consume entire broods. The young are highly vulnerable to predators until they attain flight, which requires up to 70 days. After the breeding season, Wood Ducks do not require densely wooded areas and might be seen floating with other ducks on open lakes and reservoirs.

STATUS AND DISTRIBUTION Year-round residents of the Sierra foothills, Wood Ducks have expanded their distribution upslope in response to the creation of ponds and the placement of artificial nesting boxes.

West Side. Fairly common residents and nesters in bottomlands of all the major rivers from the Feather in the north to the Kern in the south; nesting pairs frequent wooded ponds and stream courses (including beaver ponds) up to about 4,000 feet in the central Sierra, where they remain through the winter even during heavy snow storms.

East Side. Uncommon visitors from spring through fall; nesting pairs at Sierra Valley and, at least historically, at south Lake Tahoe; rarely, nonbreeding visitors observed farther south, with records from Bridgeport Reservoir and Mono Basin in spring and fall; casual to accidental in winter.

TRENDS AND CONSERVATION STATUS Extensive clearing of flooded, bottomland forests eliminated Wood Duck habitat throughout North America. According to William Dawson (1923), their California and continental populations were on the verge of extinction by 1913: "Unceasing exposure to gun-fire has brought its ruin. And for what? Simply that the pot might be kept boiling, and the great American belly might be filled . . . Because its flesh was sapid, its bridal array was stripped from it and flung on the dump, while its quivering ounce of meat went into the pot. The lord of creation has dined—but where is the Wood Duck?"

Although excessive hunting pressure and habitat loss greatly reduced their populations in the early 1900s, they are now fairly common in suitable habitats throughout California. Focused efforts by the California Department of Fish and Game, U.S. Fish and Wildlife Service, California Waterfowl Association, and Ducks Unlimited have greatly increased local populations statewide, including in the Sierra foothills. Data from Sierra Christmas Bird Count Circles and Breeding Bird Survey routes show a stunning increase, with numbers up nearly five-fold since the early 1980s. American Kestrels, European Starlings, Northern Flickers, small owls, honey bees, and wasps may compete with Wood Ducks for these artificial homes.

Gadwall • *Anas strepera*

Female

Male

ORIGIN OF NAMES Gadwall is a name of unknown origin; L. *anas* (see family account above); L. *strepera*, noisy, for the loud calls—similar to a female Mallard's.

NATURAL HISTORY Compared to other puddle ducks, Gadwalls wear subtle plumage and lack the bright, iridescent hues that characterize this group. In most places these shy ducks are outnumbered by other species, but they have one of the widest distributions of any of the world's waterfowl and occur on all continents except South America, Australia, and Antarctica.

Gadwalls are primarily freshwater birds but also visit saline and alkaline waters—even in the breeding season. They nest on dry ground near ponds and marshes surrounded by lush aquatic vegetation such as cattails and bulrushes; islands are preferred nest sites. Monogamous during the breeding season, most females are paired by November, many months before they nest in mid-April. Nests consist of shallow depressions lined with aquatic vegetation, feathers, and down. After breeding, Gadwalls roost and forage on large marshes, lakes, and reservoirs in mixed flocks with other puddle ducks and American Coots. They usually forage in deeper water than other puddle ducks, where they upend or make shallow dives for aquatic plants, seeds, and aquatic invertebrates.

STATUS AND DISTRIBUTION Gadwalls are year-round residents and migrants in California, including both sides of the Sierra.

West Side. Uncommon breeders below about 3,000 feet in the central Sierra; confirmed nesting locations include Lake Almanor, Don Pedro Lake (Tuolumne County), Millerton Lake (Madera and Fresno Counties), Lake Success, and Lake Isabella; rare in fall up to about 8,000 feet in Yosemite National Park.

East Side. Common nesters at marshes and creek deltas; in fall, large numbers congregate on large lakes and reservoirs; fairly common in winter.

Eurasian Wigeon • *Anas penelope*

Male

Female

ORIGIN OF NAMES "Eurasian" refers to the typical distribution of the species; "wigeon" from F. *vigeon,* a whistling duck; Gr. *penelope,* a weaver; possibly a mistaken form of *penelops,* duck.

NATURAL HISTORY The bright reddish heads, cream-colored crowns, and silvery flanks of male Eurasian Wigeon stand out in large flocks of wintering waterfowl. Females, however, look nearly identical to female American Wigeon. While Eurasian Wigeon do not breed in North America, some Russian Far East breeders follow flocks of the far more abundant American's on their southward journeys. Larger wintering flocks of American Wigeon (i.e., 50 to 100 birds) may contain at least one Eurasian Wigeon. The wintering behavior and ecology of Eurasian and American Wigeon are nearly identical.

STATUS AND DISTRIBUTION Eurasian Wigeon are one of the most abundant and widespread ducks in Asia and Europe and uncommon winter visitors to California.

West Side. Uncommon to rare November through March, mostly on ponds and lakes in the Foothill zone; recent winter and early spring records from large reservoirs such as Lake Almanor and Lake Isabella; has been nearly annual recently on the Springville Christmas Bird Count (Tulare County).

East Side. Uncommon to rare at large marshes and reservoirs in late fall, winter, and early spring.

American Wigeon • *Anas americana*

Male

Female

ORIGIN OF NAMES "American" to distinguish the species from Eurasian Wigeon; L. *americana*, of America

NATURAL HISTORY Most American Wigeon breed in the prairies and tundra of Canada and Alaska, but they also nest in small numbers in the Great Basin, north and east of the Sierra. Migrants arrive by mid-November to spend the winter in California. Mostly vegetarian, they feed at the surface for such aquatic plants as pondweeds, wild celery, and algae. At times, they associate with divers, such as Lesser Scaup, Canvasbacks, and Redheads, and steal aquatic vegetation after these ducks surface from dives. Wigeon are drawn to wet pastures, urban parks, and golf courses, where they graze on grasses, usually not in association with domestic ducks or geese. In winter, they consume primarily aquatic and terrestrial insects and small mollusks. On hunting days in the Central Valley, American Wigeon and other dabbling ducks sometimes flock to protected ponds and wetlands of the western foothills to escape hunting pressure.

STATUS AND DISTRIBUTION Abundant winter visitors to the Central Valley and coastal California, American Wigeon occur regularly on both sides of the Sierra.

West Side. Fairly common to locally abundant fall, spring, and winter visitors to larger reservoirs such as Lake Isabella, Lake Almanor, and Millerton Lake, and fairly common visitors to ponds and lakes in the Foothill zone; in fall, rare or casual at higher elevations.

East Side. Fairly common in winter and common in spring and fall migration at marshlands, ponds, and slow-moving creeks, with especially large numbers concentrating at Sierra Valley, Honey Lake, Bridgeport Reservoir; breeds on rare occasions in Mono County (most recently at Crowley Lake in 2007).

Mallard • *Anas platyrhynchos*

ORIGIN OF NAMES Mallard was apparently derived from L. *masculus*, male, referring to the drakes; *platyrhynchos*, flat-billed, from Gr. *platus*, flat, and *rhynchos*, a bill.

NATURAL HISTORY Mallards are the most abundant, widely distributed, and best known ducks in the Northern Hemisphere. Bred in captivity since ancient times, they are the ancestors of most domestic breeds, including white "Peking" ducks as well as other barnyard ducks, save for a few breeds that are derived from unrelated Muscovy Ducks that occur in the wild from South Texas to South America.

Female

Male

Highly promiscuous, male Mallards try to mate with as many females as possible—often very aggressively with several males chasing and trying to copulate with a single female on land or while pushing her under water until she nearly drowns, called "rape attacks" by some. Although wild Mallards rarely mate with other species, captive birds have produced fertile offspring with more than 40 different species of ducks and geese. Domestic flocks of ducks in urban parks and barnyards often feature the odd-looking progeny of such mixed-matings.

In flight, Mallards have large, broad bodies and slower wingbeats than most other ducks. They walk proficiently on land and fly directly up from the water when flushed. They mostly feed in shallow water by tipping-up but also make shallow dives for food, unlike most puddle ducks. They primarily consume submerged aquatic plants such as pondweeds, smartweed, and bulrush. In summer they also take animal foods, including the larvae and nymphs of mayflies, stoneflies, and midges.

These hardy and adaptable ducks will nest almost anywhere adequate food supplies exist. Similar to Canada Geese, they adapt rapidly to artificial habitats and are frequent visitors to ponds at farms, golf courses, and urban parks. They hide their ground nests under tall grasses or dense marsh vegetation. Females care for their young for about two months until they are capable of flight.

STATUS AND DISTRIBUTION Mallards are the most abundant ducks in the Sierra, and their distribution has expanded during the past decades with the creation of ponds, reservoirs, and other artificial wetlands, allowing them to colonize almost anywhere.

West Side. Common nesters up to about 4,000 feet in the central Sierra; uncommon at higher elevations, a few pairs nest annually up to the Subalpine zone above 8,000 feet in the central Sierra; in late summer and fall, small flocks of migrants found at all elevations.

East Side. Common nesters and fairly common year-round residents in suitable marsh and lake habitats.

TRENDS AND CONSERVATION STATUS Data from Sierra Breeding Bird Survey routes show a steady increase in numbers and expansion of breeding range throughout the region since the 1970s.

Cinnamon Teal • *Anas cyanoptera*

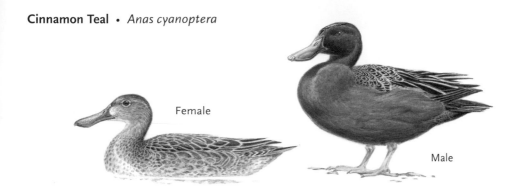

Female

Male

ORIGIN OF NAMES "Cinnamon" refers to the body coloration of males in breeding plumage; *cyanoptera*, blue-winged, from Gr. *chyaneous*, blue, and *pteron*, wing.

NATURAL HISTORY In low light conditions, male Cinnamon Teal appear as small, all-dark ducks, but full sunlight transforms them into birds of breathtaking beauty. As they wheel low over a marsh with rapid, twisting flight, both males and females reveal baby-blue wing patches. Cinnamon Teal perform courtship and form seasonal pair-bonds on their southern wintering grounds, prior to migrating northward to breed. After arriving on their breeding grounds, pairs search for nest sites in wetland habitats; both small ponds and large marshlands are used. Females tunnel under dense, matted marsh plants and create shallow scrapes for nests that are lined by soft grasses and down. Both parents may tend the young until they can fly after about seven weeks.

In the nonbreeding season, these bright-colored ducks occur in small flocks, often in association with other puddle ducks or American Coots. Cinnamon Teal forage primarily at the edges of ponds and sloughs for bulrushes, pondweed, and sedges. They also consume a limited amount of animal food when breeding, mostly bugs, beetles, and snails.

STATUS AND DISTRIBUTION Unlike most puddle ducks, Cinnamon Teal winter primarily in Mexico and fly north to breeding grounds in fresh and brackish marshes in the western United States and southern Canada.

West Side. Uncommon nesters at ponds, reservoirs, and marshes of the lower foothills, spring migrants arrive by early March and remain through mid-September; nesting confirmed at Lake Isabella and Lake Almanor; rare in winter in Kern River Valley and southern foothills; rare visitors above 3,000 feet in the central Sierra during spring and fall migration.

East Side. Common nesters in marshlands from Honey Lake south into the Owens Valley; rare or casual in winter.

Northern Shoveler • *Anas clypeata*

ORIGIN OF NAMES "Shoveler" refers to the species' shovel-shaped bill; L. *clypeum*, shield, another reference to bill shape.

NATURAL HISTORY Called "Spoonies" by hunters, Northern Shovelers have large, flattened bills that resemble spoons. These specialized bills have comb-like teeth (called "lamellae") on the mandibles, through which food is strained from shallow water or mud. They primarily feed in shallow water, where aquatic prey float near the surface; unlike most other puddle ducks, they never forage

Female

Male

on land. Shovelers are highly carnivorous and primarily consume crustaceans, snails, tadpoles, caddisflies, damselflies, and the nymphs and larvae of other aquatic insects. They also use their shovel-like bills to scoop up organic material from muddy pond bottoms. Their highly sensitive tongues make it easy to accept or reject potential food items. Shovelers tend to be monogamous and mated pairs can stay together for several years or more.

Shovelers occur in a variety of wetland habitats ranging from small ponds to large reservoirs and are especially fond of sewage treatment ponds. The primary requirement appears to be the presence of large quantities of invertebrate animal food. They congregate on large bodies of open water for roosting to avoid predators. They prefer ice-free lakes that provide access to pond bottoms, and for this reason they are among the first waterfowl to depart their breeding grounds in Alaska and Canada in late summer.

STATUS AND DISTRIBUTION Northern Shovelers begin to arrive in California by late August, and most depart for northern breeding grounds by early May.

West Side. Uncommon visitors to the lower foothills, migrants observed to above 3,000 feet in the central Sierra; possible nesting at Lake Isabella and Lake Almanor.

East Side. Uncommon but becoming more regular in winter; common during spring and fall migrations at large wetlands, including Sierra Valley, Bridgeport Reservoir, Mono Lake, and Crowley Lake; possible nesters in Sierra Valley.

TRENDS AND CONSERVATION STATUS In the fall of 1948, almost a million Northern Shovelers were recorded at Mono Lake, demonstrating this was once a major staging area for the species in the eastern Sierra. However, after more than 50 years of water diversions from its major tributary streams, Mono Lake's surface area dropped and the lake became too alkaline and salty for most puddle ducks, including shovelers. With the return of freshwater flows to the lake since the mid-1990s, waterfowl are starting to return to the lake, and hundreds of shovelers can now be seen, especially near the springs and creek deltas along the eastern lakeshore. North American breeding populations appear to be stable to increasing. Christmas Bird Counts on both sides are recording higher numbers since the early 1990s.

Northern Pintail • *Anas acuta*

Female

Male

ORIGIN OF NAMES "Pintail" refers to the species' long, pointed tail feathers; L. *acuta,* pointed.

NATURAL HISTORY Among the most stylish waterfowl, male Northern Pintails are large, elegant ducks. Like other puddle ducks, they are primarily surface feeders that dabble in shallow water. Pintails mostly consume vegetable matter such as pondweeds, sedges, and grasses. They are drawn to agricultural crops, especially flooded rice fields where waste grain is abundant. Organic rice fields, especially those with wild rice are preferred. In summer they consume small fish, frogs, mollusks, and the larvae and nymphs of aquatic insects.

Northern Pintails have a wider global distribution than any duck, and breed or winter on all continents except for Australia and Antarctica. A few Northern Pintails nest in the Sierra, but most of their population migrates north or east to breed. They are among the first migrant waterfowl to arrive in California, usually by mid-September; many continue south to winter in Mexico, Central America, or northwestern South America. Pintails nest in open, wetland habitats with low emergent vegetation such as cattails and tules. Nesting sites are usually on islands or small berms surrounded by water. The nest is a small scrape or depression lined with plant material, feathers, and down.

STATUS AND DISTRIBUTION Abundant winter visitors to the Central Valley and coastal California, Northern Pintails occur regularly on both sides of the Sierra but rarely in large numbers.

West Side. Uncommon fall, winter, and spring visitors to ponds and reservoirs of the foothills up to about 3,000 feet in the central Sierra; possible nesters have been observed at Lake Almanor and Lake Isabella.

East Side. Fairly common spring and fall migrants; a few pairs nest annually in marshes of Honey Lake, Sierra Valley, and Crowley Lake; winter status is variable, with few to none some winters and many in others.

TRENDS AND CONSERVATION STATUS In contrast to most species of waterfowl, Northern Pintails have shown generally negative population trends in recent decades. Data from California Christmas Bird Counts show that winter numbers are down significantly. Data from Sierra Christmas Bird Counts are insufficient to determine trends, but most circles are reporting smaller numbers than those seen in the 1970s and 1980s.

Green-winged Teal • *Anas crecca*

Female

Male

ORIGIN OF NAMES "Green-winged" refers to the bright wing patches of males and females; *crecca*, possibly a Latinized word to describe the species cricketlike calls.

NATURAL HISTORY Green-winged Teal are the smallest North American dabbling ducks and weigh less than a third of Mallards. These tiny ducks tend to stay at pond margins and in dense vegetation, sometimes making them hard to see. Like other teal, they have a rapid turning flight, low over water. Nesting sites are in sedge meadows or grasslands near ponds or other sources of permanent water. Females excavate small bowls lined with soft plant material, feathers, and down hidden in dense vegetation surrounded by water. Primarily vegetarians, they forage in shallow water by wading or swimming. Preferred foods are seeds of aquatic plants such as bulrushes, sedges, wild rice, smartweed, and wild millet. In summer they consume mostly animal foods such as beetles, bugs, and dragonfly and damselfly nymphs.

STATUS AND DISTRIBUTION Most of the Green-wing Teal wintering population migrates north and east to breed, but a few remain to nest and stay in California through the summer.

West Side. Uncommon visitors to marshes, ponds, and reservoirs of the foothills in fall, winter, and spring; possible breeding at Lake Almanor, and confirmed breeding at Lake Isabella.

East Side. Common spring and fall migrants to lakes and reservoirs, where they congregate in large flocks in fall and winter; uncommon breeders throughout the region.

Canvasback • *Aythya valisineria*

Male

Female

ORIGIN OF NAMES "Canvasback" for the whitish backs of breeding males; Gr. *aythya,* a water bird; Gr. *valisineria,* named for a genus of water celery *(Vallisneria),* a preferred food of the species.

NATURAL HISTORY Unlike most wide-ranging waterfowl, Canvasbacks are entirely restricted to North America. With whitish backs and sides, males stand out among other ducks on the water or in flight. One of the fastest waterfowl, they can attain flight speeds greater than 70 miles per hour. After running across water to become airborne, they fly in V-shaped flocks powered by direct, rapid wingbeats, often at high altitudes. They are also accomplished divers, capable of frequent, sustained dives in water up to 30-feet deep. However, they usually forage in water 10 feet deep, or less, where they search for preferred foods including clams, snails, fish, tadpoles, and wild celery.

While most Canvasbacks breed in the northern prairies and tundra, a portion of their population nests in marshes of the Great Basin, including the northeastern Sierra. Preferred nesting areas are dense freshwater marshes, where females conceal bulky nest platforms of marsh plants under tall emergent plants such as cattails and tules, over or adjacent to water.

STATUS AND DISTRIBUTION Primarily winter visitors to California, Canvasbacks are common in coastal waters and inland on large, open water bodies.

West Side. Rare winter visitors to foothill lakes and reservoirs but significant numbers sometimes found at Lake Almanor; no breeding records.

East Side. Uncommon spring and fall migrants and summer residents; usually scarce in winter but occasionally present in double-digit numbers on Honey Lake or South Lake Tahoe Christmas Bird Counts; rare breeders, with nesting confirmed only in Sierra Valley.

TRENDS AND CONSERVATION STATUS Canvasback populations have declined from their historical abundance across North America in response to changes in land use, loss of wetlands, and hunting pressure. Prized by hunters, the flesh of birds that have consumed wild celery or other aquatic plants (instead of animal foods) is highly sought after. Canvasback populations have remained fairly stable in recent decades (higher numbers in wet years) in response to wetland protection and restoration on their breeding grounds and the enforcement of hunting limits. Numbers from Central Valley Christmas Bird Counts show a significant positive trend since the 1970s.

Redhead • *Aythya americana*

Male   Female

ORIGIN OF NAMES "Redhead" refers to the head color of breeding males; L. *americana,* of America.

NATURAL HISTORY Female Redheads are considered nest parasites because they often lay their eggs in the nests of other ducks—especially those of other Redheads, Canvasbacks, and some puddle ducks. However, some females lay eggs only in their own nests or are partially parasitic. Because of parasitism, their clutch size is difficult to determine but probably averages about 9 eggs; "dump" nests may have more than 40 eggs laid by several different hens. Reproductive success is generally low in this species, resulting from a variety of causes including interference and desertion by parasitic hens, flooding or drying of active nests, and predation by mammals and predatory birds.

Redheads prefer to nest in fresh emergent wetlands where dense stands of cattails and tules are interspersed with areas of open water. For nesting habitat, they select relatively deep wetlands (three feet or deeper) of at least one acre, with about 75 percent open water and vegetation up to about three feet in height. They also nest in somewhat alkaline marshes and potholes of the interior. Nests are built from marsh plants and secured to tall emergent vegetation; they are usually over water but occasionally on islands or dry ground.

In winter and migration, Redheads forage and rest on large, deep bodies of water and may form rafts far from shore. Food is taken mostly by diving in deep water, but they also forage in shallow water. Unlike most diving ducks that prefer animal foods, Redheads consume mostly submergent, aquatic plants such as pond weeds, wigeon grass, and duckweed, but they also take aquatic insects, grasshoppers, larvae of midges and caddisflies, small clams, and snails.

STATUS AND DISTRIBUTION Redheads are year-round residents in California, but the small breeding population is supplemented by northern migrants in fall and winter. They winter primarily along the coast and at large, inland water bodies.

West Side. Rare or uncommon spring and fall migrants and winter visitors; breeding suspected at Lake Almanor but has been confirmed only at the Kern River Preserve (i.e., Prince's Pond) near Lake Isabella.

East Side. Fairly common spring and fall migrants and summer residents, especially in wet years when freshwater marshes provide large areas of suitable habitat; occasionally present in fair numbers in winter on lakes and reservoirs; breeding documented at Honey Lake Wildlife Area, Sierra Valley, Mono Basin, Crowley Lake, and other Mono County sites.

TRENDS AND CONSERVATION STATUS Once considered the most abundant diving ducks in California, the state's breeding and wintering Redhead population has been greatly reduced since the early 1900s, primarily due to the loss of permanent wetlands throughout the Central Valley and historical hunting pressure. Called "fool ducks" by some hunters, they are easy to decoy and hunt compared to most other waterfowl. Based on declines in the Redhead's overall population size and range, and continuing threats (e.g., ongoing wetland losses, nesting failures, and historical and possibly recent hunting pressure), the state's breeding population was added to the California's list of Bird Species of Special Concern in 2008.

Ring-necked Duck • *Aythya collaris*

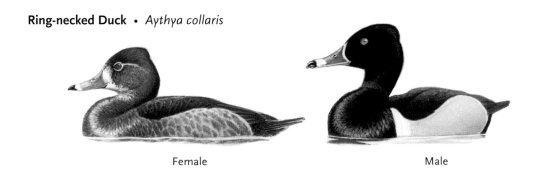

Female   Male

ORIGIN OF NAMES "Ring-necked" for the chestnut-colored neck bands of breeding males; L. *collaris,* collared.

NATURAL HISTORY Despite their name, male Ring-necked Ducks have chestnut collars so faint they are difficult to see even at close range. From a distance their dark heads and backs, vertical white stripes on the leading edge of the flanks, and white ring around the bill tip make much better field marks—they might be more properly named "Ring-billed Ducks."

Ring-necks tend to flock with members of their own kind on farm ponds and reservoirs; they almost never occur on salt water. Like most divers, they must run across the water surface for a considerable distance before taking flight. Wintering flocks frequent large, open bodies of water. In contrast, nesting pairs select smaller, secluded lakes and ponds—often surrounded by forest. Females build nests of grasses and aquatic plants, usually near or over water. Primarily vegetarians, their main foods include pondweeds, smartweeds, and occasionally insect larvae.

STATUS AND DISTRIBUTION Ring-necked Ducks are visitors and migrants to California, including both sides of the Sierra.

West Side. Fairly common fall and winter visitors to low-elevation lakes, ponds, and reservoirs the length of the West Side; rare nesters with breeding only documented at Buck's Lake but probably breeds at other deep lakes of the northern Sierra; pairs in suitable Sierra breeding habitats have been seen in midsummer as far south as Madera County south of Yosemite National Park, suggesting a recent southward-range extension.

East Side. Uncommon fall visitors to larger lakes and reservoirs; hens with broods at secluded lakes in the Tahoe Basin, Meiss Lake (Alpine County), and at Crowley Lake.

Lesser Scaup • *Aythya affinis*

Female   Male

ORIGIN OF NAMES "Scaup" was modified from OE. *scalp,* shellfish, a food of this species; L. *affinis,* allied with or related to, refers to close relationship to the larger Greater Scaup *(A. marila),* a rare species in the Sierra (see Appendices 1 and 2).

NATURAL HISTORY As their name suggests, Lesser Scaup are slightly smaller than their close relatives, Greater Scaup. The two species are often lumped together by hunters who call them "blue-bills," or "bluies," instead of scaup.

Lesser Scaup are restricted to the Western Hemisphere, where they breed in subarctic wetlands of Alaska and Canada and winter as far south as northern South America. While they also occur in coastal waters, Lesser Scaup are much more numerous in California's interior than Greater Scaup. In winter they primarily consume animal foods including crustaceans, mollusks, and aquatic insects obtained by diving. Lesser Scaup tend to feed in shallower water than some other divers, usually about 5 or 6 feet deep, but sometimes they dive up to 20 feet.

STATUS AND DISTRIBUTION Lesser Scaup are winter visitors and spring and fall migrants to California, and there are no breeding records for the Sierra region.

West Side. Uncommon visitors to most large foothill lakes and reservoirs up to about 3,000 feet in the central Sierra.

East Side. Uncommon visitors to most large lakes and reservoirs, annual records from Honey Lake, Sierra Valley, Lake Tahoe, Topaz Lake, Mono Lake, and Crowley Lake.

Harlequin Duck • *Histrionicus histrionicus*

Female

Male

ORIGIN OF NAMES "Harlequin" for this species' multicolored plumage, reminiscent of the brightly colored characters of the pantomime stage; L. *histrionicus,* related to L. *histro,* actor.

NATURAL HISTORY Nesting Harlequin Ducks prosper amid swirling torrents and rapids of mountain streams. According to William Dawson (1923): "A baby Harlequin is as thoroughly at home in wild waters as a baby trout. The trout we may seduce with worm or fly, but until we have devised an equally interesting method for attracting young Harlequins, our meetings are likely to be infrequent." As Dawson implied, Harlequins were considered rare in the state by the early 1920s.

In the late 1870s, Lyman Belding (1891) reported: "I have noticed many of these ducks on the principal streams of Calaveras and Stanislaus counties in the summer. . . . I find young broods from about 4000 feet upward, the earliest apparently hatched about the first of June, or earlier, and have often surprised the mother ducks with their broods hidden in Saxifrage . . . when I approached within a few feet of the brood . . . all would hurriedly swim from me, vigorously using both feet and wings to propel themselves against or with the rapid currents." While no nests of these hardy ducks have been described in California, Harlequins elsewhere nest on the ground, under the shelter of driftwood or rocks, and always beside swift, flowing rivers. They sometimes nest on cliff ledges and in cavities in trees and stumps lined with conifer needles, mosses, or leaf litter. Nest building begins from early May to early June, and they only produce one brood per year. Females care for their precocial young alone, when they often move to slower stretches of nesting streams.

Harlequins are adept underwater swimmers, and they seek clear, cold rapids, where they search rock crevices for aquatic insects including the adults, nymphs, and larvae of caddisflies, mayflies, and stoneflies. They use their huge feet to navigate the bottoms of rushing steams over wet, polished stones—much like American Dippers, which share their summer haunts and their favored prey.

STATUS AND DISTRIBUTION Most Harlequins currently wintering along the California coast are from northern breeding populations, but a few pairs probably nest along remote rivers of the Sierra each year in greatly reduced numbers from their historical population.

West Side. Rare, all historical records from turbulent headwaters from the Stanislaus River south to the upper San Joaquin River and specific nesting localities included Griswold Creek (tributary to the Stanislaus River), South Fork of the Tuolumne River, Cherry Creek (tributary to the Tuolumne River), South Fork Merced River, Lake Ediza (9,300 feet, near the headwaters of the San Joaquin River), and the South Fork Merced River in Yosemite Valley; also observed along the South Fork Kaweah and the South Fork Kings Rivers, but nesting there was not confirmed. Currently rare, casual, or absent from most of the historical breeding range in the Sierra; confirmed breeding records since the early 1970s include above Salt Springs Reservoir on the Mokelumne River, on the Feather River near Thermalito Forebay, and, most recently, on the Merced River in Yosemite National Park in 2002; recent breeding season observations of Harlequins (breeding status not confirmed) include North Fork Feather River, North Fork American River, Rubicon River, Silver Fork of the South Fork American River, North Fork Mokelumne River, Tenaya Creek and the South Fork Merced River in Yosemite National Park, the upper San Joaquin River, and below Friant Dam on the lower San Joaquin River.

East Side. No historical or recent records.

TRENDS AND CONSERVATION STATUS Harlequin Duck was added to California's list of Species of Special Concern in 2008. Despite its rarity and declining status, this species has not been listed as either Threatened or Endangered—mostly because it is seen so infrequently and so little is known about its recent breeding status in the Sierra. The exact cause of the Harlequin's decline is unknown. Increased disturbance from human recreational activities and damming of historical nesting streams have also reduced the suitability of many Sierra rivers for nesting Harlequins. Historical gold mining had severe but unmeasured effects on their riverine habitats and breeding populations.

Bufflehead · *Bucephala albeola*

Male   Female

ORIGIN OF NAMES "Bufflehead" derives from Fr. *buffle,* buffalo, referring to this species' large head; *bucephala* from Gr. *bous,* ox or bull, and *kephale,* head; L. *albus,* white.

NATURAL HISTORY The rounded heads of these dapper little ducks reminded early ornithologists of a buffalo's profile, hence the name Bufflehead. At a distance males look black and white, but close views reveal a bright purple and green iridescence on their heads. They sit buoyantly high on the water and can fly up directly from the surface rather than having to skitter across the water to gain speed like most diving ducks. Buffleheads fly rapidly, flashing white patches on their whirring wings. Much like their close relatives, the Goldeneyes, they dive for small fishes or search the bottom oozes for shellfish, aquatic insects, and other prey; less commonly they feed at the surface like puddle ducks.

Most of California's Bufflehead population breeds in forested mountain lakes of the Cascades and farther north, but recently they have been confirmed nesting in the Sierra. In April, drakes begin to actively court females and threaten other males by swimming at them with heads lowered and wings flapping. Unlike most ducks, Bufflehead pairs usually remain together for years. Their preferred breeding habitats are small ponds lined by conifers or aspens. Similar to Wood Ducks, they nest in tree cavities, usually larger ones excavated by Northern Flickers or Pileated Woodpeckers. Other cavity-nesting birds such as Western Bluebirds and European Starlings are considered competitors for suitable nest sites. Breeding activities begin in early May, and peak nesting extends from mid-May until late July. Females line their nest holes with down and feathers before laying eggs; the young remain in the nest for about a day before jumping to the ground and joining the hen on the nearest water and become independent after about 50 days.

STATUS AND DISTRIBUTION Similar to Ring-necked Ducks and Hooded Mergansers, Buffleheads are primarily boreal ducks that have recently extended their breeding range southward into the Sierra.

West Side. Uncommon breeders; prior to the mid-1990s, Sierra nesting confirmed only at Lake Almanor and near Buck's Lake; recent observations have confirmed localized nesting at secluded, tree-lined lakes in Sierra, El Dorado, Placer, and Alpine Counties—most of these localities are above 6,500 feet; fairly common winter visitors and spring and fall migrants to deeper reservoirs, lakes, and ponds, mostly in the foothills below about 3,000 feet in the central Sierra.

East Side. Recent breeding confirmed in Sierra, El Dorado, Alpine, and Inyo Counties; fairly common spring and fall migrants and winter visitors to larger lakes and reservoirs.

Common Goldeneye • *Bucephala clangula*

Male   Female

ORIGIN OF NAMES "Goldeneye" for the species' bright golden eyes; *clangula*, the diminutive form of L. *clangor*, noise, a reference to the whistling sounds of the species' rapid wingbeats.

NATURAL HISTORY Common Goldeneyes were once called "Whistle-wings" or "Whistlers" for the distinctive whistling or squeaking sounds made by their wings in flight. They breed at boreal lakes and marshes throughout the Northern Hemisphere. In winter, they can be found along the coast and in all but the desert regions of the United States, where they frequent large bodies of water. Migratory flocks consisting mostly of juveniles begin to arrive in California by mid-October, but the bulk of their population remains on the breeding grounds until freezing conditions force them southward. While they form large rafts along with other diving ducks in deep water, Common Goldeneyes usually forage near shorelines in water less than 12 feet deep. Preferred foods include mollusks, crustaceans, aquatic insects, seeds, and tubers of aquatic plants. On Sierra lakes and rivers, they sometimes associate with Common Mergansers.

STATUS AND DISTRIBUTION Common Goldeneyes winter at higher elevations of the Sierra than most other diving ducks, except for Common Mergansers and Mallards; there are no breeding records.

West Side. Fairly common winter visitors and spring and fall migrants to most large lakes and reservoirs, as well as slow-moving stretches of major rivers from the low foothills to the Upper Conifer zone.

East Side. Fairly common visitors from November through March from Sierra Valley south; more common in recent decades than in the past.

Barrow's Goldeneye • *Bucephala islandica*

Male   Female

ORIGIN OF NAMES "Barrow's" for Sir John Barrow (1764–1848), an Englishman who promoted Arctic exploration; L. *islandica,* of Iceland, one of the species' breeding areas.

NATURAL HISTORY In the Sierra, Barrow's Goldeneyes are usually found in association with their much more numerous relatives, Common Goldeneyes, on large, deep lakes and rivers. Barrow's Goldeneyes probably have always been rare nesters in California, as most of their population breeds at secluded lakes of the Cascades, Rockies, and mountainous portions of Alaska and Canada. Similar to Buffleheads and Wood Ducks, Barrow's Goldeneyes nest in tree cavities such as abandoned woodpecker holes. Birdwatchers in the Sierra should be alert for these strikingly beautiful birds when scanning flocks of Common Goldeneyes.

STATUS AND DISTRIBUTION Historically at least, Barrow's Goldeneyes were observed breeding in the Sierra, but there are no recent nesting records, despite extensive systematic and incidental surveys in formerly documented nesting areas.

West Side. Uncommon, before the 1940s a few nesting records from the Lassen region south to Fresno County, mostly from high-elevation lakes bordered by forests providing tree cavities for nesting; currently uncommon but regular in winter at Lake Almanor, on the Feather River at the De Sabla Reservoir, and at a few forested lakes and rivers of the central Sierra up to the Lower Conifer zone, with records from the Sonora and Valley Springs wastewater treatment ponds and Moccasin Reservoir (Tuolumne County); casual at higher elevations and south of the Yosemite region.

East Side. Fairly common along the Truckee River west of Reno and at the Truckee gravel ponds (Nevada County), with several birds present every winter; casual in fall and winter at Sierra Valley, and Lake Tahoe, but accidental farther south; careful searching of large goldeneye flocks on larger rivers, lakes, and reservoirs could reveal more records of this species.

Hooded Merganser • *Lophodytes cucullatus*

Female   Male

ORIGIN OF NAMES "Hooded" refers to the species' distinctive, crested head; L. *merganser*, "a diving goose"; *lophodytes* from Gr. *lophion*, crest, and *dytes*, a diver; *cucullatus* from L. *cucullata*, a crest.

NATURAL HISTORY Of the three species of mergansers that occur annually in the Sierra, only Hooded Mergansers are restricted to North America, as Common and Red-breasted Mergansers also occur in Europe and Asia. Hooded Mergansers are the smallest and most dramatically colored of the three. At a distance, males might be confused with Buffleheads but their sides are tan, instead of white, and their delicate bills are long and pointed. Unlike the other mergansers that consume primarily fish, Hooded Mergansers also dine extensively on crayfish, amphibians, and aquatic insects—dragonfly nymphs are especially preferred. Their habitat preferences also differ from the other mergansers since they avoid the large lakes and rivers frequented by Commons and salt water preferred by Red-breasted Mergansers. Instead, Hoodeds in the Sierra seek secluded ponds and lakes in winter. Seldom pursued by hunters due to their fishy-tasting flesh, they are still wary ducks that dive or swim for cover when humans approach.

Most of the Hooded Merganser's western population breeds along the Pacific Coast from southeastern Alaska south to Oregon. Historically they were not considered a breeding species in California, but they have expanded their range southward into the state in recent decades. Hooded Mergansers nest in tree cavities and make use of old woodpecker holes as well as artificial nesting boxes that have been installed widely in California for Wood Ducks—most of the state's recent nesting attempts have been in boxes.

STATUS AND DISTRIBUTION Grinnell and Miller (1944) described the Hooded Merganser's status in California as: "Winter visitant, relatively rare . . . never within history appreciably more numerous than now." Since the mid-1960s, however, more than 100 Hooded Merganser nesting attempts have been documented in at least 20 California counties (Pandolfino et al. 2006). The number of wintering birds in the state has also increased in recent decades.

West Side. Fairly common and widespread winter visitors and spring and fall migrants to wooded lakes and ponds of the foothills regularly up to about 4,000 feet in the central Sierra; most recent nesting records from Plumas, Sierra, and El Dorado Counties, rare or casual south of the Yosemite region.

East Side. Common winter visitors to the Tahoe region, where they are regularly observed Lake Tahoe, Donner Lake, Glenshire Pond (Nevada County), the Truckee River west of Reno; otherwise uncommon migrants and rare in winter; no breeding records.

TRENDS AND CONSERVATION STATUS As noted above, the breeding range of this species has expanded into northern California in recent years. Data from California Christmas Bird Counts show a dramatic increase in the winter population as well. Sierra counts show an even greater positive trend, with numbers up nearly six-fold from the early 1980s.

Common Merganser • *Mergus merganser*

Female   Male

ORIGIN OF NAMES "Merganser" from L. *mergus,* a diver.

NATURAL HISTORY These expert underwater hunters require fairly clear water to see their prey. Their long, thin bills have horny tooth–like projections that prevent slippery fish from escaping their grasp. This feature has earned them the epithet "sawbill." Due to their large size, they are also called "Goosanders" in Europe.

Although they are mostly fish-eaters, Common Mergansers also forage for a diversity of aquatic insects, crayfish, mollusks, amphibians, small mammals, birds, and aquatic plants. These mergansers are often accused of depleting fisheries, but their total impact on fisheries is negligible since they rarely occur in large numbers in any given area. Like other diving ducks, Common Mergansers are clumsy on land and must run across water to take flight. Once flying, their long, straight bodies knife through the air just above the water's surface, flashing white wing patches. They can be told from all other Sierra ducks by their sleeker profiles and long, narrow bills.

Common Mergansers nest in tree cavities or rock crevices near large streams or lakes; rarely they use artificial nest boxes. Breeding begins in April, and the young can navigate the swiftest streams within a day of hatching. One merganser duckling attempting to cross the Merced River at flood stage was sucked under the waves only to appear about 100 yards downstream, unhurt, and bobbing like a tiny cork. Females tend their young alone until they are independent after about five weeks.

STATUS AND DISTRIBUTION Aside from the ubiquitous Mallards, Common Mergansers are the most abundant and widespread breeding ducks in the Sierra, where they reside year-round.

West Side. Fairly common along most large creeks and rivers offering clear water and plentiful supplies of fish; most Sierra breeding records are from forested lakes and streams below about 8,000 feet; postbreeding birds can range to above 9,000 feet in the central Sierra in late summer and fall; flocks of 5 to 30 individuals remain on ice-free lakes and rivers at the highest elevations year-round, but most move to lower-elevation lakes, rivers, and reservoirs for winter; thousands winter on Lake Isabella.

East Side. Common at Lake Tahoe; fairly common to common in winter and during spring and fall migration at most large rivers, lakes, and reservoirs.

Red-breasted Merganser · *Mergus serrator*

Female   Male

ORIGIN OF NAMES "Red-breasted" for the coloration of breeding males; L. *serrator,* a sawyer, a reference to the species' saw-toothed bill.

NATURAL HISTORY Primarily winter visitors to coastal estuaries and nearshore coastal waters, Red-breasted Mergansers rarely stray inland in winter. Like all mergansers, they are adept divers that forage for small fish; they also take mollusks, crustaceans, aquatic insects, and amphibians—captured in shallow water, but they are capable of diving up to about 30 feet. This species is often overlooked inland, especially when associating with the similar-appearing Common Mergansers. Male Red-breasted Mergansers are easily distinguished by their crested heads, grayish sides, and white neck rings set off by a reddish breast. Females, juveniles, and males in nonbreeding plumage are more difficult to identify but, compared to Common Mergansers, they are smaller, with more slender necks, thinner bills, and lack the distinct white throat patches of female Common Mergansers.

STATUS AND DISTRIBUTION Red-breasted Mergansers are winter visitors and spring and fall migrants, and there are no breeding records for the Sierra region.

West Side. Casual at Lake Almanor and Lake Isabella, mostly single birds seen far from shore.

East Side. Uncommon but regular fall, winter, and spring visitors to Lake Tahoe; rare at Boca Reservoir and Prosser Creek reservoirs, and from Crowley Lake south to Owens Valley.

Ruddy Duck · *Oxyura jamaicensis*

Male Breeding

Female   Male Nonbreeding

ORIGIN OF NAMES Ruddy describes the bright reddish plumage of breeding males; Gr. *oxyura,* sharp-tailed; L. *jamaicensis,* of Jamaica, where the type specimen of the species was collected.

NATURAL HISTORY Stout, chunky birds with thick heads and necks, Ruddy Ducks sit low in the water, and breeding males often cock their stiff tail feathers upward. Along with other "stiff-tailed ducks," they differ from other waterfowl in many aspects of their biology. Males wear grayish-brown plumage for most of the year and do not molt into their bright chestnut or "ruddy" breeding plumage until March, much later than other ducks. Males display startling blue bills only when breeding, and these are used as part of elaborate and distinctive courtship displays to attract prospective females. Nest building occurs from mid- to late May, and nests are large mats of aquatic plants gathered into loose, floating platforms that rise or fall with changing water levels; most are screened from above by domes of overhanging plants. Females begin laying eggs by late May, often when nests are still under construction.

Ruddy Ducks become airborne by running across open water with their short wings beating furiously. They sink out of sight rather than flying to escape danger, much like grebes and loons. Unable to walk on land, they never stray from water, and larger lakes and ponds are preferred. They forage by skimming the water's surface and by diving in shallow water up to about three feet; favored foods include aquatic invertebrates, zooplankton, and some aquatic plants—mostly seeds and roots.

STATUS AND DISTRIBUTION Ruddy Ducks may dot the surfaces of deep lakes, reservoirs, and sewage ponds of the Sierra in spring, fall, and winter, and nesting has been documented on both sides.

West Side. Fairly common spring and fall migrants and winter visitors, mostly in the foothills; rare individuals observed in large lakes as high as the Subalpine zone; nesting confirmed at Isabella Lake and possibly Lake Almanor; a family group was seen at 9,100 feet elevation at Upper Chain Lakes in Yosemite National Park.

East Side. Common to abundant in winter, fall, and spring, sometimes the most common ducks on large lakes and reservoirs; nesting confirmed in Sierra Valley, Tahoe Basin, Mono Basin, and Bridgeport Reservoir.

QUAIL • Family Odontophoridae

New World quail are medium-sized birds only distantly related to the quail of the Old World, but both are named "quail" for their similar appearance and habits. The Western Hemisphere species are in their own family and range from the cold, high deserts of Canada to the rainforests of southern Brazil. Members of this family generally have relatively short wings and tails, and short, powerful legs. While capable of short bursts of fast flight, especially when pursued or disturbed, they mostly travel on foot. Female quail usually lay large clutches of 10 to 20 buffy-white or brownish eggs and incubate them alone for 20 to 25 days. The precocial young can run after hatching and fly in about two weeks. This family includes 32 species but only 2, California and Mountain Quail, occur in the Sierra. The family name is derived from Gr. *odontophoros*, tooth bearing—a possible reference to the sharp toothlike beaks of juveniles.

Mountain Quail • *Oreortyx pictus*

ORIGIN OF NAMES "Mountain" for the species' preferred habitat; quail, OF. *quaille*, possibly from the croaking or quacking sounds of a European species; L. *ortyx*, a quail; *pictus* from L. *picta*, painted, for the species' bright colors.

NATURAL HISTORY Mountain Quail are the largest members of their family north of the tropics. Unlike other New World quail, males and females are nearly identical, with the latter having

Male   Female

slightly shorter head plumes. These large, handsome quail are easy to miss but may be detected from April through early June by the loud, mellow *wook* or *crow* calls uttered by the males only. These ventriloquial calls are difficult to locate and at a distance might be mistaken for the single whistled notes of Northern Pygmy-Owls. These secretive birds seldom fly but instead run to dense cover when disturbed. The young are less cautious and may sometimes be seen at close range as their parents call anxiously, or even feign injury to distract human and other predators. Unlike California Quail, they rarely perch in trees, although scattered trees are usually present in their habitats.

Courting males often strut on the ground or on fallen logs to attract the attention of prospective mates. Females hide their nests in well-concealed hollows on the ground. Both parents care for the young after they hatch in late June or early July. Family groups remain together through the winter and are often joined by nonbreeding adults. Unlike California Quail, they do not typically band together with other families and winter coveys average 5 to 10 individuals.

A striking feature of the Mountain Quail's annual cycle is the attitudinal migration in spring and fall, sometimes covering over 20 miles each way over a period of days or weeks, nearly all of it on foot. During migration, coveys may traverse atypical habitats such as dense coniferous forests and open, rocky areas. They winter below the heavy snow line down to the foothill chaparral, where some coveys live side-by-side with California Quail. When the heavy snow begins to melt in spring, Mountain Quail move upslope again.

Mountain Quail's feeding habits are much like those of California Quail, suggesting that competition between these two species may have contributed to their altitudinal segregation while breeding. Foods of both species consist mostly of plant materials such as berries, seeds, flowers of perennial plants, along with a few insects—especially grasshoppers and ants.

STATUS AND DISTRIBUTION Mountain Quail visit foothill chaparral in fall and winter, but they mainly breed at higher elevations. Mountain chaparral and open forests with shrubby understories are preferred habitats, especially on steep slopes with dense thickets of manzanita, ceanothus, huckleberry oak, and other shrubs, interspersed with rocks or grassy openings.

West Side. Fairly common breeders from the Lower Conifer zone up nearly to tree line; in winter, most individuals descend below about 4,000 feet to foothill chaparral habitats to avoid heavy snow.

East Side. Uncommon residents of scrub habitats and open conifer stands up to the Subalpine zone; localized breeders in open Desert zone habitats south of the South Fork Kern River Valley, especially in the vicinity of desert oases where water is available; relatively easy to find (and to see) near Butterbredt Spring from April well into May; uncommon winter visitors to sagebrush flats along the eastern flank of the Sierra and the western Great Basin.

TRENDS AND CONSERVATION STATUS While the range of the Mountain Quail has contracted from the east, leaving the species reduced or extirpated in parts of Idaho, eastern Oregon, and Nevada, California populations appear to be stable.

California Quail · *Callipepla californica*

Male

Female

ORIGIN OF NAMES "California," the state where the species was first collected; *callipepla*, beautifully dressed, from Gr. *kallos*, beauty, and *peplos*, a robe; L. *californica*, of California.

NATURAL HISTORY This exquisitely plumaged quail is the "state bird" of California. California Quail were a favorite game bird of native Sierra tribes, who snared them along their runways; head plumes and other feathers were used to adorn clothes and head-dresses. Back-road travelers in the Sierra foothills sometimes startle large coveys of California Quail, sending them running for cover or erupting into whirring flight in all directions. Rarely found above the foothills, they take shelter in chaparral, open oak stands, and streamside thickets but mostly feed in grassy openings. They do not migrate and may spend their entire lives in areas of only about two square miles. During the long, dry summers, they seldom venture far from streams, springs, or seeps that supply their daily water. At night, quail roost in heavily foliaged trees but will use dense shrubbery if necessary.

In fall and winter, California Quail feed and roost in large coveys numbering from about 25 to 60 birds. They make a variety of clucks and calls for courtship, aggression, alarm, and maintaining contact. When separated visually, covey members utter a three-note assembly call, *chi-ca-go*, with the second note higher than the others. This call is given frequently in spring by pairs that are separated and by single birds seeking mates. Hollywood film-makers often use these sounds for background in almost any setting, including movies shot in Africa and Australia, hoping their viewers will not notice this gross biological error. Based on old episodes of *Star Trek,* they may be the most widespread birds in the galaxy!

Males and females form monogamous pairs in late winter, gradually leaving the covey in late April or early May to find nesting sites—later than most resident birds. At this time, older unmated males establish small "crowing territories" to attract unpaired females. Their *cow* calls, similar to the last note of the assembly call, are repeated several times per minute, usually from exposed perches in shrubs or trees. Each mated male defends his mate from other suitors but does not hold a nesting territory. Females construct nests by lining a small, well-hidden, ground depression with plant stems and grasses. Young hatch in early June and stay with both parents through their first winter. In late summer, two or more of these family groups band together, along with nonbreeding individuals, to form a covey once again.

The staple foods of California Quail are seeds and leaves of herbaceous plants, especially clover, lupines, tips of grasses, but acorns, fruits, wild berries, and large insects are also important. They usually forage on the ground, picking food from plants or scratching like chickens. Sometimes they move up into shrubs or even leap off the ground to snatch insects or seeds from plants.

STATUS AND DISTRIBUTION California Quail occur on both sides of the Sierra; while still more widespread on the West Side, they have become abundant in many East Side areas.

West Side. Common and widespread in the low foothills from annual grasslands and oak savannas to the Lower Conifer zone; primarily found in foothill chaparral, open oak woodlands, and riparian woodlands near a source of water and grassy areas for foraging; avoids dense conifer stands and is rare or absent above about 3,500 feet in the central Sierra; common in open Desert zone habitats, in the vicinity of oases, as in the South Fork Kern River Valley.

East Side. Uncommon to locally abundant throughout; recorded up to 8,400 feet at Aspendell (Inyo County); their range has expanded from presettlement times likely due to a combination of habitat changes, artificial water sources, and introductions; locally common near desert oases such as Butterbredt Spring.

TRENDS AND CONSERVATION STATUS Professional market hunters killed enormous numbers of California Quail during the 19th century. Quail have very high reproductive rates, so while they are probably less numerous now than before that era, this decline is due mostly to habitat changes, fire suppression, and land development. With enforced hunting limits, these gentle birds are once again common in suitable foothill habitats. Data from Sierra Christmas Bird Counts show a significant positive population trend in the past 30 years due largely to increases from nearly all the East Side count circles.

FOWL-LIKE BIRDS • Family Phasianidae

These stout-bodied birds with strong feet and legs are well suited for dwelling on the ground. Though capable of short, swift flights, they mostly run to escape danger. The turkeys, pheasants, ptarmigans, and chukars of the Sierra were introduced for hunting. The two native grouse, Greater Sage and Sooty, also are considered game birds, despite their declining populations. All Sierra species nest on the ground, making them vulnerable to predators, but the drab, mottled females are difficult to spot when motionless on their cryptic nests. Females usually lay 5 to 15 yellowish, buff, or reddish-brown eggs marked with purple or dark brown spots and incubate them alone for about 20 to 25 days. Their precocious young are well developed upon hatching, with strong legs, open eyes, and thick coats of down. Within hours the young leave the nest and begin feeding while one or both parents lead them to food and provide protection until they are capable of flight in about two weeks. This large, diverse family occurs worldwide except in the Antarctic and contains about 145 living species; 2 of these are native to the Sierra and 4 species have been successfully introduced to the region. The family name is derived from Gr. *phasianos*, a pheasant.

Chukar • *Alectoris chukar*

ORIGIN OF NAMES "Chukar" may refer to the *chuck-chuck-chuck* vocalizations of the species; *alectoris* from Gr. *alektor*, a cock.

NATURAL HISTORY First introduced from northern India (now Pakistan) to North America in the late 1800s, these medium-sized game birds are now established in many areas of the arid West. Subsequent introductions included birds native to mountainous areas from the Alps to the Himalayas, and these were mixed in game farms so that most wild populations now have an uncertain heritage.

Chukars thrive only on steep, rocky hillsides, in open areas with less than 20 inches of precipitation per year, and they avoid cultivated agricultural fields. To escape danger, they run up steep slopes and only take flight when necessary to avoid hunters and their dogs and other predators. At night they roost among rocks or shrubs. Green leaves, small seeds, and a few insects make up their diet. Little is known about the nesting habits of this species in California, since their nests are hard to find and few have ever been discovered in the state. In other areas they create simple grass-lined scrapes concealed under rocks or brush. Courtship begins in mid-March, and pairs with failed first nesting attempts may build second or third nests until at least mid-August. During the dry summer months, coveys of up to 40 or more birds concentrate near sources of water such as springs, natural seeps, and along brushy desert creeks. Similar to Mountain Quail, they descend below the heavy snow line for the winter, mostly on foot.

STATUS AND DISTRIBUTION First introduced to California in 1932, Chukars are now established on both sides of the Sierra, with the largest populations east of the crest.

West Side. Uncommon; breeding populations in the foothills of Fresno County up to about 12,000 feet in Tulare and Kern Counties (including the Kern Plateau), and possibly in the foothills below Yosemite National Park, where first introduced in the 1960s at the Red Hills south of Chinese Camp, and Horseshoe Bend Mountain near Lake McClure.

East Side. Fairly common and widespread below the pine forests along Walker Creek and in the Bridgeport Valley and south to the Mono Basin, and just east of the Sierra in the Bodie Hills; locally fairly common in eastern Lassen, Sierra, and Plumas Counties; fairly common to common in the vicinity of several desert oases south to at least Butterbredt and Jawbone Canyons; has been recorded in Owens Valley.

Ring-necked Pheasant • *Phasianus colchicus*

Female

Male

ORIGIN OF NAMES "Ring-necked" for the male's bright white neck ring; pheasant from Fr. *faisan*, a pheasant; Gr. *phaisianos*, a pheasant; *cholchius* for the Colchis region near the Black Sea in central Asia, where the species is native.

NATURAL HISTORY Ring-necked Pheasants are native to Asia and have been widely introduced to suitable habitats throughout North America, where they are now the most widely hunted upland game bird. Pheasants prefer grain or hay fields, often those with weedy margins or wetland vegetation nearby, where they forage for cultivated grain seeds, fruits, nuts, and insects. Highly adapted to life in cultivated fields and other disturbed areas, Ring-necks probably do not compete with native species of grouse or quail for food or habitat.

The bold plumage of adult males makes them easy to spot in open fields and sparse grasslands, but they can disappear entirely in even the sparsest cover when pursued by hunters or their dogs. Pheasants "freeze" until almost stepped on before erupting into flight with an enormous racket of beating wings and cackling calls. The subtle brown plumage of females enables them to stay mostly hidden, even in plowed fields, but they are almost as noisy and conspicuous as males when flushed. In early April, females begin to select their nest sites in tall vegetation near their winter foraging areas. Hens excavate shallow depressions where they toss nearby twigs, leaves, or grasses together. Chicks can fly and forage independently at about two weeks of age.

STATUS AND DISTRIBUTION Ring-necked Pheasants were introduced into California as early as 1855. They are now abundant residents in agricultural areas of the Central Valley but uncommon or absent from most of the foothill areas.

West Side. Uncommon in the Kern River Valley, the only well-established population; birds are found consistently on a couple of low foothill Christmas Bird Counts in Placer and Tulare Counties.

East Side. Uncommon residents of agricultural fields near Honey Lake with numbers on recent Honey Lake Christmas Bird Counts lower than those seen in the 1980s and 1990s; released into the Owens Valley in the early 1900s with numerous other introductions by the California Department of Fish and Game until 1977; confirmed breeding near Lone Pine (Inyo County) in 2003, but most introductions have been unsuccessful.

Greater Sage-Grouse • *Centrocercus urophasianus*

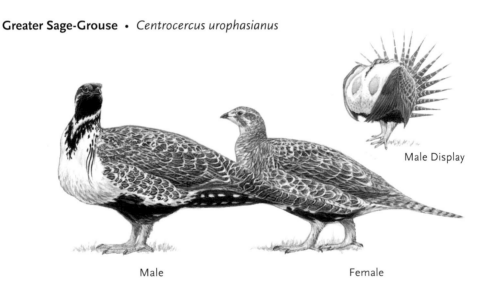

Male Display

Male   Female

ORIGIN OF NAMES "Sage-Grouse" for their dependence on sagebrush; Gr. *centrocercus*, sharp-tailed; *urophasianus* from Gr. *oura*, tail, and L. *phasianus*, a pheasant.

NATURAL HISTORY Perhaps the most inspiring Sierra birding experience is waking up in the predawn chill to view the elaborate and spectacular strutting displays of the Greater Sage-Grouse, the largest native North American grouse. As soon as the snow starts to clear in early March, males gather at traditional "lek" sites to attract females on open hillsides, grassy swales, recently burned areas, or dry lakebeds, surrounded by sagebrush—the species' preferred nesting habitat. Leks are collective display grounds composed of small, individual territories defended by different males.

Strutting displays involve a complex sequence of stereotyped postures characterized by vertical fanning of the long, sharp tail feathers, lowering of wings, and inflating of air sacs to display two olive-green skin pouches. Individual displays last about three seconds but are repeated many times during the early dawn hours. Breeding displays also include an array of strutting calls, fighting calls, low grunts, and loud popping sounds.

Displays and vocalizations serve as an alternative to physical aggression, and displaying males rarely engage in direct physical combat with other males. Leks consist of 20 or more competing males and the oldest ones, "master cocks," are the most successful. Typically one or a few males achieve more than 90 percent of all copulations. Clusters of hens gather together and serve as a sexual stimulus for other females that engage in precopulatory squatting before accepting the advances of the dominant males. After hours of preparation, the copulation only lasts a few seconds. Birders and photographers should exercise extreme caution when viewing Sage-Grouse strutting displays and should only view them from well-concealed sites at a distance of 100 yards—or more.

Once fertilized, the hens leave the strutting grounds and seek out nesting sites under the overhanging cover of sagebrush or tall grasses to provide thermal cover and protection from predators. Hens care for their precocial young for at least several weeks. Young are capable of foraging on their own within a few hours of hatching, and insects such as ants and beetles are an important component of their diet. Preferred foods of adult Sage-Grouse include the fresh leaves and buds of sagebrush as well as grasses, flowers, fruits, and a few insects; in winter, Sage-Grouse subsist entirely on sagebrush leaves, a diet unique among North American birds.

STATUS AND DISTRIBUTION Greater Sage-Grouse are residents of sagebrush flats in the Great Basin, and their range extends into the eastern Sierra.

West Side. No records.

East Side. Uncommon to rare and highly localized; some lek sites in the region persist at isolated locations near Honey Lake, west of Bridgeport Valley, Mono Basin, and Long Valley (Mono County); also just east of the Sierra in the Bodie Hills and White Mountains (Mono and Inyo Counties).

TRENDS AND CONSERVATION STATUS Due to their dependence on sagebrush, Sage-Grouse are adversely affected by land management activities that degrade their habitat, such as overgrazing by cattle and wild horses, unnatural colonization of sagebrush habitats by Utah junipers and pinyon pines, uncontrolled and destructive wildfires, and fragmentation by roads and housing subdivisions. Historical hunting pressure was also a factor in the species' decline, but strict controls since the 1980s suggest that hunting is not a current threat to their population. Such recreational activities as off-highway vehicles and uncontrolled viewing and photography of active leks can result in reproductive failures. "Green energy" projects, such as solar and wind, that destroy thousands of acres of pristine desert or sagebrush habitat, are also a significant but unmeasured threat. For these reasons, the species was included on the 2008 list of California Bird Species of Special Concern.

White-tailed Ptarmigan • *Lagopus leucurus*

ORIGIN OF NAMES "White-tailed" for the species' distinctive field mark; "ptarmigan" is probably derived from a Gaelic word "tarmachan," the "p" is silent and based on historical misspelling; Gr. *lagopus*, hare-footed, a reference to the species' feathered tarsus and feet; Gr. *leucurus*, white-tailed.

NATURAL HISTORY Brown and white in summer, these Arctic grouse molt into pure white plumage in winter to match their snowy surroundings. White-tailed Ptarmigan are silent most of the

Male, Breeding

time, but both males and females utter loud "flight screams" when flushed, and territorial males emit "gobbling" calls to attract females. Males are generally monogamous, but some are polygynous and court two or more females. Pair formation begins from late April to mid-May. After copulation, females use their feet and bills to excavate shallow scrapes on the ground near rocks in snow-free areas. These are later filled with nearby stems and grasses to form a shallow bowl. Typically females lay one egg per day, but during severe snow storms up to eight days may lapse before another egg is laid. The young are brooded by both parents until they can fly and forage alone; broods remain together until the following spring.

White-tailed Ptarmigans are native to Alaska, western Canada, and isolated pockets of the Rocky Mountains—but not the Sierra. The California Department of Fish and Game introduced White-tailed Ptarmigans to the Sierra simply because the Alpine zone had no game birds for hunters. This is especially ironic because nearly all currently inhabited sites are in National Parks where hunting is prohibited. Although there is no documentation of detrimental effects of this introduction, ecologists have long known that non-native animals can have unpredictable, and sometimes disastrous, effects on native plants and animals, which have coevolved over thousands or millions of years.

STATUS AND DISTRIBUTION All White-tailed Ptarmigans in the Sierra originated from a transplant of about 70 birds from Colorado released at two sites near Eagle Peak and Twin Lakes, Mono County, in 1971 and 1972 (Frederick and Gutiérrez 1992). Since then, the species has spread to about 50 miles north, 20 miles west, and 70 miles south of the release sites.

West Side. Uncommon residents; recorded in Humphrey's Basin and Dusy Basin in Kings Canyon National Park and as far north as Carson Pass, El Dorado County; breeding populations occur near the crest in the Yosemite Sierra near Gaylor Lakes, Granite Lakes, Mono Pass, and Dana Plateau, and as far west as Mount Hoffman, near the geographic center of Yosemite National Park.

East Side. Uncommon residents, mostly confined to the Alpine zone along the crest, but spring sightings at about 8,000 feet in Lundy Canyon, Mono County, suggest downslope movements in winter to avoid heavy snow; also observed near Pine Creek, Green Lake, Brown Lake, and Piute Pass, Inyo County.

Sooty Grouse • *Dendragapus fuliginosus*

ORIGIN OF NAMES "Sooty" for dark plumage; formerly called Blue Grouse, now considered a distinct species from Dusky Grouse *(D. obscurus)*; see "Status and Distribution" below; Gr. *dendragapus,* tree loving; *fuliginosus* from L. *fuligo,* sooty.

NATURAL HISTORY Sooty Grouse are more often heard than seen. Males of this species congregate in loose "hooting groups" from late April to early July to proclaim their territories with a series of resonant, booming calls. These calls are lower in pitch than those of any other Sierra bird, so low that some people cannot hear them. Two yellow air sacs on the neck, usually hidden, are inflated to produce these unmistakable notes. Finding a hooting male is seldom easy, because he makes

Male Display

Female

Male

ventriloquial calls from a hidden perch high overhead in the foliage of a dense conifer. Visitors to higher forests may hear, and feel, these haunting calls echoing through the woods for hours at a time.

Most Sooty Grouse in the Sierra breed in the Upper Conifer zone, where they can be found in open or partly brushy slopes with nearby stands of densely foliaged conifers for roosting and hooting. Large trees are preferred—especially Douglas-firs, white firs, red firs, and Jeffrey pines. Adult males arrive on breeding territories by early March, followed shortly thereafter by females and yearling males. Throughout the nesting season, the promiscuous males remain on their territories and attempt to court and mate with receptive females attracted by their calls. Once mated, a female prepares a nest by lining a shallow depression on the ground under a shrub, tree, or log with leaves and grasses.

The young eat mostly insects, a source of protein, for at least ten days after hatching then switch to conifer needles and seeds. By late summer, mothers with young disperse both upslope and downslope from their nesting grounds, feeding in meadows and brushy hillsides, where they are often surprisingly tame. An encounter with a family group inspired John Muir (1901) to write: "One touch of nature makes the whole world kin; and it is truly wonderful how love-telling the small voices of these birds are, and how far they reach through the woods into one another's hearts and into ours." During spring, summer, and fall, Sooty Grouse forage mostly on the ground, eating seeds, berries, green shoots, and insects. Adults also fly up into conifers to browse on needles and buds. These hardy birds survive high-elevation winters by taking shelter in dense firs, pines, and hemlocks and eating needles, buds, and pollen cones.

STATUS AND DISTRIBUTION In 2006 the American Ornithologists' Union split Blue Grouse into Sooty Grouse occurring in the Sierra, Cascades, and coastal ranges of California, Oregon, and Washington, and Dusky Grouse that live in the Rocky Mountains and Great Basin. Sooty Grouse undertake both upslope and downslope movements in spring and fall. In the Pacific Northwest they are known to migrate upslope for the winter, and birds at Sequoia National Park have also been observed to undertake such "backward" migrations.

West Side. Uncommon residents of the Upper Conifer zone, males and some females wander up to tree line after breeding where they remain until late fall; rare in Lower Conifer zone in fall and winter.

East Side. Uncommon residents of ponderosa and Jeffrey pine forests up to the Subalpine zone; populations at Sagehen Creek (Nevada County) remained at their breeding elevations through the winter but similar studies are lacking from other areas.

TRENDS AND CONSERVATION STATUS The "Mount Pinos" Sooty Grouse *(D. f. howardi)* is a distinct subspecies restricted to the southern Sierra, south of Kings Canyon National Park, including the Piute Mountains, as well as ranges to the south of the Sierra including the Tehachapi Mountains and Mount Pinos. Recent surveys suggest that they may be extirpated from most of the southern portion of their historical range. Extensive surveys in 2002 in the Greenhorn Mountains (northernmost Kern County) found only eight birds in what is likely now the southern edge of their range. Continuing threats include habitat degradation due to harvest of large trees, fire suppression and uncontrolled wildfires, and livestock grazing. This subspecies was added to the list of California Bird Species of Special Concern in 2008.

Wild Turkey • *Meleagris gallopavo*

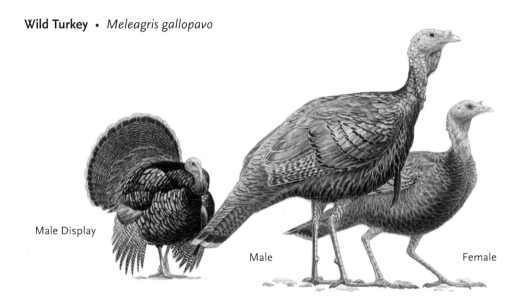

Male Display

Male

Female

ORIGIN OF NAMES Turkey, considered foreign and exotic when the species was first introduced to Europe in the 16th century; a possible reference to Asia Minor, a source of exotic goods at that time; Gr. *melagris,* spotted, for the species' patterned plumage; L. *gallopavo,* from *gallo,* a cock, and Sp. *pavo,* a turkey.

NATURAL HISTORY With their dazzling, iridescent bronze plumage, Wild Turkeys are elegant versions of common barnyard turkeys, a breed domesticated during pre-Columbian times in Mexico. Wild Turkeys are not native to California, but birds captured in Mexico were first introduced to mainland California for hunting in 1908. While most early introductions failed, subsequent releases over many decades were more successful and the species is now widespread in suitable habitats throughout the state. Most of the successful introductions to California were from wild populations in Mexico, Virginia, and Texas. Attempted introductions by the California Department of Fish and Game, U.S. Forest Service, and the National Wild Turkey Federation at high elevations of Sierra, Nevada, and Placer Counties as recently as 1997 should be discouraged because this species may compete with native species of quail and grouse with unknown consequences. They have also been observed to scour entire hillsides for snakes, lizards, and invertebrates having obvious but unmeasured impacts on those species.

Turkeys roost in trees but forage on the ground in forest openings, eating insects, green plants, seeds, acorns, and fruit. Throughout the year Wild Turkeys associate in groups, which consist during the breeding season of a "gobbler" male and his harem of four to six hens. In the Sierra the breeding season extends from late March until August, with a peak in May and June. Prior to pair formation, three to six males may strut together to attract a large group of females. The polygynous males perform courtship, a series of gobbling and strutting displays to attract females to be added to their harems. Conspicuous during courtship, fertilized hens sneak away from the immediate breeding area and are secretive when preparing to nest. They create shallow scrapes for nests and line them with leaves and grasses. Young are tended by hens alone through the winter until the following breeding season—female poults tend to remain with their mothers longer than males. With the inclusion of juveniles, winter flocks are larger than breeding flocks, and up to 100 birds might be seen foraging together on an oak-studded hillside.

STATUS AND DISTRIBUTION In the Sierra, Wild Turkeys spend most of the year in the Foothill zone, but they can be fairly common in the Lower Conifer zone in summer; postbreeding birds have been seen up to 10,000 feet on the Kern Plateau of the southern Sierra. In winter, they move as much as 50 miles downslope in search of areas with plentiful food and little snow, especially in foothill oak woodlands and chaparral.

West Side. Common to abundant in most areas of the foothills.

East Side. Uncommon; established populations near Honey Lake, in Alpine County, and near Topaz Lake.

TRENDS AND CONSERVATION STATUS Data from Breeding Bird Surveys and Sierra Christmas Bird Counts confirm the stunning increase in numbers and expansion of range for this species. In less than 30 years, Wild Turkeys have gone from unrecorded to abundant throughout the western foothills. Noted on only a couple of Sierra Breeding Bird Survey routes prior to the late 1980s, Wild Turkeys are now found consistently on more than 40 percent of the routes. Impacts to other species from this invasion are not yet evident.

LOONS • Family Gaviidae

Called "divers" in Europe, loons are sleek-bodied and superbly adapted to aquatic habitats but are almost helpless on land. Laterally flattened legs, set far back on their bodies, do not enable them to walk or stand, so loons can only move on land by sliding on their bellies. Unlike grebes, which have lobed toes, loons have webbed feet. They are also strong fliers but cannot take off from land; instead, they take flight after a long running start over open water, often into a stiff wind. Once aloft, they fly with quick, shallow strokes on slender, pointed wings. On a dive, their large, paddle-like feet propel them to great depths. Genetic studies indicate that loons are unlike any other living order of birds, not related (as once thought) to grebes. Fossil evidence shows that they have an ancient lineage dating back more than 65 million years. Males and females share identical breeding and winter plumages, but males are usually larger. There are only five species of loons worldwide, and all of them breed in North America. Two species, the Pacific and Common Loons, regularly occur in the Sierra. The family name is derived from L. *gavia*, a seabird.

Pacific Loon • *Gavia pacifica*

Breeding

Nonbreeding

ORIGIN OF NAMES "Loon" may be a reference to the species' wild calls, "crazy as a loon"; L. *gavia* (see family account above); L. *pacificus*, of the Pacific.

NATURAL HISTORY Pacific Loons consume mostly small fish captured by diving in deep, open water. They often swim with their heads and necks partially submerged, "snorkeling" for fish that they pursue at depths up to 70 feet. Behavior and habitat preference in the Sierra are very similar to Common Loons (see account below).

STATUS AND DISTRIBUTION Pacific Loons breed in Alaska and northern Canada and migrate south in small flocks to spend the winter along the Pacific Coast. In California most of the population winters in protected bays and coastal estuaries not far from shore, but migrants stray to large, inland water bodies including in the Sierra.

West Side. Uncommon fall, winter, and spring visitors; observed in most years at foothill reservoirs; rare above the Foothill zone but almost annual fall and winter records from Lake Almanor and other large mountain lakes and reservoirs up to the Subalpine zone.

East Side. Uncommon visitors to Lake Tahoe and large reservoirs such as Stampede Reservoir (Sierra County), Boca Reservoir, and Bridgeport Reservoir and Lake Crowley, and June Lake from November through May.

Common Loon • *Gavia immer*

Nonbreeding

Breeding

ORIGIN OF NAMES Sw. *immer,* ember, blackened ashes, a reference to the dark upperparts of breeding adults.

NATURAL HISTORY Common Loons have not been recorded as a California breeding species since the mid-1920s, when a few pairs and young birds were observed at Eagle Lake and other lakes near Lassen Volcanic National Park. Among the world's most accomplished diving birds, Common

Loons make foot-propelled dives that may exceed depths of 200 feet. Small fish are their preferred prey, but they also consume crustaceans as well as aquatic insects and plants, including algae. Although capable of extremely deep dives, Common Loons usually forage in relatively shallow, nearshore waters. While swimming along the water's surface, they may partially submerge their bills and cock their heads to "snorkel" for fish underwater. Upon spotting a prey item, they roll forward slowly in a smooth arc, with wings closed, and are capable of long underwater pursuits. These large-bodied loons ride low in the water when actively foraging, but float higher when resting or preening. Usually observed in winter plumage in the Sierra, Common Loons molt into breeding plumage by mid-April prior to migrating north to breed. Although rarely heard, these birds do give eerie yodeling calls from time to time, even in midwinter.

STATUS AND DISTRIBUTION Visitors from northern breeding grounds, most Common Loons in California remain in coastal waters. Each year, however, many regularly appear on large lakes and reservoirs of the Sierra.

West Side. Fairly common and regular fall and winter visitors to such large reservoirs as Lake Almanor, Millerton Lake, Lake Success, and Lake Isabella; rare nonbreeding individuals remain through summer; casual in late summer and fall at large Subalpine lakes, such as Tenaya Lake in Yosemite National Park.

East Side. Fairly common in fall, winter, and spring at large water bodies, such as Lake Tahoe, Topaz Lake, Stampede Reservoir (Sierra County), Boca/Prosser Reservoirs, and Bridgeport Reservoir, Crowley Lake, and June Lake but usually depart when substantial areas of open water freeze.

TRENDS AND CONSERVATION STATUS While late-19th century accounts documented Common Loons nesting just north of the Sierra at Eagle Lake, and at a few other large lakes above 5,000 feet east of Mount Lassen, they were probably extirpated as a breeding species in California by the early 1900s. The following account by C. H. Townsend (1887) is one of the only descriptions of Common Loons nesting from the Lassen region: "I waded out to a narrow sand bar . . . upon which a (Common) Loon had been sitting, and found her nest or rather egg . . . which was lying on the sand." The cause of the Common Loon's disappearance as California breeders is unknown, but it may have been related to human disturbance of nest sites. Studies in other parts of North America have shown that the mere presence of boats, even a single canoe passing too close, can cause Common Loons to abandon their nests and young. Formerly, Common Loons were included on the list of California Bird Species of Special Concern (Remsen 1978), but they were removed from this list since extirpation as a breeding species in California happened more than 100 years ago.

GREBES • Family Podicipedidae

Wedded to water at all seasons, grebes feed, sleep, court, and nest on water. At a distance, they look somewhat like ducks but tend to sit higher on the water and have shorter bodies, more slender necks, and sharp, pointed bills. Rather than webbed feet, they have individually lobed toes that fan out when pushed through water, as do coots and phalaropes. Grebes can dive 20 feet or more and stay submerged for up to a minute. Similar to loons, grebes swim underwater with great strength and agility, propelled by legs set so far to the rear that they cannot walk upright on land. They are rarely seen in flight, and even when pursued by predators, they usually flee by diving. Most grebes migrate at night, and a few species (such as Eared Grebes) travel thousands of miles between their breeding and wintering grounds. Female grebes usually lay three to six whitish eggs and cover them with moist nesting material when leaving their nests, causing eggs to become stained reddish-brown before they hatch. Both adults participate in incubating the eggs for about 20 to 25 days. Information on the behavior of nestlings is lacking for many species, but most young grebes attain flight after about 10 weeks.

All grebes have soft, thick plumage. Feathers are consumed incidentally while preening and form into balls in grebes' stomachs, which may function to protect the linings of their stomachs and intestines from being punctured by fish bones. Feather balls also serve as the nuclei of pellets that are regurgitated to eliminate bones and other indigestible materials. Of about 20 grebe species in the world, 7 occur in North America, and 5 of those occur regularly in the Sierra. The family name was derived from L. *podicis,* rump, and *pedis,* a foot, and roughly means "rump-footed."

Pied-billed Grebe • *Podilymbus podiceps*

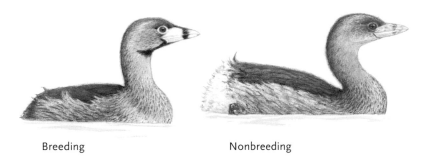

Breeding   Nonbreeding

ORIGIN OF NAMES *Podilymbus,* a diving bird, from L. *podiceps,* a grebe, and Gr. *kolumbos,* a diving bird; *podiceps* (see family account above); "grebe" is an OF. word of unknown origins.

NATURAL HISTORY Generally solitary, with subdued brownish plumage, Pied-billed Grebes may go unnoticed when swimming along the dense vegetation at a pond's margin. They dive for aquatic insects, crustaceans, or small fish in shallow waters and consume smaller quantities of tadpoles, mollusks, and aquatic plants. They also frequent deeper, open waters where they rest with other grebes or with larger coots and ducks. Most water birds take flight when predators approach, but Pied-billed Grebes sink quietly beneath the surface and reappear amid floating plants with only their heads out of water. They are typically found in quiet ponds, backwater sloughs, lakes, and reservoirs. Although they occur on water bodies from one-half to over 100 acres in size, Pied-billed Grebes are most often seen on ponds ranging from 1 to 5 acres. They prefer open water bordered by dense cattails, bulrushes, and other emergent vegetation.

Most grebes are colonial when nesting, but Pied-billed pairs are solitary. Both parents gather dead and rotting plant materials to build a thick mat nest that occasionally floats free but is usually anchored to plants growing from the bottom ooze. Their breeding season extends from about mid-March to mid-September, though peak breeding activity occurs from late June to early July. Each pair defends a small territory around the nest, where the young are fed and reared. Newly hatched chicks often ride on the adults' backs or hide beneath their wings during dives.

STATUS AND DISTRIBUTION Unlike other grebes, most Pied-billed Grebes do not migrate to salt water for the winter and prefer to reside year-round in freshwater habitats.

West Side. Fairly common residents and nesters throughout the foothills, they also nest rarely up to the Upper Conifer zone but are scarce anywhere above about 5,000 feet; a few breeding pairs have been found above 7,000 feet in the Subalpine zone of Plumas County, and postbreeding birds occur up to about 8,000 feet in Yosemite and Sequoia National Parks during late summer and fall; flocks of up to 300 individuals have been seen at Salt Spring Valley Reservoir (Calaveras County) in winter.

East Side. Fairly common permanent residents and nesters below about 7,000 feet, rare fall migrants have been seen at Alpine lakes up to about 9,000 feet in Yosemite National Park; numbers increase noticeably in October and November when they congregate at large reservoirs such as Crowley Lake and Bridgeport Reservoir; rare in winter, most depart before their preferred freshwater habitats freeze over.

Horned Grebe • *Podiceps auritus*

Breeding   Nonbreeding

ORIGIN OF NAMES "Horned" for the tufts of feathers on the heads of breeding adults; L. *auritus,* eared; for unknown reasons taxonomists applied the scientific name to this species instead of the Eared Grebe, which was named instead for its neck color.

NATURAL HISTORY Foraging Horned Grebes dive after small fish, crustaceans, and aquatic insects. Similar to other grebes, they also dive to escape danger and can swim for long distances before coming to the surface. Although they sometimes share open waters of large lakes and reservoirs with Eared Grebes, Horned Grebes tend to be much less gregarious and are less likely to occur in large flocks on their wintering grounds or during migration.

STATUS AND DISTRIBUTION Horned Grebes do not breed in California, and most of the state's visitors are found in coastal waters from October through April. A few, however, are drawn to lakes and reservoirs of the Sierra. This species is probably overreported, especially in fall when transitional plumages of Eared Grebes can blur the distinction between these species.

West Side. Uncommon fall, winter, and spring visitors to Lake Almanor, Millerton Lake, Avocado Lake (Fresno County), Lake Isabella, and other large lakes and reservoirs.

East Side. Uncommon to rare spring and fall visitors to such lakes as Boca Reservoir, Lake Tahoe, Topaz Lake, Grant Lake (Mono County), and Crowley Lake.

Eared Grebe • *Podiceps nigricollis*

Breeding   Nonbreeding

ORIGIN OF NAMES "Eared" for the feather tufts; *nigricollis,* black-necked, from L. *niger,* black; L. *collum,* neck.

NATURAL HISTORY These small, thin-billed grebes congregate at Mono Lake in enormous numbers during fall migration to feed on the superabundant brine shrimp that thrive in the lake's highly alkaline and saline waters. At the peak of their fall migration, they may consume more than 60 tons of brine shrimp daily (Winkler et al. 1977)! They also consume alkali fly larvae in shallow, nearshore waters, and the flies represent about 10 percent of their diet. Huge quantities of food are needed because the grebes are molting and rebuilding their fat reserves before continuing their migration. At freshwater habitats elsewhere in the Sierra, Eared Grebes consume mostly insects (adults and larvae), crustaceans, mollusks, and occasionally small fish and tadpoles. They usually forage by diving and capturing prey underwater but sometimes take insects from the water's surface.

Mono Lake is the Eared Grebe's largest fall staging area in North America. From breeding grounds across the western United States and Canada, birds begin to arrive at the lake in June, and the summer flock may contain more than 25,000 birds, mostly juveniles and nonbreeders. Postbreeding Eared Grebes arrive at Mono Lake at rates of up to 10,000 birds per day in late summer and fall, and numbers may peak at more than a million birds in September and October. Large numbers of Eared Grebes usually remain at Mono Lake until late November, when the brine shrimp population collapses and the grebes continue their migration on to the Salton Sea, San Francisco Bay, or the Gulf of California.

Nesting activities extend from late April until late September, with a peak from June through mid-August. Both sexes participate in building the nest, a hollowed-out mound of rotting vegetation hidden by similar material and located on the ground or floating and anchored to submerged, upright plants. Juvenile Eared Grebes can fly about 45 days after hatching and promptly begin their fall migration.

STATUS AND DISTRIBUTION Eared Grebes nest in freshwater habitats, but most seek highly alkaline and saline waters during migration through the Sierra and on their wintering grounds.

West Side. Fairly common year-round residents only at Lake Almanor and at Lake Isabella, where hundreds or thousands overwinter and a few remain through the summer; elsewhere, uncommon and irregular visitors to large lakes and reservoirs in fall, winter, and early spring; rare migrants have been seen at large Subalpine and Alpine lakes of Yosemite and Kings Canyon National Parks in late summer and fall.

East Side. Common nesters in marshes of the Great Basin, just east of the Sierra, and small numbers nest locally in shallow wetlands near large waters in the eastern Sierra, including Honey Lake, Crowley Lake, and Bridgeport Reservoir; uncommon to absent in winter except for Mono Lake, where the local Christmas Bird Count tallies hundreds or even thousands some years, and the Reno area, where double-digit numbers are not unusual.

TRENDS AND CONSERVATION STATUS A historic 1994 decision by the State Water Resources Control Board should ensure that Mono Lake's waters will not become too alkaline and salty to support abundant brine shrimp and alkali flies for Eared Grebes and other water birds. While their habitat appears secure in the eastern Sierra, mortality in other parts of the Eared Grebe's range, including the unexplained, recent die-offs of up to 150,000 birds at the Salton Sea, Imperial County, could affect the number of birds at Mono Lake and at other important western staging areas.

Western Grebe • *Aechmophorus occidentalis*

ORIGIN OF NAMES *Aechmophorus* from Gr. *aichme*, spear, and *phoreus*, bearer; L. *occidentalis*, western.

NATURAL HISTORY These largest of North American grebes (along with Clark's Grebe, see account below) are also the most widespread grebes in the Sierra. Scanning across an expanse of open water, birders often spot these large diving birds as specks of white bobbing in the distance. A closer look reveals the bright white breasts and long, snake-like necks of Western Grebes. They may appear headless, because they often sleep with heads and necks drawn back with bills pointing forward. They feed by diving and pursuing their prey under water, and they usually forage in water at least four feet deep. Westerns (and Clark's) have a neck mechanism, unique among grebes, that permits them to thrust their heads forward like spears (a fact painfully learned by one of the authors when rescuing a grebe tangled in fishing tackle). A similar mechanism has been well described for herons but has not been studied in grebes. Westerns consume more fish than most other grebes, but they also eat mollusks, crustaceans, insects, and rarely amphibians and aquatic plants. They dive to avoid predators and can remain submerged for up to a minute.

In the breeding season, Western Grebes require a large extent of open water for feeding and resting, a good supply of fish, and cattails, tules, or flooded riparian trees for securing their floating nests. Pairs of Western Grebes perform elaborate courtship displays, swimming side by side then running rapidly with their bodies erect, heads and bills pointed skyward, and only their feet touching water. In the Sierra, courtship mainly occurs from April to May, and nests can be occupied from May until August; fall nesting has been documented at Lake Isabella, with downy young riding on their parents' backs. They are monogamous breeders, nesting in dense colonies that may contain hundreds of closely spaced nests. Nests are composed of decaying aquatic plants and some green plants mounded on mud or submerged plant root masses in shallow water. They also construct floating platforms of dead plant material and anchor these to emergent plants. Parents take turns incubating the eggs, and pairs will renest if their nests are lost to predators. The precocial young leave their nests at hatching and ride on the backs of both parents. Young can swim and feed themselves soon after hatching, but their parents still feed them for four or five weeks, until they are nearly full-grown.

STATUS AND DISTRIBUTION Outside the breeding season, Western Grebes are often found in significant numbers on most large lakes and reservoirs.

West Side. Common nesters at Lake Almanor in the north and at Lake Success and Lake Isabella in the south but apparently do not breed elsewhere on the West Side; fairly common nonbreeding visitors to nearly all large foothill reservoirs; rare visitors to large Subalpine and Alpine lakes in summer and fall, especially if fish are present.

East Side. Common from April through November at Boca Reservoir, Prosser Creek Reservoir, Lake Tahoe, Topaz Lake, June Lake, Bridgeport Reservoir, and Crowley Lake; common nesters at the latter two reservoirs; casual to rare in winter.

TRENDS AND CONSERVATION STATUS Western Grebes, like all colonial water birds, are vulnerable to development around lakes and reservoirs and to other human disturbance near their nesting colonies. Personal watercraft and motor boats have been observed near breeding colonies in flooded riparian forests at Lake Isabella and could be a serious disturbance factor there and at other Sierra nesting locations. Drawdown of water for hydroelectric generation during the breeding season has marooned nests in the past at Lake Almanor. Several thousand Western and Clark's Grebes died at Lake Isabella in August and September 2005; the suspected cause was a bacteria bloom that produced neurotoxin that poisoned fish and secondarily poisoned grebes as well as other water birds such as herons and ducks. A similar die-off occurred at Lake Success in 2006, and mortality probably occurs at lower levels at these and other reservoirs in most years.

Clark's Grebe · *Aechmophorus clarkii*

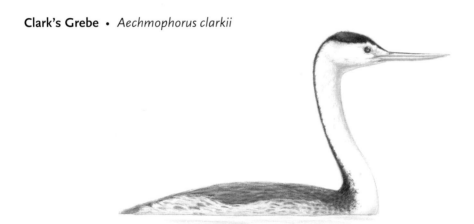

ORIGIN OF NAMES "Clark's" for J. H. Clark (1830–?) a surveyor and naturalist who collected the first specimen of this species (not the Clark of the Lewis and Clark Expedition, see Clark's Nutcracker account).

NATURAL HISTORY First described in 1858, Clark's Grebes were not considered a full species by the American Ornithologists' Union until 1985. Previously they were classified as a subspecies of the more common and similar-appearing Western Grebe. The breeding and wintering distribution of these two species overlap throughout their ranges in western North America, and they often flock together throughout the year, even during the breeding season when the two species may nest in mixed colonies.

STATUS AND DISTRIBUTION Little is known about the historical status and distribution of this species because most sightings before 1985 were recorded as "Western Grebes." Clark's typically occur in smaller numbers than Western Grebes in the Sierra and elsewhere in California.

West Side. Uncommon migrants and winter residents at foothill reservoirs; similar to Westerns, they breed only at Lake Isabella in the south and at Lake Almanor in the north.

East Side. Fairly common breeders at many of the same lakes and reservoirs as Westerns; uncommon in winter.

TRENDS AND CONSERVATION STATUS See Western Grebe account, above.

CORMORANTS · Family Phalacrocoracidae

Cormorants must leave the water periodically to dry in the sun because their feathers are not fully waterproofed like a duck's. The structure of cormorant feathers decreases their buoyancy and eases their underwater pursuit of fish. Six of the world's approximately 30 cormorant species occur in North America. Three species—Brandt's, Pelagic, and Double-crested Cormorants—occur commonly at bays and protected estuaries of the California coast, but only the Double-crested regularly ventures inland. The family name is derived from Gr. *phalakros,* bald, and *korax,* a raven.

Double-crested Cormorant · *Phalacrocorax auritus*

Adult

1st year

ORIGIN OF NAMES "Cormorant," a sea crow, from OF. *cormaran; phalacrocorax* (see family account above); L. *auritus,* eared, for the long, upturned feathers (crests) on their heads that are worn for only a short time during the breeding season.

NATURAL HISTORY With wet wings spread, Double-crested Cormorants perch like awkward black gargoyles on snags or branches of riparian trees. When they swim, their heads and bills are typically cocked upward. Cormorants frequent large bodies of water that provide ample room for their labored takeoffs and small fish, their favored prey. Diving from the water's surface, they may pursue fish to depths of 5 to 25 feet and can stay submerged for up to 30 seconds. They are equally at home on salt and fresh water, as long as fish are plentiful.

Double-crested Cormorants are colonial breeders that nest on islands or in large trees along lakes and large rivers, often in association with Great Blue Herons, Great Egrets, and other colonial species. In the breeding season, mated cormorants display to each other by stretching up their heads and necks, showing off bright orange throat patches and turquoise blue mouth linings. Both members of the pair participate in building their large nests, up to two feet in diameter and composed of twigs, sticks, and other debris placed in large trees. Nesting in the Sierra generally extends from April until late July. Females usually lay three to four pale blue eggs, and both adults incubate them for about 28 days until hatching. Adults feed and fend for their young until they are independent at about 10 weeks. During the nestling and fledgling period, both adults provide their young with regurgitated fish, crustaceans, frogs, or aquatic insects.

STATUS AND DISTRIBUTION Double-crested Cormorants frequent large lakes, reservoirs, and rivers at low and mid-elevations of the Sierra.

West Side. Fairly common nonbreeding visitors in fall, winter, and spring at most lakes and reservoirs but apparently nest only at Butt Valley Reservoir and Lake Almanor (Plumas County).

East Side. Uncommon but regular visitors except in winter; in some years pairs have nested at Hartson Reservoir, part of the Honey Lake Wildlife Area, but their numbers have fluctuated greatly there; fairly common nonbreeding visitors in spring and fall to other large reservoirs, including Topaz Lake, Crowley Lake, and at Bridgeport Reservoir, where a few pairs nested in the mid-1970s; uncommon but regular nonbreeding visitors to Mono Lake and Sierra Valley in spring and fall.

TRENDS AND CONSERVATION STATUS Once only rare visitors to the Sierra, they now occur regularly in the lower foothills due, at least in part, to construction of numerous reservoirs stocked with fish. Across North America, Double-crested Cormorant populations have increased significantly since the 1970s. A decline in historical persecution and banning of DDT are likely contributors to this increase. Although the species occurs regularly only on a few West Side Sierra Christmas Bird Counts, numbers from those counts have increased approximately four-fold since the 1980s.

PELICANS • Family Pelecanidae

With enormous, pouched bills and massive wingspans, pelicans are among the most spectacular and unmistakable of Sierra birds. Throughout the world there are six species of pelicans but only two, American White Pelicans and Brown Pelicans, occur in North America. White Pelicans occur regularly away from the coast and are likely to be seen in the interior. Brown Pelicans are common visitors to California's coastal waters and offshore islands but casual visitors to the Sierra. The family name is derived from Gr. *pelekon,* an ax, a reference to the pecking habits of a similar kind of water bird.

American White Pelican • *Pelecanus erythrorhynchos*

ORIGIN OF NAMES *Pelecanus* (see family account above); *erythrorhynchos,* red-billed, from Gr., from *erythros,* red, and *rhynchos,* a bill, as the species' bill turns bright orange-red in the breeding season.

NATURAL HISTORY Spiraling high above river canyons and mountain passes, large flocks of American White Pelicans cross the Sierra every year, unnoticed by most Sierra visitors. Among the largest of the world's water birds, they weigh up to 17 pounds and soar on 9-foot wingspans. From late winter until after the eggs are laid, adult males and females display a thin, horny plate on their upper bill. The lower halves of their unique bills are equipped with naked skin pouches that can hold up to three gallons of water and fish when expanded. After excess water has been expelled from the pouch, the fish are swallowed and temporarily stored in the esophagus; they are never stored or carried in the pouch.

Unlike coastal Brown Pelicans, White Pelicans do not dive from the air to catch fish. Instead, they forage at the surface in shallow water, thrusting their heads and bills forward to scoop up fish in their pouches. Flocks of 20 or more White Pelicans often swim and wade together in shallow water, flapping and splashing to herd fish to nearshore areas where they can be captured more easily. White Pelicans prefer to eat fish, but they also consume crayfish and amphibians. They require long expanses of open water to take flight, their wings and feet beating noisily against the water. When at last underway, flocks of White Pelicans fly with grace and agility, climbing to lofty heights and maneuvering together, flapping a few times then soaring in long lines that weave up and down together. When sun flashes across their backs and wings, they suddenly disappear and reappear, as flocks bank in unison.

STATUS AND DISTRIBUTION White Pelicans do not breed in the Sierra regularly, but migrants pass over in spring (early March through May) and fall (early August to early November) in transit between the Central Valley and Great Basin breeding lakes and marshlands. However, not all birds make unidirectional flights; one individual equipped with a radio transmitter at Stillwater National Wildlife Refuge, Nevada, made 7 round trips to the Central Valley (14 crossings of the Sierra) in 1996, probably as part of a nonbreeding flock (Yates 1999).

West Side. Uncommon but regularly observed crossing the Sierra, usually following river drainages, perhaps because they offer reliable navigational signposts and steep canyon walls provide updrafts to assist their flight; migrant flocks might be seen over the American, Feather, and Yuba River drainages, and above Yosemite Valley; fairly common nonbreeding visitors to Lake Isabella and Lake Almanor, Antelope Lake (Plumas County), and other larger northern Sierra reservoirs; regular in winter at Lake Success and Lake Isabella.

East Side. Fairly common during both eastward and westward migrations and when they stop to rest and forage at East Side water bodies from March until May, and a few nonbreeding birds may remain until September; common visitors to Sierra Valley marshes, Lake Tahoe, and such large reservoirs as Crowley Lake, Bridgeport Reservoir, and June Lake; in 1976 close to 2,000 pelicans nested at Honey Lake for the first time, but they have not attempted breeding there since; recorded as far south as Lake Tinemaha in the Owens Valley.

TRENDS AND CONSERVATION STATUS At the turn of the century, White Pelicans nested at large lakes the length of California from the Klamath Basin, through the Central Valley, to the Salton Sea. Today California's breeding population is restricted to islands at Clear Lake and Tulelake National Wildlife Refuges, near the Oregon border. The largest breeding colony near the Sierra is at Pyramid Lake, north of Reno. White Pelicans have experienced long-term declines in their statewide breeding population due to destruction of their island nesting colonies by inundation or land bridges that permit humans and other predators to disturb their breeding colonies. However, on a continental basis, Breeding Bird Survey data show a steady increase since the 1960s.

HERONS AND RELATIVES • Family Ardeidae

These long-legged waders have long necks that extend their reach and sharp, pointed bills for spearing and grasping prey. They are the only North American birds, other than Western and Clark's Grebes, having neck mechanisms that permit sudden, spearlike head thrusts toward their prey. Although they may appear awkward while landing or taking off, all members of this family exhibit strong, graceful flight. Bitterns, herons, and egrets can be recognized in flight by their folded necks and long legs trailing behind.

Female herons and egrets usually lay three to five greenish-blue, unmarked eggs (except for bitterns, which lay reddish-brown eggs), and both parents incubate them for about 25 to 30 days before they hatch. The young leave their nests one to two weeks after hatching and are tended by both parents for an unknown period thereafter. This large family includes 63 species that occur worldwide except for the Arctic, Antarctic, and some oceanic islands; 6 species occur regularly in the Sierra. Breeding plumages of most male and female herons and egrets are adorned with plumes on their lower rumps or bright colors on their bare facial skin. Aside from Great Blue Herons, most species are rare above the foothills on both slopes. The family name is derived from L. *ardea,* a heron.

American Bittern • *Botaurus lentiginosus*

ORIGIN OF NAMES Named for its call, "bittern" from Fr. *butor,* which derives from L. *butire,* meaning to "boom like a bittern"; called *butorius* because its calls resemble the bellowing of a bull; L. *lentiginosus,* freckled, referring to spotting of the species' plumage.

NATURAL HISTORY The combination of cryptic earth tones, blacks, and whites of the American Bittern's plumage provides a perfect camouflage against a backdrop of cattails or bulrushes. Freezing with heads and necks pointed skyward aligning their vertical throat stripes with surrounding vegetation, they stand motionless until the source of their alarm has passed. Usually solitary, they rely on stealth to capture aquatic insects, amphibians, crayfish, fish, and small mammals. American Bitterns typically remain hidden in marsh vegetation, where they communicate by means of resounding, ventriloquial calls which Ralph Hoffmann (1927) has described as "either wooden, like to blows of a mallet on a stake, or liquid like the gurgling of a pump." They inhabit shallow wetlands dominated by tall, emergent vegetation near lakeshores, ponds, meadows, and canals in agricultural areas.

Their nesting season extends from late March to late August, with a peak in May and June. Usually solitary breeders, American Bitterns sometimes nest in groups with one male guarding several females within one large territory, suggesting that at least some males may be polygynous. Females alone select nest sites and build nests: large platforms of matted aquatic vegetation, over water, in tall, dense cover.

STATUS AND DISTRIBUTION American Bitterns are uncommon residents and nesters in isolated foothill wetlands of the Sierra.

West Side. Formerly uncommon residents of marshes near Lake Isabella; rare and infrequent visitors and probable nesters at foothill marshes elsewhere on the west slope; a few spring and fall migrants have been found at mid- and high-elevation marshes near tree line.

East Side. Common spring and summer visitors and suspected nesters in Sierra Valley; extirpated as a breeding species from many former areas in the eastern Sierra, including Lake Tahoe and Mono Basin; uncommon spring and fall migrants in near Tahoe Keys (south Lake Tahoe), but breeding has not been confirmed; since the 1970s, only casual migrants have been observed in dense marshlands near Mono Lake and the upper Owens River, with only one record from those areas since the early 1990s.

TRENDS AND CONSERVATION STATUS The secretive nature of this bird makes monitoring populations difficult, but data from U.S. Breeding Bird Survey and Christmas Bird Counts suggest long-term steady declines. California Christmas Bird Counts also show declines since the 1970s, with the most dramatic trends seen in coastal areas.

Great Blue Heron • *Ardea herodias*

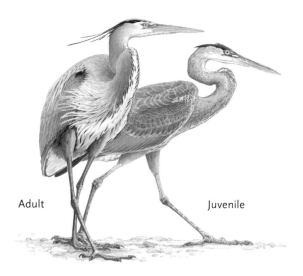

Adult   Juvenile

ORIGIN OF NAMES "Great Blue" refers to large size and overall coloration; "heron" is from OF. *hairon*; *ardea* (see family account above); Gr. *herodias*, heron.

NATURAL HISTORY Few Sierra birds are as majestic as North America's largest heron. Standing fully four feet tall, with wingspans of seven feet, Great Blue Herons dwarf most other water birds of the region. Although typically silent, Great Blues utter deep, harsh croaks when startled or when interacting with others of their species. They are solitary feeders and forage both night and day but are most active around dusk and dawn. They stalk fish, frogs, crayfish, and aquatic insects in the shallow margins of lakes, ponds, marshes, and rivers where fish and aquatic prey are plentiful. In winter and spring, Great Blues hunt for mice, gophers, and other small animals in wet pastures, recently irrigated agricultural fields, and even dry land. When hunting, they may stand motionless for several minutes or inch forward stealthily, with necks outstretched. Suddenly they strike, grasping small prey in their bills or spearing larger animals. Strong fliers, these birds may travel up to 10 miles from their nests to find food.

Great Blues nest colonially, and more than 100 nests may be present in large rookeries, which are often shared with Great Egrets and sometimes with Snowy Egrets, Black-crowned Night-Herons, or Double-crested Cormorants. In the Sierra most rookeries are in the tops of tall riparian trees that are relatively protected from wind and human disturbance. The same trees may be used year after year until they fall or are removed by humans. Less commonly, Great Blues nest on cliffs or amid marsh vegetation. Adults usually arrive at their nesting colonies in February, and nesting activities often extend through July. Nesting platforms may exceed three feet in diameter and are built from large sticks and lined with smaller twigs and grasses. Both parents tend the young, and one adult is almost always present at the nest for the first few weeks. During the nestling period, the hoarse, monotonous clacking calls of juveniles resonate for long distances.

STATUS AND DISTRIBUTION This successful, adaptable, and widespread species is the most common Sierra heron and the only one regularly seen above the foothills.

West Side. Fairly common year-round near lakes, reservoirs, and rivers up to about 4,000 feet in the central Sierra; most breed in lowlands of the Central Valley, but a few rookeries have been observed along lower portions of the North and South Forks of the Kern River, South Fork Kaweah River, lower American River, and Yuba River; the highest elevation west-slope breeding colonies are at Lake Almanor (about 4,560 feet); postbreeding birds sometimes follow river drainages into the high Sierra to Subalpine lakes and meadows above 8,000 feet in late summer and fall.

East Side. Fairly common year-round residents, the cold-hardy Great Blues outnumber all other herons and egrets; they nested, at least formerly, near Honey Lake, Boca Reservoir, and in the Lake Tahoe Basin, and one colony has been active since the 1990s in Long Valley (Mono County); nonbreeding Great Blues can often be found at Sierra Valley, Honey Lake, Lake Tahoe, Bridgeport Reservoir, Mono Basin, Crowley Lake, and along the upper Owens River.

Great Egret • *Ardea alba*

ORIGIN OF NAMES "Egret" from OF. *egrette (aigrette)*, egret, from Gr. word for "heron" with diminutive *-ette* added; L. *ardea,* heron; L. *alba,* white.

NATURAL HISTORY Great Egrets are one of the most striking and conspicuous members of their family in North America. Their brilliant white plumage and large size (slightly smaller than Great Blue Herons) make Great Egrets stand out in any landscape. They forage along the edges of marshes, lakes, canals, and slow-moving streams as well as irrigated croplands and pastures, where they often associate with Great Blue Herons and Snowy Egrets. Standing motionless or stalking slowly, they suddenly strike their prey with daggerlike bills. Their diverse diet includes fish, amphibians, reptiles, snails, crustaceans, insects, small mammals, and such small birds as rails (including Threatened Black Rails) in coastal areas.

In the breeding season, extravagant plumes grace the plumage of both males and females. Just prior to breeding, each bird grows 40 to 50 plumes that extend from their backs to well below

their tails. They fan these plumes while performing elaborate courtship displays; plumes are lost soon after breeding. In the Sierra, Great Egrets usually nest in tall riparian trees near water, often in association with Great Blue Herons and Double-crested Cormorants. Both parents collect large sticks and construct large platform nests up to two feet in diameter. Young hatch with a soft covering of down but begin to grow feathers within a week. They remain in their nests for about three weeks and then begin taking experimental flights to nearby branches. Both adults feed and tend the young for about two months until they gain independence.

STATUS AND DISTRIBUTION Great Egrets are mostly confined to low elevations on both sides of the Sierra.

West Side. Fairly common at foothill wetlands and rare visitors to lakes and reservoirs up to the Lower Conifer zone; in fall, a few individuals have been spotted at Subalpine meadows such as Tuolumne Meadows (about 8,600 feet) in Yosemite National Park, and near the Sierra crest above the North Fork American River; uncommon year-round residents at Lake Almanor in the north and at Lake Isabella in the south, where they are also irregular nesters.

East Side. Uncommon spring and fall migrants, they apparently do not remain to nest; each year migrants visit marshlands in Sierra Valley, Bridgeport Valley, Mono Lake, Lake Tahoe, and the upper Owens River but seldom in large numbers; casual to rare in winter.

TRENDS AND CONSERVATION STATUS The millinery trade reached its heyday in the late 1800s, as feathers were in great demand for women's hats and garments. Plume hunters targeted Great and Snowy Egrets because of their large size, abundance, and long plumes (called "aigrettes" in the millinery trade). In less than 20 years, Great and Snowy Egrets were nearly exterminated across a large swath of North America, including California. None were seen in the state for many years, and both species were considered rare until the 1920s. Great Egrets became a symbol for a growing conservation movement, led by the National Audubon Society and the American Ornithologists' Union, aimed to educate men not to shoot egrets and women not to adorn their hats with aigrettes. By the early 1900s public opinion and fashions had changed, and these egrets and most other native birds received protection under the federal Migratory Bird Treaty Act. With strict protection Great Egret populations have largely recovered and the species is now common across most of its former range, including California. Like all colonially nesting birds, Great Egrets are susceptible to human disturbance around their nesting colonies. They may be increasing in winter in the western foothills based on results from some lower elevation West Side Christmas Bird Counts.

Snowy Egret • *Egretta thula*

ORIGIN OF NAMES "Snowy" for the species' bright white plumage; *egretta* from OF. *aigrette,* egret; *thula,* a Chilean name for this bird.

NATURAL HISTORY Some birders call Snowy Egrets "golden slippers" because of their bright yellow feet. Similar to Great Egrets, these birds sport brilliant white plumage but they are small and dainty compared with their more common relatives. Snowy Egrets frequent shallow waters along the edges of marshes, slow-moving streams, and canals. They also forage in irrigated fields and pastures with other egrets and herons. They are more active foragers than other members of their family in the Sierra, and often dash through shallow water using their golden slippers and wings to flush unsuspecting prey. Like other egrets and herons, they also stand motionless or stalk slowly while searching for preferred foods, including small fish, crustaceans, and large aquatic insects. On occasion they also consume amphibians, reptiles, worms, and small mammals.

STATUS AND DISTRIBUTION Common residents of the Central Valley, Snowy Egrets rarely venture into the Sierra.

West Side. Rare and mostly confined to the lowlands below about 1,000 feet, Snowy Egrets have been observed at Lake Almanor, Lake Isabella, Lake Wildwood (Nevada County), and in Yosemite Valley in spring, summer, and fall; a single bird was observed in late May at Tuolumne Meadows (about 8,600 feet) in Yosemite National Park, but there are few other records above 4,000 feet; rare to casual in winter at lowest elevations.

East Side. Rare spring and fall visitors to east side marshlands, with individual records from Sierra Valley, Boca Reservoir, Honey Lake, Lake Tahoe, Mono Basin, Grant Lake (Mono County), and Crowley Lake; a pair nested at Mono Lake in the early 1990s, but there are apparently no other nesting records; accidental in winter.

TRENDS AND CONSERVATION STATUS See Great Egret account, above.

Green Heron • *Butorides virescens*

ORIGIN OF NAMES "Green" for the dark green upperparts of adults; L. *butorides* probably means "bittern-like"; L. *virescens,* becoming green.

NATURAL HISTORY Perched on a log or branch at water's edge with their necks folded, these chunky, crow-sized birds may be difficult to recognize as herons. Green Herons inhabit secluded lakes, ponds, marshes, and slow-moving foothill rivers and creeks that are shaded by riparian trees and often clogged with floating wood. They nest and roost in willows, oaks, and other trees and prefer to forage in their shade. Extremely agile, they climb among wooded tangles while stalking small fish, crustaceans, insects, and rarely small mammals. They also wade or stand in shallow water, much like Great Blue Herons. Sometimes Green Herons hunt from logs or streambanks and may even dive headfirst at their prey. They are known to fish by placing "lures" such as feathers on the water's surface, thereby attracting their prey close enough to catch.

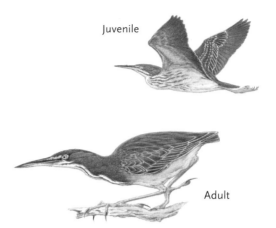

Juvenile

Adult

Unlike most other members of their family, Green Heron pairs are solitary while nesting. In the Sierra their nesting season extends from early March until late June. Both members of mated pairs participate in building nests, which can range from a simple pile of sticks to a more elaborate, woven structure that is reused in successive years. Their nests, up to 18 inches in diameter, are positioned in shrubs and trees up to about 50 feet above the ground. Both parents tend the young until they can fly, about 25 days after hatching.

STATUS AND DISTRIBUTION Green Herons are mostly confined to low elevations of the Sierra.

West Side. Uncommon year-round residents in the lower foothills, usually below about 3,000 feet; rare at higher elevations although Sequoia Lake at 5,300 feet (Fresno County) has two records; possibly present at all seasons except winter, and they are suspected nesters at Lake Isabella, Yosemite Valley, and Lake Almanor.

East Side. Rare spring and summer migrants through the Mono Basin, June Lake, and riparian habitats, with most records from March through June; a few records from Sierra Valley and from Carson Valley.

Black-crowned Night-Heron • *Nycticorax nycticorax*

ORIGIN OF NAMES "Black-crowned" to distinguish it from the similar Yellow-crowned Night-Heron *(Nyctanassa violacea)*; *nycticorax,* a night bird, from Gr. *nyctos,* night, and *corax,* a crow.

NATURAL HISTORY As their name suggests, Black-crowned Night-Herons forage primarily at night and in the twilight hours. When feeding young, however, they are active all day because food sources may be many miles

Juvenile

Adult

from their nesting sites. When not nesting, they spend most diurnal hours in large, colonial roosts in dense riparian trees and shrubs or marsh vegetation. Night-Herons forage and roost along densely vegetated lake and pond margins; wet, brushy meadows; and riparian vegetation lining slow-moving streams and canals. They are attracted to fish hatcheries, which provide an abundant source of easily captured prey. Opportunistic foragers, they consume a varied diet of small fishes, snakes, crayfish, and other crustaceans, aquatic insects, and other invertebrates. Night-Herons are also voracious predators of birds' eggs and nestlings, especially those of colonial-nesting species such as other herons and egrets, White-faced Ibis, and Tricolored Blackbirds. Night-Herons also consume ducks' eggs and nestlings and may destroy entire clutches of Wood Ducks, Mallards, and Cinnamon Teal.

Highly gregarious, Night-Herons nest in dense rookeries with members of their own species as well as other herons and egrets. Both members of mated pairs participate in building their large nests, which average about 2 feet in diameter and 18 inches in depth. Nests are built from locally available materials including sticks in riparian areas or cattail and bulrush thatch in marsh habitats. In the Sierra their courtship behavior begins in February and nesting continues into July. The young remain in their nests for at least two or three weeks and are tended by both parents. Juveniles attain complete feathering within about a month and can fly within about six weeks. Gregarious even outside the breeding season, large day roosts of many birds are found in marshes and dense riparian areas.

STATUS AND DISTRIBUTION Black-Crowned Night-Herons are mostly confined to low elevations on both sides of the Sierra.

West Side. Rare to locally uncommon in wetlands within annual grassland, foothill chaparral, and oak woodland habitats up to about 1,000 feet in the Yosemite region and somewhat higher in the southern Sierra, as at Lake Isabella (about 2,600 feet), where they reside year-round; casual at Lake Almanor and at Subalpine lakes and marshes in the Yosemite region up to about 9,000 feet in late summer and fall; the probable Sierra altitude record-holder was at Saddlebag Lake near Tioga Pass, just outside Yosemite National Park (10,200 feet) in October 2007.

East Side. Fairly common from March until October in marshlands of Sierra Valley, Honey Lake, and rare visitors to Lake Tahoe; locally common farther south, with colonies at Mono Lake and Laurel Pond (Mono County); they may also nest at Bridgeport Reservoir, Crowley Lake, and in dense emergent marshlands along Hot Creek and the upper Owens River (Mono and Inyo Counties).

IBIS • Family Threskiornithidae

This family includes both the ibises and spoonbills, representing more than 30 species worldwide; five of these exist in North America but only one, the White-faced Ibis, regularly occurs in the Sierra. White-faced Ibis can be distinguished from most other wading birds by their long, decurved bills, distinctive glossy plumage, and highly gregarious habits in all seasons. The family name is derived from Gr. *threskiornis,* a sacred bird.

White-faced Ibis • *Plegadis chihi*

ORIGIN OF NAMES "White-faced" for the white facial feathers of breeding birds; "ibis" is an Egyptian name taken from Sacred Ibis *(Plegadis aethiopicus)* that was mummified and depicted on tombs; Gr. *plegas,* sickle, for curved the bill; *chihi* is probably a native South American name for this bird.

NATURAL HISTORY White-faced Ibis are strong fliers that depend on ever-changing seasonal wetlands, requiring them to constantly move in search of productive feeding areas and secure nesting sites. They forage by probing their long, decurved bills into shallow water or moist soil

searching for aquatic and moist-soil insects, crustaceans, and earthworms. They prefer shallow marshes with islands of cattails or tules but also forage in upland soils, including agricultural fields—especially rice.

While males are slightly larger, both sexes have identical plumage and display beautiful metallic green and bronze hues in the breeding season. A thin white border of feathers surrounds their bills, hence their common name "white-faced." These white markings, as well as their iridescent plumage highlights, are lost when they molt into winter plumage.

These long-legged waders emit soft grunts and *oink* calls at their nesting colonies or when flushed. They fly in long, undulating lines, constantly shifting positions with strong, rapid wingbeats and alternating glides. They normally fly fairly close to the ground, but flocks also climb to great heights before making dramatic dives—pulling up just before hitting the ground.

In California, ibis initiate nesting in shallow, freshwater marshes from mid-April until late May. Both sexes participate in nest building, bending stems of partially submerged cattails or bulrushes over to form a woven platform just above the water's surface. Nests may be hard to see, hidden deep within dense patches of emergent vegetation, but their colonies are obvious with ceaseless traffic of dark birds entering and leaving the marsh. Nesting colonies move to new sites when water levels or other conditions change. Females lay three to four greenish or bluish eggs, marked with dark spots or scrawls that both sexes incubate for about 22 days. By about three weeks after hatching, fledglings can move a short distance to acquire food from their parents, and by five weeks the adults start leading the young on a series of short flights through the colony, feeding them at each stop.

STATUS AND DISTRIBUTION White-faced Ibis are now common residents of the Central Valley (see "Trends and Conservation Status," below), and they occasionally stray into the Sierra.

West Side. Rare postbreeding wanderers can be seen at low-elevation marshes; regular but uncommon in the Kern River Valley in spring and fall.

East Side. Nesting has been observed at a few isolated marshes north of Truckee and intermittently at Honey Lake since the mid-1970s, with the largest colony at Sierra Valley (hundreds of pairs in wet years but absent during dry years); they wander widely during migration and after the nesting season and might be seen in large wetlands or agricultural areas; accidental in winter.

TRENDS AND CONSERVATION STATUS Formerly common nesters the length of California, White-faced Ibis were reduced to very low numbers by the 1960s and were included on the list of California Bird Species of Special Concern in 1972. The principal cause of their dramatic decline was draining of Central Valley and coastal marshes; eggshell thinning due to DDT was also implicated. In the late 1970s ibis began returning to the state in numbers, possibly in response to inundation of historical marsh nesting areas at the Great Salt Lake. Their breeding population has increased steadily since. The species was rarely recorded on Central Valley Christmas Bird Counts in the 1970s and 1980s before numbers began to grow exponentially from the early 1990s and reach a plateau in recent years. For example, the species was nearly unseen in the 1980s at Lake Almanor, where it is now common in mid- to late summer. Flocks numbering in the tens of thousands now roam marshlands and agricultural fields of the Central Valley and northeastern California to the extent that in 2008 they were removed from California's list of Bird Species of Special Concern. Increases in Central Valley managed wetlands and winter flooding of rice fields may have contributed to the recovery of the species in California.

NEW WORLD VULTURES • Family Cathartidae

New World vultures are large scavengers with long, broad wings enabling them to soar great distances in search of carcasses. These carrion-eaters have extraordinary vision and can find dying or recently dead animals from high in the air. They often feed by thrusting their featherless heads into the body cavities of rotting animals, so baldness avoids the problem of chronically soiled head feathers. Of the seven species in this family, Turkey Vultures are fairly common to abundant in the Sierra, and California Condors historically nested there. The family name is derived from Gr. *kathartes*, a cleanser or purifier, for their habit of cleaning up carcasses.

Adult

Juvenile

Turkey Vulture • *Cathartes aura*

ORIGIN OF NAMES "Turkey" for the species' resemblance to the similarly naked-headed Wild Turkey; "vulture" from OL. *vuellere,* to pluck or tear; *cathartes* (see family account above); *aura* may be related to a South American Indian name for the bird.

NATURAL HISTORY Spiraling almost effortlessly on warm updrafts, Turkey Vultures soar gracefully over open grasslands, chaparral, and oak-lined canyons of the Sierra foothills. Their long wings, slightly smaller than those of Golden Eagles, are held in a flattened "V" with grayish flight feathers spread almost like fingers. Exceptionally light for their size, they wobble unsteadily from side to side, only rarely flapping their wings. So finely tuned are their abilities that vultures may soar for many hours simply by moving from updraft to updraft.

Vultures are ever on the lookout for carrion, which they identify both visually and by detecting drifting molecules of decay. Their keen sense of smell, one of the most refined of all birds, is so acute that Turkey Vultures can locate carrion buried under leaves in forests or dead animals as small as insects. They usually forage individually, but several may gather to feast communally on an especially desirable carcass, where they waste no energy fighting among themselves. While

Turkey Vultures are capable of subduing very small prey, poet Kathleen Lynch (2000) has it pretty much right:

Of all the birds
watching from winter-stripped
trees. Vultures are
kindest, killing
nothing.

Dependent on unpredictable and sparsely distributed carrion, vultures are the undisputed masters at conserving energy between meals. A sunny morning will find them perched with spread wings, warming up without using calories. Raising and lowering the ruff of feathers around their neck is another means of efficiently regulating body temperature. Vultures expend extra energy during courtship flights, when determined males may spend up to an entire day flying close behind a prospective mate. In California breeding activities usually begin in mid-April, when pairs seek out dark recesses in cliffs or rocky slopes. These can sometimes be found by smell, as the stench of regurgitated carrion at nest sites is almost unbearable. Both parents help incubate two eggs that hatch in about a month. It takes 70 to 80 days before the young birds leave the nest.

STATUS AND DISTRIBUTION Small numbers of Turkey Vultures reside year-round in northern California, but most migrate south to spend the winter in more productive foraging grounds from central California to northern South America. Fall migration can be extremely impressive when groups of hundreds or even thousands of birds float together over Sierra ridgetops and river canyons. Recent studies suggest the foothills of the West Side comprise a major vulture migratory route drawing birds from the Pacific Northwest. Especially high numbers have been recorded in the South Fork Kern River Valley, where nearly 30,000 are counted each fall, but thousands of migrants can also be seen over most Sierra river canyons from late September until mid-October.

West Side. Fairly common winter residents of the foothills, their numbers increase dramatically from early March until late April, when migrants and breeding birds return to the region; common from early May until mid-September in oak savanna and annual grasslands below the Lower Conifer zone, especially where there is a mix of intact forest and open grazing land; uncommon up to the Upper Conifer zone and rare up to the Subalpine zone in late summer and early fall.

East Side. Fairly common in both spring (mostly April) and fall (mostly September) migration; uncommon during summer and rarely nest; casual in winter.

California Condor • *Gymnogyps californianus*

ORIGIN OF NAMES "Condor" is a Spanish rendering of an indigenous Peruvian name for these birds; Gr. *gymnos,* naked, and *gyps,* vulture; L. *californianus,* of California.

NATURAL HISTORY California Condors once soared over open expanses of western North America searching for decaying carcasses provided in abundance by vast herds of bison, elk, deer, and pronghorn. Prior to European colonization, they occurred from British Columbia to Baja California, but shooting, lead poisoning, nest disturbance, decline in food supply, and other human intrusions have caused dramatic declines, almost to the point of extinction (see "Trends and Conservation Status," below).

With nine-foot wingspans, California Condors are the largest land birds in North America. These giants routinely make round trips of 70 miles or more between roosting and feeding sites.

They always feed on the ground in open areas such as grasslands and oak savanna, where there is sufficient space and suitable air currents for their labored landings and takeoffs. Both fresh and decomposing carcasses of cattle, sheep, deer, and ground squirrels are the mainstays of their diet. Lacking a keen sense of smell, condors locate the majority of their food by watching other scavengers such as Turkey Vultures.

Condor reproductive rates are low and a successful pair will generally hatch only one egg every other year. Their chicks take about 100 days to gain adult size but do not fledge for another two months. They depend on the adults for food and protection for at least another six months, an extraordinarily long period of dependency. Nests are simple depressions on cliff ledges or potholes in caves, although in the Sierra they are known to have nested in cavities in giant sequoias. Female condors do not breed until at least their sixth year, and mated pairs often remain together for long periods. Prior to their first mating, immature birds, either singly or in groups, will spend a year or more visiting and becoming familiar with the foraging range and nesting territories of the adult population.

STATUS AND DISTRIBUTION According to Grinnell and Miller (1944), California Condors were "formerly (within historic times and up to about 1870) common throughout that portion of the State lying west of the Great Basin and desert territories, and north from the Mexican line to the Oregon line." These authors noted that condors were known from the foothills of Kern, Tulare, Fresno, Madera, and Tehama Counties as well as the Owens Valley, mainly below 6,000 feet.

West Side. The last known nesting pair of condors in the Sierra was recorded in 1984 nesting in a giant sequoia in Sequoia National Forest (Tulare County); as of 2011, there were no records of wild condors present in the Sierra, although rugged and remote areas in the southern Sierra may be prime candidates for reintroduction efforts in the future and are the likely destinations of wandering birds from current reintroduction efforts in the mountains of southern California. There have been a few records from the Kern River Canyon in recent years.

East Side. No modern records.

TRENDS AND CONSERVATION STATUS By 1950 the global population of California Condors numbered only about 150, and the species was listed as Endangered by the U.S. Fish and Wildlife Service in 1967 and by California in 1971. Their numbers dropped to around 15 birds by 1987, when all remaining wild birds were captured and taken into captive breeding programs to prevent the species' extinction. Since 1988, captive-bred birds have been produced and, as of 2010, more than 380 had been released into suitable habitat areas in coastal southern California, Utah, Arizona, and Baja California. These releases have met with variable success, with relatively high mortalities of both adult and juvenile birds due to the failures of captive-reared birds to exhibit wild behaviors and from consumption of lead fragments from bullets in deer and other game. Establishment of self-sustaining, wild populations may not be realistic until the problem of lead contamination from bullet fragments can be addressed on a range-wide basis.

OSPREY · Family Pandionidae

In recent decades Ospreys were considered a subfamily of the family Accipitridae, along with hawks, kites, and eagles. However, they are again considered a distinct family, represented by only a single species. This is supported by genetic studies and physical characteristics including having toes of equal length and rounded, rather than grooved, talons. Ospreys and owls are the only raptors with reversible outer toes, allowing them to grasp prey with two toes in front and two behind. The family name is thought to be derived from *Pandion,* a king of Greek mythology whose two daughters were turned into birds; this name may also derive from L. *pan,* all; and L. *dio,* god.

Osprey · *Pandion haliaetus*

ORIGIN OF NAMES "Osprey" from L. *ossifraga,* bone breaker, for a European vulture, but how the name got transferred to this fish-eating bird is unknown; *pandion* (see family account above); *haliaetus,* a sea eagle, from Gr. *hals,* sea; *aetos,* eagle.

NATURAL HISTORY Among the world's most widespread birds, Ospreys occur on six continents, although they are rarely common in any single area. Their favored habitats are large lakes, reservoirs, rivers, and coastal areas with clear water and ample fish, and they are seldom observed away from such sites. The commonly used term "fish eagle" is appropriate, as they are the only North American raptor dependent almost solely on fish. They watch for fish from perches on rock outcrops or snags or hunt by flying low over the water. They prey on many types of fish but prefer those 6 to 12 inches long that school near the surface. If fish are not available, they may take other prey, such as turtles, snakes, frogs, shorebirds, and ducks on rare occasions.

Ospreys capture fish by diving headfirst, swinging their feet forward past their heads and extending their wings back behind their bodies just before striking the water. Almost instantly they surge back out of the water with one powerful beat of their wings. Their long toes have tiny spikes that help them grip and subdue slippery prey. They usually fly back to a favorite perch to eat, holding the fish tightly with its head pointing forward to reduce air resistance or to prevent theft by other raptors.

In their second year Ospreys may form pair bonds and build a nest, but their first nesting attempt is usually during their third or fourth year (up to five years for some pairs). Elaborate male courtship flights are rewarded with as many as 400 copulations in the three weeks prior to egg-laying. Nests are huge, bulky structures built mostly of sticks and other debris by males, with the female lining the inside with grasses or other soft material. They are constructed on snags, on cliff faces, telephone poles, high tension towers, and other artificial structures—always near large rivers or lakes. Nests are repaired and added to each year, and the same pair (or later generations) might use the same nest for decades; nests may eventually grow to huge proportions—more than 6 feet wide and 12 feet deep. Two to three large whitish eggs are incubated for just over a month, mostly by the female. Nestlings remain in the nests for about 40 days and are capable of flight in about 2 months. Canada Geese, which begin nesting earlier than Ospreys, sometimes take over Osprey nests, which results in fierce battles after Ospreys return to discover the intruders.

STATUS AND DISTRIBUTION Primarily summer residents though many remain in California through the winter, with the largest numbers along the coast and in the southern half of the state. Breeding pairs are scattered throughout the Sierra.

West Side. Nesting pairs occur near most of the large lakes and reservoirs and many large rivers; representative breeding locations include Lake Almanor, Lake Oroville, New Bullards Bar Reservoir (Yuba County), Bass Lake (Madera County), and Lake Isabella, where fairly common. During both spring and fall migrations, they are more widespread and might be observed over any habitat type below the Upper Conifer zone, though rarely far from water.

East Side. Uncommon and scattered; several pairs nest each year on tufa towers at Mono Lake, flying to distant lakes to find fish; they also breed in the Tahoe Basin and other lakes in Mono County; breeding season observations in Alpine and Inyo Counties.

TRENDS AND CONSERVATION STATUS Common until the second half of the 20th century, the introduction of DDT and other chemicals led to a dramatic decline in Osprey populations throughout their North American range. Since DDT was banned in the 1970s, Osprey numbers have steadily increased. The creation of numerous reservoirs and artificial nesting structures facilitated an increase in population and expansion of their range in the Sierra.

HAWKS AND RELATIVES • Family Accipitridae

All species in this large and diverse family are daytime predators with keen eyesight, sharp talons, and hooked beaks for dismembering and devouring prey. Like owls, they often consume whole animals and digest them in their highly acidic stomachs. Fur, feathers, and bones are then passed on to the gizzard (a muscular stomach), compressed into pellets, and regurgitated. These pellets, or castings, provide useful information about diets. Females are larger than males, but most species do not show noticeable plumage differences between the sexes.

Female kites and harriers lay four to six eggs, while most hawks only lay two to four eggs. All Sierra members of this family have buffy-white eggs that are variably splotched with brownish or maroon streaks. Female kites, harriers, and hawks incubate the eggs for 25 to 35 days, with occasional help from the males, who provision food to the females and nestlings. Eagles only lay two eggs, which both parents incubate for 35 to 45 days. All raptors begin incubating their eggs after the first one is laid (unlike waterfowl and many other birds), so they hatch asynchronously and the youngest nestlings may starve or be eaten by their older siblings if food is in short supply. Immatures require at least one and, in some species, two or more years to achieve adult plumage.

Golden and Bald Eagles take four or five years, respectively, to reach maturity. Worldwide this family is represented by about 205 living species, and 13 of these occur annually in the Sierra. The family name is derived from L. *accipiter,* to grasp or seize.

White-tailed Kite • *Elanus leucurus*

Juvenile

Adult

ORIGIN OF NAMES "Kite" may derive from an Indo-European root meaning to shoot or move quickly; *elanus,* from Gr. *elauno,* to harass or drive forward; *leucurus,* from Gr. *leukos,* white, and *oura,* tail.

NATURAL HISTORY Attired in bold shades of white, gray, and black, hovering White-tailed Kites grace open landscapes of the Sierra foothills. Some long-term residents call them "Angel Hawks" for this behavior. From hovering positions 20 to 100 feet above fields and pastures, kites scan intently for the movements of meadow voles. Nearly their entire diet consists of these small rodents, making kites an effective regulator of rodent populations. When voles or other prey are spotted, kites have the peculiar habit of making slow, vertical descents by pulling their wings up into a steep "V" and parachuting gently downward feet-first. It's a surprisingly effective strategy, with capture rates approaching 50 percent. This degree of specialization has its drawbacks, however, for kite populations fluctuate broadly in close synchrony with the ebb and flow of vole populations.

Kites take advantage of peak rodent numbers in late March and April to ensure the best chance of raising their young in the face of a fluctuating food supply. By rearing clutches of four or five eggs, and double-brooding in good years, they end up having one of the highest reproductive success rates of all local hawks. Pairs build nests among the dense outer foliage of trees, exceedingly well concealed from below but open to the sky above so they can scan for approaching danger. Females incubate the eggs for about 30 days while the males provide food for mates and young throughout the entire nesting period, sometimes transferring food to the female in midair exchanges. The young attain flight in about 40 days, but they return to the nest to receive handouts from their parents for up to six months.

STATUS AND DISTRIBUTION Year-round residents that nest below 2,000 feet, White-tailed Kites scarcely make it into the western fringe of the Sierra foothills. They favor wet meadows, grasslands, oak savannas, irrigated pastures, and alfalfa or hay fields, dropping out at elevations where trees start forming a continuous cover. By all accounts, they remain close to water, probably because voles and other prey are most abundant there. They prefer to nest in or near open areas where they can readily hunt. After the breeding season, kites may wander over a wider range, sometimes appearing at higher elevations and on the East Side.

West Side. Uncommon and localized year-round residents below the Lower Conifer zone but casual as high as the Sierra crest, especially in late summer; outside their few known Sierra breeding areas, kites might be seen dispersing high overhead.

East Side. Generally quite rare except for a spate of activity in the mid-1970s, when population numbers were at a high and West Side birds were wandering, probably to avoid a drought when a handful of birds showed up around Reno, Sierra Valley, and Honey Lake in fall with some into winter or were observed in transit elsewhere; other fall and winter records from Honey Lake, Sierra Valley, Mono County, and Owens Valley attest to the ability of this species to disperse widely outside the breeding season; a few spring records from the north end of the Owens Valley are exceptional.

TRENDS AND CONSERVATION STATUS These elegant birds suffered a precipitous decline in the first half of the 20th century at the hands of hunters and egg-fanciers. Fully protected in California in the 1960s, well before the state or federal Endangered Species Acts were passed, kites began a dramatic recovery that lasted well into the 1970s. Their success is a function of their unusually high reproductive rate and ability to tolerate some habitat fragmentation and human disturbance, but their population growth has also been linked to a 40 percent expansion of irrigated agricultural land between 1944 and 1978. Since then, their population may be declining, perhaps as a result of residential and commercial development of former agricultural lands or conversion to more intense agriculture like vineyards and orchards in the Central Valley and Sierra foothills.

Adult

Subadult

Bald Eagle • *Haliaeetus leucocephalus*

ORIGIN OF NAMES "Bald" for white head color of adults; both Fr. *aigle* and OE. *eagle* are derived from L. *aquila*, eagle; similar to *aquilo*, the north wind, a symbol of Roman military prowess; *haliaeetus*, sea eagle, from Gr. *hals*, sea, and *aetos*, eagle; *leucocephalus*, white-headed, from Gr. *leukos*, white, and *kephale*, head.

NATURAL HISTORY Bald Eagles are familiar to most people as the country's national symbol. They are also the second largest bird of prey in North America, after the California Condor, with wingspans exceeding seven feet. They obtain much of their food by scavenging animal carcasses or using their formidable size to steal freshly caught items from other more efficient predators like Ospreys and Great Blue Herons. All types of carrion and prey are consumed, with fish being preferred and forming the bulk of their diet, although they also take many American Coots and waterfowl during the winter. Along Highway 395 in the northeastern Sierra, road-killed deer carcasses are especially preferred and injured waterfowl are pursued in hunting areas in the Central Valley and Sierra foothills. Bald Eagles capture fish and other animals by swooping down from perches or from flight.

For successful nesting, Bald Eagles require sizable bodies of water with dependably productive fisheries, suitable nest sites, and little human disturbance. Recent studies show that younger Bald Eagle pairs are nesting closer to areas of human activity than older pairs, suggesting habituation to human presence. Peak nesting in the Sierra takes place from March to June, and pairs often reuse the same nests for years, gradually building into massive structures. The largest nests in North America have measured up to 18 feet tall and nearly 2 tons in weight. Tall, older trees within half a mile of water are most frequently used, but occasionally nests are built on cliff ledges, especially along the coast. Nests are usually well shaded by foliage or rock overhangs. Nestlings are capable of flight in about 11 weeks but often return to the nest to receive food from the adults for several months after that. Young birds wander until attaining adult plumage at about five years, when they may breed for the first time.

STATUS AND DISTRIBUTION Historically, Bald Eagles probably nested throughout the Sierra. As their population steadily recovers from mid-20th-century lows, their nesting range continues to expand as new sites are occupied each year. From October to March, they can be seen regularly though rarely throughout the Sierra with many at a time occasionally observed at some sites; such large winter gatherings may increase in the future as Sierra populations continue to grow. In the nonbreeding season, they travel great distances over many types of habitats but occur most often near large rivers, lakes, and reservoirs. During the breeding season (April through September), Bald Eagles are still rare in the central and southern portions of the Sierra.

West Side. Fairly common to uncommon, pairs have nested as far south as Tuolumne County but will likely expand further southward in the future; large reservoirs along the full length of the range attract visiting Bald Eagles during the nonbreeding season, creating the possibility that new pairs will become established where conditions are favorable.

East Side. Uncommon in the breeding season, breeding records south to Long Valley (Mono County) with possible breeding further south; fairly common nonbreeding visitors to large bodies of water with plentiful fish.

TRENDS AND CONSERVATION STATUS After World War II, the use of DDT and other organochlorine compounds became widespread as insecticides to control mosquitoes and other insect pests. Breakdown products from these compounds accumulated in Bald Eagles and other birds of prey, causing eggshell thinning and almost complete reproductive failure in many species. In addition, Bald Eagle nests have been abandoned due to disturbances from nearby human activities. For these reasons the U.S. Fish and Wildlife Service and the California Department of Fish and Game listed the species as Endangered in 1978. A ban on use of DDT in the late 1970s, and greater protection of known nesting territories by federal and state agencies, allowed Bald Eagles to dramatically increase their population and expand their range in California and throughout the United States. Because of this recovery, the U.S. Fish and Wildlife Service reclassified Bald Eagles as Threatened in 1995 and delisted them entirely in 2007. They are still listed as Endangered and considered Fully Protected by California and are also protected under the federal Bald Eagle and Golden Eagle Protection Act. Historically huge Sierra salmon runs are now a thing of the past due to dam construction and water diversions, and this may limit the extent of the Bald Eagle's recovery in this region, although construction of reservoirs has undoubtedly benefited the species.

Northern Harrier • *Circus cyaneus*

ORIGIN OF NAMES "Harrier" from "hen-harrier," a British name for the species' habit of harrying poultry; *circus* from Gr. *kirkos,* a hawk, refers to the species' circling flight; L. *cyaneus,* blue, for the male's grayish-blue upperparts.

NATURAL HISTORY One often sees these low-flying hawks hunting low over grasslands and marshes of the Sierra Nevada. Northern Harriers have an owl-like facial disk that funnels sound into large ear openings. While not hunting like owls in the dark, they do work very close to the ground in a "low patrol" fashion, where their

Female

Male

excellent hearing helps them locate prey, even in dense cover. These hunting flights can be quite dramatic, as harriers hunt almost solely through active flight: "It oftenest moves in a huge zigzag course, quartering its territory like a hunting dog" (Dawson 1923). When prey is detected, the harrier may suddenly stop, perform an acrobatic downward turn, and plunge suddenly onto an unsuspecting vole, bird, or reptile.

While many breeding pairs are monogamous, some male harriers maintain harems of two to five females, a rare behavior among hawks. In the Sierra the breeding season begins in late March or early April, when males are in frequent flight, performing energetic roller-coaster courtship displays that may include dozens of U-shaped dips. Both sexes carry grasses and other materials, but the females do the nest building alone. During incubation, males bring a steady supply of skinned and beheaded prey to the female, with the exchanges frequently occurring in flight with the male dropping the item and the female snatching it in midair. Their nests are well hidden in wet meadows or marshes, and recently fledged birds spend much of their time hiding in hidden pathways in the vegetation. Both parents feed the young, but males with extensive harems may ration their efforts to benefit just one or two nests. Juveniles can fly and can hunt independently about two months after hatching.

In winter, the larger and more aggressive adult females defend territories with the best habitat, forcing males and juveniles to forage in less productive areas. Unlike most other hawks, sexes differ in appearance, with females brownish and adult males gray, black, and white. Both male and female harriers are brownish during their first year of life.

STATUS AND DISTRIBUTION Found around the edges of the Sierra, harriers favor open areas of wetland, pastures, and grassland but can be seen foraging over sagebrush flats as well. After young birds fledge, until the arrival of snows they readily wander into the mountains and can be found over open slopes, rarely up to the Alpine zone. Local populations often increase in winter with the influx of birds from the north and east.

West Side. Fairly common to uncommon, primarily associated with Central Valley wetlands, pasture, forage crops, and rice (in winter), but their breeding range extends from the lowest reaches of the Foothill zone up into open oak savanna in the north; in the south known nesters only in the Kern River Valley, at about 2,600 feet; during spring and fall migration they are frequently observed soaring in-transit over all habitats up to the Subalpine zone.

East Side. Fairly common to uncommon, pairs breed locally throughout the Great Basin and scattered populations range to the base of the Sierra; common in low-elevation wet meadows, but historical overgrazing and development have impacted and fragmented their best habitats; sometimes breed at higher elevations (up to more than 6,000 feet) in wet mountain meadows.

Sharp-shinned Hawk • *Accipiter striatus*

ORIGIN OF NAMES "Hawk" from Teutonic root *hab,* to grasp or seize; "sharp-shin" refers to the raised ridge on the front of the bird's lower leg; *accipiter* from L. *accipere,* to take or seize; L. *striatus,* striped.

NATURAL HISTORY Whether scanning from a hidden perch or dashing through forests, Sharp-shinned Hawks are constantly on the alert for small songbirds, their primary prey. Finch-sized birds are captured in midair or plucked from the ground or from foliage with the aid of the hawk's long central toes and talons. Quarry range in size from Anna's Hummingbirds to California Quail, and they occasionally take other small vertebrates, such as mice and lizards, or large insects. These adept hunters attack from low, chasing flights or by sneaking in from behind concealing obstructions (which also means that they do not see prey until the last instant).

Juvenile

Adult Male

The relatively short, rounded wings and long, rudderlike tail of the "sharpie" make it exceedingly agile as it pursues elusive, darting birds into dense thickets. Chases are short and quickly aborted if the surprise attack fails. If successful, the hawk returns to a favorite perch to pluck the prey fastidiously before it is eaten or fed to hungry nestlings. During the breeding season, a male will bring beheaded prey to a perch near the nest, where he calls softly for the female to come and take the food to the nestlings.

In the Sierra, Sharp-shins nest in groves of oaks and conifers or in riparian woodlands, where they hide their loosely constructed nests high on horizontal branches near the trunks of densely foliaged trees. While both sexes collect twigs, sticks, and bark for the nest, the much larger female does the actual construction. Nest sites (but not the actual nests) may be reused in the following year, although it is unclear if the pairs that return are the previous occupants.

STATUS AND DISTRIBUTION Sharp-shinned Hawks can be found in the Sierra year-round, with significant numbers of migrants moving through the region in spring and fall and an influx of wintering birds at lower elevations. Spring migration begins in early February, but peak numbers move through the region in late March; breeding pairs are likely to establish nesting territories in April during late spring and summer at scattered locations in the northern half of the range primarily between the Lower and Upper Conifer zones. Numbers pick up again from August to the end of October, corresponding with the addition of wandering juvenile birds and the arrival of migrants from the north. During late summer and fall, they range above their normal nesting elevations and follow flocks of songbirds to the Sierra crest but retreat back to lower elevations before the heavy snows of winter set in.

West Side. Rare nesters but fairly common visitors in winter and during migration; rare in the southern Sierra, and they rarely breed south of the Tahoe region; in winter, they are most readily observed around open oak woodlands of the Foothill zone.

East Side. Thinly distributed and scarcely observed, most reliably found in September and October during fall migration, often near the Sierra crest; nesting observed in the Tahoe Basin and Mono County and may breed in the Owens Valley.

Cooper's Hawk • *Accipiter cooperii*

ORIGIN OF NAMES "Cooper's" for William C. Cooper (1798–1864), a generous naturalist who helped start the New York Lyceum of Natural History in 1818 and was the father of James C. Cooper (1830–1902), a noted California ornithologist for whom the Cooper Ornithological Society was named.

NATURAL HISTORY Stealthy hunters within riparian groves, urban forests, oak woodlands, and conifer forests, Cooper's Hawks always use the element of surprise when attacking prey. Typically they wait patiently on hidden perches then make quick, powerful flights to take unsuspecting birds and mammals. Birds up to the size of American Crows are their primary victims, but

Juvenile

Adult Female

Cooper's Hawks also attack rabbits, squirrels, mice, and reptiles. These powerful hawks avoid eating feathers or fur and prefer to pluck their prey at favored perches. Like Sharp-shinned Hawks, they readily pursue birds through dense thickets, maneuvering with ease due to their relatively short wings and long, rudderlike tails. Occasionally they land on the ground in search of prey hiding in cover, resuming the chase when their quarry takes flight.

The smaller size of the Cooper's Hawk renders it a somewhat less formidable and aggressive nest defender than the Northern Goshawk. However, both deserve caution and respect, and their nesting territories should be avoided. Near the North Fork American River in Placer County, a pair of Cooper's Hawks built their nest in a dense mixed conifer forest near a well-traveled trail. Despite the frequency of human traffic, the stealthy adults were never seen flying from their nest to distant feeding grounds. Their stick platform was finally discovered high in a white fir, lodged between the trunk and a large branch. Excrement or "whitewash" splattered the foliage and ground. The plucked remains of a chipmunk and Steller's Jay dangled from nearby branches. Three juveniles peered from the nest, but the adults were gone. The parents returned suddenly and attacked the human intruders with frenzy—driving them about 100 yards through the forest.

Cooper's Hawks begin nest building or reconstruction of old nests as early as March, and most pairs have produced eggs by May or June. Both adults take part in nest building, but males do most of the work. The young become fully feathered and can fly within about six weeks; they can hunt without help from their parents after about eight weeks. Scattered, dense stands of live or blue oaks and lowland riparian forests are preferred breeding habitats, but some pairs build their nests in pine and oak forests up to the Upper Conifer zone. They tend to nest in slightly older and more open stands than Sharp-shinned Hawks. Small streams are often nearby, but nests have also been found in arid forests and foothill canyons.

STATUS AND DISTRIBUTION Occurring year-round in the Sierra, Cooper's Hawks are fairly common breeders in a variety of closed canopy woodlands with an open understory from the Foothill zone into the Upper Conifer zone. From mid-August through October, their numbers increase as juveniles begin wandering and migrants push through the area. During this season it's not uncommon to see them up to or above tree line. Raptor migration surveys show that Cooper's and Sharp-shinned Hawks occur in roughly equal numbers and are the two most abundant migrating raptors in the Sierra. Band recovery data for all three *Accipiter* species suggest that migrants found on either side of the Sierra crest are from distinct populations and there is little crossover between flyways.

West Side. Fairly common but widely scattered and relatively hard to find in breeding season; more readily observed during spring or fall migration, or in winter, when widespread and fairly common.

East Side. Uncommon to rare breeders, Cooper's Hawks are even more sparsely distributed than on the West Side and, likewise, more likely to be found during fall migration and in winter, especially in residential areas with large numbers of bird feeders.

In the past few decades Cooper's Hawks have increased across much of their range. They have become more common in winter and even as breeders in residential areas, where bird feeders provide benefits further up the food chain. Sierra Christmas Bird Counts show a significant positive trend for this species throughout the region.

Northern Goshawk • *Accipiter gentilis*

Juvenile

Adult

ORIGIN OF NAMES "Northern" for the species' distribution; "goshawk" or "goose-hawk" from AS. *gos*, goose, and *havoc*, hawk; L. *gentilis*, gentle or noble.

NATURAL HISTORY Harsh, strident *kak-kak-kak* calls issuing from the deepest conifer forests announce the presence of nesting Northern Goshawks. If approached too closely, these largest and most powerful of North American *Accipiters* will defend their nesting territories like demons. They peer down at intruders with defiant red eyes and fly boldly at their targets with talons spread. The best strategy when attacked is to protect your head and eyes and quickly leave the nesting territory. William Dawson's (1923) experience is not unusual: "It was late June and the ornithologist was not aware that a certain stretch of woods which the trail cleft belonged to a highly virtuous pair of Goshawks, until Whoof! Buff! the blue terror struck a blow from behind and sent the bird-man sprawling." The National Park Service has on several occasions temporarily closed the popular Four-Mile Trail from Glacier Point to Yosemite Valley to protect visitors from a pair of nesting Northern Goshawks.

Goshawks nest in a variety of mature forest types, usually in dense stands with complex structure (variety of trees sizes and ages) near openings for foraging. Their use of extensive forest stands makes goshawks especially vulnerable to logging practices that fragment forests and remove large trees. However, in East Side pine forests they will nest in small patches of trees. Males and females often work together to build bulky stick nests on the lowest large branches of mature conifers. Nests are two or three feet across and may be conspicuous or concealed by foliage. Nest sites generally have an open midstory, allowing clear access to the nest. Banding studies suggest that they do not mate for life. As early as February or March, goshawks begin to build a new nest or reline an old one with bark, small conifer branches, and needles. From courtship through early nestling stages, the male will provide nearly all the food for the female. At about 40 days after hatching, young birds are ready to leave the nest, and spend 4 to 6 weeks in the nest area learning to hunt before dispersing.

These cunning predators glide silently and swiftly beneath the forest canopy in search of prey, or else they pause to sit and watch for short spells before moving on to another perch. A chorus of alarm calls from chipmunks, Douglas squirrels (their favored prey in the Sierra), and Steller's Jays is often the first clue to their presence. Even the silhouette of a goshawk in its distinctive flap-glide flight is sufficient to trigger a flurry of nervous activity from small forest animals. Goshawks also hunt in clearings or along meadow edges. They attack from a perch or while in flight, pursuing birds in midair at tremendous speeds. Foliage of trees and shrubs are brushed aside or deftly avoided as they career through close behind a bird or rabbit. Goshawks also attack on the ground; one in the Sierra hopped around a small lake for more than an hour chasing Mallard ducklings and eventually caught three.

STATUS AND DISTRIBUTION Northern Goshawks mainly occur in mature conifer stands during all seasons but are more often observed hunting in meadows or other openings, especially near water where prey may be abundant. On rare occasions, migrating goshawks may be seen circling high overhead. Estimates of the Sierra breeding population range up to 700 pairs. They are sparsely distributed on both sides of the Sierra, with the majority apparently resident year-round on their breeding territories. In late summer, some individuals travel above tree line and soar over open ridgetops in search of Belding's Ground Squirrels and other prey in the Alpine zone.

West Side. Uncommon breeders from the Lower Conifer zone up to the Subalpine zone; uncommon in winter, some goshawks may move downslope into foothill woodlands for brief periods, but this movement does not seem to be related to the severity of winter conditions or prey availability.

East Side. Resident at mid- to high elevations in stands of large conifers but even more sparsely distributed than on the West Side; also found down to the base of the Sierra in appropriate habitat (e.g., Jeffrey pine stands north of Mammoth Lakes, Mono County). They also breed in mature aspen stringers within pine and scrub habitats over a wide range of elevations extending out into the Great Basin.

TRENDS AND CONSERVATION STATUS Much of the habitat of this species is within public lands such as National Forests and National Parks. While there is comparatively little threat in the way of habitat destruction in National Parks, logging of older forests on private lands and in National Forests is a significant threat. Due to this threat and the relatively small California breeding population, the species is on the list of California Bird Species of Special Concern, and it is considered a Sensitive Species by the U.S. Forest Service.

Red-shouldered Hawk • *Buteo lineatus*

ORIGIN OF NAMES "Red-shouldered" for the wing patches of adults; L. *buteo,* a hawk; L. *lineatus,* lined, for the species' bold white-and-black barring on the wings and tail.

NATURAL HISTORY Because of their ability to maneuver adeptly through dense forests with rapid wingbeats, these brilliantly patterned woodland hawks are more like *Accipiters* than typical soaring *Buteos.* Historically, Red-shouldered Hawks nested primarily in wooded river bottoms and large tracts of contiguous oak forest. They have now adapted to living in fragmented, suburban settings of the foothills, where they often nest near houses or farm buildings and their piercing *kee-yer* calls are a common sound. Wherever they nest near Great Horned Owls or Red-tailed Hawks, those larger raptors often usurp or prey upon the smaller Red-shoulders.

The distinctive nests are located in the main forks of large tree trunks in riparian habitats or oak woodlands near water. Red-shouldered Hawk nests are compact but substantial and fill tree crotches to considerable depths, comfortably lined with fine pieces of bark, lichens, or mosses. In early spring, both parents help to build new nests or refurbish old ones. Within appropriate habitat, Red-shoulders are resident and are thought to stay on the same territory for their entire lives. Young birds fledge around six weeks and are quickly able to catch small prey items for themselves.

These hawks mainly hunt from a perch, searching for small prey (rodents, birds, reptiles, and large insects). Though generally associated with woodlands, they are often seen perched on roadside utility wires, staring down between their feet into the roadside grasses.

STATUS AND DISTRIBUTION Historically a low-elevation species, in the past 20 years Red-shouldered Hawks have moved up from the Central Valley (where the population has nearly tripled) to previously unrecorded elevations on the West Side; they have even become regular visitors and breeders on the East Side.

West Side. Mainly associated with dense riparian forests or adjacent upland slopes up to the Lower Conifer zone and have recently extended their breeding range above 3,000 feet; rare individuals may wander into the high mountains or cross over the Sierra crest in late summer, fall, and even winter.

East Side. Fairly common but localized; in the 1970s Red-shoulders first appeared near Honey Lake, near Lake Tahoe, and in the Owens Valley, and they are now increasingly regular residents in many lower-elevation areas, mostly associated with cottonwood riparian zones and such residential areas as Susanville (Lassen County), Loyalton (Sierra County), and Reno.

Swainson's Hawk • *Buteo swainsoni*

ORIGIN OF NAMES "Swainson's" for the prolific 19th-century naturalist and author William Swainson (1789–1855), a friend and contemporary of John James Audubon; nine species are named in his honor, including a thrush, a warbler, and this hawk that breed in North America, as well as six tropical species.

NATURAL HISTORY Swainson's Hawks have one of the most interesting life histories of any Sierra raptor. Arriving in the region in late February or early March from their tropical winter range, they breed in open, lowland valleys and agricultural areas. Historically they preferred grass-dominated landscapes, but now they usually forage in agricultural areas wherever crops are not too tall for spotting prey—mowed or flooded alfalfa fields are preferred foraging areas.

Their large, flimsy nests (perhaps the most insubstantial of all *Buteos*) are built by both sexes in riparian woodlands or even in isolated trees. Wherever they coexist alongside Red-tailed Hawks, they tend to use smaller trees or smaller clumps of trees than their more common cousins. During the breeding season, they hunt low over open country in search of rodents, rabbits, and reptiles. Swainson's Hawks frequently gather to hunt communally, especially after the breeding season or around peak food events such as when a farmer floods or mows a field or a fire flushes prey from the grass. They are extremely agile on the ground, walking easily and running expertly in pursuit of prey. After the breeding season, they switch over to a diet that is almost entirely composed of grasshoppers, dragonflies, and other large insects, and a single hawk may eat up to 100 per day.

A female incubates her eggs while the male provides food and takes over incubation duties after leaving food for the female near the nest. About 40 days after hatching, the young birds begin taking their first flights. Swainson's Hawks depart by September or October and undertake marathon flights to wintering grounds. Until recently, most were thought to fly all the way to the pampas of Argentina for winter, but radio-telemetry studies have shown that most nesting birds in the Central Valley and Sierra foothills winter in central Mexico instead (Bradbury et al. in prep.).

STATUS AND DISTRIBUTION Swainson's Hawks are almost entirely restricted to lowland valleys and open areas of the foothills on both sides of the Sierra. In California they breed in the Central Valley, on the Modoc Plateau, and along the East Side of the Sierra.

West Side. While fairly common breeders in the Central Valley, they are rare visitors to the lowest portions of the Foothill zone as far south as Fresno County; nesting pairs seem to drop out as soon as the habitat changes from irrigated pastures and annual grasslands to oak savanna; however, recent breeding records just below 500 feet (Amador County) and just above that elevation (Mariposa and Madera Counties) may indicate an expansion of their breeding range into the western foothills; a few postbreeding birds may wander upslope to forage over mountain meadows prior to their southward migration; less than a half-dozen are observed each fall in the Kern River Valley vulture counts (see Turkey Vulture account).

East Side. Locally fairly common breeders near irrigated fields and open country around Honey Lake, and in Sierra Valley and Carson Valley; a separate population of about 20 pairs nests near alfalfa fields in the northern Owens Valley; uncommon in spring or fall migration and might be observed over wet meadows or irrigated pastures.

TRENDS AND CONSERVATION STATUS Swainson's Hawks were listed as Threatened by California in 1983. Central Valley populations are centered in Sacramento, San Joaquin, and Yolo Counties. During historical times (ca. 1900), Swainson's Hawks may have maintained a California population in excess of 17,000 pairs. Based on a study conducted in 1994, the statewide population was estimated to be approximately 800 pairs. However, anecdotal observations suggest that numbers have increased in recent years. The loss of agricultural lands to various residential and commercial developments is a serious threat to these birds in California. Additional threats are habitat loss due to riverbank protection projects, conversion from agricultural crops that provide abundant foraging opportunities (e.g., alfalfa) to crops such as cotton, vineyards, and orchards that provide almost no foraging opportunities. On their wintering grounds in Mexico and Argentina they are threatened by shooting and pesticide poisoning.

Red-tailed Hawk • *Buteo jamaicensis*

ORIGIN OF NAMES "Red-tailed" for the most visible field mark of adults; L. *jamaicensis,* of Jamaica, where the species was first collected.

NATURAL HISTORY More often than not, a large raptor seen wheeling above the landscape almost anywhere in the Sierra is a Red-tailed Hawk. Red-tails frequent a greater range of habitats than any other Sierra raptor, though they prefer open terrain such as annual grasslands, river canyons, mountain meadows, granite outcrops, chaparral, broken forests and Alpine fell-fields where their ground-dwelling prey can be most easily spotted. When not hunting, they spend much of their time soaring on broad, outstretched wings and fanned tails. While hunting, Red-tails search for prey from exposed perches and attack from low, quartering or hovering flight. Unlike the agile *Accipiters* that chase songbirds through woodland thickets, Red-tails hunt in open habitats, using their extraordinary eyesight to detect ground squirrels, gophers, rabbits, mice, snakes, or lizards that stray from cover; occasionally they attack larger birds such as American Kestrels and Western Meadowlarks. They kill prey in a direct dive, stunning their unsuspecting victims with powerful legs and sharp talons.

Red-tails are usually monogamous, and courting pairs fly high above their nesting territories, shrieking and diving at each other with legs dangling. The sexes can often be distinguished by size when seen flying together because females, like most raptors, are slightly larger than males. Pairs build their sturdy nests between forked limbs of tall trees or, less often, on cliff faces well above the reach of ground predators such as coyotes or raccoons. Nests may be reused year after year unless they are appropriated by Great Horned Owls. Breeding activities may extend from January to September. Chicks leave the nest a month and a half after hatching but may stick around to be fed by their parents through the next winter.

Red-tailed Hawks exhibit tremendous variation in color, ranging from mostly light to entirely dark, chocolate-brown individuals. The full range of color variation can be seen in the Sierra, but light-morph birds are decidedly more common.

STATUS AND DISTRIBUTION Up to 16 subspecies of Red-tails have been recognized in North America. This extremely widespread species is commonly observed in all types of open country. Individuals seen flying over forested areas are likely in transit or venturing out from an adjacent breeding territory. They are year-round residents at lower elevations, but by midsummer they are readily observed in the high mountains, where they remain until driven downslope by heavy snows. Due to an influx of Red-tails from farther north and east, numbers typically increase dramatically in winter wherever snow cover is minimal and prey are readily available.

West Side. Common resident from the Foothill zone into the Upper Conifer zone, and migrating Red-tails also readily observed at higher elevations—below the snow line in spring and as high as the crest in fall; common to abundant among the oak savannas and annual grasslands of the lower foothills during fall and winter.

East Side. Fairly common breeders in appropriate open habitats and winter concentrations can be quite high in valleys that support agriculture from Honey Lake south to the Owens Valley.

Ferruginous Hawk • *Buteo regalis*

ORIGIN OF NAMES "Ferruginous" derived from L. *ferrugo,* rust, a reference to the rusty-red color of thighs, backs, and shoulders of adults; L. *regalis,* royal.

NATURAL HISTORY This largest and most powerful of our soaring hawks was considered by A.C. Bent (1937), author of the classic series *Life Histories of North American Birds,* to have more in common with Golden Eagles than other soaring hawks, since both have feathered legs, bulky nests, and many other similar traits. Ferruginous Hawks are sit-and-wait predators of open habitats where they feed on rabbits, ground squirrels, pocket gophers, or whatever other small mammal is most readily available in a given locale. From perches on the ground, fences, or on nearby trees or rocky outcrops, these hawks swoop in with low, coursing flight to catch prey. Sometimes they will stand and wait to grab ground squirrels or gophers that are clearing out burrows and pushing

dirt up to the surface. In winter, groups of six to twelve birds at a time may be observed hunting and roosting together, a curious behavior about which little is known.

STATUS AND DISTRIBUTION Ferruginous Hawks breed mainly in the Great Plains, upper Midwest, Great Basin, but many head for the valleys of California in winter where they can be common around favored habitats with ample prey. Except during migration, Ferruginous Hawks are nearly always associated with open terrain in and around valleys, plateaus, desert edges, and agricultural areas. Migrating birds apparently cross all habitat types, including extensive coniferous forests. Wintering birds occur at lower elevations around the Sierra periphery in annual grasslands and irrigated pastures.

West Side. Uncommon and localized from late September until early April in the Foothill zone; while locally fairly common along the Central Valley edge, locally uncommon to rare above 500 feet in large areas of annual grassland and open oak savanna.

East Side. Fairly common to locally abundant fall through winter in lowland areas; often abundant in Sierra Valley, where as many as 65 have been tallied on a single Christmas Bird Count, and they can be fairly common in many other valley locations from Honey Lake south to the Owens Valley.

Rough-legged Hawk • *Buteo lagopus*

ORIGIN OF NAMES Both common and scientific names refer to the bird's legs, which are feathered to the toes; *lagopus* from Gr. *lago*, a hare, and *pous*, foot, a reference to the fur on a hare's foot.

NATURAL HISTORY Each winter, hawk enthusiasts eagerly anticipate the arrival of these unusual far northern breeders, which are common some years and scarce or absent in others. Not only are these stocky *Buteos* highly variable in their plumage patterns, but they are also notable for their hunting technique of often kiting in midair while scanning the ground below for voles, mice, shrews, or carrion. This strategy is useful in areas where perches are limited, and it opens up hunting grounds that are not accessible to *Buteos* that hunt from perches. Rough-legged Hawk numbers fluctuate widely in response to the boom-and-bust cycle of Arctic lemming populations, and in winter their numbers are even more variable because they seek areas with abundant food supplies and can be absent from areas where they were common in previous years. During peak years they may rival Red-tails in total numbers in some locations.

STATUS AND DISTRIBUTION Being well adapted for open country, these hawks spend their nesting season on Arctic tundra then in winter seek out similarly open habitats—annual grasslands, pastures, wet meadows, fields, and sagebrush steppe—mostly at low elevations and with relatively level or slightly rolling topography. They are widespread in California from late October to early April, with individual birds sometimes arriving early or leaving late.

West Side. Rare in lower-elevation grassland and oak savanna habitats; there is a scattering of November and December records from high mountain meadows and barren slopes, especially in years when the snowpack is not extensive.

East Side. Fairly common to absent in valleys and sagebrush flats; sometimes abundant around agricultural areas with fields and pastures as near Honey Lake, in Sierra Valley and Carson Valley; Christmas Bird Counts at Honey Lake and Sierra Valley have recorded totals of nearly 100 birds in a single day.

TRENDS AND CONSERVATION STATUS In recent decades the winter range of Rough-legged Hawks has shifted with many more birds wintering farther north, mainly in the northern Great Plains. This shift correlates with generally milder winters and decreasing snow cover in these areas. Winter numbers in the Sierra and the Central Valley have decreased, as they have in all the eastern and southern parts of this species' winter range in North America.

Golden Eagle • *Aquila chrysaetos*

ORIGIN OF NAMES "Golden" for the napes of adults; L. *aquila,* eagle, and Gr. *chrysos,* golden, and *aetos,* eagle.

NATURAL HISTORY Gliding silently among granite cliffs, forested canyons, and sagebrush-covered hills, the keen-eyed Golden Eagle betrays little of the awesome power of its great bulk and talons. It inhabits the entire length of the Sierra wherever undisturbed nesting sites, open terrain for foraging, and dependable food supplies are to be found. This majestic bird has a wingspan exceeding seven feet; the only larger North American raptors are Bald Eagles and California Condors.

Golden Eagles build huge stick nests on steep cliff ledges or in tall trees with commanding views of open country. These bulky structures can be five or six feet across and may weigh a ton or more, especially if they are used repeatedly over many years. Nests are typically reused by mated pairs in successive years but may be abandoned if humans disturb the nesting pair. One pair in Yosemite Valley deserted a nest with eggs because too many people climbed up for closer views. Golden Eagles have a prolonged nesting period, extending from January through September, with the young birds taking more than two months to make their first flights. Early in the season, adults mount high in the air and make up to 20 successive courtship dives in a row. Whether gliding above lowland grasslands or Alpine meadows, Golden Eagles search for unwary squirrels, rabbits, marmots, or pikas. They occasionally scan for prey while perched but more often hunt from low, quartering flight. At times, a hiker can be alerted to the approach of an eagle by a string of alarm calls from terrified ground squirrels. On rare occasions they take prey as large as Sandhill Cranes or Tundra Swans and will consume carrion, though not as readily as Bald Eagles. Pairs are sometimes observed hunting in what appears to be a cooperative fashion, with one bird flushing prey that is captured by the other bird.

STATUS AND DISTRIBUTION Golden Eagles might be seen at any elevation of the Sierra, particularly over open habitats where prey animals are easier to see and where updrafts ease their heavy

Juvenile

Adult

flight. Winter storms generally drive eagles below the snow zone to lowland regions where hunting is more productive. Eagles from the Great Basin and northern locations invade the Sierra foothills in winter, noticeably swelling the local population.

West Side. Most often found in the Foothill zone up to the lower edges of the Upper Conifer zone but also regularly seen in the higher mountains—especially above rugged river canyons and near the Sierra crest, where updrafts are particularly strong; most pairs nest in the Foothill zone, but breeding has been observed to the Sierra crest with vertical cliffs above river canyons being their favored nest sites.

East Side. Uncommon, eagles soaring high above are visible from nearly all elevations, making it difficult to determine which habitats are favored; many breeding pairs are located below the pine forests where open, sagebrush-foraging habitats are located close by.

TRENDS AND CONSERVATION STATUS While Golden Eagles are not listed as Threatened or Endangered by state or federal agencies, they are protected under the federal Bald Eagle and Golden Eagle Protection Act. Due to their large size and relatively slow flight, they are one of the most susceptible raptors to collisions with wind turbines. High mortalities of this species have been observed at wind energy sites such as Altamont Pass in the Coast Range, and expansion of this industry in the Sierra will require careful planning to avoid similar impacts to Golden Eagle populations.

FALCONS • Falconidae

More than any other birds of prey, falcons are built for speed, streamlined and compact. Small heads, short necks, long, pointed wings and narrow tails are all adaptations for rapid flight. When pursuing prey, they rocket through the air at enormous velocities and kill with powerful blows from outstretched talons. Their hooked beaks are sharply notched about halfway between the tip and base, useful for tearing flesh or breaking the necks of their victims. Like many other predatory birds, falcons dissolve food in their highly acidic stomachs and then regurgitate indigestible feathers or fur as compressed pellets.

Falcons do not make nests but instead use rock ledges, natural cavities, or the abandoned nests of other large birds. Females usually lay two to four eggs at two- to three-day intervals, meaning that the youngest nestlings may starve or be eaten by their siblings when food is in short supply. Females incubate their eggs (heavily spotted or splotched with brown or chestnut patterns) for 28 to 30 days, while the males return to the nests with prey items at frequent intervals and intermittently to help incubate the eggs. Young falcons are covered with white down at hatching, making them somewhere in between "altricial" and "precocial" compared with other birds. Nestlings can move about the nest site on their own after 20 to 25 days, and they are fully feathered and can fly after about 30 to 40 days (longer for Prairie and Peregrine Falcons than for American Kestrels).

This large family is represented by almost 60 species that occur worldwide except for Antarctica and some oceanic islands, but only 4 of these occur regularly in the Sierra region. Recent genetic studies suggest that falcons may actually be more closely related to parrots than to other birds of prey. The family name is derived from L. *falx*, a sickle, for the bird's claws.

American Kestrel • *Falco sparverius*

ORIGIN OF NAMES Kestrel apparently derives from OF., a reference to a ringing bell; L. *sparverius*, relating to a sparrow, either for the species' size or preferred food items.

NATURAL HISTORY Along almost any country road in the valleys and low foothills of the Sierra, American Kestrels can be seen perched on utility lines, poles, or fences. These small, colorful falcons thrive in open terrain like annual grasslands, woodland edges, and chaparral. The majority of their hunting is done while perched patiently, but they also search for prey while hovering with rapid wingbeats.

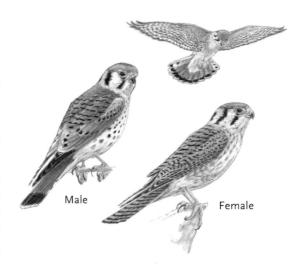

Male

Female

Kestrels typically catch insects, reptiles, and small mammals on the ground, although some individuals learn to specialize on catching large insects or small birds in flight. After catching a prey item, they return to preferred perches with tails bobbing. Although birds are not a large part of their diet, some still refer to them as "Sparrow Hawks."

Nearly all of a kestrel's life is spent in open areas with short ground vegetation. These open patches are critical for finding prey, and kestrels will use patches of nearly any size. Habitat type seems a secondary consideration because kestrels use parks, schoolyards, lawns, and other open urban habitats as readily as they use such wild habitats as meadows, grasslands, or deserts. During the breeding season, they require large snags with old woodpecker cavities for nesting, but in the absence of woodpecker holes, they seek out other natural cavities. Because cavities are a limited resource, people putting out nest boxes have had great success encouraging these small confiding raptors to take up residence in areas that lack natural nest sites.

Starting in April, male kestrels locate and inspect all potential nesting cavities on their territories, later escorting their mates to them so that the females can select ones to their liking. No special preparations are made to the nest cavity, and the eggs are sometimes simply laid on bare wood. About 30 days after hatching, the chicks leave the nest, though they may stay with the parents in family groups until September. After breeding, pairs split up with the larger females occupying the best habitat in winter and males relegated to less-productive areas.

STATUS AND DISTRIBUTION From April to September, American Kestrels are fairly common breeding residents in the lowlands of the Sierra, rarely occurring as high as the Upper Conifer zone. After breeding, some travel upslope to forage over open hillsides, mountain meadows, and talus slopes near the Sierra crest. A few remain in the high country until winter storms force their return to more productive lowland hunting grounds. In winter, numbers in the valleys and Foothill zone increase with the arrival of northern migrants.

West Side. Common to fairly common residents of open terrain below the Upper Conifer zone but occasionally found nesting higher; in winter, mostly restricted to open areas in the low Foothill zone.

East Side. Common breeding residents in open areas up to the pine forests; in winter, fairly common near Honey Lake, and in Sierra, Carson, and Owens Valleys; numbers variable elsewhere from year to year.

TRENDS AND CONSERVATION STATUS As noted in the chapter "Recent Trends in Sierra Bird Populations and Ranges," Breeding Bird Survey and Christmas Bird Count data show widespread, long-term declines in American Kestrel populations throughout their range. The causes for these downward trends remain uncertain.

Male

Juvenile

Merlin • *Falco columbarius*

ORIGIN OF NAMES Merlin from OF. *esmerillon*; L. *columbarius*, a pigeon keeper.

NATURAL HISTORY Perhaps the wizardry of its flight earned this small northern falcon the title of Merlin, the magician. Certainly its powerful, swift flight is a wonder to behold, especially as they pursue small birds or dragonflies and snatch them in midair with astonishing agility. Unlike their smaller relatives American Kestrels, Merlins fly with direct, powerful flight, never stop to hover, and do not bob their tails after landing. If they were more common, the expression for following a straight path would be "as the Merlin flies," rather than "as the crow flies."

Visiting the Sierra region only as fall and winter visitors, Merlins typically linger in open habitat with abundant prey. They are often observed along lakeshores where shorebirds gather on mudflats or around grain elevators and farm buildings where small birds congregate to feed on spilled grain. Hunting flights usually originate from adjacent perches like trees, fence posts, or rock outcrops that command broad views of the surrounding area. Unlike American Kestrels, Merlins rarely perch on wires.

STATUS AND DISTRIBUTION Small numbers of Merlins visit the state each fall and winter. Of the three subspecies that occur in the Sierra, the "Taiga" Merlin *(F. c. columbarius)* is by far the most common. This subspecies breeds primarily in Canada and Alaska. In coloration this subspecies is intermediate between the beautiful, pale "Prairie" subspecies *(F. c. richardsonii)* from the Midwest and the hauntingly dark "Black" subspecies *(F. c. suckleyi)* from the Pacific Northwest—both rare visitors to the Sierra. Arriving in early fall, Merlins are widespread and may be seen over barren high mountain slopes as readily as in low foothill valleys. With the arrival of winter storms, they descend below the snowline and most leave by the end of April.

West Side. Uncommon fall-winter visitors to the Foothill zone as far south as the Kern River Valley, primarily in low-elevation oak savannas and annual grasslands, but they can be seen over a variety of habitats up to the Alpine zone.

East Side. Rare in open areas below the pine forests, most easily found in such agricultural areas as Honey Lake, Carson Valley, and the upper Owens Valley; might be seen near shorebird concentrations at Mono Lake, Bridgeport Reservoir, and the Owens Valley.

Peregrine Falcon • *Falcon peregrinus*

Adult

Juvenile

ORIGIN OF NAMES "Peregrine" refers to wandering for the species' wide-ranging behavior; *peregrinus*, from L. *per*, through, and *agri*, a field.

NATURAL HISTORY Peregrine Falcons, among the world's fastest fliers, reportedly attain speeds over 200 miles per hour when "stooping" with wings pulled tight against their bodies. From lofty, exposed perches or from searching flights, these powerful predators survey the landscape for flying birds such as doves, pigeons, jays, woodpeckers, gulls, ducks, shorebirds, or even darting swifts. When they spot likely victims, they hurtle down and kill or stun prey in midair with a single blow or piercing grab of their spread talons. They usually pluck and eat prey on the ground or on cliff ledges but sometimes consume smaller items in flight. Peregrines are highly versatile in their hunting behaviors. Most individuals attack from above, but they sometimes come up from below to snatch prey in midair. Some pairs work cooperatively to decoy and flush other birds, chasing them into water and grab them when they swim ashore. Peregrines often seem to pursue birds for the sheer joy of the chase, and from the perspective of their prey, they are indeed "terror incarnate" (Hoffmann 1927).

In March or April females lay their eggs in "eyries," or nesting sites, usually simple recesses on tall, inaccessible cliffs that offer expansive views of the surrounding landscape, often near lakes or rivers. Peregrines typically mate for life and frequently reuse the same eyries year after year. During the breeding season, large amounts of food may be cached as a safeguard against bad weather and both sexes may visit the cache to feed. After about 40 days the young birds fledge and spend at least two months learning how to hunt, initially following the parents and catching prey items dropped to them in midair, then making their own tentative playful chases that eventually lead to successful captures.

STATUS AND DISTRIBUTION Far-ranging, Peregrine Falcons can be expected in any location from the lowest valleys to the highest peaks, though in winter they either leave much of the range or move down into the foothills. They have been steadily recolonizing the Sierra and new nesting locations are noted nearly every year.

West Side. Uncommon and sparsely distributed breeders and easily missed because many flights are at great altitudes; up to four nesting pairs have been observed in Yosemite Valley at one time, and nesting has occurred in major river canyons such as the North Fork of the American River; pairs have used cliffs created by gravel-mining operations in Yuba, Nevada, and El Dorado Counties; nonbreeding Peregrines might be seen anywhere high cliffs provide good views of productive foraging areas.

East Side. Uncommon, breeding pairs are widely scattered on open cliffs; birds might be seen flying over almost any terrain, similar to the West Side.

TRENDS AND CONSERVATION STATUS Similar to Bald Eagles and many other sensitive raptors, Peregrine Falcons declined after World War II, when the use of DDT and other organochlorine compounds became widespread (see Bald Eagle account above). The U.S. Fish and Wildlife Service and the California Department of Fish and Game listed the species as Endangered in 1978. A ban on use of DDT in the late 1970s and an aggressive program of captive breeding allowed Peregrine Falcons to dramatically increase their population and expand their range throughout the United States. In 1999 the U.S. Fish and Wildlife Service removed the Peregrine Falcon from the Endangered Species list. However, they are still listed as Endangered and considered Fully Protected in California.

Prairie Falcon • *Falco mexicanus*

ORIGIN OF NAMES "Prairie" from the species' preferred, open habitats; L. *mexicanus,* of Mexico, where first collected.

NATURAL HISTORY These large, powerful falcons typically hunt medium-sized mammals and birds by flying fairly low to the ground and ambushing unwary prey with a combination of speed and stealth. Other times they search for prey while soaring or sit and wait while perched on a tree, utility pole, or rock outcrop. Open-country specialists, they terrorize winter flocks of American Pipits and Horned Larks in Sierra grasslands. Ground squirrels, pocket gophers, and other such rodents are also favored prey. In any season they favor grasslands, agricultural areas, and shrub-steppe habitats. Breeding begins in March with a month of courtship and nest site selection. A pair will visit numerous cliff faces and rock outcrops, investigating ledges, cavities, and abandoned ravens' nests while performing stylized courtship behaviors. Clutches are usually timed to coincide with the emergence of juvenile ground squirrels in spring. This ensures a steady and easily obtained source of food. Young birds leave the nest after about 40 days but are dependent on their parents for another couple months.

STATUS AND DISTRIBUTION Prairie Falcons wander nomadically in response to seasonal shifts in food supply. In mid- to late summer, after ground squirrels start to aestivate (summer dormancy) in lowland areas, these birds move into higher elevations. At this time, Prairie Falcons search for prey around high mountain meadows and slopes up to 14,000 feet in the southern Sierra. Some remain there until the first heavy snows drive them back into the foothills on both slopes.

West Side. Uncommon residents the length of the Foothill zone, where they forage primarily in open grasslands and rarely above the most open oak savannas, even though breeding sites may be at much higher elevations; numbers increase noticeably in the lowlands from fall through winter as migrants from the north and east arrive.

East Side. Uncommon year-round residents, where they nest on steep, rocky cliffs near scrub habitats; some birds, mostly juveniles, wander upslope to the Alpine zone in late summer and fall.

RAILS AND RELATIVES · Rallidae

With the exception of the gregarious American Coot, marsh-dwelling members of this family are generally secretive, poorly known, and seldom seen—most often detected by their loud and distinctive calls. Thin bodies and strong legs allow them to slip effortlessly through dense marsh vegetation. When absolutely forced to fly, they make short, quick flights to cross open spaces. Although they have a very high ratio of leg muscles to flight muscles, some rails manage to fly long distances during migration. Despite their short stubby wings, migrating Soras have been clocked at almost 60 miles per hour!

These birds can be subdivided into three groups based on foraging styles. Long-billed rails, like Virginia Rails, have long, curved bills to probe for invertebrates in the mud. Crakes (such as the Black Rail and the Sora) and gallinules (the Common Gallinule) use their shorter bills to pick food from surfaces. As a result, they tend to eat more plant foods. American Coots have found a ducklike niche in open water, where they paddle around in large flocks and dive for food. Female rails usually lay 8 to 12 creamy or pinkish eggs with sparse spots or splotches. Both adults incubate the eggs for 15 to 20 days, and the young are independent after about a month; they can fly after four to six weeks.

More than 130 species in this family exist worldwide, but only 6 have been observed in the Sierra (including the Yellow Rail, casual in the Sierra, see Appendix 2). Although the common expression "thin as a rail" does refer to the narrow proportions of these birds, the name more likely derives from their calls, possibly from OF., *rale*, a throat rattle, imitative of their distinctive vocalizations.

Black Rail · *Laterallus jamaicensis*

ORIGIN OF NAMES "Black" for the species' overall plumage coloration; rail (see family account above); *laterallus* from L. *lateo,* to lurk or hide, and *rallus,* suggesting either a hidden bird or its hidden calls; *jamaicensis,* from Jamaica, where the species was first collected.

NATURAL HISTORY When Black Rails were first discovered in the foothills of Yuba County in 1994 (Aigner et al. 1995), it drastically changed our knowledge of the species' distribution in California. They were formerly known only from tidal marshes around San Francisco Bay and marshes along the lower Colorado River. No biologist would have remotely suspected that one of the state's largest populations resided in the foothills of the western Sierra.

These smallest of North American rails (smaller than many sparrows) are so secretive that few people have ever seen one. However, their loud *kik-kee-do* calls are so distinctive, it makes one wonder how Black Rails eluded ornithological fieldworkers until recently. In the Sierra, Black Rails are almost always found in wet meadows with emergent vegetation or shallow freshwater marsh habitats, though the birds shift regularly between sites in response to local conditions. They readily colonize new marshy areas that form near existing colonies, forming a clumped rather than uniform distribution. Recent genetic studies suggest that the foothill population is strongly linked with the Bay Area population, that interchange occurs between the populations, and that most movement appears to be from the foothills to the Bay (Girard et al. 2010).

They apparently remain year-round on very small territories less than one acre in size. Their favored niches in the foothills are places where water depths are about one-inch deep, a condition created at spring-fed seeps, where water from irrigated pasture runs off into gentle swales, or downslope from leaky irrigation canals and ditches, where water spreads into adjacent annual grasslands and marshes. If moist conditions persist into the spring, rails will begin calling and courting. Their nests are small, deep cups of sedges and grasses hidden on the ground under dense vegetation.

STATUS AND DISTRIBUTION Black Rails were recently discovered to be uncommon year-round residents of the Sierra foothills.

West Side. Uncommon, recent intensive surveys have detected Black Rail populations using more than 200 distinct locations (ranging from 300 feet to nearly 1,000 feet elevation) from year to year in just 4 counties: Butte, Yuba, Nevada, and Placer; future surveys in nearby counties could possibly reveal additional populations in marshes, swales, and irrigated pastures of the Foothill zone.

East Side. No records.

TRENDS AND CONSERVATION STATUS California listed the "California" Black Rail *(L. j. coturniculus)* as Threatened in 1971 due to historical and ongoing loss and degradation of its known habitats along the coast and lower Colorado River. Primary threats to these populations are water diversion and flood-control projects, land-use changes, and agricultural runoff. While the recent discovery of the large population in the Sierra foothills is an encouraging sign, populations here are also vulnerable to the same threats as other populations in the state. In addition, their foothill habitat is almost entirely dependent on human sources of year-round water. Efforts to seal leaking canals and decreases in irrigated pasture will reduce available breeding sites.

Virginia Rail • *Rallus limicola*

ORIGIN OF NAMES "Virginia," where the species is common, but this is a poor name because they have reside in all the lower 48 states; *limicola*, mud dweller, from L. *limus*, mud, and *colo*, to inhabit.

NATURAL HISTORY Few visitors to the Sierra actually see these chunky, long-billed "marsh hens," since they rarely emerge from dense marsh cover. Like owls, however, Virginia Rails can be easily recognized by their distinctive calls, most often uttered at night or dawn and dusk. They give a loud and descending, hoarse, piglike *oink, oink, oink, oink* as well as several shorter calls, none of which are likely to be confused with the high-pitched whinny sounds of Soras that often share their wetland habitats.

Highly adapted for life amid dense marsh vegetation, Virginia Rails would rather run, dive, or swim than fly to escape danger. Special adaptations that allow them to slip between dense cattails and bulrushes include laterally compressed bodies, flexible vertebrae, modified feathers on the head that resist abrasion, and valves that close to protect nostrils from sharp vegetation. Pairs cooperate in building up to five dummy nests in addition to their main nest, a cuplike structure made from marsh plants. Both adults also share in incubation duties, trading off every couple hours. The young are mobile and active soon after hatching and able to move quickly if danger threatens or if water levels suddenly rise or decline.

Despite living in isolated wetland areas, Virginia Rails are sometimes locally abundant. They seem to prefer younger marshy sites where the density of living and accumulated dead vegetation is less than in older marshes, and they quickly move into newly formed marshes as soon as the vegetation provides adequate cover. Their specific habitat requirements seem to include a mix of robust emergent vegetation (mainly cattails and bulrushes) and patches of open mudflats or shallow pools. Where they co-occur with Soras, they typically use drier portions of the same marshes. With proportionately longer bills than other Sierra rails, Virginia Rails are capable of probing into mud, where they search for submerged foods such as aquatic and terrestrial insects and their larvae as well as the seeds and tubers of aquatic plants.

STATUS AND DISTRIBUTION Widespread in California, Virginia Rails are found throughout the Sierra. While their true distribution and numbers are poorly documented, it is suspected that they are more common and widely distributed than currently known. In winter, those at higher elevations may descend into the foothills or migrate south.

West Side. Locally fairly common in marshes with patches of dense, emergent vegetation from the low Foothill zone up to the lower edges of the Upper Conifer zone from about 3,500 feet in Tehama County south to the Kern River Valley; higher-elevation birds probably descend down to the Lower Conifer zone or lower during winter, as there are few winter records above the Foothill zone.

East Side. Fairly common in marshy lowland sites below the pine forests, where they primarily occur from late April through September, but a few may remain at low elevations for winter rather than migrating, most consistently in the Owens Valley.

Sora • *Porzana carolina*

ORIGIN OF NAMES "Sora" may be a Native American name of uncertain origin; *porzana*, from It., a crake; L. *carolina*, of Carolina, where the species was first collected.

NATURAL HISTORY Like other rails, Soras reveal their presence mainly by distinctive calls and rarely leave the dense cattails or tules. They make an unmistakable whinny—a rapidly descending series of clear-toned notes—as well as a harsh *keek* and a clear, whistled *er-weeeee* that rises in pitch. A patient observer may be rewarded with a clear view of this small rail as it steps gingerly along a muddy marsh edge or swims across a small stretch of open water.

Similar to Virginia Rails, Soras require only a small patch of marsh plants for their survival. The two species often occur together, but Soras seem to require sites with at least some standing water. Habitats with a high density of floating residual vegetation and a mix of tall cover along with shorter seed-producing plants are favored, although Soras regularly use areas with low cover and do not seem to require the tall vegetation that Virginia Rails do. Here they search for seeds and a diversity of invertebrates by raking floating vegetation with their feet or pulling aside vegetation with their bills and looking for food. During the breeding season, pairs will aggressively defend high-quality territories by chasing and fighting with neighboring Soras.

Courtship and nesting peaks in May in many parts of their range, with eggs laid in a crude cup of vegetation constructed over standing water, usually hidden beneath dense marsh plants with a runway leading to the nest. Males bring nesting material to females, who continue to improve and fortify their nests throughout the egg-laying period.

Like other rails, distribution of the secretive Sora is poorly documented, especially on the west slope, where little is known about their elevational limits; found at some high-elevation lakes with a large fringe of grassy wetlands.

West Side. Uncommon localized, resident breeders the full length of the foothills in scattered ponds and marshes from the Central Valley at least up to the Lower Conifer zone and probably higher; cold conditions (and lack of appropriate habitat) probably defines their upper breeding limits and drives them downslope in the winter.

East Side. Locally common, nests near Lake Tahoe and at some other high-elevation locations (Meiss Lake, Alpine County) just over 8,300 feet; also known to nest in marshes in Sierra Valley, where they arrive in mid-April and leave by the end of October; some winter in the Owens Valley but casual to accidental further north in winter.

Common Gallinule · *Gallinula galeata*

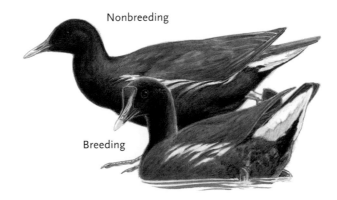

ORIGIN OF NAMES "Gallinule" from L. *gallinula,* a diminutive form meaning "little hen"; *galeata* from L. *galeatum,* helmeted, a reference to the bills of breeding adults; formerly known as Common Moorhen until 2011, when it was concluded the New World species was distinct from the Old World one.

NATURAL HISTORY The gentle soundscape of a marsh at dawn may be broken by the startling lunatic cackle of a gallinule, hiding just out of view in the vegetation. Common Gallinules use the same types of habitat as American Coots, slow-moving or standing waters with patches of dense emergent vegetation, but gallinules tend to keep more to cover and are seen less often in open waters. The relationships between gallinules and coots ("mud-hens") was aptly described by ornithologist William Dawson (1923): "It is a common misfortune of men to be overshadowed by the presence of others neither more deserving nor more clever, only a little more self-assertive . . . the accident of association, together with the still greater accident of similarity to the multitudinous 'Mud-hen' has completely obscured this bird's claim to public recognition."

Gallinules may nest in small colonies or solitarily, each pair building multiple floating platforms for incubating eggs, brooding chicks, and resting. They are unusual among birds in that females compete for males, fighting for their top choice—small, fat males. Because females lay up to three or four clutches a year, males take on a larger share of the duties of nest building, incubation, and care of the chicks than most aquatic birds, so females have more time to feed and conserve energy. The chicks in each brood are able to feed themselves after three weeks and are soon independent, although older chicks frequently remain with the parents and help feed or brood younger chicks or even help incubate eggs. Gallinules feed on a wide variety of plant and animal materials, ranging from seed and roots to snails, tadpoles, and eggs of other birds. They feed while swimming, walking across floating vegetation, or on upland areas adjacent to water.

STATUS AND DISTRIBUTION Gallinules are year-round residents in warmer, lowland areas throughout California west of the Sierra but are among the least likely members of this family to be found in the Sierra proper.

West Side. Uncommon to rare residents or migrants at the lowest elevations, especially in the central and southern Sierra; locally fairly common in the Kern River Valley.

East Side. Rare, migrants are reported on occasion at large marshes, such as at Sierra Valley, Mono Basin, Owens Valley, and around low-elevation reservoirs.

American Coot · *Fulica americana*

ORIGIN OF NAMES "Coot" is of unknown origin used loosely for various waterfowl, possibly related to the word *scoter,* a name still used for sea ducks; L. *fulica,* coot; perhaps from *fuligo,* soot, for the species' dark-grayish color; L. *americana,* of America.

NATURAL HISTORY Coots and ducks provide a good example of convergent evolution; over the millennia they have come to resemble each other superficially in adapting to similar aquatic lifestyles. Both are excellent swimmers and divers, due largely to their specialized feet. While ducks have webs connecting their toes, coots have "lobed" feet with separate webbing around each toe. This adaptation, lacking in other members of the rail family in the Sierra, allows coots to swim and feed efficiently in deeper water. They dive as much as 25 feet below the surface or tip up like Mallards to feed on plant stems, tubers, and leaves or the occasional small animal. Coots jerk their heads backward and forward while swimming and they have a fleshy structure called a frontal shield on their foreheads that flushes reddish during courtship.

Coots frequent the edges of ponds, lakes, and slow-moving streams, where they prefer deeper waters bordered by large patches of cattails or tules. During the breeding season, they are conspicuous when pairs engage in vigorous and vocal territorial fights with other birds. Prominently displaying their white undertails and fluffing up their feathers to look impressive, coots will fight viciously and birds are often injured or killed. Pairs build up to nine floating platforms, adding a nest cup to one and using the other for resting or brooding sites. As the eggs start hatching, one parent cares for the precocial young while the other incubates the remaining eggs. Within a couple of days the chicks start swimming out to greet parents bringing food or start accompanying parents on foraging expeditions. After about eight eggs hatch, the remaining ones are either abandoned or dumped out of the nest, frequently being replaced with a second clutch. In late summer, coots molt all their flight feathers at once, rendering them flightless for about four weeks during which time they gather nervously in groups in the center of larger bodies of water.

STATUS AND DISTRIBUTION Coots are fairly common year-round at low elevations throughout the Sierra, occurring only rarely in the higher mountains and almost always during spring and fall migration. Augmented by migrants from the north, foothill populations increase (sometimes dramatically) during fall and winter.

West Side. Common nesters at ponds and marshes throughout the lower foothills below the Lower Conifer zone; they can be abundant on large bodies of water starting from early May through early November, and rare observations have been made during summer and fall as high as the Subalpine zone.

East Side. Common nesting birds at many low-elevation marshes and wetlands up to the pine forests; numbers can increase significantly during migration, with many birds lingering into winter until they are pushed out of the region when lakes and marshes start to freeze over; many winter on Lake Tahoe.

CRANES • Family Gruidae

Represented by 14 living species (and almost 40 extinct species) worldwide, cranes can be recognized by their behavior of flying with their necks extended with feet trailing behind, as compared to herons or egrets, which always crook their necks. On the ground, cranes can also be distinguished by the "bustle" of feathers over their rumps. Two species of crane, the Whooping and Sandhill, regularly occur in North America, but only the latter has been seen in the Sierra. The family name was derived from L. *grus,* a crane.

Sandhill Crane • *Grus canadensis*

ORIGIN OF NAMES Named for the Sandhills region of Nebraska, where many cranes gather during migration; "crane" derives from an Indo-European root that means to "cry out"; L. *canadensis,* of Canada, where much of the population breeds.

NATURAL HISTORY Even the most oblivious nonbirder will turn their head skyward when they hear the haunting calls of a flock of migrating Sandhill Cranes. These loud bugling vocalizations are sometimes called "voices of the Pleistocene" because fossil evidence shows that their syrinx, or "voice box," has not changed over a million years. Perhaps that explains the general sense of ancient familiarity so many feel with this sound.

In northeastern California cranes nest in wet meadows and marshes on large mounds built using a variety of marsh vegetation. Females lay one to three eggs (usually two), and these are incubated by both adults for about 30 days. Pairs stay together year-round, with young birds waiting two to seven years to begin breeding. Although two eggs are usually laid, a pair will rarely raise more than one chick to fledging. Even then, parents are kept busy defending their chick from repeated approaches by hungry coyotes and other large predators, at least until the young are capable of independent flight after about 70 days. Juveniles stay with their parents for at least a year. Cranes use their long, stout bills to forage for a wide variety of plant and animal materials, although they feed extensively on cultivated grains where available.

STATUS AND DISTRIBUTION Large numbers of "Greater" Sandhill Cranes *(G. c. tabida)* wintering in the Central Valley migrate over the Sierra to and from breeding grounds in southeastern Oregon, northeastern California, and northwestern Nevada. On their Central Valley wintering grounds, they are joined by "Lesser" Sandhill Cranes *(G. c. canadensis)*, a slightly smaller form that breeds in the high Arctic and into Siberia. The migration is a prominent and spectacular sight (and sound) over the Sierra, usually north of the Yosemite region from late February until late April in the spring and from mid-September until late November in the fall.

West Side. Fairly common flying overhead in migration, but finding a crane on the ground away from some Plumas County locations such as Indian Valley is unlikely; a few birds breed from time to time near Lake Almanor and other sites in Plumas County and as far south as Kyburz Meadow (Sierra County).

East Side. Common to fairly common in Sierra Valley, a significant stopping ground during migration, with peak numbers in late February and early March, when "courtship" dances are a common sight; a few pairs remain to nest at Sierra Valley and other Sierra and Plumas County locations each year; a least one pair has nested at Bridgeport Reservoir in most years since 2007 and nesting has occurred in Carson Valley; a few individuals may winter as far south as the Owens Valley.

TRENDS AND CONSERVATION STATUS Due to historical and ongoing conversions of native grasslands and marshes to agricultural and urban development, California designated the "Greater" Sandhill Crane a Threatened species in 1983. While they readily use corn, rice, and other agricultural fields in winter, the species has been adversely affected on its breeding rounds by plowing and human and livestock disturbance. Collisions with transmission lines cause high mortalities of juvenile birds in winter and during migration, especially during periods of dense fog in the Central Valley.

PLOVERS · Family Charadriidae

While sandpipers and most other shorebirds have long bills that allow them to find prey by touch under water or mud, plovers are visual, surface hunters. They have a distinctive feeding strategy of walking in short bursts, stopping to scan for movement, then pecking at the surface. This pattern makes plovers immediately recognizable at a distance or when they are mixed in with other shorebirds. Plovers can also be recognized by their short bills and stout bodies. Large eyes aid plovers in spotting prey and help them hunt at night. Unlike most other shorebirds, plovers seldom wade but work the areas above the water's edge or even forage in completely dry habitats.

Only two plovers breed in the Sierra, Killdeer and Snowy Plovers. Females usually lay a clutch of four buffy eggs covered with dark specks and splotches that make them blend with the sand or mud. Both sexes incubate the eggs for 24 to 26 days, with females taking the day shift and males replacing them at night. Chicks leave the nest within hours of hatching, but they stay near their parents for at least a month. More than 60 plover species exist worldwide, but only 4 occur in regularly the Sierra. The family name was derived from Gr. *charadra,* ravine, a name chosen by Aristotle for a water bird that nests in clefts.

Breeding

Nonbreeding

Black-bellied Plover · *Pluvialis squatarola*

ORIGIN OF NAMES Both "plover" and *pluvialis* derive from L. *pluvial,* rain; no known reason for this association; *squatarola* is thought to be a Venetian name that Linnaeus coined for this bird.

NATURAL HISTORY Primarily birds of ocean shorelines, a few Black-bellied Plovers appear at inland sites during their long flights between coastal wintering areas as far south as southern Chile and breeding grounds in the high Arctic. Most inland birds gather at sandy and muddy flats at the margins of large lakes and reservoirs, with each bird well spaced from its nearest neighbor. During migration and in winter, they often visit agricultural fields. Inland birds feed on relatively large prey, including aquatic and terrestrial insects such as moths and grasshoppers, earthworms, and snails. Individual birds send out warning calls to other shorebirds at the first hint of nocturnal danger. By late March some Black-bellieds begin to show dramatic black-and-white breeding plumage, males brighter than females, and early returning adults in August may also retain much of their summer finery. These plovers may often give away their presence with a vaguely mournful *pee-oh-weee* call.

STATUS AND DISTRIBUTION During spring and fall migration, Black-bellied Plovers can occur almost anywhere, but nearly all sightings in the Sierra are on the East Side. Though many winter in California along the coast, they are accidental anywhere in the Sierra in that season.

West Side. During spring and fall migration casual at Lake Almanor and rare in the Kern River Valley; otherwise, casual or accidental.

East Side. Uncommon in spring (March–April) and fall (August–October) migration with most records from the Owens Valley, Crowley and Mono Lakes; rare in migration elsewhere.

Snowy Plover • *Charadrius nivosus*

ORIGIN OF NAMES Named for its overall pale coloration compared to other plovers that display bold patterns; *charadrius* (see family account above); L. *nivosus,* snowy.

NATURAL HISTORY Snowy Plovers are wedded to barren, salty beaches and dry lakebeds where their sand-colored plumage provides perfect camouflage. Here, in a land of no shade and little fresh water, conditions are particularly harsh for incubating adults, untended eggs, and newly hatched chicks. Common Ravens are especially effective predators of their eggs and young, and they are known to fly systematically over occupied areas searching for nests; coyotes probably also take their share when they find nests.

Adult birds rely on their light, reflective color to help keep cool, but to keep eggs from overheating, they often wet their breasts and stand shading the nest. They construct simple nests by lying on their bellies in the sand and kicking back with their feet while turning in a circle. Parents are not known to feed their chicks, though they lead the fledglings to prime feeding areas and watch carefully for predators and other dangers. Snowy Plovers consume a wide variety of terrestrial and aquatic insects that can be found around the water's edge; at Mono Lake they are fond of *Ephydra* alkali flies and *Artemia* brine shrimp, which occur in extreme abundance.

STATUS AND DISTRIBUTION In California some Snowy Plovers live year-round on the coast, but a portion of this population migrates inland to San Joaquin Valley marshes and large playas and alkali mudflats on the western edge of the Great Basin. They arrive at inland sites in March to early April and remain until late September, when they migrate back to the coast.

West Side. Rare visitors during spring and fall migration to the shores of Lake Isabella; casual at other lakes and reservoirs with a few summer and fall records from Lake Almanor.

East Side. Fairly common isolated breeding pairs or populations reside around Great Basin playas and alkaline mudflats where they favor seeps and springs; the largest breeding populations (with more than 100 breeding pairs) are at Honey Lake, Mono Lake, and Owens Lake; smaller numbers are also known to breed near Bridgeport Reservoir and Crowley Lake; rare migrants may appear at south Lake Tahoe from late July through early September; accidental in winter with at least one Mono County record.

TRENDS AND CONSERVATION STATUS Due to declining populations and poor reproductive success, the U.S. Fish and Wildlife Service designated the coastal population of the "Western" Snowy Plover *(C. n. nivosus)* as Threatened in 1993. Major threats include disturbance of nesting pairs on beaches frequented by humans and dogs, recreational vehicles, as well as direct habitat losses resulting from development. Increasing numbers of Common Ravens and gulls near areas of human habitation can devastate nesting colonies. Interior populations are not listed, but they were included on California's list of Bird Species of Special Concern in 2008. Since most of these inland populations breed in remote areas, they have a lower risk of disturbance by humans and their pets and vehicles than coastal breeders, but they are still vulnerable to predation and possible changes in water management.

Semipalmated Plover · *Charadrius semipalmatus*

Breeding

Nonbreeding

ORIGIN OF NAMES "Semipalmated" refers to the webbing that extends between the toes on this species; *semipalmatus,* half-palmed, from L. *semi,* half, and *palma,* a hand.

NATURAL HISTORY These plovers are readily found inland because a large percentage of them migrate from northern breeding grounds over the interior of North America on their way to coastal wintering areas from central California nearly to the southern tip of South America. At migration stopover points they are generally approachable and easily observed and tend to remain in the same areas for a week or more. Studies have shown that individuals use the same migration pathways and stopover points year after year, making them fairly predictable. Their sharp *zhrrr-ick* calls can often be heard over the din of more numerous shorebirds.

Semipalmated Plovers favor many types of muddy shorelines or beaches, alkaline bodies of water, and flooded fields. Here they feed on invertebrates of beach and shoreline habitats, including fly larvae, beetles, and spiders. Specific habitat requirements are open, even surfaces where the plovers can run unhindered in their active search for prey.

STATUS AND DISTRIBUTION Semipalmated Plovers can appear at almost any inland wetland where they favor wet sandy or muddy edges. Spring migration is a rather abrupt passage in late April and early May, while fall migration tends to be more drawn out because juvenile birds lag behind the adults.

West Side. Rare migrants except at Lake Almanor where uncommon.

East Side. Common to fairly common in migration; during spring migration dozens to hundreds of individuals are possible at Honey Lake, Mono Lake, and Crowley Lake; fall migration peaks in August, but some juveniles linger to late September or even early October; rare but regular spring and fall migrants at Boca and Prosser Creek Reservoirs, south Lake Tahoe, and Bridgeport Reservoir.

Killdeer • *Charadrius vociferous*

ORIGIN OF NAMES L. *vociferous,* vocal; older common names have included "Chattering Plover" and "Noisy Plover."

NATURAL HISTORY Named for their loud, plaintive *kill-deer* cries, these ring-necked shorebirds can be heard at any hour, day or night. William Dawson (1923), having his attempts to photograph other birds frequently foiled by this one, concluded that the Killdeer "is the noisemaker extraordinary, the professional scold, the yap yap artist, the irrepressible canine of the bird world . . . her (or his) shrill cries arouse the countryside to attention, and in nine cases out of ten the object of her vociferous spite is a human being."

Killdeer commonly live along the edges of lakes and streams, but unlike many shorebirds do not require a substantial body of water. They will settle for a small seep, wet meadow, irrigated field, even a golf course or schoolyard, so long as the terrain is flat and free of tall plants that would obstruct their view or impede their movements. There must also be an ample supply of ground-dwelling insects, spiders, snails, and worms. Typical for plovers, Killdeer run quickly along the ground then stop short, scanning motionlessly before suddenly pecking at their prey.

Nesting mainly occurs in April and May in the Sierra. Killdeer scrape small depressions in gravel, turf, or bare earth in areas with little or no vegetation—often on river bars or seldom-used farm roads. The downy young leave the nest soon after hatching, depending on their parents for protection but finding food by themselves. Because their eggs and young are in the open and vulnerable to predators, Killdeer have become particularly adept at luring enemies away from their nests. Before a predator approaches too closely, the incubating parent sneaks quietly away then makes a great racket from a different location. Another tactic is to feign a broken wing and flap awkwardly along the ground to entice predators farther and farther from the nest or young. Many ground-nesting birds have evolved such distraction displays, but the Killdeer's performances are Academy Award–worthy.

Year-round residents throughout the lower elevations of California, Killdeer are widespread on both sides of the Sierra.

West Side. Common, readily observed in open areas below the Lower Conifer zone, but a few breed in open, wet meadows up to the Subalpine zone, where the species is uncommon; in winter, may be abundant at low-elevation pastures, city parks, athletic fields, and golf courses; generally less common in the southern Sierra.

East Side. Fairly common at lower elevations, though in colder or snowier winters may disappear altogether; uncommon nesters at higher elevations such as Lake Tahoe; can be locally common in fall.

STILTS AND AVOCETS · Family Recurvirostridae

While they are considered shorebirds, these large, long-legged, long-billed waders are placed in a separate family. Compared with other shorebirds, stilts and avocets have small heads, long necks and wings, and short, square-tipped tails. Front toes are partially webbed, hind toes rudimentary or absent, and they walk with long, graceful strides. The long legs and bills of stilts and avocets give them access to deeper water and submerged food items that cannot be reached by smaller shorebirds.

In the Sierra, stilts and avocets usually nest from April through August with a peak in June. Clutches usually contain four light-brownish eggs covered with lots of black spots and blotches. Both parents incubate the eggs for about 25 days until their precocial young hatch.

Four species of avocets and six species of stilts occur in temperate and tropical regions world-wide except for some oceanic islands. Stilts, however, show extreme geographic variation and the actual number of species remains in question. The family name is derived from L. *recurvus,* bent backward, and *rostrum,* a bill.

Black-necked Stilt · *Himantopus mexicanus*

ORIGIN OF NAMES "Stilt" refers to the species' outrageously long legs; *himantopus* from Gr. *himantos,* a leather strap, and Gr. *pous,* foot; L. *mexicanus,* of Mexico.

NATURAL HISTORY Stilts stride elegantly along the margins of shallow marshes atop improbably long, coral-red legs. Among North American birds, only flamingoes have such lengthy legs relative to their body size. When disturbed, or during courtship and social displays, stilts further accentuate their legs by dangling them conspicuously in flight, usually while calling incessantly.

Stilts often frequent the same wetlands as their close relatives, American Avocets, but they prefer fresher water, stay closer to shore, and are less gregarious. Stilts are dainty feeders, pecking lightly at the surfaces of water, mud, or low vegetation in search of various aquatic or terrestrial invertebrates and, on occasion, small fish. Around open or sparsely vegetated marshes they gather to nest, frequently near avocet nesting colonies. Female Black-necked Stilts excavate shallow nesting scrapes in loose sand or on mud that they may line with sparse amounts of dead vegetation. Though the newly hatched chicks are at first wobbly on their long legs, the parents lead them away from the nest in search of food and shelter on their first day of life. The young gain independence from their parents after about 30 days.

STATUS AND DISTRIBUTION   Black-necked Stilts are found in the Sierra mainly on the East Side from late March through September. Many head south to winter in tropical America, but large numbers remain to winter on the California coast and, increasingly in recent years, in the Central Valley. However, they are casual to accidental in winter anywhere in the Sierra.

West Side. Uncommon spring and fall migrants to Lake Almanor and Lake Isabella but generally rare elsewhere.

East Side. Common to fairly common breeders throughout in lower-elevation wetlands; much more common in the Owens Valley as migrants than as breeders; migrants are occasionally seen at high-elevation lakes.

American Avocet • *Recurvirostra americana*

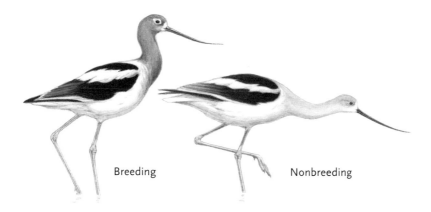

Breeding   Nonbreeding

ORIGIN OF NAMES   "Avocet" from It. *avosetta,* a word of unknown origins that could mean "graceful"; *recurvirostra* (see family account above); L. *americana,* of America.

NATURAL HISTORY   These gregarious birds of shallow marshes and ponds may be recognized even at a great distance by their flashy black-and-white patterns. At close range their most distinctive feature is a curiously upturned bill, more pronounced in females, that at first looks unsuited to the task of feeding. Avocets step forward at a half-run, swing their bills from side-to-side in an elegant, scythe-like motion, and capture small aquatic invertebrates with lightning speed. Among the most aquatic of all shorebirds, they readily swim, submerge their heads and shoulders like dabbling ducks, and on rare occasions resort to diving. At times, groups of up to 300 feed cooperatively, lining up and walking in a phalanx to herd their aquatic prey into shallow water.

Even courtship involves cooperation, as pairs join into a circle and display by leaning forward together and stepping around the circle. Mating is followed by the male draping one wing over the female's back as the pair runs side by side with bills crossed. Nesting occurs in loose colonies adjacent to open wetlands. Though avocets sometimes choose freshwater habitats, optimal nesting locations appear to be islands surrounded by alkaline or brackish waters. Adults and recently fledged juveniles gather in large groups after the breeding season.

STATUS AND DISTRIBUTION Very like Black-necked Stilts, American Avocets occur mostly on the East Side from spring through fall. However, small numbers are more likely to linger into late fall in the Sierra, and they are slightly less rare in winter than stilts (though still quite rare). Also similar to stilts, recent trends show increasing numbers breeding and wintering in the Central Valley. Avocets occur in all types of shallow waters but favor alkaline marshes at low elevations. During fall migration they have been observed on the shorelines of mountain lakes, with a high-elevation record close to 10,000 feet near Tioga Lake, just east of Yosemite National Park.

West Side. Uncommon spring and fall migrants at Lake Almanor and Lake Isabella, and casual at other lower-elevation ponds and reservoirs.

East Side. Common to fairly common from late March through September at low-elevation wetlands from Honey Lake through Mono County; numbers in Owens Valley highest in spring and fall migration.

SANDPIPERS AND RELATIVES • Family Scolopacidae

While the name "sandpiper" is often restricted to species in the *Tringa* and *Calidris* genera, it more broadly includes snipe, phalaropes, and other relatives. Most of these slim-bodied birds are far-ranging visitors to beaches, mudflats, and shorelines. Capable of mind-boggling annual round-trip migrations of as much as 20,000 miles, most North American sandpipers breed on the Arctic tundra then migrate across the continent or along the coasts as far as the southern tip of South America. Some species only rarely visit inland sites, where they are highly sought after by bird-watchers. A large part of their appeal is knowing that a bird seen in California today, might have been in the Arctic several days ago and might be on a beach in Chile in another week.

Sandpipers have long, slender bills ideally suited to probing for invertebrates hidden in soft mud or sand. Longer-legged and longer-billed species are able to wade into and forage in deeper water than smaller shorebirds, resulting in mixed flocks breaking into discrete feeding groups according to the length of each species' bills and legs. Because most sandpipers breed in the high Arctic, few people witness their striking breeding season alternate plumages, loud songs, or elaborate courtship displays. Fortunately, five species (Spotted Sandpiper, Willet, Long-billed Curlew, Wilson's Snipe, and Wilson's Phalarope) breed in the Sierra.

Their nests may be on bare ground or hidden in grasses or marsh vegetation. Females usually lay four eggs that vary in color from whitish to greenish, or even pink, with a generous helping of purplish-brown spots and streaks. Both parents usually incubate the eggs for 20 to 28 days and tend their precocial young for several weeks or until they can fly after 20 to 25 days. Worldwide there are about 90 species of sandpipers, of which 11 occur regularly in the Sierra. The family name derives from Gr. *skolopos,* something pointed—in this case the bills of this group.

Spotted Sandpiper · *Actitis macularia*

Breeding

Nonbreeding

ORIGIN OF NAMES "Spotted" for the extensive spotting on the underparts of breeding adults; *actitis* from Gr. *aktites,* a dweller of the sea coast; *macularia* from L. *macula,* spot.

NATURAL HISTORY Between late summer and winter, Spotted Sandpipers lose their dark polka dots but retain characteristic mannerisms, making identification easy in all seasons. Over short distances they fly with stiff, shallow wingbeats, interrupted by frequent, brief glides on down-curved wings. On land they almost incessantly teeter, holding heads low while tail ends vigorously bob up and down. Highly vocal, even outside the breeding season, their piecing, rolling *pee-de-de-wheet, pee-de-de-wheet* songs can be clearly heard over the tumbling roar of mountain streams. In the Sierra they feed on insects and other invertebrates along sandy or pebbly shores of lakes and streams.

Spotted Sandpipers usually lead a solitary existence except while breeding. Parents display an interesting reversal of the typical avian sex roles, with males incubating eggs and caring for young with little help from their mates. In the Sierra they start nesting in April at lower elevations but often not until the end of June higher up. They nest in simple grass-lined depressions along shore, often hidden among low plants, rocks, or logs, and always near water. The polyandrous females often mate and lay clutches with two or more males and defend large territories where the males raise their broods. Females are particularly aggressive toward other females during courtship battles and injuries are common. Soon after hatching, the young leave their simple nest near the water's edge; although they must find their own food, the males (with occasional help from the females) brood and protect them for about three weeks, until they are capable of flight and independent living.

STATUS AND DISTRIBUTION More common in the northern Sierra than in the south, Spotted Sandpipers are the only shorebirds that nest regularly above the Foothill zone on the West Side.

West Side. Fairly common at all elevations along streamside gravel bars and lakeshores, though they nest mostly from the Upper Conifer zone up to the Subalpine zone; most leave the higher mountains soon after the breeding season, and some remain in the Foothill zone through winter.

East Side. Fairly common, similar to West Side but perhaps more common as a nesting species along creeks and rivers.

Greater Yellowlegs • *Tringa melanoleuca*

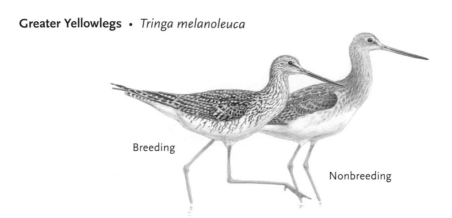

Breeding

Nonbreeding

ORIGIN OF NAMES "Greater" to distinguish the species from its close relative Lesser Yellowlegs; both are named for their brightly colored legs; *tringa* from Gr. *tryngas,* a sandpiper or water bird; *melanoleuca* from Gr. *melanos,* black; *leucos,* white, a reference to the contrasting pattern on the upperparts.

NATURAL HISTORY These graceful, medium-sized shorebirds frequent marshes, lake margins, and flooded fields. Single birds or small groups scour mudflats in search of insects, worms, small fish, frogs, and occasionally seeds and fruits. At all times they are alert and active, moving forward with a characteristic style of quick, staggering steps and sudden chases after prey. Yellowlegs are relatively wary, and among a group of mixed shorebirds they are often the first to flush and give their *tu-tu-tu* alarm calls at the approach of an intruder.

STATUS AND DISTRIBUTION During migration Greater Yellowlegs are a common and widespread species throughout California, and they are often found in the Sierra. Spring migration peaks from late March through April, fall migration from early August through September. On occasion, nonbreeders remain to summer in California while the rest of the population heads north to nest in the spruce muskeg of Canada and Alaska. In winter, they are common in the Central Valley but only rarely encountered in the Sierra.

West Side. Fairly common to uncommon in spring and fall along slow-moving rivers, ponds, and lakes; most reliable locations are Lake Almanor and the Kern River Valley; records of fall migrants up to 8,000 feet in Yosemite National Park suggest that this species could appear at freshwater sites at almost any elevation; uncommon to rare in winter except at lower Foothill zone ponds and reservoirs.

East Side. Common migrants in the Owens Valley; fairly common to uncommon elsewhere; rare in winter.

Willet • *Tringa semipalmata*

ORIGIN OF NAMES "Willet" is imitative of the bird's call; *semipalmata* from L. *semi,* half; *palma,* a palm, in reference to the partial webbing between this species' toes, similar to the Semipalmated Plover.

NATURAL HISTORY For those who only know this bird from the coast in winter, you must experience Willets on the breeding grounds. They don a somewhat more distinctive alternate plumage for summer, but the real show begins when they take to the air in courtship displays, gliding over

Breeding

Nonbreeding

wet meadows and shallow marshes on quivering wings while loudly singing *pill-will-willet, pill-will-willet*. Even when simply flying from point-to-point, this is a bird that transforms from a relatively unassuming appearance to one of our most conspicuous shorebirds when they open their wings and flash that vivid black-and-white wing pattern. They are also prone to perch conspicuously on fence posts and utility poles in breeding areas.

Nests are shallow, grass-lined depressions that may be located up to several hundred yards from the nearest water. Pairs cooperate in tending their eggs and raising young, though their parental stint is short. Adults leave while chicks are still flightless in order to gather in premigratory staging areas before flying south or west. Young remain on the breeding territories until they attain flight after about 30 days. Willets use long, stout bills to peck and probe in search of mud-burrowing insects and crustaceans. They bob their heads while walking, frequently wade, and may even swim.

STATUS AND DISTRIBUTION The very southwestern edge of the "Western" Willet's *(T. s. inornatus)* breeding range extends to the northeastern Sierra. At a few locations they are fairly common breeders, otherwise Willets are rare migrants in the region. Northbound migrants can be observed from early April to mid-May. Postbreeding adults begin returning in late June and are followed by juveniles in mid-July. The largest numbers are typically seen from late July to early September. This species winters along the coast from central California south to Central and South America.

West Side. Rare spring and fall migrants through the Kern River Valley and at Lake Almanor; otherwise casual elsewhere.

East Side. Common to abundant nesters in April, May, and June at Honey Lake and Sierra Valley; a small breeding colony also exists at Crowley Lake, and Willets probably breed in the Washoe and Carson Valleys of Nevada and near Bridgeport (Mono County); a nesting record at Lake Tahoe and one possible nesting record near Big Pine (Inyo County); fairly common in migration at most wetlands and lakeshores.

Long-billed Curlew • *Numenius americanus*

ORIGIN OF NAMES "Long-billed" for the species' long, down-curved bill; "curlew" is echoic of the bird's call; *numenius* from Gr. *neos,* new, and *mene,* moon, as the species' bill resembles a half moon; L. *americanus,* of America.

NATURAL HISTORY Largest of North America's shorebirds and sporting outrageously long bills, Long-billed Curlews draw attention to themselves with loud, clear *cur-lee* calls and spectacular roller-coaster flight displays on the breeding grounds. On wet meadows where they nest, males engage in territorial behavior that includes fighting or bouts of stylized posturing. To attract females, males make a series of nest scrapes, one of which is later lined with grasses and used as a nest. Both parents incubate the eggs and tend the young, though females leave when the chicks are only two to three weeks old. Compared with most other members of this genus, Long-billed Curlews are short-distance migrants and only move a few thousand miles between their breeding and wintering grounds. They are early to arrive on their breeding grounds and early to leave, a schedule that seems in keeping with the tendency of grasslands to dry out earlier than most other habitats. While at inland sites, curlews eat a varied diet of insects, worms, spiders, snails, crayfish, and berries, using their bills to make long reaches or deep probes.

STATUS AND DISTRIBUTION During the nonbreeding season, curlews are locally common on the coast and in the Central Valley. By April most head north and east toward the Great Basin and Great Plains to begin breeding, leaving some nonbreeding birds behind. In the Sierra they breed only in the northeast portion of the region. Fall migration begins in late June with movements continuing through August.

West Side. Rare to uncommon spring and fall migrants through the Kern River Valley, and on grasslands of the southern Foothill zone (Mariposa County south), where some linger through winter.

East Side. Fairly common breeders at Honey Lake and Sierra Valley and some apparently breed in the Carson Valley and at the north end of the Owens Valley; Mono County's first breeding record confirmed near Crowley Lake in 2011; common in spring and fall migration near Owens Lake, where some occasionally winter.

TRENDS AND CONSERVATION STATUS In contrast to most shorebirds, Long-billed Curlews have a very restricted range and a relatively small population (approximately 20,000). The U.S. Shorebird Conservation Plan designated this species as "highly imperiled" because of its small range and population, contraction of its historic range, and ongoing threats to breeding and wintering habitats.

Western Sandpiper · *Calidris mauri*

Nonbreeding

Breeding

ORIGIN OF NAMES "Western" for the species' distribution and preferred habitat; Gr. *caladris,* a name used by Aristotle for a gray sandpiper; L. *mauri* in honor of Ernesto Mauri (1791–1836), a distinguished Italian naturalist.

NATURAL HISTORY During their brief visits to the Sierra in migration, Western Sandpipers are intent on refueling for the next leg of their journey. They primarily consume adults and pupae of flies, beetles, and other insects by foraging at or near the edge of lakes and reservoirs, or by wading into the shallows. They prefer areas largely devoid of vegetation. While they often occur in association with their smaller relatives, Least Sandpipers, their slightly larger size and longer bills allow them to use this habitat in a noncompetitive fashion. They tend to wade more often than Leasts, and they can probe more deeply into the mud for morsels.

STATUS AND DISTRIBUTION While hundreds of thousands of Western Sandpipers winter in the San Francisco Bay Area, they only visit the Sierra during spring and fall migration when they can be locally abundant from late April to early May on their way to the Arctic breeding grounds and again in fall heading south. Fall migration is more prolonged and complex, with adults coming through from mid-July to early August, then mostly juveniles making their first trip in late August through September. Small numbers of stragglers can be found through October.

West Side. Rare to casual in migration with the exception of Lake Almanor in fall, where they occur regularly; otherwise, a surprisingly difficult bird to find; rare in winter in Kern River Valley, casual to accidental elsewhere.

East Side. Common to locally abundant in spring and fall migration from Mono County through the Owens Valley; fairly common near Reno through Carson Valley, Sierra Valley, Lake Tahoe, and Honey Lake; casual to accidental in winter.

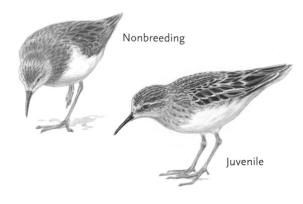

Nonbreeding

Juvenile

Least Sandpiper • *Calidris minutilla*

ORIGIN OF NAMES "Least" for being the world's smallest shorebirds; L. *minitulus,* very small.

NATURAL HISTORY Least Sandpipers make annual round trips from breeding areas in Alaska and northern Canada to as far south as central South America. Often found in large flocks mixed with Western Sandpipers, Leasts mainly forage just away from the water, harvesting a variety of invertebrate prey from the sandy or muddy surface or probing just below the surface. One of the most vocal shorebirds outside the breeding season, they sound rolling, high-pitched trills as flocks flush from one spot to the next.

STATUS AND DISTRIBUTION In spring and fall migration, Least Sandpipers visit a variety of inland wetlands—from valley marshes to high-elevation lakeshores. They are the default *peep* (small sandpipers like Westerns, Leasts, and Dunlin) anywhere in the Sierra in winter and at any season on most of the West Side. Compared with Westerns, they arrive earlier in the spring, and many stay for winter at lower elevations and more southerly parts of the region. Spring migration begins in late March and peaks in late April, fall movements begin with adults in late July and early August, with highest numbers in late August but fairly large flocks can be found in some places through October.

West Side. Uncommon but widespread in spring and fall migration; in winter, sizable flocks can be found in the Kern River Valley, and small numbers are seen around lakes and ponds of the lower Foothill zone.

East Side. Common to locally abundant in spring and fall migration at Mono Lake, Crowley Lake, and the Owens Valley; fairly common to uncommon elsewhere; rare in winter except in the Owens Valley, where they can be locally common some years.

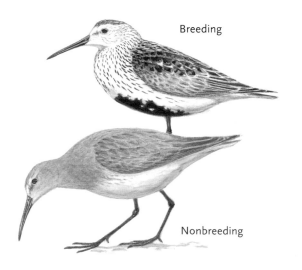

Breeding

Nonbreeding

Dunlin • *Calidris alpina*

ORIGIN OF NAMES "Dunlin" literally means "little dun-colored one"; L. *alpinus,* alpine, for Arctic breeding grounds.

NATURAL HISTORY Dunlins undergo one of the more dramatic transformations from basic plumage in winter to bold and colorful alternate plumage in spring. To see these birds in fancy dress in California, you need to catch them in April on their way to the high Arctic because by the time they return in September, they have molted into their winter grays. In many shorebirds male and females have different length bills, with females generally having longer bills. This difference is particularly pronounced in Dunlins, and the females of our western subspecies *(C. a. pacifica)* have bills averaging more than 10 percent longer than the males. This adaptation probably allows pairs to reduce competition for prey in common foraging areas. Dunlins tend to be gregarious, especially when they are snoozing during the day, and one can see tightly packed flocks all with bills tucked under their wings. As with many other shorebirds, they do much of their feeding at night.

STATUS AND DISTRIBUTION In California, Dunlins mostly winter along the coast and in the Central Valley and are mainly spring and fall migrants to the Sierra, with the largest concentrations usually seen in April. They are typically the last shorebird species to arrive in fall, with most showing up in late September.

West Side. With the exception of Lake Almanor and Kern River Valley, where they are uncommon in migration and rare in winter; otherwise casual.

East Side. Fairly common in spring and fall migration in the Owens Valley; uncommon but regular farther north at Honey Lake, Lake Tahoe, and Mono County; some winters large numbers seen in the Owens Valley and Honey Lake but generally rare in winter.

Long-billed Dowitcher • *Limnodromus scolopaceus*

Breeding

Nonbreeding

ORIGIN OF NAMES "Long-billed" compared to its close relative, the Short-billed; "dowitcher" may be derived from the Iroquoian word for the bird; *limnodromus,* marsh runner, from Gr. *limne,* a marsh, and *dromos,* runner; *L. scolopaceus,* from Gr. *skolopax,* a snipe or woodcock.

NATURAL HISTORY Long-billed Dowitchers are the wind-up toys of the shorebird guild as they walk along probing at an impossibly rapid pace with their very long bills, much like little sewing machines. Nearly indistinguishable in winter from Short-billed Dowitchers, they can be identified by their high-pitched *keeks* given singly with pauses when they are calm but delivered in a rapid series when flushed. In a clear demonstration that looks can deceive, these two dowitchers are genetically quite distinct, much more so than other shorebirds that look much different from each other. As is true with many sandpipers, those long bills are actually quite flexible, and the tip is highly sensitive and able to detect tiny prey buried deep in the mud. Without these adaptations, feeding would be like a human trying to pick up grains of rice with four-foot chopsticks.

STATUS AND DISTRIBUTION Long-billed Dowitchers, in spite of large winter concentrations on the California coast and Central Valley, are mainly spring and fall migrants to the Sierra. Compared to Short-billed Dowitchers, Long-bills migrate south later in fall and have more prolonged migrations, with spring migrants seen from late March into early May, and fall birds present in good numbers from August well into October. An adult dowitcher in July has a good chance of being a Short-billed, and distinctively plumaged juvenile Short-billed Dowitchers are most likely to be found in mid- to late August.

West Side. Uncommon migrants in fall and spring at Lake Almanor, Lake Isabella, and some lower Foothill zone reservoirs, ponds, and wastewater treatment plants; rare in winter, though fair numbers sometimes found on the Kern River Valley Christmas Bird Count.

East Side. Common to locally abundant migrant with largest concentrations usually in fall from Mono County south; spring numbers often higher than fall at Honey Lake and Sierra Valley; uncommon in Tahoe Basin; in winter, generally rare except in Reno area and Owens Lake, where fair numbers winter some years.

Wilson's Snipe · *Gallinago delicata*

ORIGIN OF NAMES Recently renamed from "Common" back to "Wilson's" after the Scotsman Alexander Wilson (1766–1813), the foremost 19th-century ornithologist who published his nine-volume *American Ornithology* between 1808 and 1814; while Audubon had the greater gifts for art and self-promotion, Wilson was the true "father of North American ornithology"; "snipe" from AS. *snite,* meaning to snip with shears, a possible allusion to the species' bill; *gallinago* from L. *gallina,* a hen; perhaps in reference to the species' size; L. *delicata,* paramour or favorite, unclear why the name was applied to this species.

NATURAL HISTORY This chunky, long-billed shorebird will often remain motionless even when closely approached, making it difficult to spot until it suddenly flushes from dense ground cover with a loud *skype* call. Then it bursts forth in rapid zigzag flight, suddenly dropping into another hiding place or climbing high into the sky and out of sight. Snipes are conspicuous only during the breeding season when either sex may perform an impressive display known as "winnowing." After circling high above its nesting territory, a snipe dives steeply, veering slightly to one side and forcing air through outspread tail feathers. This produces an accelerated series of hollow *hu-hu-hu-hu-hu-hu-hu-hu* notes that are a distinctive summertime sound in lush, wet valleys where they nest. This display is used both to attract mates and to establish and defend territories. A steady *tuk-a-tuk-a-tuk-a-tuk-a* vocalization is another common sound heard during breeding season.

Wilson's Snipe nests are nearly impossible to find, hidden in dense grassy hummocks within or at the edges of marshes and wet meadows. Their parenting strategy is unusual among shorebirds, as only the female incubates but the two mates split the brood after hatching with each parent caring for their own contingent of young. With their extremely long, slender, sensitive bills, Wilson's

Snipes probe in mud for worms and insect larvae and pick adult insects from the ground. Feeding behavior is a rapid series of deep probes while the bird walks forward slowly, often up to its belly in water. Many prey items are swallowed while the bird's bill remains submerged in wet mud; only large or awkward items, like worms, require that the bird extract its bill and visibly swallow.

STATUS AND DISTRIBUTION Wilson's Snipe are capable of long migrations; however, Sierra birds are present year-round and appear to make only short movements between different elevations or across the crest. Their secretive nature makes it hard to know their true abundance at any location during the nonbreeding season.

West Side. Uncommon and very local breeders in the Foothill zone, and rare breeders to the Upper Conifer zone; fairly common in appropriate habitat outside the breeding season, suggesting some migration to these areas; fall migrants have occurred as high as 9,000 feet.

East Side. Common, though local, breeding species south to the north end of the Owens Valley; in winter, fairly common to uncommon from Reno area south; rare further north.

Wilson's Phalarope • *Phalaropus tricolor*

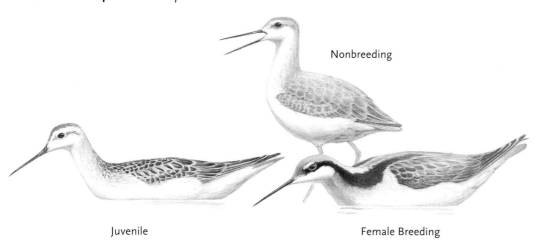

Nonbreeding

Juvenile

Female Breeding

ORIGIN OF NAMES Wilson's (see Wilson's Snipe account); *phalarope,* a coot foot, from Gr. *phalaris,* a coot; *pous,* foot, for the phalarope's semipalmate (partially webbed) feet and lobed toes; L. *tricolor,* three colors, for the species' white-black-red breeding plumage.

NATURAL HISTORY Phalaropes defy our expectations in many ways. First, unlike most other shorebirds, they are much more likely to be seen swimming than wading or walking. They peck insects or small aquatic invertebrates from the surface or spin rapidly to create a vortex, bringing prey from the mud below to within reach. That brightly colored bird we see in spring is not the male, it is the female, and her fancy plumage is only one example of the reversed sex roles exhibited by these shorebirds. The polyandrous females fight each other for access to males, but egg-laying is the only parental duty they perform. Once the clutch is laid, she is off to either mate with another male or fly to a productive spot where she can molt and put on fat to fuel her migration south. The male usually covers the nest site, located in or near a marsh or wet meadow, with vegetation to conceal it and does all the incubation, brooding, and tends to the young until they can fly. Unlike the other two phalaropes, Wilson's do not winter on the open ocean but seek alkaline lakes in South America, where little is known about their biology.

The breeding range for Wilson's Phalaropes just reaches the eastern edge of the Sierra. However, Mono Lake is a site of paramount importance to this species. In July and August tens of thousands gather to molt and put on the fat reserves needed to fly thousands of miles to their South American wintering locations. They gather to gorge on the local superabundance of brine shrimp and alkali flies found at this lake. This concentrated source of food is not just important to fuel onward migration but also to allow this species to undergo one of the most rapid molts of any bird. All body feathers and most flight feathers are replaced in just over a month during this fall staging period.

West Side. Fairly common spring and fall visitors to Lake Almanor, where they may nest in some years; uncommon in the Kern River Valley in fall; uncommon to rare elsewhere but can appear at almost any large water body, with foothill wastewater treatment ponds the most likely sites; spring migration from mid-April through May, fall movements mid-July through August; nesting records on the Central Valley floor and recent expansion into eastern North America attest to the ability of this species to colonize new locations.

East Side. Fairly common nesters at low-elevation wet meadows and marshes around Honey Lake, Sierra Valley, Mono Basin, and Owens Valley as well as at scattered locations in the higher mountains, including the south end of Lake Tahoe; huge numbers stage at Mono Lake, with high concentrations of adult females in July, gradually replaced by adult males and juveniles in August; mostly gone by late September but a few may linger into October.

Red-necked Phalarope • *Phalaropus lobatus*

Male Breeding

Juvenile

Female Breeding

ORIGIN OF NAMES "Red-necked" for the plumage of breeding adults, females are brighter; L. *lobatus,* lobed, a reference to webbing on toes.

NATURAL HISTORY Red-necked Phalaropes are dainty shorebirds, the smallest of the phalaropes, but they have an exotic life, breeding in the Arctic and spending the entire winter well out in the open ocean. In the Sierra we see them only when they pause during their flights between these habitats. Just like Wilson's Phalaropes, typical sex roles are reversed and females are the more brightly colored birds. Both phalaropes forage in a similar fashion, plucking prey from the surface as they swim or spin in place to draw food from the bottom. However, Red-necks are much less likely to be seen foraging on land than Wilson's. Also unlike Wilson's Phalaropes, which consume huge quantities of brine shrimp at Mono Lake, recent studies have shown that Red-necks consume almost entirely alkali fly pupae and reject shrimp unless they are near starvation weights. After morning feeding bouts near shoreline tufa, flocks of Red-necks move offshore to roost for the day or fly briefly to freshwater sites where they bathe and drink.

STATUS AND DISTRIBUTION Red-necked Phalaropes are strictly spring and fall migrants to the Sierra, and spring is the time to see them in their lovely alternate plumage. It is possible to see both phalaropes together in late April through early May. Red-necks also stage in large numbers at Mono Lake in fall, with adult females arriving in numbers in late July and males and juveniles in August and September. Unlike Wilson's Phalaropes, Red-necks molt only body feathers during this period, waiting to replace flight feathers until out at sea.

West Side. Rare to uncommon migrants to low-elevation ponds, Lake Almanor and Kern River Valley, with scattered records up to 10,000 feet.

East Side. Common to abundant from late July through September at Mono Lake, when tens of thousands may be seen in a single day; elsewhere common to fairly common in spring and fall.

GULLS AND TERNS • Family Laridae

Gulls are widespread scavengers whose hooked bills allow them to eat almost anything, and many species are adept at begging a meal from picnickers and tourists. Terns have long sharply pointed bills, and they specialize on catching small fish and insects in fast, adroit hunting maneuvers. Gulls and terns tend to be gregarious at all times of year, and during the breeding season generally nest in large, noisy colonies. Some species remain on the coast year round, while others fly inland to nest in marshes or on islands in large lakes and rivers. Pairs cooperate in building nests of twigs and plant materials on the ground next to low plants or rock outcrops. Nests are often closely associated with those of other colonial water birds. Females lay two to three eggs beginning in early to mid-May; eggs are olive and covered with irregular splotches and spots of black. Both parents incubate the eggs for 20 to 25 days, and the juveniles are able to fly after about one month.

All species are migratory, and some tern species have the longest migrations of any of the world's birds. For this reason some individuals of even the most unusual species are spotted in the Sierra every year. Identification of these birds to species can be difficult due to their long period of growth (up to four years for large gulls) and many different, juvenile, immature, and adult plumages. Add in the proclivity of many species to interbreed, and this family poses identification challenges that some find fascinating and others frustrating. Of about 100 Larid species in the world, only 4 gulls, and 3 terns occur regularly in the Sierra; 13 additional gulls and 3 terns are rare, casual or accidental visitors to the region (see Appendices 1 and 2). The family name is derived from Gr. *laros,* a gull.

Bonaparte's Gull • *Chroicocephalus philadelphia*

ORIGIN OF NAMES "Bonaparte's" for Charles Bonaparte (1803–1857), a famous 19th-century ornithologist in Europe and North America and nephew of Emperor Napoleon of France; gull, from OE. *gullan,* possibly imitative of their loud calls; *chroicocephalus,* colored head, from Gr. *chroa,* color, and *cephalus,* head; L. *philadelphia,* for the city where the type specimen was collected.

NATURAL HISTORY Bonaparte's Gulls make long-distance migrations to the northern boreal forests of Canada and Alaska each spring, where they have the distinction of being the only gulls that nest regularly in trees. They walk with a slight waddling gait but fly with the buoyant grace and dexterity of terns. They float lightly on the water and peck at food on the surface, similar to phalaropes, leading one of the authors to refer to them as "Greater Phalaropes." In the Sierra, Bonaparte's Gulls are usually seen alone or in small groups passing through during migration when they may stop briefly at various lakes, reservoirs, and wastewater treatment ponds in search of small fish and insects.

STATUS AND DISTRIBUTION Occurrences of Bonaparte's Gulls in the Sierra are erratic and unpredictable, but most records in spring migration are in April, and most fall observations are from mid-August through October.

West Side. Uncommon, seen at Lake Almanor almost annually (sometimes in numbers), otherwise rare.

East Side. Uncommon to fairly common visitors to Lake Tahoe, Mono Lake, and the Owens Valley during spring and fall migrations in some years (e.g., during fall 2007 double-digit numbers were seen in many locations); occasionally remain through summer in Mono County (most notably 1991, when 120 summered at Crowley Lake); rare visitors to Boca and Prosser Creek Reservoirs, Stampede Reservoir (Sierra County), and Indian Creek Reservoir (Alpine County); rare to casual in winter.

Breeding

Nonbreeding

Ring-billed Gull • *Larus delawarensis*

ORIGIN OF NAMES "Ring-billed" for the species' most distinctive field mark; *delawarensis*, L. of Delaware, for the Delaware River where the species was first collected.

NATURAL HISTORY Easily dismissed as "parking lot" birds, Ring-billed Gulls provide interesting subjects for study. Their lack of shyness allows one to closely examine the pristine whites and tasteful grays and blacks of their plumage and the complex interactions between individuals as they beg for a handout. More assertive birds throw back their heads and call to establish dominance while others hunch submissively and defer. Opportunistic foragers, they eat insects, worms, fish, rodents, grains, and human garbage. Pretty much anything that can be caught and swallowed is gull food.

The open, relatively barren locations of most nesting colonies makes the eggs, young, and adults susceptible to attack by a host of ground and aerial predators. Both parents share in nest-building, incubation, and caring for the young until they can fly. This cooperation is essential and, if one mate is killed, the nest or young are almost invariably abandoned. This may explain why numbers of female-female pairs have been documented at several Ring-bill breeding locations. These pairs lay their clutches in the same nest and cooperate in all aspects of parenting. This may occur because of insufficient numbers of males or because one or both females lost their mates.

STATUS AND DISTRIBUTION The breeding range of Ring-billed Gulls stretches across north-central North America and into the northeast corner of California, just reaching the Sierra in Lassen County. Ring-bills are present in the Sierra at all times of year and large numbers congregate from mid-March to mid-May prior to nesting. As with some other gull species, numbers may linger through summer in nonbreeding areas. During migration they may appear at any large lake or reservoir as lone birds or in flocks of dozens, as evidenced by records above 10,000 feet just east of Yosemite National Park. Wintering birds may linger in any body of water that provides food and does not freeze over.

West Side. Fairly common and widespread throughout the year, though typically in low numbers at most locations except during spring and fall migration and in winter at Lake Almanor and Lake Isabella.

East Side. Fairly common to uncommon year-round, usually in higher numbers than on the West Side; locally common during fall and spring migration; in the breeding season, up to 2,000 may gather to nest at Honey Lake from mid-May until late July, depending on lake levels and availability of food; common most winters near Honey Lake, Lake Tahoe, and from the Reno area south into the Carson Valley; generally uncommon elsewhere.

California Gull • *Larus californicus*

ORIGIN OF NAMES Named in honor of California, where it was first collected; however, the largest breeding colonies are at the Great Salt Lake, Utah, where this species is recognized as the "state bird."

NATURAL HISTORY California Gulls nest on isolated islands in large lakes scattered around the arid interior West, and the Sierra is home to one of the most important breeding sites. Since the early 1900s, naturalists have been aware of a large California Gull breeding colony at Mono Lake's Negit Island. When the first systematic censuses in 1976 estimated the size of this colony at 50,000 breeding adults,

Breeding

Nonbreeding

we learned that it was the second largest colony in the world (Winkler et al. 1977). The size of Mono Lake's breeding population varies greatly from year to year (from about 18,000 to 65,000 adults since the early 1980s), and annual totals seem to be most strongly influenced by the availability of their primary food (brine shrimp), local and regional weather patterns, and winter survival. Since they do not breed until their fourth year, changes in reproductive success usually are not apparent until four years later.

Mated pairs build simple nests on the ground lined with small sticks and feathers and bones of birds that died the year before. Most nests are placed near rocks, logs, or shrubs to shelter the eggs and chicks from intense sun and strong winds. Precocial young are fed regurgitated food by both parents until they are capable of flight, about six weeks after hatching. As described earlier for Ring-billed Gulls, California Gulls are common denizens of shopping center parking lots, especially where fast food restaurants enhance the local habitat quality. Their food preferences are likewise highly flexible and their behaviors are equally fascinating. Time spent watching them can reveal subtleties of behavior and appearance well worth noting. Of course, the occasional French fry tossed out the window is always appreciated.

STATUS AND DISTRIBUTION California Gulls are widespread in California year-round, where they winter mainly along the coast and in the Central Valley and breed at isolated colonies at large lakes and wetlands in northeastern and eastern California. They are occasional visitors to high mountain lakes, especially in summer.

West Side. Common to abundant in winter at Lake Isabella and Lake Success, fairly common at Lake Almanor and locally fairly common at large rivers and reservoirs of the Foothill zone; fairly common throughout in spring and fall migration.

East Side. Abundant at Mono Lake from early April through mid-October; small breeding colonies are also found at Honey Lake and Laurel Pond (Mono County), and they have attempted nesting recently at Crowley Lake; a small nesting colony is well established at Virginia Lake in Reno, and in 2011 a few pairs nested successfully within the confines of a fenced parking lot of a closed shopping mall in Reno, far from any saline waters; elsewhere fairly common in spring and fall migration and generally uncommon to rare in winter except at Lake Tahoe and from Reno into the Carson Valley, where large numbers winter.

TRENDS AND CONSERVATION STATUS Until recently, breeding California Gulls were included on the list of California Bird Species of Special Concern, due mostly to threats to Mono Lake's island breeding population. Diversion of the Mono's principal freshwater streams began in the early 1940s, causing the lake to lose more than half its volume and to become twice as salty and alkaline over the ensuing four decades. In 1979 a land bridge formed between Negit Island and the mainland, allowing coyotes and other predators to invade the primary gull breeding colony with a devastating effect on gull eggs and chicks. In the following years, many breeding pairs relocated to small "islets" near Negit and Paoha Islands. This predation event triggered major conservation efforts, legal actions, and extensive media coverage to "Save Mono Lake" and its California Gulls, and subsequent actions have allowed water levels to steadily increase (see the chapter "Bird Conservation in the Sierra").

Herring Gull • *Larus argentatus*

ORIGIN OF NAMES "Herring" for a preferred food of the species; L. *argentatus*, ornamented with silver, for the color of the adult's upperparts.

NATURAL HISTORY Despite increasing attention to gull identification and some excellent field guides in recent years, Herring Gulls still epitomize the challenges that lead some birders to quickly scan by the gull flocks. This gull takes four years to obtain adult appearance, and Howell and Dunn (2007) use the term "notoriously variable" to describe both first- and second-cycle plumages. Besides the variation in plumages with age, up to nine different subspecies (or species, depending on the source) have been recognized worldwide, and the taxonomic status is further complicated by its willingness to hybridize with other large gulls. Like other gulls, Herrings prefer a diet of fish and invertebrates but will take advantage of other sources of food, such as grains and human garbage.

STATUS AND DISTRIBUTION In California, Herring Gulls primarily winter along the coast, but some visit the Sierra from late fall through winter.

West Side. Uncommon from late October to early March from Lake Almanor south to the Kern River Valley; one mid-July record at Tenaya Lake in Yosemite National Park.

East Side. Uncommon during spring migration and uncommon to locally fairly common in late fall; in winter, uncommon to rare at Honey Lake, Lake Tahoe, and in the Reno area south into Carson Valley; rare to casual elsewhere.

Caspian Tern • *Hydroprogne caspia*

ORIGIN OF NAMES "Caspian" for the Caspian Sea, where the species was first collected; "tern" from L. *terni*, a water bird; *hydroprogne*, a water swallow, from L. *hydro*, water, and Gr. *procne*, for a Greek goddess who was turned into a swallow by her father; L. *caspia*, as above.

NATURAL HISTORY The largest and one of the most widespread of the world's 44 tern species, Caspian Terns are larger than some gulls. Unlike the smaller terns, they use direct and powerful flight and hover only when pursuing surface-feeding fish. In many ways these large terns are more "gull-like" than the very "tern-like" Bonaparte's Gulls described earlier.

Caspian Terns nest in colonies on islands in large rivers and lakes scattered throughout the interior of North America. The quality of breeding sites depends on fluctuating water levels, which means that this species may nest in a given location in some years and not in others. Caspian Terns typically nest with California Gulls or other large water birds starting in mid-May. Nesting near gulls can be a two-sided coin—they offer protection from many nest predators, but the gulls themselves often prey on tern eggs and young. Pairs of terns prepare a simple scrape in the sand, sometimes lining it with vegetation and creating a shallow rim around the nest. Unlike most other members of this family, young Caspian Terns remain dependent on their parents well after they are able to fly. This means that observing a young tern being fed does not necessarily confirm local breeding.

STATUS AND DISTRIBUTION Caspian Terns winter commonly along the Pacific coast north to Marin County and breed at isolated lakes and reservoirs of the Great Basin, including some sites in the Sierra.

West Side. Uncommon in spring (mid-March to early June) and rare in fall (July to September) migration with records from most large reservoirs; accidental in winter.

East Side. Uncommon to fairly common spring and fall migrants; locally common at breeding colonies at Honey Lake (more than 100 nesting pairs in some years), Bridgeport Reservoir, and Mono Lake from April through August; rare in fall after mid-September and accidental in winter anywhere.

Black Tern • *Chlidonias niger*

Juvenile   Adult Breeding

ORIGIN OF NAMES "Black" for the striking plumage of breeding adults; *chlidonias* is based on a historical misspelling of Gr. *khelidon*, a swallow; L. *niger*, black.

NATURAL HISTORY Unlike other members of their family, Black Terns forage for flying insects as well as for surface-feeding fish. These small terns fly with unequaled gracefulness and dexterity, able to match the sudden changes of direction of one their favorite foods—dragonflies. They sometimes nest in colonies and build floating rafts of vegetation for their eggs. In flooded cattle pastures another nesting substrate is sometimes used, as delicately described by William Dawson

(1923): "One cannot forebear to mention the frequency with which these birds are beguiled by the attraction of floating cakes of cow dung for use as nesting sites." Young hatch in mid- to late June, and colonies are largely devoid of birds by mid- to late July, when all the terns move on to favored feeding areas. Juveniles typically migrate south about a month behind adults.

STATUS AND DISTRIBUTION Black Terns are present in the Sierra from early May until mid-September, when they depart for traditional wintering grounds on the northern coast of South America.

West Side. Fairly common some years only in the vicinity of known breeding areas at Mountain Meadows Reservoir (Lassen County); uncommon to rare in migration from May to early August at Lake Almanor and Lake Isabella, and generally rare elsewhere during spring and fall migration with most records from spring.

East Side. Uncommon and intermittent breeders at Honey Lake and Sierra Valley; apparently extirpated from former breeding areas at south Lake Tahoe; rare to uncommon spring and fall migrants elsewhere (including Lake Tahoe).

TRENDS AND CONSERVATION STATUS Black Terns were once common breeders throughout the Central Valley and northeastern California. Loss of wetland habitat in the valley has reduced their range and abundance to a tiny fraction of what it once was. Their numbers and the extent of their range have also decreased in the Sierra as water diversions have altered seasonal flooding and many wetlands have been converted to agricultural use. As a result, the species was designated a California Bird Species of Special Concern in 2008.

Forster's Tern • *Sterna forsteri*

ORIGIN OF NAMES "Forster's" for Johann Reinhold Forster (1729–1798), a German naturalist who published *A Catalogue of Animals of North America* in 1771 and sailed around the world with Captain Cook in 1772 when he did pioneering ornithological studies at Hudson Bay; Thomas Nuttall (see Common Poorwill account) named this tern in Forster's honor.

NATURAL HISTORY As elegant in flight as they are in appearance, Forster's Terns are a great treat to watch as they cruise over the marsh, suddenly stopping to hover then plunging with complete abandon into the water, often coming out with a wriggling bit of silver grasped in the bill. The only members of their genus that do not undertake long migrations to South America each year, they spend the winter mainly in coastal areas of North and Central America.

Forster's Terns breed in marshes, where they may nest on floating vegetation, the mound of a muskrat lodge, or use a simple scrape on a protected rocky beach. The young grow quickly on a diet of regurgitated, then whole small fish, and are able to fly within a month of hatching.

STATUS AND DISTRIBUTION In California, Forster's Terns winter along the coast from Sonoma County south. Other than small numbers that remain to breed on the San Francisco Bay, most migrate to nesting areas at isolated lakes and marshes of the Great Basin and Great Plains.

West Side. Uncommon in spring migration from late April to early May, when possible at any large, low-elevation reservoir or lake; rare in fall migration from mid-August to late September; accidental in winter.

East Side. Fairly common only near breeding colonies including those at Honey Lake, Sierra Valley, Lake Tahoe, and possibly at Bridgeport Reservoir and Crowley Lake; Inyo County's first nesting attempt occurred at Tinemaha Lake in the Owens Valley in 2004; otherwise uncommon in spring and fall migration and accidental in winter.

PIGEONS AND DOVES · Family Columbidae

There are no technical differences between "pigeons" and "doves," except that the former tend to be larger, and these names are often used interchangeably. Most pigeons and doves are stocky, fast-flying birds with short legs, long tails, and small heads and bills. Their flight muscles may make up more than 40 percent of the bird's body weight to power strong, fast flight. Wing shape is often a good indicator of the species' migratory behavior, and pigeons with the longest wings tend to fly the greatest distances. Pigeons and doves have soft skin at the base of their bills, and a ring of bare skin around their eyes that can be red, blue, or white in Sierra species.

Pigeons and doves usually lay two glossy white eggs, and both parents share incubation duties for 14 to 18 days. Downy, altricial young are fed and cared for by both parents in the nest for about two weeks until they fledge; most are capable of independent flight after about 35 days. Pigeons and doves have unique, bi-lobed crops (the expanded lower portions of the esophagus) with specialized cells for producing "pigeon-milk" that they regurgitate into the gullets of their young.

All North American species forage primarily on seeds and nuts but occasionally eat fruits, leaves, buds, flowers, and insects. Seed-eating birds must drink a lot of water to digest their food, and all members of this family drink by submerging their bills and sucking, unlike most birds, which scoop water and lift their heads to swallow. More than 300 species exist worldwide, except in the polar regions and some oceanic islands, but only 4 species (2 native and 2 introduced) are found in the Sierra. The family name is derived from L. *columba*, a pigeon or dove.

Rock Pigeon • *Columba livia*

ORIGIN OF NAMES "Rock" for the species' preferred native habitat on rocky cliffs; "pigeon" from L. *pipire*, a young bird that chirps or peeps; L. *columba* (see family account above); L. *livia*, blue, the dominant color of wild birds.

NATURAL HISTORY Native to Europe, Asia, and northern Africa, Rock Pigeons were introduced to North America in the early 1600s. Feral populations adapted to a free-living existence in cities and agricultural areas across the continent from Central America to southern Alaska. Despite their large distribution, wild populations seem to always depend on humans for food and shelter, and are seldom observed in wildland settings. Rock Pigeons are known for fast flight and accurate homing abilities, and they were among the first birds to be domesticated—perhaps for sending long-distance messages in medieval times or earlier. They are also the best known and most intensively studied of the world's birds and have been the subjects of many pioneering studies of navigation, orientation, endocrinology, mechanics of flight, and genetics of color. They were also key subjects of Darwin's analysis of evolution by natural selection.

Feral Rock Pigeons nest on the sheltered ledges of abandoned and occupied buildings, church steeples, and bridges—all structurally similar to their native nesting habitats on rocky ledges in the Alps and Himalayas. A few twigs or sticks suffice for a nest, preventing eggs from falling off. Worldwide, Rock Pigeons nest throughout the year, but the Sierra breeding season primarily extends from February to September. Adult birds come in a variety of colors and patterns, but almost all have white rumps. This visible and consistent mark reliably distinguishes them from their native relatives, Band-tailed Pigeons. Rock Pigeons primarily forage on the ground for seeds but sometimes make awkward landings in trees or shrubs to pluck ripening fruits.

STATUS AND DISTRIBUTION Rock Pigeons are primarily birds of the foothills on both sides of the Sierra and are mostly confined to cities, ranches, and farms, where human-made structures and foods are readily available.

West Side. Common to abundant residents in settled areas throughout the Foothill zone up to about 3,000 feet.

East Side. Common residents of most lower-elevation towns and some ranches; a few rare records up to the higher passes such as Tioga Pass in Yosemite National Park and near Donner Pass in Placer County.

Band-tailed Pigeon • *Patagioenas fasciata*

ORIGIN OF NAMES "Band-tailed" for their distinctive tails; *patagioenas* from Gr. patagion, gold edging; L. *fasciata*, banded.

NATURAL HISTORY While they sometimes visit towns, ranches, and farms in search of productive foraging areas, Band-tailed Pigeons are most often seen in undeveloped oak woodlands. Aside from raptors, some water birds, and Common Ravens, they are the only large Sierra birds that regularly make long flights above the treetops. For this reason, they are among the most favored prey of Peregrine Falcons, which specialize on capturing birds in flight.

Fruits and berries are important components of their diet when available, especially those of dogwoods, chokecherries, madrone, and toyon. However, for most of the year Band-tails depend on acorns, which comprise about half of their annual diet. They sometimes forage on the ground but more often in tall trees that produce fruits, pine seeds, and acorn mast. Similar to most seed-eating birds, Band-tails consume small bits of gravel that are stored in their muscular gizzards to grind acorns and seeds. Similar to most finches (family Fringillidae), they are drawn to mineral deposits and springs.

Due to their dependence on acorns, Band-tailed Pigeons seldom stray far from oak forests during years of good mast production, and it seems that any kind of oak will do—blue, interior live, canyon live, or black. Once a dependable acorn source is depleted, Band-tails roam nomadically in search of other foods. In lean years, when few acorns are produced, they may forsake oak woodlands entirely in favor of other hardwood trees and pines or resort to taking grain from agricultural fields and farms in the foothills. Like all pigeons and doves, they require a reliable water source and are drawn to mineral waters especially in fall and winter.

Even during the breeding season, Band-tails feed and drink in flocks, but circumstantial evidence suggests that some nesting pairs may defend nesting territories of several acres or more. In the Sierra most pairs initiate courtship and breeding by early May, but active nests have been found as late as September. Males advertise breeding territories with a series of owl-like *hoot-whoo* sounds with the second phrase louder but shorter than the first. These calls end abruptly after females begin to lay eggs. Both parents participate in constructing a shallow platform of twigs criss-crossed on high branches, lined with grasses or mosses.

STATUS AND DISTRIBUTION In the Sierra, Band-tailed Pigeons are associated with oaks year-round, and with the exception of the Tahoe Basin, are generally uncommon above the Lower Conifer zone. After breeding, however, some wander as high as the Subalpine zone, but they descend to winter below the heavy snow, even reaching the Central Valley in some exceptional years with poor oak mast production.

West Side. Common, locally abundant, residents in the Foothill and Lower Conifer zones in areas with ample supplies of acorns; otherwise rare or absent; fairly common in the Upper Conifer zone and rare up to the Subalpine zone.

East Side. Uncommon, localized breeders along the east shore of Lake Tahoe, Mount Rose, and the Carson Range; fairly common in some years, depending on the local productivity of pine nuts and acorn production on the West Side; casual to accidental in winter.

TRENDS AND CONSERVATION STATUS Band-tailed Pigeons were favored targets for hunters, and they were nearly extirpated in California before hunting was regulated in the early 1900s. While the population has largely recovered, data from Breeding Bird Surveys and Christmas Bird Counts suggest negative population trends both continent-wide and in the Sierra. However, the patchy distribution and flocking behavior of this species make the data difficult to interpret, and these trends are not statistically significant.

Eurasian Collared-Dove • *Streptopelia decaocto*

ORIGIN OF NAMES "Eurasian" since native to Europe and Asia; *streptopelia* from L. *strepto*, to turn, and *peleia*, a dove; Gr. *decaocto*, eighteen, phonetically similar to the species' repeated cooing calls.

NATURAL HISTORY Similar to Rock Pigeons, Eurasian Collared-Doves were introduced from Europe and have spread across most of North America at an unprecedented rate. Thought to be originally native to India and a few surrounding countries, they began to colonize much of Asia and Europe in the 1600s, either by natural range expansion or by human introduction. However, unlike Rock Pigeons, which have lived in North America for centuries, Eurasian Collared-Doves are recent arrivals. While some think these are descendants of caged birds from the Bahamas that escaped in the mid-1970s, others speculate that they may have arrived here under their own power from Europe or western Africa. Since then, they have made an amazing population increase and now occupy urbanized areas across the continent.

Currently in our region, they are mainly found in suburban and rural residential settings, rarely in purely natural or highly urbanized areas. While they mostly feed on the ground, Eurasian Collared-Doves roost and nest in suburban trees, usually planted evergreens; some pairs also nest on building ledges up to 40 feet above the ground. This species is known to breed year-round, but in the Sierra most pairs breed from March to August, and they usually produce three broods per season. Females build loose platform nests of twigs and grasses, while males supply food.

STATUS AND DISTRIBUTION Eurasian Collared-Doves were first recorded in California in the late 1990s, but it is unclear whether these first records were from local escapes or releases, or from range expansion from the east. In any case, this species is now an established breeder in the Sierra.

West Side. Rare visitors or transients in foothill towns throughout and population expanding annually.

East Side. First observed in about 2000, now common residents throughout in all areas with significant human development.

TRENDS AND CONSERVATION STATUS Both the rates of population growth and of range expansion for this species have been completely unprecedented, and numbers are still increasing in most areas at a nearly exponential rate. It remains to be seen if this expansion will have a negative effect on any other species, with Mourning Doves appearing to be likely candidates.

Mourning Dove • *Zenaida macroura*

ORIGIN OF NAMES "Mourning" for the species' melancholy calls; "dove" from AS. *dufan*, to dive; *zenaida* for Zenaide, wife of Charles Bonaparte (see Bonaparte's Gull account); *macroura*, long-tailed, from Gr. *makros*, long, and *oura*, a tail.

NATURAL HISTORY Among the most visible birds in the Sierra foothills, Mourning Doves are a familiar sight in towns and ranches as well as in wild areas. Sleek and streamlined in flight, they are favorite targets for hunters, and millions are shot annually across North America. The male's plaintive *cooing* notes given throughout the day in spring and summer earned these gentle birds' their name, but these calls are motivated by territoriality and not by mood. The loud, whistling sounds of Mourning Doves erupting into flight are made by their wings and may serve as a type of alarm call to others when predators or hunters approach.

Mourning Doves require open ground for foraging, and preferred foods include such agricultural grains as corn, millet, and wheat. They also consume large quantities of grass seeds, mustard, and occasionally acorns and pine seeds, and rarely snails and a few insects. Similar to other pigeons and doves, they also take small bits of gravel to grind their food in muscular gizzards. Water is an essential habitat component that they visit at least twice per day, often by flying many miles in transit between foraging and drinking sites.

In the breeding season, males defend small territories around their chosen nest sites. They prefer trees for nesting, especially along woodland edges, but will also nest on the ground in treeless areas. Once a suitable nest site has been selected, males deliver small twigs and grasses to females for constructing flimsy platforms on horizontal branches. Sometimes nests are built on top of old nests, including those of other species. In the Sierra, nesting begins in early March, but active nests can be found until late August. Mourning Doves nest two or three times per season (and more in some locations), possibly accounting for their overall abundance. After nesting, they aggregate into large flocks of 50 or more and feed and roost communally. While normal life span is only about two or three years, one exceptional banded bird survived in the wild for more than nineteen years.

STATUS AND DISTRIBUTION Mourning Doves are widespread and common residents of the foothills on both sides of the Sierra, where they frequent open woodlands, sparse chaparral, rural residential and suburban areas, and occasionally open conifer forests.

West Side. Abundant from March to September in the Foothill zone, especially in open forests, towns, and ranches where human-produced foods are available; uncommon breeders in the Lower Conifer zone, and rare at higher elevations; most leave the Sierra entirely in winter except in more temperate urban areas of the Foothill zone, where feeder-dependent birds tend to remain year-round.

East Side. Abundant in scrub habitats and uncommon in open pine forests from March through September; formerly rare in winter, now uncommon to abundant, with large numbers found in more urbanized, lower-elevation settings (see below).

TRENDS AND CONSERVATION STATUS Population trends for Mourning Doves show a complex pattern. Breeding Bird Surveys for the western region (and for California specifically) suggest a stable population, but U.S. Fish and Wildlife monitoring shows a decline in western populations. California Christmas Bird Count data show a strong positive trend, and this is reflected by increased numbers on recent Sierra Christmas Bird Counts, especially in areas that have seen significant urbanization over the past 30 years, such as Reno and the Carson Valley. In contrast, we see a significant negative trend in numbers on Sierra Breeding Bird Surveys, with the western foothill routes showing steady declines. For the Sierra this suggests a pattern with fewer breeding birds but increasing numbers of wintering birds, possibly driven by access to good winter foraging in urbanized areas. It is too early to say if the recent expansion of the Eurasian Collared-Dove will impact Mourning Dove numbers (see previous account).

CUCKOOS AND ROADRUNNERS • Family Cuculidae

It may seem strange that the tree-dwelling cuckoos and ground-dwelling roadrunners are in the same family. Their similarities are mostly anatomical and best seen by examining their feet—all species in this family have their two inner toes pointing forward, while the two outer ones point backward. Roadrunners make their own nests and incubate their own eggs, as do Yellow-billed Cuckoos. However, Yellow-billed Cuckoos are occasional brood parasites, laying their eggs in other cuckoo nests (and rarely those of other species). Cuckoos migrate for long distances to winter in tropical rainforests of northern South America, but roadrunners fly infrequently and remain in the same general areas year-round. Approximately 125 species of cuckoos, anis, and roadrunners exist worldwide but only 2 species, the Yellow-billed Cuckoo and Greater Roadrunner, occur regularly in the Sierra. The family name is derived from L. *cuculus*, a cuckoo.

Yellow-billed Cuckoo • *Coccyzus americanus*

ORIGIN OF NAMES "Cuckoo" is imitative of the sounds made by some European species; L. *coccyzus* from Gr. *kokkuzo*, a reference to the species' calls; L. *americanus*, of America.

NATURAL HISTORY Hidden in the dense foliage of the tallest cottonwoods and willows, Yellow-billed Cuckoos are more often heard than seen. Their loud, drawn-out calls sound something like *ka-ka-ka kuk-kuk-kuck* followed by *kowlp-kowlp-kowlp* and echo for hundreds of yards through riparian woods. The latter calls are probably only given by males, usually those seeking mates and

defending their breeding territories from other competing males. Calling birds sit motionless in the treetops and can only be found by careful and methodical searching. In flight, their loud, rapid wingbeats and flashing white tail feathers give them a superficial resemblance to flying Mourning Doves.

Large, hairy caterpillars are their preferred foods, and cuckoos search for them with slow, deliberate movements, often turning their heads from side to side; they also consume a variety of other large insects such as grasshoppers, crickets, and beetles, and rarely fruits and seeds. Many studies have shown that cuckoos' breeding behavior is triggered by the local abundance of caterpillars, and in years of reduced food supplies, they may not breed successfully or at all. Among the latest of our neotropical migrants, cuckoos do not arrive on their breeding grounds until about mid-June; most pairs depart for their South American wintering grounds by mid-August.

More than any other Sierra bird species, Yellow-billed Cuckoos require large, wide tracts of riparian forest for successful breeding, and most breeding pairs select unbroken stands of at least 40 acres. Compared to other birds of similar size, their breeding cycle is extremely rapid and only takes about 17 days from the initiation of nest building until the eggs are hatched. Both adults participate in building the nest by collecting nearby twigs and forming a loose platform in the forked branch of a tall tree. Females usually lay three or four bluish-green eggs and incubate them for about 10 days with some help from the males. Altricial young are almost completely naked when hatched, and they leave the nest after about a week, compared with most similar-sized birds, which remain in their nests for a month or more. Both parents feed fledglings for about three weeks until they are capable of independent flight and can forage on their own.

STATUS AND DISTRIBUTION Greatly diminished in numbers throughout California, breeding Yellow-billed Cuckoos are now confined to a few remaining large tracts of lowland riparian forest in the state.

West Side. Uncommon nesters only in the Kern River Valley; casual or accidental migrants elsewhere, with early June records from Butterbredt Spring (probably birds in transit to the South Fork Kern River).

East Side. Casual or accidental, possible at most large riparian habitats during spring and fall migration; occasional mid-June and early fall records from scattered locations (Honey Lake, Tahoe Basin, near Mono Lake, and the Owens Valley); most published records prior to the 1970s and a few date to the 19th century.

TRENDS AND CONSERVATION STATUS In the late 19th century, Yellow-billed Cuckoos were fairly common summer residents of riparian forests throughout the lowlands of California. During the 20th century the state's breeding population suffered catastrophic losses with the destruction of riparian habitats due to flood control, agriculture, and urbanization. Breeding cuckoos are now restricted to a few remaining tracts that are large enough to support breeding pairs along the lower Colorado River, the Sacramento River (Sutter, Butte, and Glenn Counties), and in the Kern River Valley. Due to dramatic population declines, California listed Yellow-billed Cuckoos as Threatened in 1971 and as Endangered in 1988.

Greater Roadrunner • *Geococcyx californianus*

ORIGIN OF NAMES "Greater" for their larger size compared to the Lesser Roadrunner *(Geococcyx velox)*, a closely related species restricted to Mexico and Central America; "roadrunner" for primary means of locomotion; Gr. *geococcyx*, a ground cuckoo; L. *californianus*, of California.

NATURAL HISTORY These large, brownish, ground-dwelling cuckoos are voracious and indiscriminate predators, suggestive in many ways of the raptorial theropod dinosaurs from which birds are believed to have descended. Their varied diet includes large insects, large spiders (such as tarantulas and wolf spiders), scorpions, birds' eggs and young, rodents, lizards, and snakes, including rattlers—which they beat to death on rocks.

Although not nearly as fast as the cartoon character, Roadrunners can maintain running speeds of 18 miles per hour or more for considerable distances. They can fly from danger but usually sprint instead. Roadrunners are wary and easy to overlook. The best time to locate them is in spring, when males reveal their presence with a short series of whiny, cooing notes rather like a descending whimper of a puppy. They call most frequently around sunrise from elevated perches, including fence posts, snags, or tall cacti, and will respond to imitations of their calls.

They live in open, arid chaparral and deserts where good-sized thickets of large shrubs or cacti are separated by ample open ground. Specific adaptations for life in hot and dry environments include reabsorption of most consumed water through their digestive tracts and cloacas and secretion of salts from nasal glands. Behavioral techniques include panting and opening wings widely to expose lightly feathered areas underneath to the cooling effects of desert winds. On cold mornings they also expose their dark rumps to the sun to warm up.

Pairs are apparently monogamous and stay together in the same territory for years, possibly for life. Their multipurpose territiories are used for breeding, roosting, and foraging and average about a half mile in diameter. In the Sierra most pairs initiate breeding in April. Females do most of the nest building while males deliver sticks to the nest site, usually situated in a somewhat isolated patch of thorny trees or shrubs or a tall cacti. Nests are sturdy, compact platforms composed mostly of sticks and lined with softer materials like leaves, grasses, and feathers. Females usually lay four whitish, chalky-looking eggs over a several-day period, and both adults share incubation duties for about 20 days. Eggs hatch asynchronously, depending on when they were laid, and the first-born young are noticeably larger than the others. The young are altricial but active and strong soon after hatching. They remain in the nest for almost two weeks until coaxed away by the adults and can fly and feed themselves about three weeks later.

STATUS AND DISTRIBUTION Fairly common in deserts of the Southwest, Roadrunners are uncommon and seldom seen residents of the Sierra foothills, except in the southern desert areas, where they are more often seen.

West Side. Fairly common in desert areas south of the South Fork Kern River, including in the foothills of the Piute Mountains along Kelso Valley Road; rare to uncommon, highly localized residents of mixed chaparral/grassland habitats up to about 3,500 feet from Tehama to Kern Counties, and most frequently found near Highways 49 and 41 from Tuolumne to Madera Counties; possibly extirpated from the mid-Sierra foothills (e.g., Nevada and Placer Counties).

East Side. Formerly rare residents in Mono Basin scrublands but currently only found east of the region in Mono County; uncommon south to Owens Lake; fairly common in desert regions of eastern Inyo and Kern Counties, and in the extreme south along Highways 14 and 58.

BARN OWL • Family Tytonidae

There are two families of owls in the Sierra: Tytonidae, represented only by the Barn Owl and "typical" owls (family Strigidae) with 10 other Sierra species. Barn Owls are medium-sized owls that differ from typical owls in details of their bone structure and feet but share most other physical and behavioral characteristics (see below). They are the only North American representative of this widely distributed family, which includes 16 living species worldwide, except the polar and desert regions. The family name is derived from Gr. *tyto*, an owl.

Barn Owl • *Tyto alba*

ORIGIN OF NAMES "Barn" for their habit of roosting in barns and other open structures; *tyto* (see family account above); L. *alba*, white, for the species' pale underparts.

NATURAL HISTORY Barn Owls have the largest global range of any owl and can be found on every continent except Antarctica. Since they are among the world's most widespread vertebrate species, more is known about their natural history than most other land birds. Many foothill residents know these owls because they nest and roost in old barns, silos, church towers, and other buildings. At dusk or when caught in light from streetlamps, their pale, ghostlike forms appear flying softly like giant moths. While nest sites in human structures have allowed their numbers to increase in the past century, Barn Owls also roost in densely foliaged trees and nest on protected ledges on cliffs or sometimes in cavities in oak trees or occasionally streambank cavities.

Unlike other Sierra owls, they regularly raise two broods per year. The first nesting effort begins in early February with display flights and harsh, unearthly screams given by males as they wander their territories aloft (note that Barn Owls screech, Western Screech-Owls do not). Mated pairs usually remain together for life and will use the same nest site for as long as they live.

Females usually lay four to six eggs from late February to early March and perform all the incubation duties alone, while males provide all the food for females and young, at least until they are capable of independent flight after about two months. Second broods, if attempted, are usually started in May or June in the Sierra, but this species has been observed nesting in all months of the year throughout its global range.

Barn Owls frequent open country, where they forage in agricultural areas, marshes, open grasslands, deserts, and woodland edges. A breeding pair raising young can consume prodigious numbers of gophers, voles, and other small mammals in a season. Surprisingly, pellet castings in one foothill area showed them eating substantial numbers of crayfish but leaving the question unanswered as to whether the crayfish were caught after leaving the water or whether the Barn Owls plunged in after them. Because they require open terrain for hunting, nearly all their habitat use is centered in, or around, meadows and fields. Using asymmetrically placed ear openings and the most acute hearing of any animal ever tested, they can pinpoint and capture prey even in total darkness (verified under laboratory conditions).

STATUS AND DISTRIBUTION Barn Owls reside year-round near low-elevation grasslands throughout the Sierra foothills, especially around ranches and farms with buildings for roosting and nesting. Some seasonal movement in response to cold or heavy snow is likely but difficult to verify.

West Side. Fairly common in open habitats of the Foothill zone; rare visitors to mountain meadows of the Upper Conifer zone.

East Side. Rare to fairly common below the pine forests, with localized populations centered mainly in agricultural areas such as near Honey Lake and Bridgeport Valley.

TYPICAL OWLS • Family Strigidae

Fearsome hunters of the night, owls hold a mysterious fascination for humans. Their direct stares and stolid upright postures have earned them a reputation for wisdom. Because most species are nocturnal, casual observers only rarely see more than an owl's silent, fleeing silhouette. Owls reveal their presence by calls, distinctive for each species and usually given at night.

With large eyes and broad facial disks, an owl's gaze can be piercing and intimidating. Owls have excellent vision, not only at night but also in daytime. Both eyes point straight ahead, permitting depth perception for capturing prey. While their eyes do not move, many species can turn their heads up to 270° to search for prey (Peeters 2007). Facial disks aid in collecting and focusing sound toward their extraordinarily acute ears (asymmetrically placed in some species) enabling them to sense distance and direction precisely. Other adaptations for hunting include soft, velvety feathers permitting silent flight, powerful feet and talons for capturing prey, and strong, hooked beaks for tearing flesh.

Owls regurgitate pellets consisting of indigestible residues such as bones, fur, and feathers, which provide biologists with valuable evidence of their food habits. Hawks and falcons tear apart their prey to get at the edible parts, while owls swallow their victims whole or in large pieces and so produce a large volume of pellets compared with other birds of prey. Their pellets also contain more bones because they are less efficient at digesting those hard tissues than hawks and falcons.

All members of this family are nocturnal, but several species often hunt by day, especially Short-eared, Northern Pygmy, Burrowing, and Great Gray Owls. Since most owls nest in well-concealed locations where females constantly guard the nests, they have little reason to camouflage their eggs. Most species lay two to four uniformly white to cream-colored eggs with no dark splotches or scrawls. For most species, incubation is performed mostly by females for about 25 to 30 days. Females brood the young while the males bring food until the young can fly after about a month. This large family is represented by more than 140 species worldwide. The family name is derived from Gr. *strigx*, to utter harsh sounds.

Flammulated Owl • *Otus flammeolus*

ORIGIN OF NAMES "Flammulated," reddish, for the species' reddish-gray plumage; "owl" may be derived from AS. *ule*, an owl; Gr. *otus*, a horned owl; *flammeolus* from L. *flammeus*, flame-colored.

NATURAL HISTORY "If a Martian in black livery were to sidle up on the dark side of our planet, all on a moonless night, to spy upon us, he could scarcely keep his business so well concealed as this ghoulish avian mystery, the Flammulated Owl." As William Dawson (1923) implied, these diminutive and mysterious owls were long regarded as rare because of their irregular calling, secretive behavior, and poorly known habitat needs. Even when calling, they are hard to locate due to their low frequency but loud, ventriloquial hoots. Most Sierra birders never see this species because they perch near tree trunks, deep within the canopy, where their matching coloration blends with bark and branches—especially at night. Surprisingly, surveys based on a better understanding of their behavior have revealed that in local areas of optimal habitat, Flammulated Owls can be found in higher densities than any other Sierra owl. They seem to prefer dry, open stands of older ponderosa or Jeffrey pines and incense cedars mixed with black oak; they generally avoid the dense, moist fir forests favored by Spotted Owls. In a given area, Flammulated Owls may be found on dry, south-facing slopes and ridgetops and be replaced by Spotted Owls in shaded draws or north-facing slopes nearby.

Unlike most Sierra owls, Flammulated Owls migrate long distances, spending the winter in southern Mexico and Central America. They return to the Sierra by mid-April, though they rarely call before late May. The call is a single- or two-three-note mellow hoot, given every two to three seconds that, under ideal conditions, can be heard for up to a half mile. In the Sierra they typically nest in the abandoned holes of Pileated Woodpeckers or Northern Flickers. These owls reuse the same sites year after year, as long as both members of the pair return. Prior to egg laying, the female will cease foraging and let the male bring her food. By the time the first egg is laid in late May, the female may have gained so much weight that she has difficulty flying.

Dedicated insectivores, Flammulated Owls are particularly fond of large moths but also eat nocturnal spiders and flying insects, often capturing the latter in midair. The lack of flying insects in Sierra pine forests during winter explains their long migration to the tropics. By early October most have departed to the south, making their long journeys in the dark of night.

STATUS AND DISTRIBUTION Flammulated Owls are largely restricted to dry ponderosa or Jeffrey pine forests between 2,000 and 5,500 feet in the Lower Conifer zone. They mostly breed in, or migrate through, patches of pines interspersed among various conifers, but their status and distribution are poorly known in the Sierra except in a few areas where researchers have worked extensively.

West Side. Common in isolated pockets of suitable habitat, with scattered records from the length of the Sierra, but many seemingly suitable habitats are not occupied; some known nesting locations include Henness Ridge, just west of Yosemite National Park; Mosquito Ridge Road (Placer County); and San Juan Ridge (Nevada County).

East Side. Uncommon or rare; nests in pine forests in Plumas County, east rim of the Tahoe Basin, and possibly elsewhere; late-season banding studies east of Lake Tahoe suggest there may be significant numbers of migrants moving down the East Side.

Western Screech-Owl • *Megascops kennicottii*

ORIGIN OF NAMES "Screech" describes some of the sounds of the Eastern Screech-Owl but not the Western; named for Robert Kennicott (1835–1866), a founder of the Chicago Academy of Sciences, and a bird and snake collector for the Smithsonian Institution in the northwestern United States and Canada; L. *megascops*, a small owl with ear tufts.

NATURAL HISTORY Western Screech-Owls were poorly named, for none of their typical calls even remotely resembles a screech. Rather, their calls are a short series of low, whistled notes on one pitch, speeding up like a bouncing ball at the end. They have an extensive vocabulary of other calls, including barks, steady trills, and even a rising wail not unlike a Spotted Owl contact call. Courtship extends from January into April and includes mutual preening and feeding. Pairs maintain and defend territories year-round, which may explain why this owl is vocal at any time of year. Young birds remain largely dependent on their parents for about five weeks, then begin wandering more and more, developing skills to catch insects, small mammals, and small birds.

They are the most abundant members of their family in foothill oak woodlands of the West Side. They occur in high densities in riparian woodlands but are also frequently encountered in dense oak and oak-conifer stands on dry slopes and ridgetops. Screech-Owls nest in either natural tree cavities or those left behind by Pileated Woodpeckers or Northern Flickers (occasionally taking over an already active woodpecker nest). They readily use artificial nest boxes, such as those placed for Wood Ducks.

STATUS AND DISTRIBUTION Screech-Owls, while strongly associated with the Foothill oak zone, also occur in groves of cottonwoods and willows near water and within the Lower Conifer zone in mixed stands with oaks.

West Side. Common and widespread residents almost anywhere there are medium-large oaks below the Upper Conifer zone; in late summer, juveniles wander widely, rarely up to the Subalpine zone.

East Side. Rare to uncommon residents in low-elevation woodlands; casual to rare from the southern Sierra, in contrast to the adjacent Great Basin, where the species can be fairly common and widespread in riparian and dense pinyon-juniper habitats.

Great Horned Owl • *Bubo virginianus*

ORIGIN OF NAMES "Great horned" for the species' large size and ear tufts resembling horns; L. *bubo* from Gr. *buzo*, to hoot; L. *virginianus* of Virginia, where the first specimen was collected.

NATURAL HISTORY Eagles may get all the press, but Great Horned Owls are really the top predator of the North American bird world. They feed primarily on rodents and rabbits but will kill almost any animal they can carry, including birds, snakes, fish, and even animals other predators avoid completely, like porcupines and skunks. They can fly away with prey as much as three times their own weight and are known to kill house cats, small dogs, geese, and Great Blue Herons on occasion. These magnificent birds weigh more than any other owl in the Sierra, and only Great Gray Owls exceed them in length and wingspan.

Great Horned Owls are extremely versatile in their choice of habitats and nest sites. Their preference, however, is for areas with large clearings or open forests for hunting, with sheltered roosting sites nearby. Pairs are highly territorial and proclaim their domain with persistent nightly hooting. They will attack and kill other owls or raptors and are, in turn, mobbed enthusiastically by crows, jays, and other birds when a day roost is discovered.

Like other owls, Great Horns do not build their own nests. Instead, they settle in old nests of Red-tailed Hawks and other large birds or, less often, in caves, cliffs, or tree cavities. One of the earliest-nesting birds of the Sierra, they begin courtship in winter and egg-laying as early as January. Young owls make their first flights after about seven weeks but may remain with their parents until October, many months after leaving the nest. Individuals may live for more than 20 years.

STATUS AND DISTRIBUTION Great Horned Owls are widespread throughout most of the Sierra. Their large territories keep densities low and they are frequently found in areas of human settlement, where they take advantage of open habitats and higher concentrations of rodents.

West Side. Fairly common in all forested areas below the Alpine zone; avoids large stands of dense coniferous forest but otherwise ubiquitous; most common from the Upper Conifer zone down into the Central Valley.

East Side. Similar to the West Side, fairly common in all habitats below the Subalpine zone; in open desert settings, most activities centered around homesteads and groves of tall trees where they often roost during the day.

Northern Pygmy-Owl • *Glaucidium gnoma*

ORIGIN OF NAMES "Northern" for northern distribution compared with other pygmy-owls that occur at lower latitudes; "pygmy" for their small size; *glaucidium* from Gr. *glaukidion*, glaring, or a little owl that glares; Gr. *gnoma*, reason or opinion, related to gnome, a diminutive and intelligent goblinlike creature that guards the earth's treasures.

NATURAL HISTORY Do not be misled by the fact that Northern Pygmy-Owls are the smallest Sierra owls by weight. Gram for gram, they are the most deadly beasts in the forest. Fierce and apparently fearless, they prey on a variety of birds and readily attack species much larger than themselves, like quail or Mourning Doves. They have even been observed killing domestic chickens more than 60 times their own weight! By this standard, if Great Horned Owls were equally aggressive, they could kill a 200-pound man. Probably because of their hunting prowess, no other owls—not even Great Horned Owls—are mobbed so fiercely and frequently by songbirds. By calling and diving at them, small birds can drive owls from their neighborhood entirely, or at least disrupt their hunting. Many birders imitate the steady *toots* of pygmy-owls to try to attract other birds into view, although the owls (if present) readily respond as well. These owls are active much of the day and hunt most often around dawn and dusk.

Year-round residents, pygmy-owls live in open woodlands and in forests bordering meadows, chaparral, and other large openings, and steep hillsides with widely scattered oaks are especially preferred. They nest in medium-sized cavities like those left by Hairy Woodpeckers or Northern Flickers. Females usually lay eggs in the same cavities year after year. Young pygmy-owls are extremely voracious feeders, requiring males to hunt almost ceaselessly, bringing prey to the nest cavity for females to pluck and tear into small bites for the young. Young owls leave the cavity after about a month but remain dependent on their parents for an unknown period, probably several more weeks.

STATUS AND DISTRIBUTION Grinnell and Miller (1944) described this species' status as "usually, where detected at all, 'common,' for an owl." Northern Pygmy-Owls are resident in most woodlands of the Sierra except the higher elevations, very dense forests, and deserts; there is likely some movement in response to local weather conditions in winter.

West Side. Fairly common in oak habitats from the Foothill zone up to the Lower Conifer zone; rare in the Upper Conifer zone.

East Side. Uncommon to fairly common residents in and below the pine forests as far south as northern Inyo County, rare to absent farther south.

Burrowing Owl • *Athene cunicularia*

ORIGIN OF NAMES "Burrowing" for the species' preferred underground nest burrows; Gr. *Athene*, goddess of wisdom; L. *cunicularius*, a miner or burrower.

NATURAL HISTORY People unfamiliar with Burrowing Owls may be startled to see these small raptors disappear into holes in the ground. These owls often perch by their burrows all day or on nearby fence posts, making them easy to spot. They prefer large expanses of relatively level, treeless

Adult   Juvenile

grassland kept nearly barren by ground squirrels, grazing, or roadside mowing or weed control. Burrowing Owls also colonize fields in agricultural settings as well as golf courses, cemeteries, and vacant lots in residential and industrial areas, where they make use of rock piles, artificial levees, and berms. They hunt mainly in the early evening and at night, hawking insects in midair and attacking rodents from perched or hovering positions. Sometimes they alight and run or hop after prey on the ground, making good use of their unusually long legs. Snakes, lizards, and small birds may also be on the Burrowing Owl's menu.

Though they almost always use burrows abandoned by ground squirrels or other burrowing rodents, Burrowing Owls are effective diggers in their own right and can excavate a burrow up to nine feet deep in two days. Loose colonies of up to 12 families may nest in the same area year after year unless their habitat is destroyed. Each pair lines their burrow with manure to mask the scent of their nest and youngsters from any of the numerous predators, such as gopher snakes, that find these ground nesters easy prey. As an added defense, both adult and juvenile owls can utter a rattlesnake-like sound that would keep most bird or mammal predators from sniffing further into the burrow entrance.

STATUS AND DISTRIBUTION Burrowing Owls are associated with valley lowlands adjacent to Sierra slopes, where they are restricted to dry valley bottoms or open foothill habitats. Migrants arrive in March or April and begin heading south in September or October. Some populations are almost certainly resident since they are found year-round in the Kern River Valley and in some Foothill zone localities. Burrowing Owls tend to wander widely during migration and after the breeding season.

West Side. Uncommon, scattered colonies exist in grasslands and open agricultural areas, rarely as high as the upper edge of the Foothill zone; rarely wanders to the Sierra crest in late summer and early fall as evidenced by odd sightings on top of El Capitan in Yosemite Valley at the end of October and one at 12,000 feet on the Dana Plateau at the end of September.

East Side. Rare, small breeding colonies have been present from time to time near Honey Lake and in Sierra Valley; rare or casual elsewhere.

TRENDS AND CONSERVATION STATUS Although still found in scattered locations in the Sierra, Burrowing Owls have declined throughout California because of habitat loss to development or more intense agriculture, elimination of ground squirrels and their burrows, and ingestion of poisons used for rodent control. They are especially vulnerable to disturbance in agricultural areas where the weedy areas along field margins are treated with rodenticides. For these reasons, Burrowing Owls were included on the 2008 list of California Bird Species of Special Concern. Repeated petitions to list this species as Threatened or Endangered in the state have been denied, largely because of sizable populations in the Imperial Valley and despite the dependence of those populations on agricultural activities that could be abandoned or altered at any time.

Spotted Owl · *Strix occidentalis*

ORIGIN OF NAMES "Spotted" for the breast coloring of adults; Gr. *strigx*, screech-owl, harsh or strident; L. *occidentalis*, western.

NATURAL HISTORY Nocturnal and confined to dense, mature forests, Spotted Owls are known to most observers as elusive shadows with an impressive array of calls. Besides the typical *hoo hoo-hooo, hoooo . . . hoooo* territorial call, these owls have a remarkable variety of vocalizations, including rising screams, barks, and conversational chortles. Roosting by day next to the trunks of tall trees high overhead, these owls are very difficult to see against the dappled texture of tree bark. Once located, however, they are extremely easy to watch due to their relatively tame nature and apparent lack of fear of humans. So calm is this species that females lifted off their nests do not protest or show interest but merely look around with big, dark eyes.

They breed almost exclusively in mature conifer forests, especially those with older ponderosa and sugar pines, incense cedars, and Douglas-firs, or in moist stands where black oaks, madrones, or other hardwoods form a well-developed understory. Spotted Owls are much more general in their choice of foraging habitat, using a wide variety including open second growth forest and recently burned areas. Sensitive to overheating, they seek shady stands for daytime roosting in summer. They usually hunt from elevated perches, feeding primarily on small mammals, especially woodrats and flying squirrels, but also on bats, birds, and insects. Curiously, they may also feed on garbage around dumps or picnic areas. Like most owls, Spotted Owls do not build their own nests but occupy those built by other species; most pairs in the Sierra nest in tree cavities or broken-top snags. Parents defend large breeding territories from February through September during the period when they are raising and defending young.

STATUS AND DISTRIBUTION "California" Spotted Owls *(S. o. occidentalis)*, the subspecies found in the Sierra, are resident in appropriate habitat throughout the West Side.

West Side. Fairly common in mature forests but mainly confined to the Upper Conifer zone, where they live year-round; some adults descend to Foothill zone in October to spend the winter; breeding status in the Foothill zone is uncertain but probably occurs rarely if at all.

East Side. Uncommon to rare breeders in Plumas County, the Tahoe Basin, and possibly at Mammoth Creek (Mono County) and at a few locations in Alpine County; elsewhere casual visitors to pine forests.

TRENDS AND CONSERVATION STATUS "California" Spotted Owls are vulnerable to habitat disturbance and are likely declining in the Sierra despite urgent measures to safeguard the species. A recent 15-year study of population dynamics indicated that while the population is not in dramatic decline, it is vulnerable, especially where forest management practices permit more habitat disturbance. Each pair requires several hundred acres of mature forest, and such large stands are rapidly succumbing to logging. Logging practices such as clear-cutting, salvage logging, and group selection cuts that remove most or all large trees or snags from forest stands are particularly harmful to this species. Habitat degradation due to residential and commercial development, and large, intense wildfires also destroy and fragment their preferred breeding habitats. The recent invasion of California by the Barred Owl *(Strix varia)* also threatens the Spotted Owl's future survival because Barred Owls will displace, attack, and kill, as well as hybridize with Spotted Owls. For these reasons, "California" Spotted Owls were included on the list of California Bird Species of Special Concern in 2008.

Great Gray Owl • *Strix nebulosa*

ORIGIN OF NAMES "Great gray" for the species' large size and overall coloration; L. *nebulosa*, cloudy, a reference to the species' gray color.

NATURAL HISTORY Their grandeur and extreme rarity south of Canada make Great Gray Owls one of the most eagerly sought birds in the Sierra. Gazing at close range into their penetrating yellow eyes, accentuated by huge facial disks, is a memorable and enchanting experience. No one forgets their first Great Gray. Although the Great Horned Owl is heavier, no other North American owl rivals the Great Gray in overall length and wingspan. While they occur in boreal forests across a wide swath of the continent, the central Sierra population is geographically isolated, and recent studies have shown that it represents a genetically distinct evolutionary lineage from other populations and should be considered a new subspecies, *S. n. yosemitensis* (Hull et al. 2010).

They require moist meadow systems of 20 acres or more for foraging—another compelling reason to preserve these teeming centers of biological activity whenever possible. When hunting, Great Grays watch and listen for prey from low perches along meadow edges, flying distances of a few feet to more than 50 yards to make a kill. Despite the owl's large size, their victims are usually small rodents like gophers or voles. Great Grays have exceptional hearing and have been observed diving for gophers through a foot of snow catching them underground by breaking through tunnels near the surface. Unlike most owls, they often forage in daylight, usually early or late in the day. They also hunt at night, as evidenced by the remains of nocturnal flying squirrels that are often found in their pellets.

A recent study of Great Grays in the Sierra suggested that they may only nest successfully in years when there is an above-average density of prey such as voles to feed their young—apparently many Sierra pairs do not even attempt to breed in some years (Powers et al. 2011). Most of the few nests found in California have been in the tops of large, broken-off snags, but elsewhere many pairs have used abandoned hawk nests. In the Sierra, Great Gray Owls call most frequently after 1:00 a.m. but may be heard at any time of night or in late afternoon. They advertise breeding territories from April through August, making low double hoots or 5 to 12 low-pitched, evenly spaced *whoos*.

STATUS AND DISTRIBUTION Great Gray Owls breed in Lower and Upper Conifer zones, dominated by large pines, cedars, and firs with meadows or other open foraging areas nearby. Some move up into the Subalpine zone in late summer and fall. In winter, most move down into the Lower Conifer zone. Based on recent sightings, some may even move into the Foothill zone. Primarily denizens of the boreal forests of Canada and Alaska, Great Gray Owls also occur in southward extensions of suitable habitat in the Rocky Mountains and Sierra. While sight and specimen records from California date back to 1857 and 1858 respectively, the state's first mated pair was not discovered until 1915, in the Yosemite region; a brood patch on the female provided the first confirmation of breeding south of Canada (Grinnell and Storer 1924). Their known Sierra breeding population continues to be mostly concentrated in the Yosemite region, but Great Grays are also observed occasionally elsewhere in the Sierra and likely breed at other undiscovered locations.

West Side. Rare to uncommon; in Yosemite National Park breeding pairs found at Crane Flat, Peregoy Meadow, Westfall Meadow, McGurk Meadow, and just west of the park at Ackerson Meadow; other recent nesting records outside the Yosemite region include near Beasore Meadow (Madera County) in 2002; near Shaver Lake (Fresno County) in most years since 1998; near Plummer Ridge (El Dorado County) in 2002; along Pliocene Road near Pike (Sierra County) in 2009; and winter records at 3,000 feet near Nevada City (Nevada County) and, most surprising, an early spring record in Yuba County below 1,200 feet in oak woodland; no evidence of nesting in Sequoia and Kings Canyon National Parks, but at least one nonbreeding record from there; and a winter record from Alpine Lake (Alpine County).

East Side. Rare; possible breeding pairs near Independence Lake and at Perazzo Meadow (both Sierra County); also a few records of apparently nonbreeding individuals exist for the northern Sierra, including a single bird at Kyburz Flat (Sierra County) in 2009; fall record at over 10,000 feet north of Yosemite National Park.

TRENDS AND CONSERVATION STATUS Since 1980, Great Gray Owls have been classified as Endangered in California, and a recent survey estimated fewer than 50 breeding pairs in the Sierra. However, recent anecdotal reports from several Sierra locations of second or third growth forests suggest the population may be larger than we suspect. Recent research suggests that much of the population may winter in areas of private land not subject to the same protections available on public land. This species is highly sensitive to human disturbance, and birders should view them from a respectable distance and never approach active nests. Because they do not breed in years of low rodent abundance, their populations are especially vulnerable to reproductive failures due to human disturbance. Large-scale commercial logging during the past 100 years has greatly reduced the extent of mature conifer forests and the number of large-diameter trees that Great Grays depend on for nest sites, and flooding of meadows to create reservoirs has eliminated prime foraging habitat. A recent radio-telemetry study in Yosemite National Park documented that Great Gray Owl hunting behavior is negatively affected by human disturbance, and even more so by bird-watcher disturbance (Wildman 1992). Since Great Grays are among the rarest, and most sensitive, Sierra birds, it is unwise, unethical, and illegal to play recordings of their vocalizations or to harass them in any way. The best chance of seeing one is to sit quietly in a concealed position at the edge of a meadow and wait for one to appear.

Long-eared Owl • *Asio otus*

ORIGIN OF NAMES "Long-eared" for the species' long, distinctive ear tufts; L. *asio*, a horned owl; L. *otus*, horned owl.

NATURAL HISTORY Few other Sierra birds are so poorly known as these extraordinarily secretive owls. Long-eared Owls are masters of camouflage and spend their days in dense thickets, elongating their bodies to resemble dead sticks when alarmed. They rarely call except during the spring breeding season and occasionally in fall. A varied repertoire of calls are heard in the Sierra, including a low, mellow, prolonged hoot, a series of five or six hoots (resembling those of the Great Horned Owl, but higher), catlike meows, and screeches.

Like Great Horned Owls, Long-ears nest in vacant nests of other birds, especially hawks, magpies, and crows. Both parents will fiercely defend the nest and young by hissing loudly then falling to the ground and struggling frantically as a distraction. If these tactics fail, they will savagely attack the face and throat of any intruder. At three weeks of age, young birds lessen the risk of being found by predators by climbing out of their nests and roosting solitarily in dense foliage nearby. They can fly within about 40 days and are largely independent of their parents after about two months but may remain in family groups into the fall and rarely into winter.

Long-eared Owls often gather in communal roosts of up to about 20 birds in dense foliage of trees, willow thickets, or even shrubs when not incubating eggs or brooding young. They need open areas for hunting near their nest or roost sites and a sufficient supply of food, especially voles and mice. Due to their narrow diet, they typically leave an area when populations of favored prey drop dramatically. Long-eared Owls nearly always hunt at night, flying low over the ground and listening for prey in the manner of a nocturnal Northern Harrier.

STATUS AND DISTRIBUTION Although seldom recorded in the Sierra, Long-eared Owls may be fairly common in some localities, and their actual status is unknown.

West Side. Uncommon breeders at low elevations in riparian habitats of the Foothill zone but recorded nesting as high as the Upper Conifer zone and may range even higher after the breeding season; in winter, most probably descend to low elevations around the Sierra or leave the area based on the few records of wintering birds.

East Side. Uncommon in pine and riparian habitats; studies just east of the region in Mono County suggest the species may be more common and widespread than is apparent from the low frequency of reports; documented nesting in Washoe Valley, just east of Highway 395; winter roosts in streamside thickets at low elevations suggest downslope movements after breeding.

Long-eared Owls apparently have declined in numbers due to human impacts on their preferred riparian habitats. Because of habitat loss and uncertainty about populations, this species was included as a California Bird Species of Special Concern in 2008.

Short-eared Owl • *Asio flammeus*

ORIGIN OF NAMES "Short-eared" for the species' short ear tufts that are rarely visible, even at close range; L. *flammeus*, flame-colored.

NATURAL HISTORY Short-eared Owls are readily observed hunting in early morning and late afternoon. They fly slowly over open country with deep, mothlike wingbeats, keeping near the ground and sometimes stopping to hover before dropping on their prey. They hunt somewhat like Northern Harriers, and the two species often share habitats with Short-eared Owls coming out just as the harriers are heading off to bed. Their most common victims are meadow voles, but parents feeding young capture other small mammals and birds as well. These items are captured mainly with acoustical clues and secondarily by vision, but curiously Short-eareds have olfactory nerves as well developed as birds that have a strong sense of smell.

Short-eared Owls inhabit broad, relatively treeless expanses, especially marshes and tall, moist grasslands. Males are famous for their dramatic "sky dance" courtship displays in which they hoot repeatedly while making acrobatic flights and loudly clapping their wings together over the females. By late March or early April, females create ground nests in slight depressions that they scrape out and line with grasses. Nests are extremely difficult to locate because they are located under tall grasses and females sit tight when approached. Due to their vulnerability, young owls disperse from the nests on foot after about 15 days but cannot fly for about two weeks after that. Short-eareds may nest in loose colonies, and while migrating or wintering they sometimes gather in communal roosts of up to 200 birds.

STATUS AND DISTRIBUTION Formerly considered regular breeders east of the Sierra crest, recent breeding season records of Short-eared Owls have been infrequent and scattered. Most occurrences are of wintering birds or postbreeding wanderers.

West Side. Rare visitors to foothills fall through winter; casual or accidental at higher elevations.

East Side. Rare or casual, former breeding records from Honey Lake, Mono County, and the Owens Valley; recent breeding season records of multiple birds (Honey Lake 2008; Sierra Valley and Bridgeport Reservoir 2011) suggest the species is not extirpated as a breeder; localized migrant or winter visitor at other marshes and wet meadow areas.

TRENDS AND CONSERVATION STATUS Short-eared Owls apparently have declined in numbers due to human impacts on their preferred marsh and grassland habitats. Because of habitat loss and uncertainty about populations, this species was included as a California Bird Species of Special Concern in 2008.

Northern Saw-whet Owl • *Aegolius acadicus*

ORIGIN OF NAMES "Saw-whet" for the breeding season calls resembling a saw being sharpened; Gr. *aigolios*, an owl, a name of uncertain origin used by Aristotle, perhaps from *aigos*, a goat, for similarity of bird's call to the bleating of a goat; L. *acadicus*, from Acadia (now called Nova Scotia), where the species was first collected.

NATURAL HISTORY Like other small owls, Northern Saw-whets must keep constant watch for their greatest enemies, the larger owls. Because these talented little ventriloquists usually call from dense cover, they are difficult to find, even when perched close by. Once they start calling, their steady string of *toots* can go on and on and on without pause for minutes at a time. Scarcely larger than thrushes, Saw-whets are hard to see—roosting by day in holes or in dense foliage, where their speckled plumage provides good camouflage. Despite their secretive habits, they are remarkably tolerant of people, allowing close approach in daylight and at night sometimes coming down to perch on people's shoulders or venturing into tents.

Saw-whet Owls eat small mammals, insects, frogs, and birds in proportion to local availability. Excellent hearing and highly asymmetrical placement of ears allow them to precisely locate and catch prey in near-total darkness. They are patient hunters and will sit for hours waiting for a prey item to reappear from a hiding place. A recent study of Saw-whets in the Tahoe Basin, one of very few Sierra-specific studies, suggests they prefer more open forests and avoid areas where white fir grows in high densities.

Saw-whets usually nest in deserted woodpecker holes, especially those left behind by large woodpeckers. Males begin calling in late February, and most Sierra nests are started by mid-March. With a high rate of metabolism, these diminutive owls require a significant effort to feed their rapidly growing brood. Saw-whets often die when snow blankets the ground, one reason they move downslope in winter.

STATUS AND DISTRIBUTION During spring and summer, Saw-whet Owls occur mainly in the Lower and Upper Conifer zones, but in winter they abandon higher elevations where snow is heavy and move down into the foothills or even to the valley floor where there are numerous records. Winter movements, either downslope or southward, begin in October and reach a peak in December.

West Side. Fairly common in summer in mid-elevation conifer forests; exact status uncertain and few Sierra nests have been found; in winter, many killed by vehicles along Highway 80 between about 3,500 and 6,500 feet and at other locations as low as 600 feet, indicating they are more common than otherwise known and/or that significant movements occur from time to time in winter.

East Side. Uncommon, the few nests found to date have been in pine forests, indicating this is their primary nesting habitat but calling birds have also been found in pinyon-juniper habitats near Mount Rose and in the Mono Basin; winter habits less well known because few birders spend time in these mid-elevation forests and sightings are extremely limited.

NIGHTHAWKS AND RELATIVES • Family Caprimulgidae

Like owls, which may be their closest relatives, nighthawks and poorwills are primarily nocturnal and have unusually soft feathers that permit silent flight. They are most active at sunrise and sunset, and when the moon is out they forage through the night. They hunt visually for nocturnal flying insects, sweeping them up with a net of "rictal bristles" that extend out from the mouth on some species. Despite their short bills, these birds have surprisingly wide gapes when they open their mouths (up to two-inches wide). Legs are short and weak and mostly serve to help perch on the ground or low branches.

All species have unmistakable, far-carrying calls for communication in the dark. These calls are often the only way to detect these species in the field. By day they remain motionless, relying on their camouflaged plumage to avoid discovery, and they will permit extremely close approach at this time. From mid-April to late June in the Sierra all members of this family breed on the open ground with no apparent "nest." Females usually lay two whitish, heavily spotted eggs on these barren sites. These are exposed to extreme heat, and females sit tight on the eggs all day to safeguard their temperature. They may lay eggs in areas shaded by rocks or low plants or temporarily roll the eggs into shade during the hottest part of the day. Mostly they stay on the eggs in the sunlight and regulate temperatures by "gular flapping" (panting by fluttering the throat) and raising feathers. Males perch close by, and when the chicks hatch after 18 to 20 days, males do most of the feeding and guarding until the young begin flying after about three weeks.

This large family is represented by almost 70 species occurring worldwide, except New Zealand and some oceanic islands; 3 of these can be found regularly in the Sierra. The family name is from L. *caper*, a goat, and *mulgere*, to milk, for an incorrect legend that members of this family fed on goat's milk—hence their historical name "goatsuckers." Nighthawks, poorwills, and their relatives are also called "nightjars," possibly for the jarring, nocturnal calls of a European species.

Lesser Nighthawk • *Chordeiles acutipennis*

ORIGIN OF NAMES "Lesser" for the slightly smaller size of this species compared with the Common Nighthawk (see account below); "nighthawk" for hunting insects in the dark; *chordeiles* from Gr. *khorde*, a musical note, and *deile*, evening; *acutipennis* from L. *acutus*, pointed, and *penna*, wing or feather.

NATURAL HISTORY Formerly known as "Trilling Nighthawks," these birds (males only) may be heard giving intermittent, whistling trills at night as they perch on the ground or swoop low over open country. By day they sleep on the ground or low branches then awake to hunt flying insects in the evening and early morning hours. Although they eat a wide variety of insects, Lesser Nighthawks travel long distances in search of large single-species swarms that appear each evening. Unlike Common Nighthawks, Lessers typically hunt low over the ground. Both species have graceful, buoyant flight with deep wingbeats and quick erratic glides on wings held in a dihedral.

STATUS AND DISTRIBUTION In California, Lesser Nighthawks are most common in the deserts of the south. In the Sierra they occupy barren or partly brush-covered areas and are often associated with dry washes or gravel bars with scattered shrubs from May through mid-September.

West Side. Rare to uncommon and highly localized breeders at the very lowest edge of the foothills (except for the Kern River Valley); most reports are from the southern half of the region.

East Side. Fairly common and localized breeders in the Owens Valley, becoming rare in Mono County and absent further north.

TRENDS AND CONSERVATION STATUS In the Central Valley and the lower Sierra foothills, Lesser Nighthawks were historically fairly common breeders up to the early 20th century. Flood control projects and resulting channelization of rivers and creeks have destroyed much of the brushy gravel bars on which Lesser Nighthawks nest, which may be responsible for the local decline of this species.

Common Nighthawk • *Chordeiles minor*

ORIGIN OF NAMES L. *minor*, as they are smaller than similar European Nightjars *(Caprimulgus europaeus).*

NATURAL HISTORY There are many good reasons to forego a tent on a summer night in the Sierra, but none better than the opportunity to awake to the booms of Common Nighthawks displaying overhead at dawn. Formerly known as "Booming Nighthawks," males courting or defending eggs perform steep power dives with a loud *whoof* or *boom* sound at the end. These unforgettable sounds are produced by air rushing through their wings and tails.

Darting erratically through the air in pursuit of high-flying insects, Common Nighthawks fill the role of swallows and swifts at night. They feed mostly between sunset and sunrise, taking prey ranging in size from mosquitoes to moths, and often hunt in groups. In summer, nighthawks come out in the day as well, perhaps because they need extra food for their young. Even when not "booming," they announce their presence with a peculiar, loud *peent* call, frequently given in flight. When not out foraging in the air, Common Nighthawks are seldom observed during the day since they will sit motionless on a branch or on the ground, allowing their remarkable camouflaged plumage to make them nearly invisible. If you are lucky enough to find one roosting, they will usually allow you to approach within inches, apparently supremely confident in their camouflage.

STATUS AND DISTRIBUTION In the Sierra, Common Nighthawks arrive late (late May to early June) and leave early (most by the end of August), probably because they require abundant insects and are intolerant of cold. Most nest from the Upper Conifer zone up to the Alpine zone, where they often forage over large mountain meadows.

West Side. Fairly common nesters and can be observed in open country from the Foothill zone to the Sierra crest; mostly occurs from Fresno County north; a few even nest at the very edge of the Central Valley floor (e.g., Placer and Sacramento Counties) in areas where Lesser would be considered the likely nighthawk.

East Side. Common from lower-elevation sagebrush flats and marshes up to the Alpine zone; generally more abundant than on the West Side.

Common Poorwill · *Phalaenoptilus nuttallii*

ORIGIN OF NAMES "Poorwill" is imitative of the species' call; *phalaenoptilus*, moth's wing, from Gr. *phalaina*, moth, and *ptilon*, feather; *nuttalli* for Thomas Nuttall (1786–1859), a noted 19th-century botanist and ornithologist who traveled and collected extensively in the West and published the first portable guide book to the birds of North America, *Manual of the Ornithology of the United States and Canada*.

NATURAL HISTORY It is often difficult to get a good look at these nocturnal insect-eaters of the chaparral. A slow drive along quiet dirt roads at night may provide a glimpse of one resting on the ground, their eyeshine glowing pink or orange or fluttering up like giant moths in the glare of the headlights. Like owls, Common Poorwills are much more easily heard than seen. On warm summer nights, beginning at dusk, they utter a mellow whistled *poor-will-lo*, the second note higher than the first and the low third note audible only at close range. Sometimes several birds call back and forth at once, creating a beautiful, haunting chorus. Unlike high-flying Common Nighthawks, poorwills flit mere feet above the ground, often leaping vertically to catch passing insects then settling immediately back on the ground. As ground-dwellers, poorwills are vulnerable to predators, but their camouflaged plumage makes them almost invisible as they perch on open rocks or ground during the day. It is thought that most pairs rear two broods in a single season.

In addition to using panting, gular flapping, and water loss from their exposed skin to stay cool during hot summer days (like nighthawks), poorwills are able enter torpor on a daily basis, or over the course of weeks, to survive cold weather and food shortages. Torpor is a state of inactivity characterized by reduced metabolic rate, slow heartbeat, and body temperatures down to about 40°F. A few other birds enter torpor overnight, but only poorwills are known to enter a prolonged state of dormancy (similar to hibernation when it occurs in winter) that may continue for more than three weeks. One wild bird remained torpid for 88 days, maintaining a body temperature of 65°F—about 40° below the normal temperature of an active poorwill. Only a few torpid Common Poorwills have been reported in California, but it is possible that some use this ability to spend the winter here.

STATUS AND DISTRIBUTION Based on reports (mostly of calling birds), poorwills appear in the southern Sierra in mid-April and remain until early October. Northern Sierra dates range mostly from May through September. A pulse of migrants is encountered in October at many low-elevation sites in the lower foothills and even on the Central Valley floor.

West Side. Fairly common and regularly observed, or heard, in open, shrubby, or rocky habitats from the Foothill zone to the Upper Conifer zone (around 6,000 feet); also possible in open forests or forest clearings, especially open stands of junipers on open ridgelines.

East Side. Fairly common breeders in sagebrush and greasewood habitats of the flatlands; uncommon on rocky hillsides of the Subalpine zone.

SWIFTS • Family Apodidae

No other birds are so lightly tethered to earth as the swifts. They spend nearly every waking moment of their lives on the wing, and even other aerial gymnasts, like nighthawks and swallows, cannot match the speed and daring maneuvers of the swifts. With their highly streamlined bodies and long, pointed wings, these are aptly named birds. Swifts are superficially similar to swallows in habits and appearance, but swallows are passerines, or "perching birds," while swifts are related to hummingbirds and share similar internal wing structures and tiny feet. Incapable of "perching," swifts can only cling. They feed, drink, gather nesting material, and even copulate on the wing. Their weak legs make it almost impossible for them to land on or launch themselves from branches or the ground; instead, they cling to vertical rock faces or the trunks of large trees.

Nesting material is difficult for swifts to gather, so their nests are usually sparse accumulations of mosses, ferns, and twigs glued together with saliva. In the Sierra, Vaux's and White-throated Swifts usually lay four or five eggs, but Black Swifts only lay one. Swift eggs are white and unmarked and elongated to roll in circles to prevent them from falling off cliffs. While information from the Sierra is lacking, elsewhere both parents incubate the eggs for about 20 to 25 days.

These highly social birds typically leave their nesting and roosting sites early in the morning and spend the entire day traveling long distances, perhaps hundreds of miles, in search of flying insects. Swifts seek out areas where updrafts along canyon walls, cliff faces, and ridges force insects high into the air. They favor swarming insects like flying ants because they contain significant quantities of fat but also eat flies, beetles, bees, and other insects. During periods of cold or rainy weather, swifts enter torpor and remain in their roosting cavities rather than draining their energy in a fruitless search for food. This family is represented by about 80 species that occur worldwide, except for the Polar regions and on some oceanic islands; 3 species are regular nesters in the Sierra. The family name means "footless" from L. *a*, without, and *pous*, a foot.

Black Swift • *Cypseloides niger*

ORIGIN OF NAMES "Black" for the adult plumage; swift (see family account above); *cypseloides* from L. *cypselus* (the original scientific name for European Swifts), and Gr. *eidos*, similar to; L. *niger*, black.

NATURAL HISTORY Few birds are so particular in their choice of nesting sites as Black Swifts or have preferences so grand and majestic. All Sierra nesting colonies (usually 5 to 15 birds each) are on moist cliffs near or behind great waterfalls such as Yosemite or Bridalveil Falls in Yosemite National Park, or Hamilton and Marble Falls in Sequoia National Park. Due to the inaccessibility of their waterfall nesting sites, the first Black Swift nest was not examined until 1901. The discoverer, a little known egg collector named A. G. Vrooman (cited in Dawson 1923), was surprised to find only one egg in the nest, and later searches of other nests confirmed that they only lay one. The single egg is large—about three times the size of a White-throated Swift's. Nestlings take an extraordinarily long time to fledge, partly because their nests are located in such cold damp places and because they must go without food each day while their parents hunt. Adults return at night to feed their young regurgitated food pellets. Fledglings take their first flights 45 to 50 days after hatching.

Black Swifts may spiral thousands of feet into the air pursuing clouds of insects. These sooty-black acrobats are larger and fly with slower, less erratic wingbeats than White-throated Swifts, which often cruise with them and can be distinguished from Vaux's Swifts by their larger size and longer bodies.

STATUS AND DISTRIBUTION While Black Swifts are widespread in California during migration, they are known to breed in only a handful of sites near large waterfalls, mainly on the West Side. Based on the few known nesting sites, the Sierra population may be fewer than 50 pairs. However, recent intensive surveys of potential habitat in the southern Rocky Mountains found many previously unsuspected nesting areas, more than tripling the number of known sites in that region and raising the possibility that many Sierra sites are yet to be discovered. Black Swifts arrive at the end of April and leave by mid- to late August, when they might be seen up to the Sierra crest. Evidence suggests the California breeding population mostly winters in northern South America, possibly the western Amazon Basin, where birds from the Rocky Mountains were very recently shown to winter.

West Side. Uncommon and highly localized breeders between 3,000 to 7,500 feet; foraging birds most likely observed from the Lower to Upper Conifer zones; the largest Sierra breeding population is in Yosemite Valley, but known and suspected breeding areas from other isolated river gorges exist from Butte County south to Tulare County.

East Side. Rare breeders and migrants; only known nesting sites are at Rainbow Falls in Devil's Postpile National Monument (Madera County), west of Woodfords (Alpine County), and probably in Inyo County.

TRENDS AND CONSERVATION STATUS Black Swifts were included on the list of California Bird Species of Special Concern (of 2008) primarily because their breeding population is so small. Their inaccessible nest sites on wet, high, vertical surfaces make them relatively immune to human disturbance. Changes in climate that might affect insect prey population and possible but unknown threats on the wintering grounds could affect the Sierra population.

Vaux's Swift • *Chaetura vauxi*

ORIGIN OF NAMES John K. Townsend (see Townsend's Solitaire account) named this swift in honor of William S. Vaux (1811–1882), who donated his huge collection of natural history materials to the Philadelphia Academy of Natural Sciences (and pronounced his name *vauks* not *voh*); *chaetura*, bristle-tailed, from Gr. *khaite*, bristle, and *oura*, a tail.

NATURAL HISTORY These smallest of Sierra swifts are also the plainest. Though originally applied to the eastern Chimney Swift, Roger Tory Peterson's (1990) description "like a cigar with wings" works for Vaux's Swifts as well. With rapidly beating wings, they look somewhat like stiff-winged bats as they dart and swoop after flying insects. During the day, Vaux's Swifts cruise over mature forests, meadows, and meadow edges. Most birds observed in the Sierra are passing through in migration to other areas, but some remain to breed. They nest in cavities in large conifers and, like other swifts, glue tiny nests to the vertical surfaces with their saliva. Few nests have been observed in the Sierra, and little is known about their breeding biology. After breeding, Vaux's Swifts sometimes gather into huge flocks and most observations are of migrating birds. In some places thousands of birds may gather in September, forming temporary roosts in chimneys, smokestacks, or other suitable cavities.

STATUS AND DISTRIBUTION Vaux's Swifts breed in coniferous forests as far north as Alaska and winter in Central and South America. Most Sierra observations of Vaux's Swifts are in spring (April–May) or fall (mostly September) migration. The majority of breeding probably occurs on the West Side, where larger, hollow snags provide the best nesting sites.

West Side. Uncommon, migrants appear at almost any elevation and over any habitat; based on observations in the breeding season, nesting appears to be limited to forested areas between 3,000 and 7,000 feet as far south as Tulare County.

East Side. Uncommon to fairly common in migration; rare in breeding season but probably breeds in pine forests in the Tahoe Basin and elsewhere in Placer, Nevada, and Sierra Counties.

TRENDS AND CONSERVATION STATUS Since they require large snags for nesting, Vaux's Swifts are adversely affected by forest practices such as sanitation cuts and fuel management activities that remove large trees and snags from mature conifer stands. Due to these practices, large, hollow stumps are becoming increasingly rare in the Sierra and in redwood forests of the Coast Range. For these reasons, Vaux's Swifts were included on the 2008 list of California Bird Species of Special Concern.

White-throated Swift • *Aeronautes saxatalis*

ORIGIN OF NAMES "White-throated" for the species' distinctive field mark compared with other swifts; *aeronautes*, an air sailor, from Gr. *aer*, air, and *nautes*, a sailor; L. *saxatalis*, rock dwelling.

NATURAL HISTORY "Swifter than swift is the White-throated Swift!" As William Dawson (1923) exclaimed, few North American birds can match the velocity and aerial virtuosity of White-throated Swifts. One was observed in Yosemite Valley outracing a Peregrine Falcon that was diving at a speed estimated at more than 175 miles per hour! These highly social swifts frequently mingle with Violet-green and Tree Swallows and other swifts, but they always stand out because of their fast, dexterous maneuvers, larger size (except for Black Swifts), and white throats contrasting brightly against a cliff or other dark background. They can often be located by listening for their distinctive, descending chittering calls.

Although closely associated with cliffs and rocky canyons, where they nest and roost, White-throated Swifts may fly many miles during their daylong foraging bouts and can be seen above almost any landscape. They best display their uncanny flying precision when returning to a roosting site, usually entering a tiny hole or crack at tremendous speed, folding their wings at the last possible instant as they disappear inside. When the young are ready to make their first flight, it is a do-or-die proposition, since once on the ground, they are incapable of getting back into the air. One author found such an unlucky young swift one day in late summer along the American River. The bird was unharmed and clung to a finger with remarkable tenacity. Once lifted up and turned into the breeze, it immediately flew off, hopefully to begin a long and mostly aerial life.

STATUS AND DISTRIBUTION The seasonal occurrence of White-throated Swifts is complex because their population includes both migratory breeders and year-round residents. Where they are resident, many can be seen on warm winter days, but they seem to disappear during long cold spells. These birds probably go into torpor on cold days, but they could conceivably leave the area and then return during warm periods. Migrants apparently begin to arrive in mid-March at low elevations, then at higher elevations in mid-April to mid-May, and remain until the end of September.

West Side. Fairly common to abundant in appropriate nesting habitats in rocky canyon walls, cliffs, and within bridges and other human structures from the foothills to about 9,000 feet but highly localized and many potentially suitable areas are not occupied; occurs year-round in the lower Foothill zone but generally rare to uncommon in winter except in the Kern River Valley, where hundreds have been recorded on Christmas Bird Counts.

East Side. Fairly common breeders in Mono County and the Owens Valley; uncommon to rare farther north; in winter, rare to casual in the Owens Valley, accidental farther north.

HUMMINGBIRDS • Family Trochilidae

With whirring wings and shimmering iridescent plumage, these tiny aerial wizards bring a sense of magic to Sierra meadows, rock gardens, and scrublands. Both hummers, and the flowers they visit, have coevolved to enhance their mutualistic relationships. Many "hummingbird flowers" have internal structures that facilitate pollination by hummers. Some, such as penstemon and paintbrush, hide their nectar from insects deep in tubular "throats" that effectively reserve it for hummingbirds. Others, like manzanita, have small flowers with narrow openings and no attractive landing places for bees or other pollinators that might compete with hummers. Their small size, long bills and tongues, and ability to hover and fly backward enable efficient foraging on nectar and pollen as well as on insects and spiders found in flowers. Hummingbird flowers are often red, explaining the well-known attraction of hummers to red hats, shirts, and feeders, which they defend aggressively from competitors.

Hummingbirds have the fastest wingbeats of all birds—50 to 75 beats per second for most small and medium-sized species—and they are the only ones capable of backward flight. An example of convergent evolution, sunbirds of Africa and Eurasia (members of the Passerine family Nectarinidae) have many of the same traits, including similar foraging behaviors and brilliant iridescent plumage, but they are unrelated to hummers and cannot fly backward.

Most Sierra hummers, except for Anna's, migrate long distances to winter in the tropics, and they store up to half their body weight in fat prior to attempting this journey—sometimes flying 500 miles in a single night. To sustain such exertion over long periods, hummingbirds have evolved the highest metabolic rates of any vertebrate; consequently they must conserve energy whenever possible. On cold nights, rather than depleting their fat reserves to stay warm, many hummingbirds enter a state of torpor, a brief hibernation, with greatly lowered body temperatures and metabolic rates. When active, most hummers feed 5 to 10 times per hour for bouts of 30 to 60 seconds.

Male hummingbirds are admired by people, and by female hummers, for their iridescent "gorgets," or throat feathers, and for their spectacular diving displays, used during aggression and courtship. These striking colors are not due to any pigment but result from a complex interference phenomenon (not unlike that seen in a film of oil on water) produced by the internal structure of the feathers. Variations in the structure from species to species produce the different colors characteristics of each.

Females exclude the males from their nesting territories. They line their tiny, deep cup-nests with plant down and wrap them in spider webs and other soft fibers. Sierra nesting species usually lay two unpatterned, white eggs that resemble jelly beans in size and shape. Females incubate the eggs alone for 14 to 18 days and assume the sole duties of caring for their naked, altricial young for about 20 to 25 days before they leave the nests.

Hummingbirds occur only in the Western Hemisphere and include a dazzling array of almost 320 species. The most spectacular species live in the tropics. Five species occur regularly in the Sierra, but eleven species have been observed, including the Ruby-throated Hummingbird (twice) and Blue-throated and Violet-crowned Hummingbirds (once each). The family name is derived from L. *trochilus*, a runner, applied by Linnaeus to this family for unknown reasons.

Male

Female

Black-chinned Hummingbird • *Archilochus alexandri*

ORIGIN OF NAMES "Black-chinned" for their dark-colored chins and throats in all lighting conditions; "hummingbird" for the sound of their whirring wings; *archilochus*, from Gr. *archos*, a chief, and *lochos*, to ambush, an unclear but possible reference to stealing pollen from flowers before other hummers find it; *alexandri* for Dr. M. M. Alexandre, who lived and practiced medicine in Mexico and was the first to collect this species, and many other birds, in the Sierra Madre in the mid-1840s.

NATURAL HISTORY Every male Sierra hummingbird has a jewel on his throat, but the Black-chinned Hummingbird's is hardest to see. Under most conditions his throat simply looks black, but the proper lighting and angle reveal a stunning blue-violet band below his truly black chin. Black-chins nest in deciduous streamside woodlands lining the bottoms of dry foothill canyons or in irrigated orchards and gardens. In general they prefer moister, riparian habitats than do Anna's, the most common of foothill hummingbirds. Females usually build their nests in low foliage over a stream or dry creek bed. Males occupy nearby sites, typically upslope in tall, broken chaparral mixed with live or blue oaks. When displaying to intruders or potential mates, they "swing" back and forth through the air in wide, shallow arcs.

Promiscuous, both sexes copulate with more than one mate and are together only for a few copulations. In the Sierra, nesting begins in mid-May and extends through mid-July; males arrive in breeding habitats before the females and also depart sooner. After about mid-July, the majority of Black-chins in the Sierra are either females or juveniles. Females construct their nests almost entirely from spider webs. They shape their nests like a potter forming a bowl at a wheel, whirring around and molding nest materials with their breasts. Their diet is mostly nectar from larkspurs, paintbrush, and sage as well as sugar water from feeders, but they also take small spiders and insects when encountered on preferred flowers.

STATUS AND DISTRIBUTION Black-chinned Hummingbirds arrive in the Sierra foothills in early April and most depart for Mexico by late August or early September. They tend to be more numerous in the southern part of the range and less common in the northern Sierra.

West Side. Fairly common in the Foothill zone in areas where riparian habitats and chaparral predominate; in July and August a few move up into the Lower or Upper Conifer zones, or even higher; most often seen at feeders at all elevations.

East Side. Rare spring and summer visitors up to about 7,000 feet (e.g., Tom's Place, Mono County, Reno area), where mostly confined to feeders, riparian areas, and landscaped gardens; casual visitors to higher-elevation meadows in late summer.

Anna's Hummingbird • *Calypte anna*

ORIGIN OF NAMES "Anna's" for Anna de Belle Massena (1806–1896), an Italian duchess whose husband, Prince Victor Massena, collected a specimen of the species; the name was given to the bird in her honor by the French naturalist René Primevère Lesson (1794–1849); John James Audubon described Anna as beautiful, graceful, and polite; *calypte*, possibly from Gr. *kalypira*, a hood, a reference to the male's iridescent cap.

NATURAL HISTORY Male Anna's Hummingbirds display their aerial prowess to the fullest when intimidating trespassing hummingbirds at flower patches or feeders. First they hover about 10 feet

Female

Male

above the intruder, orienting toward the sun to display their brilliant red crowns and gorgets. They mount slowly in wavering vertical paths to heights of 120 feet or more. Suddenly dropping into steep power dives at other hummers, they pull out just in time and loop up to the starting point to hover and buzz again. Air rushing through their specially adapted tail feathers makes a startling sound at the base of the dive, a sharp *peep* that resembles the alarm notes of California ground squirrels. Males may repeat their J-shaped dives several times in sequence, directing them at intruding males and at prospective mates alike. Apparently fearless, they attack birds many times their size, even the usually truculent Western Scrub-Jay. One naturalist observed and photographed a particularly aggressive male near Nevada City grabbing a competing male from a feeder, pulling him high into the air with his feet, then diving and body-slamming his competitor to the ground, eventually pecking his eyes out and killing him.

More Anna's breed in the Sierra than any other hummer, and they are the only members of their family to remain year-round in the foothill woodlands, chaparral, and gardens. For this reason they require plentiful supplies of winter-blooming flowers, such as manzanita, gooseberry, and currant, as well as artificial feeders. They are also drawn to many ornamental flowering plants as well as eucalyptus groves. More than other Sierra hummers, they need large numbers of small insects and spiders to sustain them during the cold winter months when most plants are dormant and not providing nectar, and when many people forget to maintain their feeders.

Anna's Hummingbirds are among the few California birds that begin nesting in winter, and egg-laying begins in late December. Females prefer dense stands of live oaks, where they build tiny nests with lichens and spider webs and camouflage them in dense shrubs or trees. Meanwhile, males establish territories in open woodlands or chaparral and advertise them with distinctive, buzzy notes, uttered from a perch, behavior that can be noted in every month of the year at lower West Side locations. Females initiate courtship by entering a male's territory to feed. Several chases and displays ensue, eventually ending up in the female's nesting area, where the pair copulates. The male soon departs, leaving the female to raise the family alone.

STATUS AND DISTRIBUTION Historically Anna's Hummingbirds were restricted to the western Sierra foothills and coastal California from Baja California north to about the Bay Area, but ornamental gardens and hummingbird feeders have enabled them to expand their breeding range north as far as southern Canada and east to at least Arizona since the 1930s. They are the only hummingbirds present year-round and the most common members of their family in the Sierra. Anna's primarily frequent oak woodlands, chaparral, and mountain meadows with plentiful flowers and blooming shrubs; they are also the most frequent visitors to feeders and landscaped gardens of the Sierra.

West Side. Common in the Foothill zone year-round, and higher in areas with hummingbird feeders; from April through September, postbreeding birds regularly move upslope to the Upper Conifer zone or even higher, usually choosing arid brushy habitats but also mountain meadows.

East Side. Uncommon summer residents of the Donner Summit region (Placer and Nevada Counties) and Reno, and rare in southern Mono County and the Owens Valley; and uncommon fall visitors to riparian woodlands up to the Subalpine zone; observed in September and October and sometimes through winter, especially where people keep feeders full.

TRENDS AND CONSERVATION STATUS Historical range expansion and population increases of Anna's Hummingbirds appear to continue up to the present. Numbers on Sierra Breeding Bird Survey routes and from Sierra Christmas Bird Counts show highly significant positive trends. The species is now routinely recorded on half the Breeding Bird Survey routes, up from less than 25 percent in the 1970s, with numbers of birds per route having also doubled.

Male   Female

Costa's Hummingbird • *Calypte costae*

ORIGIN OF NAMES "Costa's" for Louis Marie Pantaleon Costa de Beau-Regard (1806–1864), a French archaeologist and naturalist who specialized in studying and collecting hummingbirds; in 1839 his name was attached to the bird to honor his contributions by Jules Bourcier (1797–1873), who named a number of other hummingbirds, including several tropical species during his residence in Ecuador.

NATURAL HISTORY Costa's Hummingbirds are among the most diminutive hummingbirds in the Sierra, only the Calliope is smaller. In the southern Sierra they usually arrive in late February or early March when desert flowers are blooming. In the early season they consume nectar from creosote bush but later forage extensively on ocotillo, yucca, and saguaro as well as willow buds and flowers along desert washes. They also glean small insects and spiders from desert plants, and they are adept at fly-catching small insects on the wing. When courting females, male Costa's employ elaborate flight displays including numerous oval-shaped loops around the perched female, always orienting toward the sun to show their gorgets to best advantage. They punctuate these vigorous dives with high-pitched whistles near the end of the loop; these vocalizations are also given when males perch in the female's territory.

STATUS AND DISTRIBUTION Costa's Hummingbirds are almost entirely confined to hot, arid scrublands and deserts during the breeding season, and the northernmost extension of their breeding range barely extends to the southern Sierra. They are most likely to be observed during their breeding season during March and April; they are mostly absent from June through October, when they abandon their desert breeding grounds for moister coastal areas of southern California and western Mexico where flowering plants are plentiful. Since females and juvenile Costa's can be confused with Anna's and Black-chinned Hummingbirds, Costa's may be more frequent visitors to northern areas on both slopes than the few existing records suggest.

West Side. Uncommon in the South Fork Kern River and Piute Mountains; casual or accidental farther north but can show up in any season.

East Side. Uncommon to fairly common in early spring in eastern Kern and Inyo Counties, where Desert zone habitats predominate; rare from early May to early June north to at least the Mono Basin.

Calliope Hummingbird • *Stellula calliope*

Female

Male

ORIGIN OF NAMES "Calliope" from an ancient Greek rhyme meaning "a beautiful voice"; unclear where the name came from since the bird's voice is not beautiful or particularly distinctive; L. *stellula*, little star, a reference to the bird's small size and distinctive, starlike gorget.

NATURAL HISTORY Despite their tiny size, smaller than any other bird north of Mexico, Calliope Hummingbirds somehow survive the chilling temperatures of summer nights in the high Sierra. Males probably conserve energy by reducing their body temperature at night, but females on the nest must maintain normal temperatures to incubate their eggs and protect their featherless young. Females stay warm by snuggling down in their thick nest insulation and by situating their nests strategically; they place them on branches directly beneath other limbs or sprays of foliage, thus reducing radiation of heat to the cold night sky.

Calliope hummers frequent a variety of habitats, including montane chaparral, open forests near streams, seeps, or meadow edges as long as they provide good supplies of "hummingbird flowers" such as gooseberry, currant, manzanita, penstemon, and paintbrush. They often take sap from sapsucker drillings or forage in rock gardens amid scattered Jeffrey pines and junipers on sunny slopes and ridges. From early May through June, males defend a patch of flowers and chase away other intruding hummers. They make impressive flight displays in the form of a broad, vertical U, with a power dive down one side, a short, quiet buzz at the bottom, and a gradual ascent up the other side. Usually directed at females, flight displays probably play a role in both courtship and territorial defense. At feeders, visited mainly in migration, Calliopes tend to be the least aggressive hummer.

STATUS AND DISTRIBUTION Calliope Hummingbirds have the distinction of being the smallest long-distance migrants of any of the world's birds, making annual migrations from central Canada to southern Mexico. They arrive in the Sierra by mid-April, and by early July most males have departed for their wintering grounds. The females and young follow later and are mostly gone by mid-August.

West Side. Fairly common nesters mainly from the Lower Conifer to the Upper Conifer zone, the only hummers that regularly breed above the foothills; postbreeding birds move upslope to the Subalpine and Alpines zones in July and early August prior to fall migration; uncommon in spring migration in the Foothill zone.

East Side. Fairly common breeders but mostly restricted to moist meadows, aspen thickets near small streams, and brushy canyons up to 9,500 feet; rare in fall, a few pioneer up to the Alpine zone (up to 10,000 feet) in late July and early August before heading south by mid-September.

Female

Male

Rufous Hummingbird • *Selasphorus rufus*

ORIGIN OF NAMES "Rufous" for the colors of breeding males; *selasphorus*, light-bearing, from Gr. *selos*, light, and *phoros*, to bear; L. *rufus*, reddish.

NATURAL HISTORY Because they are so common in mountain meadows and around hummingbird feeders in summer, when other hummingbirds might still be breeding, early ornithologists assumed Rufous Hummingbirds nested in the Sierra. Further study revealed that they only pass through the region in migration and do not breed in California. While Calliope Hummingbirds have the distinction of being the smallest of the world's long-distance migrant birds, the slightly larger Rufous Hummingbirds are known to undertake the longest migratory journeys in proportion to their size—they travel annually from southern Alaska to southern Mexico and back.

Due to the extreme energy demands of their long migration, they can lose almost half their body mass without regular refueling on nectar. Rufous are among the most aggressive hummers at feeders and favored flower patches, and they routinely engage in flight chases of competitors. In addition to foraging on typical "hummingbird flowers" such as penstemon, Indian paintbrush, columbine, and scarlet gilia, they hawk considerable numbers of flying insects in midair as a source of protein to replenish muscle mass lost during their exertions. Migrants are slightly lighter in spring, suggesting that they encounter fewer rich nectar sources than are available on their return trips.

STATUS AND DISTRIBUTION Outnumbering all other hummers at many locations in summer and fall, these tiny migrants linger in meadows, brushfields, and rocky slopes with plenty of favored blooms, up to well above tree line. The southbound "fall" migration starting in late June is led by a wave of adult males, for they are free to leave the breeding grounds once their limited role as sperm donors is complete. By mid-July migration is in full swing when females and juveniles begin to appear, and by early August adult males are rare, but females and young are common through at least early September. The species appears to time its migration routes to take advantage of peak blooming seasons, traveling at lower elevations northward in spring and higher elevations southward in summer.

West Side. Locally common spring migrants through the Foothill zone from mid-March through April; common to locally abundant from the Lower Conifer zone to the Alpine zone from late June through August; fairly common at high elevations until mid-September.

East Side. Nearly identical to West Side in terms of timing, habitats, and relative elevational distribution.

KINGFISHERS · Family Alcedinidae

Their bright plumages and loud vocalizations make kingfishers conspicuous, as does their habit of perching on open snags or branches near streams or lakes. About 90 species exist in temperate and tropical forests of the world, but the Belted Kingfisher is the only member of this family in the Sierra. The medieval name "kingfisher" means "chief of the fishes"—the primary prey of most species in this family. The family name is derived from L. *alcedo*, a kingfisher.

Belted Kingfisher · *Megaceryle alcyon*

Male

Female

ORIGIN OF NAMES "Belted" for the bright chestnut breast band of adult females; kingfisher (see family account above); Gr. *megaceryle*, large kingfisher; *alcyon* from Gr. *halcyon*, also meaning kingfisher.

NATURAL HISTORY A harsh, grating rattle announces the passage of Belted Kingfishers as they fly with irregular wingbeats and glides above Sierra lakes, ponds, and streams. When perched, their double-crested heads and massive bills seem out of proportion to their short tails and tiny feet. Unlike most sexually dimorphic birds, females are the more brightly colored of the two, with a broad, chestnut band beneath a darker upper band across their breasts and flanks.

Kingfishers hunt by flying over water and scanning from exposed perches on snags, limbs, bridges, and wires. They prefer areas with clear, shallow water where they can easily spot prospective prey. They attack fish and other aquatic prey with dramatic dives from perched or hovering positions, often disappearing briefly underwater before returning to a favorite spot to kill and eat their catch. Kingfishers kill unruly prey with sharp blows from their daggerlike bills or by beating them against branches or rocks. Adults mainly consume small fish, but crayfish and tadpoles are also taken. Solitary hunters for most of the year, both males and females loudly and aggressively defend foraging territories.

Mated kingfishers remain together only through the breeding season, from late March through mid-July in the Sierra. They excavate nesting burrows in steep stream banks of clay or sand near productive fishing areas. Both parents work together, and the entire digging process may take up to three weeks. These horizontal holes are one to three feet below the tops of banks and six or seven feet deep, rarely up to fourteen feet. The burrows slope upward, leading to the nesting chamber, possibly to keep eggs dry if floodwaters rise. Females usually lay six to eight glossy white eggs, and both adults take turns incubating them for about 24 days. Featherless, altricial young do not open their eyes for about two weeks. Both parents feed nestlings partly digested, regurgitated fish and occasionally large insects such as dragonflies. Young kingfishers remain in the nesting burrows until fully feathered and capable of flight after about five weeks.

STATUS AND DISTRIBUTION Belted Kingfishers are year-round residents the entire length of the Sierra where water remains unfrozen on both sides of the Sierra.

West Side. Fairly common at lower elevations from the Foothill zone up to the Lower Conifer zone; uncommon higher up but a few birds follow streams up to tree line after breeding; most move downslope and southward in winter.

East Side. Uncommon or rare residents and breeders at lower elevations; some move upslope in late summer and fall to forage in Subalpine lakes and streams.

WOODPECKERS • Family Picidae

As their name suggests, woodpeckers are admirably equipped for drilling in wood. They excavate nest cavities and winter roosts with their powerful bills and drill holes or flake off bark in search of insects. Special cushioning protects their brains from shock, and an unusual bone structure secured to the back of their skulls support their extremely long, barb-tipped tongues that are used to extract insects or sap from holes. With strong feet and stiff tail feathers, they brace themselves vertically on trees while pecking.

Woodpeckers' bills are also used for "drumming," pecking loudly in specific patterns to communicate with mates or territorial rivals. In some species drumming is more important than vocalizing and is equivalent to the singing of other birds. Woodpeckers generally peck rhythmically in rapid-fire patterns when drumming and much more slowly and erratically when feeding. Woodpecker-created cavities are used by a host of other Sierra animals, such as small and medium-sized owls, nuthatches, chickadees, chipmunks, gray squirrels, Douglas squirrels, and flying squirrels as well as some reptiles and amphibians.

All Sierra woodpeckers lay about four to six white, glossy eggs. The eggs are unspotted, probably because nesting in cavities precludes the need to camouflage them. Both parents share incubation duties for 12 to 18 days, with males generally taking the night shifts. The altricial young of all species are naked and featherless when hatched. They require constant feeding of regurgitated food while in the nest cavities and often emit harsh chatters that betray their locations. Young do not leave the nests until about 25 to 30 days, and the adults feed them for at least several weeks until they gain independence.

About 200 species represent this family worldwide, and woodpeckers are found almost everywhere except some oceanic islands and the Polar regions and, curiously, Australia. Sierra forests, scrublands, and deserts support 12 nesting species, with up to 7 co-occurring in some western conifer forests. There may be no other place in North America where so many woodpecker species can be seen in a single day of birding. Most Sierra species are nonmigratory, and several remain to brave the fierce winters in the highest forests. The family name is derived from L. *picus*, a woodpecker.

Lewis's Woodpecker • *Melanerpes lewis*

ORIGIN OF NAMES "Lewis's" for Meriwether Lewis (1774–1809), a personal assistant to President Thomas Jefferson and coleader of the Lewis and Clark Expedition (1803–1806); he was the first to describe this bird in writing—apparently one of the few bird specimens from the expeditions that still exists; Gr. *melanerpes*, black creeper, from *melanos*, black, and *herpes*, a creeper.

NATURAL HISTORY Lewis's Woodpeckers are stunning but peculiar-looking woodpeckers, with rose-pink bellies, bright-red faces, and dark, iridescent-green backs and wings. Unlike most other Sierra woodpeckers that fly in undulating, up-and-down patterns, they proceed on a steady, level course, appearing more like small crows in flight. These woodpeckers are irruptive, meaning that their numbers change unpredictably from year to year in response to weather conditions and annual variations in local food supplies.

Most Lewis's Woodpecker breeding pairs reuse nest holes from previous years or use natural cavities, often situated at the bases of the lowest large limbs, where they line their nests with small chips of wood. They winter in oak savanna and open oak woodlands of the Foothill zone with good crops of acorns, their staple food at that season. Like Acorn Woodpeckers, they store acorns, but Lewis's use natural crevices rather than drilling their own holes. They defend acorn caches individually rather than in groups and may spend more than 40 percent of their time guarding these stores. They also relish such large insects as dragonflies, grasshoppers, ants, beetles, and bees, capturing them in midair or swooping down to snatch them from shrubs and grasses. Sometimes they act like large swallows, chasing one insect after another in long acrobatic flights before returning to a perch.

STATUS AND DISTRIBUTION   Curiously, Lewis's Woodpeckers nest in the interior Coast Range but not in similar habitats in the foothills of the western Sierra. They are fairly common nesters east of the crest, where they prefer open stands of ponderosa pines with at least a few shrubs, in flat, open country far from steep canyons. In spring and fall, they migrate across the Sierra between breeding grounds in open forests east of the crest and their wintering areas in the western foothills and beyond. Wintering populations swell in some years, when acorns are abundant, as others move in from the Cascades and other northern breeding grounds.

West Side. In some years common and widespread in these fall-winter irruptions in the Foothill zone, rare or absent in others; usually found in open stands of blue and valley oaks with highest concentrations generally in the northern foothills (e.g., Spenceville Wildlife Area and western Tehama County); in exceptional years many reach higher elevations during late summer and fall; migrants numbering in the thousands were sighted in September 1925 crossing an 11,000-foot pass near Foerster Peak in Yosemite National Park and again in the fall of 1932 passing through Yosemite Valley (Gaines 1992).

East Side. Fairly common nesters in open Jeffrey and ponderosa pine forests from mid-May through early September; postbreeding birds irregularly seen in numbers in pinyon-juniper and pine forests up to tree line; may be seen at ranches with clumps of trees well into October.

Acorn Woodpecker • *Melanerpes formicivorus*

Female

Male

ORIGIN OF NAMES   "Acorn" for the species' staple food; L. *formicivorus,* an ant eater, from *formica,* ant, and *vorare,* to eat or consume.

NATURAL HISTORY   The Acorn Woodpecker's familiar *whacka-whacka* calls provided the inspiration for Walter Lantz's creation of Woody the Woodpecker, although this cartoon figure more closely resembles the much larger Pileated Woodpecker. Among the most familiar of Sierra woodpeckers, these clamorous, clown-faced birds have a habit of caching acorns, which are central to their individual and social existence. They prefer areas with two or more species of oak because that makes the acorn supply more dependable. Each kind of oak has bad acorn years, but it is unusual for several species to have crop failures in the same year. Unless the acorn supply is exhausted, Acorn Woodpeckers remain on their territories year-round. They are most common in open oak woodlands of the foothills, but large numbers also live in the Lower Conifer zone in areas where black oaks predominate.

The tight-knit social groups of Acorn Woodpeckers are unique among Sierra birds. Others may nest in loose colonies or feed in flocks, but only this species forms cooperative breeding groups consisting of up to seven males competing to mate with one to three egg-laying females. Up to ten additional adults, including offspring from previous years, serve as "helpers-at-the-nest," bringing food to the young birds and sometimes helping excavate the nests. Two closely related females may use the same nest hole and will aggressively exclude other unrelated females by pecking "foreign" eggs or killing the young of intruders. In this promiscuous mating system, several males mate with the same females and vice versa; males are known to disrupt the copulations of other males, relatives or otherwise.

In the Sierra foothills Acorn Woodpeckers primarily nest from early April through early June, but breeding can extend through September. Once settled in the nest, all members of the group incubate the eggs until they hatch and feed young regurgitated insects until they leave the nest. After fledging, the young remain on the territory for at least a year, and some never leave. A central feature of each group territory is the communal granary, usually a huge old oak or conifer snag, but sometimes a utility pole or group of fence posts, where group members stash acorns and defend them against squirrels, jays, and Acorn Woodpeckers from other colonies. A single granary may contain more than 10,000 acorn-sized holes drilled by the woodpeckers and stuffed with acorns in the autumn. Although acorns are the main food from late summer through winter, insects and sap are more important in spring and early summer. Unlike most woodpeckers, Acorns rarely drill for insects but will instead take dragonflies and bees on the wing, flying up like oversized flycatchers from the tops of oaks. Rarely they take nestlings of other birds, and one family group of three adults in Yosemite Valley was observed robbing a female Western Wood-Pewee of all three of her nestlings.

STATUS AND DISTRIBUTION Acorn Woodpeckers are mostly confined to the western Sierra, but their enormous range extends from the Pacific Northwest south through Mexico and Central America to northern Columbia.

West Side. Common to abundant residents and breeders in open, oak-dominated woodlands of the Foothill and Lower Conifer zones up to about 4,500 feet in the central Sierra; stands of blue, valley, or black oaks with large conifer snags are preferred.

East Side. Resident in Lassen County where black oaks occur between Susanville and Lauffman, and in Inyo County near Independence and along Sage Flat Road; a small colony existed (at least historically) north of Mono Lake along Highway 395; casual elsewhere, but individuals may stray from the West Side areas in late summer and fall, especially in years of poor acorn crops.

TRENDS AND CONSERVATION STATUS Although data from Sierra Breeding Bird Survey routes and Christmas Bird Counts show no current trend, a combination of loss of oak habitat to development, lack of natural regeneration of blue and valley oaks, and competition with European Starlings for nest sites all suggest looming threats to this species. Acorn Woodpeckers have been extirpated from some parts of the San Francisco Bay area due to a combination of habitat loss and increasing competition with starlings. "Sudden Oak Death" syndrome, which has not yet reached the Sierra, adds another potential long-term threat.

Williamson's Sapsucker · *Sphyrapicus thyroideus*

Male

Female

ORIGIN OF NAMES "Williamson's" for Robert Williamson (1824–1882), a military engineer, Civil War veteran, and leader of a survey party to Oregon when a male of the species was first collected; "sapsucker" for the species' habit of drilling shallow holes to permit the flow of sap, a preferred food; *sphyrapicus* from Gr. *sphyra*, hammer, and *picus*, a woodpecker; L. *thyroideus*, shield-like, from Gr. *thyreos*, shield, and *eidos*, resembling, a possible reference to the female's black breast patch.

NATURAL HISTORY Unlike most other woodpeckers, male and female Williamson's Sapsuckers differ strikingly in appearance, so much that from 1852 to 1873 they were classified as separate species, and one early ornithologist even placed them in separate genera. Williamson's Sapsuckers drum in a pattern similar to other sapsuckers but distinguishable by its more rapid beginning and slower, more deliberate final taps: *ttttttt-tt-tt-tt-tt-t-t*. Like Red-breasts, they eat sap and cambium, but the Williamson's drill holes in irregular rather than neatly aligned rows. In the Sierra, Williamson's Sapsuckers often nest in old, living aspens with rotten heartwood or in dead or dying conifers such as lodgepole pines, western white pines, and red firs.

Unlike most woodpeckers that take circuitous routes to their nests to deceive predators, Williamson's Sapsuckers fly directly to their holes, making nests relatively easy to find. Nesting season extends from mid-May through late July. Newly excavated cavities are preferred as nest sites, and males chisel away for three or four weeks with some help from females before the job is done. While feeding young, adults specialize almost exclusively on ants gleaned from trees, logs, and the ground. In winter, when sap is barely flowing in Sierra conifers, they probably feed on dormant bark insects.

STATUS AND DISTRIBUTION In the Sierra, Williamson's Sapsuckers primarily frequent open Subalpine forests, especially those dominated by lodgepole pines. They generally breed at higher elevations than the Red-breasted Sapsuckers, but both species may nest in red firs of the Upper Conifer zone. Even there, most Williamson's Sapsuckers stick to forests where lodgepoles are numerous, as around wet meadows and on rocky slopes and ridgelines. The winter whereabouts of Williamson's Sapsuckers remains a mystery, partly because they are so quiet at that season. The few Sierra records come mostly from the Lower and Upper Conifer zones, suggesting a downslope migration when it gets cold. The paucity of sightings at higher elevations, however, may merely reflect the lack of winter observers there. Another possibility is that Sierra populations move south for winter; some Williamson's Sapsuckers of the subspecies that breeds in the Rocky Mountains *(S. t. nataliae)* migrate into California and south as far as northern Baja California and the Sierra Madre in northern Mexico. There are fall and winter records of this subspecies throughout California, including several Sierra records.

West Side. Uncommon residents from the Upper Conifer zone up to the Subalpine zone; rare but annual at low elevations in winter, a few representative locations include: Nevada City (Nevada County), near Mariposa (Mariposa County; 1,500 feet elevation), Tuolumne Grove (Tuolumne County), Yosemite Valley, near Bass Lake (Madera County), and Grant Grove in Sequoia National Park; in nearly all cases, it is unknown whether these low-elevation wintering birds represent the Sierra or Rocky Mountain subspecies.

East Side. Uncommon residents above about 6,000 feet and rare above 9,000 feet; breeding populations at Yuba Pass, Tahoe Basin, and canyons above the Mono Basin; winter status uncertain but rare or casual in most forested areas, suggesting that most migrate out of the Sierra and winter elsewhere; some fall records in lowlands of the Desert zone suggest migration to those southern California locations, but there are apparently no spring records from the desert.

Red-naped Sapsucker • *Sphyrapicus nuchalis*

ORIGIN OF NAMES "Red-naped" for the red patch on the backs of the species' head; *nuchalis* from Gr. *nucha*, back of neck.

NATURAL HISTORY Until 1983, and despite ongoing taxonomic controversy, Red-breasted and Red-naped Sapsuckers were lumped as subspecies, along with the eastern Yellow-bellied Sapsucker *(Sphyrapicus varius;* covered in Appendix 2) as part of a "superspecies." The three taxa have distinct distributions, and recent studies concluded that they are indeed three distinct species. Like their close relatives, Red-naped Sapsuckers are partial to deciduous woodlands that provide ample flows of sap, and they are often found in apple orchards for the same reason. The breeding and foraging behaviors of this species are nearly identical to the Red-breasted Sapsucker (see account below).

STATUS AND DISTRIBUTION Red-naped Sapsuckers primarily live in the Rocky Mountains, northern Cascades, and Warner Mountains, and they visit the Sierra irregularly, mostly in winter. They have nested in the eastern Sierra a few times, where they are known to hybridize with Red-breasted Sapsuckers as far south as Inyo County; most Sierra breeding records are from aspen groves. Because of the regular occurrence of sapsucker hybrids in the Sierra, any "Red-naped Sapsucker" should be examined carefully for evidence of hybridization.

West Side. Rare or casual fall, winter, and spring visitor to the Foothill and Lower Conifer zones; seen in most years on at least a few Christmas Bird Counts at foothill locations.

East Side. Casual in the nesting season, when mostly found in aspen groves and open ponderosa pine forests; a pair in Horseshoe Meadow, west of Lone Pine (Inyo County), provided the southernmost nesting record for the Sierra; more regular nesting just east of the Sierra in the White Mountains; casual visitors in winter and early spring with records from Sierra Valley and Mono Basin, where hybrid Red-naped × Red-breasted are found more frequently than pure Red-naped and hybrids are known to have bred.

Red-breasted Sapsucker • *Sphyrapicus ruber*

ORIGIN OF NAMES "Red-breasted" describes the overall coloration of their breasts and heads; L. *ruber*, red.

NATURAL HISTORY With their brilliant red heads and breasts, yellow bellies, white wing patches, and spotted backs, Red-breasted Sapsuckers are among the most beautiful and distinctive members of their family. Unlike most woodpeckers, males and females wear identical plumage. The resident subspecies in California and Nevada *(S. r. daggettii)* is less colorful and generally darker than their bright red northern relatives *(S. r. ruber)*.

Despite their vivid colors, Red-breasted Sapsuckers are relatively inconspicuous. They drill quietly, call infrequently, and only occasionally produce their distinctive, irregular drumming signals. They do, however, leave an unmistakable sign of their presence: small, shallow holes drilled in neat horizontal rows on living trees. Hardwoods are preferred over conifers for foraging, and apple trees are among their favorite targets. After drilling holes, sapsuckers return at frequent intervals to consume the oozing sap and any insects attracted to it. Some of their hard-earned food is "stolen" by hummingbirds, warblers, and other forest birds that have learned sapsucker holes provide a free lunch. Sapsuckers also eat considerable quantities of cambium—the soft, growing tissue under the bark of living trees. They return repeatedly to a favored tree until it is riddled with holes, occasionally girdling it to death. Still, serious damage is usually limited to a few trees. While they seek hardwoods and orchards for foraging, they are also drawn to incense cedars and giant sequoias, which produce considerable quantities of sap. In the nesting season, these woodpeckers supplement their diet by hawking insects in midair over meadows and forest openings. Their need to forage in living trees distinguishes them from other montane woodpeckers such as Hairy, Black-backed, White-headed, and Pileated that primarily frequent dead or dying trees or branches.

In the Sierra, Red-breasted Sapsuckers nest from mid-May through early August. Most nest cavities are excavated in dead conifers or in live ones with rotten heartwood or in larger aspens. After hatching, the adults care for their young for only a week or so after they leave the nest until they learn to eat sap for themselves. They often drill sap wells in small willows around mountain meadows.

STATUS AND DISTRIBUTION Many are resident in their range in the Sierra, but some individuals disperse upslope after nesting and most winter at lower elevations, below the heavy snow zone. No one knows for sure whether Sierra birds migrate south, but many winter in the foothills and Central Valley and south to Baja California, where warmer temperatures keep the sap flowing.

West Side. Fairly common and widespread residents in Lower Conifer zone and breeders into the Subalpine zone though uncommon amid the red firs of the Upper Conifer zone; usually found in rather open forests near streams or wet meadows; some postbreeding individuals move up to the Subalpine zone.

East Side. Fairly common residents of aspen groves and planted groves of deciduous trees and orchards in towns and ranches as well as conifers up to above 9,000 feet; rare to casual in winter.

Ladder-backed Woodpecker • *Picoides scalaris*

ORIGIN OF NAMES "Ladder-backed" for the distinctive black and white barring on back; *picoides* from L. *picus*, a woodpecker, and Gr. *eidos*, like or similar to; *scalaris* from L. *scale*, a ladder.

Female

NATURAL HISTORY Ladder-backed Woodpeckers of southwestern deserts closely resemble the Nuttall's Woodpeckers but have more white on their faces, a slightly different upper back pattern, and noticeably different calls. Ranges of the two species overlap slightly at the southern end of the Sierra, in the South Fork Kern River Valley, and nearby valleys and slopes where Joshua tree woodland abounds. "Cactus Woodpeckers," as some call them, is an apt name since they often forage and nest in large cacti of the Mojave desert.

Male

In the Sierra, Ladder-backs are almost entirely confined to Joshua tree woodlands, and rarely venture into riparian woodlands or pinyon-juniper forests, except when they border desert areas. They forage for insects and other arthropods on the surface of Joshua trees and excavate nests in the relatively soft inner tissues. Little is known about the natural history of Ladder-backs, as few detailed studies have been done on this species. However, from what is known, they initiate nesting in mid-March and most of the nest excavating is done by the males. Young remain in the nests for an unknown period and are tended by both parents until they emerge to forage and fend for themselves, usually by mid-July.

STATUS AND DISTRIBUTION Ladder-backed Woodpeckers occur almost exclusively in the Desert zone of the extreme southern Sierra.

West Side. Fairly common residents of Joshua tree woodlands in the South Fork Kern River watershed as well as similar habitats in the Piute Mountains.

East Side. Fairly common residents of Joshua tree woodlands in southern Kern County; rare as far north as Lone Pine (Inyo County).

Female

Male

Nuttall's Woodpecker • *Picoides nuttallii*

ORIGIN OF NAMES "Nuttall's" for Thomas Nuttall (see Common Poorwill account).

NATURAL HISTORY Nuttall's Woodpeckers are small but hard to overlook with their frequent and unmistakable rattling calls and resonant drumroll pecking in spring. Like their close relatives, Downy and Ladder-backed Woodpeckers, they use their small bills to probe in crevices and pry off bark in search of insects and only occasionally drill holes as the larger woodpeckers do. Unlike Downy's, which prefer to forage on smaller branches and twigs, Nuttall's concentrate their foraging efforts on tree trunks and large branches of oaks. In addition to gleaning bark for beetles, caterpillars, ants, and spiders, they sometimes capture insects on the wing; they also consume seeds as well as berries from elderberry and poison oak shrubs in fall.

Given a choice, Nuttall's Woodpeckers prefer to forage on oaks and to nest in riparian trees with softer wood such as willows, cottonwoods, and sycamores. Thus they are most common where both habitat types exist near each other, frequently near streams. In the Sierra, nesting begins in late March and extends through late June. Both adults excavate the nest cavity in dead trees or limbs. The young can fly after about 30 days, when they can forage for themselves.

STATUS AND DISTRIBUTION The global range of Nuttall's Woodpeckers is confined to oak woodlands and urban forests of California and northern Baja California. They are nonmigratory throughout this limited range. While nesting, this species is generally confined to the Foothill zone, a lower-elevation range than any other Sierra woodpecker. Within this zone they frequent almost all wooded habitats except pure stands of willows, a domain claimed exclusively by Downy Woodpeckers.

West Side. Common residents in the Foothill zone; uncommon breeders up to the Lower Conifer zone in areas where black oaks predominate; rare postbreeding individuals venture upslope to the Upper Conifer zone especially along watercourses; casual up to over 7,000 feet on the Kern Plateau in Kern County.

East Side. Rare in the Owens Valley and southern Mono County; casual fall visitors observed irregularly in the Sierra portions of Inyo and Mono Counties likely from the isolated breeding populations from the White and Inyo Mountains, east of the Sierra; individual winter and early spring records from the Yuba Pass region.

Downy Woodpecker • *Picoides pubescens*

ORIGIN OF NAMES "Downy," a reference to the smaller size and lack of "maturity" of this species compared with the similar but much larger Hairy Woodpecker; L. *pubescens*, entering puberty.

NATURAL HISTORY Smallest of the North American woodpeckers, Downy Woodpeckers are almost identical in coloration to Hairy Woodpeckers but have black barring on their white outer tail feathers as well as tiny bills—less than half the length of their heads. More than any other woodpecker, they stick to streamside groves of willows, alders, cottonwoods, and other broadleaved trees for feeding and nesting. Less often they forage in nearby oaks or, rarely, conifers. Orchards, especially apple, often substitute for their natural habitats. They also frequent residential areas and city parks with planted ornamental trees.

Female

In the Sierra, Downy Woodpeckers initiate breeding in late March or early April, and their breeding season extends to late July. They excavate their small nest and roost holes in heavily decayed dead trees or limbs, perhaps because firmer wood offers too formidable a challenge for their small bills. Both adults excavate the nest hole and trade off this duty after about 20 minutes of work. Their incubation period lasts for only about 12 days; the young are able to climb to the tops of nest cavities after 10 to 12 days but do not leave for another 3 weeks.

Male

When feeding, Downy Woodpeckers search for beetles, ants, spiders, and small snails by flaking off bark or prob-ing into woody crevices with their bills, rather like large, upright nuthatches. More than most other woodpeckers, they work on small branches and twigs, preferring live trees over dead ones. Downys' quiet, single-note *peek* calls, and even their hole-drilling, can be heard only from short distances. They also make a noisy rattle call, like the Hairy Woodpecker's but higher in pitch and descending at the end, and a loud, regular drumming pattern, also like the Hairy Woodpeckers. Downys produce these louder signals infrequently, however, making them difficult to track down.

STATUS AND DISTRIBUTION Downy Woodpeckers occur in deciduous forests across North America, and they have been recorded in every state and all Canadian provinces that border the United States. In California they are found year-round in riparian forests and other deciduous woodlands.

West Side. Fairly common year-round in the Foothill and Lower Conifer zones usually along stream courses with significant stands of riparian trees; more common in extensive stands; some probably move upslope to breed in higher conifer forests where they usually forage in deciduous understory trees and shrubs; some postbreeding birds, especially juveniles, may wander upslope to the Upper Conifer zone.

East Side. Rare residents typically found in aspen groves and orchards; small resident populations in Sierra Valley, Honey Lake, Tahoe Basin, Mono Basin, Mammoth Lakes (Mono County), and regular south at least to Inyo County.

Hairy Woodpecker • *Picoides villosus*

ORIGIN OF NAMES "Hairy," a reference to the species' long, white back feathering, which had a hirsute but combed appearance to early bird collectors; L. *villosus,* hairy or shaggy.

NATURAL HISTORY From the gray pines of the Foothill zone to the lodgepole pines of Subal-pine zone, Hairy Woodpeckers relentlessly chisel away the bark of conifers and riparian trees in pursuit of their prey. Even during harsh Sierra winters, when most birds have deserted higher elevations, these hardy generalists remain. Midwinter surveys of lodgepole pine forests sur-

Female

Male

rounding Tuolumne Meadows, and in similar habitats elsewhere in Yosemite National Park, showed that they, along with Mountain Chickadees, Common Ravens, and rarely a Red-tailed Hawk, were the only birds to be found after hours of concentrated searching.

Hairy Woodpeckers prefer older forests or recent burns because swarms of bark- and wood-boring beetles lay their eggs in the dead or weakened trees. Like Black-backed Woodpeckers, they move in to feast on the abundant larvae, or "grubs" and other wood-boring insects that they obtain by drilling vigorously or chipping off bark. Preferred items include larvae and adult beetles and ants, centipedes, millipedes, spiders, and aphids; they also consume considerable quantities of berries and fruits from ripening dogwoods, wild cherries, serviceberries, and poison oak; pine seeds pried from green cones supplement their diet in winter.

In the eastern United States, Hairy Woodpeckers usually nest in live trees with rotted cores, but in the Sierra most pairs select entirely dead trees. Unlike most woodpeckers, they often nest in smaller trees, 15 inches in diameter or less. They tend to remain close to their nesting territory for life unless food becomes scarce. In the Sierra, Hairy Woodpeckers begin nesting activities in early March and most pairs have raised their broods by late July. Males do most of the nest excavation work with some help from the females, and the entire process takes about three weeks. Featherless young are fed by both parents for about 30 days until they grow plumage and leave the nest; then they are fed and cared for up to 2 weeks by the adults and return to the nest at night.

Hairy Woodpeckers are noisy and conspicuous, often making sharp, squeaky, single-note calls or loud rattles. Both sexes drum frequently in loud, regular, rapid rolls. These woodpeckers are larger than the similarly plumaged Downys, but size is often difficult to judge; the Hairys' bills are larger and more robust, well over half as long as their heads. Juveniles occasionally show a yellow, rather than red, forehead patch, causing confusion among birders hoping to find a Black-backed Woodpecker.

STATUS AND DISTRIBUTION Among the most widespread birds in North America, Hairy Woodpeckers have the largest range of any member of their genus extending from central Alaska to Central America. They can be found in almost any forested area of the Sierra but reach their highest densities in conifer forests with at least some large trees; they also occur in riparian woodlands, oak woodlands, city parks, and residential settings with large ornamental trees.

West Side. Fairly common permanent residents from about 2,500 feet up to tree line; most common in older pine and red fir forests of the Upper Conifer zone; uncommon in the Subalpine zone; rare at the highest elevations in winter; uncommon down to the Foothill zone in winter, possibly in response to heavy storms.

East Side. Uncommon to fairly common permanent residents of pine and pinyon-juniper forests up to tree line; usually found in riparian habitats and isolated aspen groves in winter.

White-headed Woodpecker • *Picoides albolarvatus*

Female

Male

ORIGIN OF NAMES "White-headed," as no other North American woodpeckers have entirely white heads or faces; *albolarvatus*, white-masked, from L. *albo*, white, and *larva*, mask.

NATURAL HISTORY These striking woodpeckers often feed and nest low on the trunks of large conifers. They are familiar and characteristic inhabitants of giant sequoia groves, and perhaps more than any other Sierra woodpecker, occur among large pines, especially ponderosa, sugar, and Jeffrey. Although they seem to favor pines, they also feed on firs and may prefer incense cedars in winter. In other parts of their range, studies have shown that more than half of their fall and winter diet consists of pine seeds, extracted mainly from cones still hanging on trees. Near Lake Almanor, White-headed Woodpeckers attacked nearly a third of all sugar pine cones studied, leaving deep vertical trenches and cleaning out most seeds before the cones had opened. When mountain cabins are damaged by woodpeckers, they are the most likely culprits; they seem to specialize in drilling on redwood and cedar siding and riddling entire walls with holes during courtship.

In areas where sugar pines are absent or scarce, these woodpeckers require additional winter foods, for all the other common pines in their habitat shed their seeds in late summer and fall. At these times, White-headed Woodpeckers feed heavily on larvae of beetles and ants as well as spiders gleaned from conifer needles and by chipping bark from deep crevices of incense cedar and pine bark. Unlike Pileated and Hairy Woodpeckers inhabiting the same forests, they usually select live trees for feeding and do not drill as much for food. In recently burned forests, however, they join Hairy and Black-backed Woodpeckers to feast on the abundant bark beetle larvae in dead and weakened trees.

For nest cavities they prefer large, dead, broken-off pines or cedars with partially decayed wood that permit easy excavation but are not so rotten that the nest hole will lose its shape. Compared with other Sierra woodpeckers, they show a tendency to nest low on the trunk, averaging only eight feet above the ground. They readily use taller stumps left from logging and may use large, fallen logs for nesting. The configuration of their nest holes are distinctive from other woodpeckers, by including and excavated "spout" at the bottom of the entrance hole, presumably to drain water away from the nest. Their nesting season extends from early May when nest excavation begins through mid-August when the young are fledged. Both adults participate in creating the nest hole, sometimes with several false starts, until the nest is completed after three or four weeks. These woodpeckers make rapid, even drumroll tapping similar to Hairy Woodpeckers. Their sharp two- or three-note calls, similar in tone to the Hairys' single harsh notes, are distinctive. These calls sound like stronger, sharper versions of Nuttall's Woodpecker calls.

STATUS AND DISTRIBUTION Throughout their relatively restricted range in western North America, White-headed Woodpeckers are closely associated with mature stands of ponderosa pine. In the Sierra, however, they also frequent a variety of conifer trees, including pines firs, cedars, and giant sequoias. Very rarely they may stage movements as in fall 1996, when birds were seen in Kelso Valley, Bakersfield, and a number of southern California counties.

West Side. Fairly common and widespread permanent residents of the Lower and Upper Conifer zones; uncommon in nearly pure stands of red fir and Douglas-fir, and rare or absent from continuous stands of lodgepole pine.

East Side. Fairly common summer residents in appropriate habitat south to Mono County, where found in Jeffrey pines and red fir; rare south of Mono County and in any pinyon-juniper association; rare or casual in winter in Mono County.

Female

Male

Black-backed Woodpecker • *Picoides arcticus*

ORIGIN OF NAMES "Black-backed" for coloration; L. *arcticus*, northern.

NATURAL HISTORY Like Pine Grosbeaks and Great Gray Owls, Black-backed Woodpeckers' rarity and limited range in the United States make them a treasured find to birdwatchers. They occur primarily in cold-climate coniferous forests of Canada and Alaska and in fingerlike southward extensions of these boreal forests in the northern Rocky Mountains and Sierra, where all three species reach the southern limits of their ranges.

These woodpeckers have only three toes on each foot, instead of the usual four, but appear just as agile on vertical surfaces as other woodpeckers. Adult males sport yellow caps, unique among Sierra woodpeckers—all the others have at least some red on their heads (males only for most species), but beware of the occasional juvenile Hairy Woodpecker in July or August with yellow instead of red on the forehead. Surprisingly bold for woodpeckers, Black-backs tolerate close approach and sometimes nest in busy, higher-elevation campgrounds. They actively drill and chisel off bark with their long, stout bills, gobbling up grubs, mostly beetle larvae.

Black-backs prefer to forage low on the trunks of dead or dying conifers and choose stands with a high density of dying trees infested with bark- or wood-boring beetles. Although rare in most Sierra forests, they can be fairly common in recently burned stands at higher elevations. Although they occur in wide variety of unburned forests, more than any other woodpecker they are tied to fire-killed trees. Thus forestry practices such as postfire salvage logging and clear-cuts that eliminate recently killed trees deprive this species of essential foraging and breeding habitat.

Although often quiet, these woodpeckers can be noisy near their nests, particularly when disturbed. They make harsh, gurgling, descending rattles—unmistakable calls like no other Sierra bird's. They also produce harsh *chip* calls, somewhat like those of Hairy Woodpeckers but have a flatter and duller sound. Like Hairys, they drum in a regular, rolling patterns but Black-backs usually pick hollow limbs for drumming, creating a distinctive sound that carries farther than those of most other woodpeckers. In the Sierra dead lodgepole pines with considerable rotten sapwood are preferred nest sites, often at the edges of meadows. Nest trees are often small, sometimes less than 10 inches in diameter, and most nest holes are below 20 feet. Due to their high-elevation existence, they initiate nesting later than most Sierra birds, not until late May or early June, and young hatch by late July. Males do most of the nest excavation with some help from the females. The young leave the nest after about three weeks, and each adult tends one or two nestlings for at least several weeks.

STATUS AND DISTRIBUTION Throughout their large range, Black-backed Woodpeckers are confined to high latitudes or high-elevation forests farther south, where they remain year-round. While sometimes found dependably nesting in a particular place from year to year (especially following burns), most move about searching for dead and dying and recently burned trees with ample supplies of wood-boring beetles.

West Side. Uncommon permanent residents of lodgepole and red fir forests of the Upper Conifer and Subalpine zones; casual down to the Lower Conifer zone in some winters; the species has not been recorded in Kern County despite careful surveys of suitable habitats.

East Side. Generally rare to uncommon and irregular; low numbers are present in Tahoe Basin year-round; persist in burned areas for several years, then usually move on; no records from Inyo County.

TRENDS AND CONSERVATION STATUS The small population and wandering nature of this species makes it nearly impossible to get good data on trends. A substantial portion of the Black-backed Woodpecker's range overlaps with National Parks and designated Wilderness areas. While Black-backs may benefit from increased fire frequency in the Sierra, the widespread practice of postfire salvage logging is clearly detrimental to this, and to other fire-damaged tree specialists.

Northern Flicker • *Colaptes auratus*

ORIGIN OF NAMES "Northern" for distribution, which does not extend south of Central America; "flicker," one who strikes; *colaptes*, from Gr. *kolaptes*, a chisel; L. *auratus*, golden, a reference to the underwing and tail colors of the yellow-shafted eastern subspecies.

NATURAL HISTORY Northern Flicker's salmon-red underwings and tails, polka-dot breasts, and red "whiskers" (males only) set them apart from all other Sierra woodpeckers. Add in the black bib and white rump, and you have a bird with an overabundance of distinct field marks. Native Americans highly valued their bright feathers and wove the stiff wing and tail quills along with shells and other ornaments into their elaborate ceremonial robes and headdresses.

Northern Flickers are remarkable in their cosmopolitan taste in habitats, equaled by no other woodpecker and few other birds in the Sierra: they nest in all forest types from the Central Valley up to tree line. They prefer open stands with large, scattered trees and open ground for foraging, and most retreat below the heavy snows for the winter; at this season they may gather in small groups to feed or to roost communally. Flickers avoid snow-covered areas because, unlike other Sierra woodpeckers, they obtain most of their food on the ground. They eat mainly ants but take other insects such as beetles and grasshoppers as well as seeds and fruits—wild grapes, blackberries, poison oak, and dogwood berries are their favorites. Although they also take bark-dwelling insects, most flickers pecking on trees are engaged in other pursuits.

They communicate by drumming on trees in loud, regular patterns, and they excavate holes in dead or live trees for nests and winter roosts. Except for Pileated Woodpeckers, flickers are the largest Sierra woodpeckers and drill the biggest nest holes. These large cavities are of great importance to small owls, American kestrels, and other wildlife that cannot use the nest holes of smaller woodpeckers. Flickers frequently give a piercing *cleer!* call year-round and make loud *ca-ca-ca-ca* calls in the breeding season. The latter call may be mistaken for that of Pileated Woodpeckers, which vocalize year-round. Compared to this species, flicker calls are softer, more constant in pitch, and slower at the end.

In the Sierra, Northern Flickers breed from early May until late July, and peak activities are in June. Both sexes participate in nest excavation, but males do most of the work. They select dead conifers, or dead portions of live conifers, and trees with relatively soft wood; rarely pairs have been observed using earthen burrows for nesting, including abandoned nests of Belted Kingfishers and Bank Swallows. After hatching, adults feed the young regurgitated food for several weeks until they are capable of leaving the nest after about 28 days; males do most brooding of the young in the nests, especially at night.

STATUS AND DISTRIBUTION "Red-shafted" *(C. a. cafer)* and "Yellow-shafted" Flickers *(C. a. auratus)* were recognized as separate species, along with the Gilded Flicker *(C. chrysoides)* of the Sonoran Desert, until 1983 when they were lumped together as one species (the Gilded Flicker is once again recognized as a separate species). In fall and winter an occasional "Yellow-shafted" Northern Flicker shows up in the Sierra. Members of this eastern subspecies have yellow, rather than salmon, on their wings and tails, different head coloration, and black mustache marks (on males). Many of the Sierra "Yellow-shafted" Flicker reports are actually intergrades between the "Red-shafted" and "Yellow-shafted" forms, as these two interbreed across a long, narrow zone in the Midwestern United States and Canada. Any odd-looking flicker should be scrutinized for characters that might suggest an intergrade between these two subspecies.

West Side. Common breeders in open forests from the Foothill zone through the Upper Conifer zone; uncommon up to the Subalpine zone and rare in the Alpine zone, where they have nested as high as 10,000 feet near Silliman Creek in Kings Canyon National Park and at similar elevations at the Hall Natural Area just east of Yosemite National Park; due to frequent snow cover, rare or absent in the Upper Conifer zone in winter.

East Side. Common residents in most forested habitats up to tree line; in winter, few remain above about 7,000 feet or the local snow line; often seen in open areas near aspen groves, orchards, and other ornamental deciduous trees.

Pileated Woodpecker • *Dryocopus pileatus*

Female

ORIGIN OF NAMES "Pileated" from L. *pileatus*, capped, a reference to the species' prominent red crest; *dryocopus*, tree cleaver, from Gr. *drys*, a tree, and *kopis*, a cleaver.

NATURAL HISTORY Almost as large as crows, with jet-black bodies and flaming-red crests, Pileated Woodpeckers are among the most spectacular and unforgettable of Sierra birds. They are surpassed in size in the Americas only by Imperial Woodpeckers *(Campephilis imperialis)* and Ivory-billed Woodpeckers *(C. principalis)*, both of which may be extinct. Despite their large size and striking plumage, Pileated Woodpeckers in the Sierra may be shy and hard to see; they can also be surprisingly fearless—sometimes allowing a close approach. They can be heard at great distances, calling loudly or drumming a rolling tattoo. One call—a repeated *kak-kak-kak-kak*—is given year-round and closely resembles that of the Northern Flicker but is louder and often rises and then falls slightly in pitch.

Pileated Woodpeckers find optimal habitat in the most majestic stands of Sierra trees: the giant sequoias and towering pines and firs in mature forests of the Lower and Upper Conifer zones. They also occur in second-growth stands that support trees with sufficient girth to contain their nests and roost sites, and sufficient snags and down logs for feeding. They are normally seen nesting in live trees greater than 12 inches in diameter or in dead trees greater than 20 inches in diameter, often near a broken top.

Male

With powerful, deliberate blows, Pileated Woodpeckers chisel into soft, dead wood or strip away loose bark to extract carpenter ants, their favorite food, and the larvae of wood-boring beetles. These deep excavations, often roughly rectangular, provide unmistakable evidence of their presence in an area. In late summer and fall they supplement their diet with berries and nuts and are sometimes seen hanging like oversized chickadees from the slender branches of dogwoods, elderberries, and madrones, pecking berries and flapping awkwardly to change positions.

Most nests in the Sierra are in large dead conifers or live aspens. The oval or triangular nest holes, about four to five inches across, provide homes for many of the larger hole-nesting birds and are the only woodpecker cavities large enough to accommodate cavity-nesting species such as Wood Ducks and Common Mergansers. Males do most of the nest excavation, starting in late March or early April, and females begin laying eggs by May. Both adults tend the newly hatched young, feeding them regurgitated ants, beetles, and berries. The young are capable of flight about 28 days after hatching but remain in family groups with their parents through the summer.

STATUS AND DISTRIBUTION Widespread in forested areas across North America, Pileated Wood-peckers find the southern limits of their western distribution in the Sierra. They are usually associated with large stands of mature forest but also occur in areas with a mixture of stand ages (including near houses), if suitable nest sites and foraging habitat is available. Densities in the Sierra are much lower than in the southeastern United States, and possibly as low as in any part of their range.

West Side. Uncommon residents of the Lower and Upper Conifer zones; rare in the highest red fir forests; nearly absent from lodgepole pine forests of the Subalpine zone; recorded south to the Greenhorn Mountains in Kern County.

East Side. Uncommon residents in the Tahoe Basin but casual farther south with individual late summer and fall records from larger stands of ponderosa pine near the Mono Basin; has nested east of Yuba Pass in Sierra County, and there are occasional winter records in Sierra Valley and in East Side portions of Lassen County.

TRENDS AND CONSERVATION STATUS Although not included on any "official" lists of declining species, Pileated Woodpeckers in the Sierra have undoubtedly declined due to logging, especially clear-cutting and other harvest methods that remove all or most large trees on which this species depends. Each pair requires several hundred acres of mature forest, and such stands are adversely affected by management practices that remove the snags and logs this species requires for feeding and nesting.

TYRANT FLYCATCHERS • Family Tyrannidae

Tyrant flycatchers form a group of birds taxonomically intermediate between oscine passerines (songbirds) and nonpasserines (all other nonperching birds). Although tyrant flycatchers sing somewhat like songbirds, their songs are not complex or musical because their vocal apparatus (syrinx) is not well developed, hence they are known as suboscine passerines. Also, their songs are mainly hard-wired from birth, rather than learned as in oscine passerines. These flycatchers are the only North American representatives of this large and diverse group of mostly New World, tropical, suboscine families.

Tyrant flycatchers characteristically feed by swooping from perches and snatching insects either in the air, from foliage, or on the ground. Most species have broad-based bills with hooked tips that help them grab and hold small prey, while "rictal bristles" that arise from the base of their bills may help flycatchers pinpoint their prey or else protect their eyes when flying insects are missed or make their "catcher's mitts" larger. Depending on their size, flycatchers eat almost every type of flying and crawling insect, from tiny gnats to heavily armored wasps or cicadas, but what is more surprising is that many flycatchers eat fruit, especially during migration and winter.

Except for the brilliantly colored Vermilion Flycatcher, most members of this family in the Sierra are rather plain and some are difficult to identify. However, the uniformity of their coloration is offset by distinct differences in their calls or songs, which can be learned with practice. Some species may also be recognized by the habitats in which they occur, though such distinctions are difficult to make when nonsinging birds are migrating through a variety of habitats. Behavioral cues like foraging style and tail-dipping can also be very helpful for some species. Identification of some of the "Empids" (members of the genus *Empidonax*) can be especially difficult.

Female flycatchers usually lay three or four white or cream-colored eggs with brown or purplish speckling and splotches at their larger ends (most Empid eggs are entirely white) and incubate them alone for 12 to 15 days. Males bring food to females during incubation, and both parents feed and care for their young until they leave the nests after 15 to 20 days. Fledgling departure from the nest is drawn out because young birds spend time on nearby branches or return to the nest after their first flights. This delay may be possible because Tyrannids are so pugnacious in defending against possible nest predators.

Tyrant flycatchers are widespread throughout the Americas, and with more than 400 living species they are the largest bird family in the World; 13 species occur regularly in the Sierra and an additional 11 rare, casual, or accidental species also have been recorded (see Appendices 1 and 2). The family name suggests their sometimes tyrannical or aggressive behavior toward other birds, and it is derived from either L. *tyrannus*, king, or Gr. *tyrannos*, lord or ruler.

Olive-sided Flycatcher • *Contopus cooperi*

ORIGIN OF NAMES "Olive-sided" for the species' most conspicuous field mark; "flycatcher" for the foraging habits of this family; *contopus*, short-footed, from Gr. *kontos*, short, and *pous*, foot; L. *cooperi* (see Cooper's Hawk account).

NATURAL HISTORY Olive-sided Flycatchers lead enviable lives, calling and hunting from the tallest treetops, overlooking broad Sierra landscapes. From such commanding viewpoints, these large, stocky flycatchers sally out to make long acrobatic flights after flying insects—mostly large bees, wasps, and flying ants. The larger stinging insects are then beaten against a branch to subdue them and to break off the stinger. Their loud calls—sometimes interpreted as an insistent request for refreshment, *quick-THREE-beers*—can be heard at great distances, as can their *pip-pip . . . pip-pip-pip* calls.

Olive-sideds typically avoid dense forests in favor of those with clearings and edges, and perch on tall, mature trees towering over open spaces with younger trees, mountain meadows, or chaparral. Pairs of breeding Olive-sided Flycatchers will aggressively defend large territories of 100 acres or more. Nesting begins in earnest in late June, and nest sites are usually placed well out on horizontal branches anywhere from eye level to over 60 feet above the ground. Young birds may linger with their parents until migration.

STATUS AND DISTRIBUTION Olive-sided Flycatchers primarily breed from boreal forests of northern Alaska and Canada south to northern Mexico, and winter in the northern Andes and western Amazon Basin, much farther south than most Sierra birds. They are summer residents of conifer forests the length of the Sierra, where they breed in most forest types except the upper reaches of the Subalpine zone. They arrive in late April or early May, later than many other neotropical migrant songbirds, and most depart by mid-August.

West Side. Fairly common on their breeding grounds from the Lower Conifer zone up to the Subalpine zone and in the Foothill zone during spring migration (mostly in early May).

East Side. Uncommon breeders in pine forests only where large, mature trees are present; spring and fall migrants can appear anywhere where there are trees, including desert riparian habitats and suburban gardens.

TRENDS AND CONSERVATION STATUS As noted in the chapter on "Recent Trends in Sierra Bird Populations and Ranges," Breeding Bird Survey data show long-term, highly significant, range-wide declines in Olive-sided Flycatchers over the past several decades. Possible reasons for the Olive-sided Flycatcher's decline in the Sierra include forestry practices (such as postfire salvage logging, which remove all or most large snags from forest stands). Declines may also be linked to historical fire suppression efforts that have reduced the extent of open canopy forests and destruction of tropical forests where Olive-sideds winter. For these reasons, the Olive-sided Flycatcher was included as a California Bird Species of Special Concern in 2008.

Western Wood-Pewee • *Contopus sordidulus*

ORIGIN OF NAMES "Western" for range; "pewee" perhaps imitative of the calls of Eastern Wood-Pewees, but not Western's; L. *sordidus*, dirty, either named for its dingy color or by someone who had the unfortunate task of distinguishing this bird from museum specimens of the nearly identical Eastern Wood-Pewee—without the benefit of hearing both species call.

NATURAL HISTORY Western Wood-Pewees are the most conspicuous and widespread of all Sierra flycatchers. Their vaguely mournful, down-slurred *pee-er* calls are heard from before dawn, through the heat of midday when other birds have gone silent, and well into the evening. Nearly as tolerant of close approach as the Black Phoebe, they perch in plain view at the base of the forest canopy and near habitat edges, burned or cleared forests, wet meadows, and riparian forests; then they sally out to catch a flying insect and almost invariably return to the same perch. This reuse of a perch is consistent enough to be helpful as an aid to identification.

Equipped with long wings, wood-pewees are fast, powerful fliers and cover more distance when chasing prey than the similar *Empidonax* flycatchers (see accounts below). Those long wings are also very useful to birders as a field mark to distinguish them from the Empids. Masters of aerial pursuit, wood-pewees can match the dodges and darts of even the most agile flying insects. Female Wood-pewees build cups of tightly bound plant fibers, grasses, and moss held together with spiderwebs that straddle the forks of open horizontal branches 15 to 75 feet high. Nests are found in wide variety of tree species, but aspens, cottonwoods, and pines account for most nest sites.

STATUS AND DISTRIBUTION Western Wood-Pewees breed from northern Alaska to Central America, and they winter farther south than any other Sierra flycatcher, save the Olive-sided. Due to their long migrations compared with most other passerines, they usually do not arrive in the Sierra until mid-May and remain until late August, with a few stragglers into September.

West Side. Common to abundant summer residents breeding in all but the densest forest types or heavily logged or burned areas that lack tall snags or trees; common in spring migration in the Foothill zone.

East Side. Common summer residents and breeders in pine forests, as well as in aspen and riparian habitats; arrival and departure dates similar to the West Side.

Willow Flycatcher · *Empidonax traillii*

ORIGIN OF NAMES "Willow" for the species' preferred breeding habitat; *empidonax*, king of gnats, from Gr. *empis*, gnat, and *anax*, a king; John James Audubon named this bird in honor of his friend, the Scottish zoologist Thomas Stewart Traill (1781–1862), a founder of the Royal Institution of Liverpool, where Audubon first met him.

NATURAL HISTORY No name could be more appropriate for these small, brownish flycatchers that nest almost entirely in patches of willows. Few other Sierra birds are so specialized in their choice of habitats. Willows in wet meadows and beside streams and lakes provide foraging lookouts and roost sites as well as concealment for nests. Like most of the Empids, Willow Flycatchers tend to hunt from low perches, spending more time in the tree or shrub canopy and less time perched in plain view than most other flycatchers. Liquid *whit* calls and a characteristic *fitz-bew* song often reveal their presence. Females build tight, compact cup nests in vertical forks of willows with nest-building starting in May and eggs very rarely laid before June. Both parents feed nestlings, but the female makes the larger contribution. Where Brown-headed Cowbirds are present, rates of brood parasitism tend to be high and breeding success very poor.

STATUS AND DISTRIBUTION Three subspecies of Willow Flycatchers visit the Sierra annually including the "Little" *(E. t. brewsteri),* "Southwestern" *(E. t. extimus),* and "Mountain" *(E. t. adastus),* and these subspecies occupy distinct breeding ranges and habitats. "Little" Willow Flycatchers breed in dense scrubby willow thickets in mountain meadows and riparian woodlands of the Lower and Upper Conifer zones across most of the length of the Sierra and typically place their nests within 5 to 10 feet from the ground. "Southwesterns" are restricted to much taller black willow riparian habitats, and their range in the Sierra does not extend north of the Kern River Valley, where 15 to 50 active nests have been found annually since 1989. "Mountains" migrate through the East Side, and some breed there at the Great Basin edge and in the Owens Valley. A late-arriving summer resident (generally the last of the Empids to arrive), they begin showing up on their nesting territories from mid-May (South Fork Kern River) to early June (mountain meadows). Most leave their nesting territories by early August, with a few stragglers still nesting until mid-August.

West Side. Uncommon, the largest breeding population found along the South Fork Kern River ("Southwestern"), smaller, localized populations of "Little" Willow Flycatchers breed in wet meadows near Shaver Lake (Fresno County), Beasore Meadow (Madera County), Donner Summit (Nevada and Placer Counties), and in flooded willows near Lake Almanor and Buck's Lake; for such an uncommon breeder, the species can sometimes appear fairly common in spring migration (mid-May) through riparian areas of the Foothill zone.

East Side. Uncommon, "Little" Willow Flycatchers breed at a few mountain meadows, including small, stable populations at Perazzo Meadows and Weber Lake (Sierra County) and the Tahoe area; small breeding populations also found among willow thickets along the lower reaches of tributary creeks to Mono Lake south to the upper Owens River; migrants can appear in riparian habitats almost anywhere, especially from May through early June.

Grinnell and Miller (1944) described the status of Willow Flycatchers as "where conditions are right, common." Historically they nested in willow-dominated riparian habitats and wet meadows throughout California, including both sides of the Sierra up to about 7,000 feet. They now survive in the Sierra in perilously low numbers at widely scattered riparian and wet meadow sites. There may be only 200 nesting pairs left in the state, mostly in the Sierra. A recent finding that the species appears to have been extirpated as a breeder from Yosemite National Park is particularly worrisome as this park includes some of the best and most well-protected habitat in the Sierra.

In addition to nest parasitism by Brown-headed Cowbirds, Willow Flycatcher nests may fail when browsing cows knock nests and eggs down or expose the nests to predators by eating away protective leaves. Historical grazing practices in mountain meadows altered the hydrology by causing streams to down-cut and become dry so that most no longer support good habitat for this species. All Willow Flycatcher subspecies in are listed as Endangered in California, and "Southwestern" Willow Flycatchers are also listed as Endangered by the U.S. Fish and Wildlife Service. This isolated population is heavily impacted by cowbirds, with up to 80 percent of the flycatcher nests failing before cowbird trapping efforts were initiated in the early 1990s; parasitism rates have now declined to 20 to 40 percent of active nests in most years. All Willow Flycatchers in the Sierra are highly sensitive; therefore it is unwise, unethical, and illegal to play recordings of their vocalizations or to harass them in any way.

Hammond's Flycatcher • *Empidonax hammondii*

ORIGIN OF NAMES "Hammond" for William Alexander Hammond (1828–1900), surgeon general of the United States Army between 1862 and 1864; the prominent ornithologist John Xantus (1825–1894) named this flycatcher in Hammond's honor.

NATURAL HISTORY Hammond's and the flycatchers Gray and Dusky (see accounts) exemplify the identification challenges posed by birds in this genus. Theodore Roosevelt, an avid and skilled birder, asked John Muir how to distinguish between the songs of Hammond's and "Wright's" (now Dusky) Flycatchers. Muir's answer is unrecorded, but it was likely unsatisfactory because Roosevelt was unimpressed with Muir's bird identification skills, especially as applied to bird song. The common name Roosevelt (1913) used for Dusky Flycatcher, "Wright's," is further evidence of the identification difficulties. The original type specimen used to name what is now the Dusky Flycatcher was later determined to actually be a Gray Flycatcher (hence the scientific name for Gray Flycatcher, *Empidonax wrightii*).

With Dusky and Hammond's Flycatchers having broadly overlapping ranges, they offer a frequent identification dilemma, best resolved by waiting until they call or sing, though many can be correctly identified by structure with practice and very good views. According to many accounts, Hammond's Flycatchers forage higher in trees than Duskys, but this distinction is less consistent than their choices of breeding habitats. Hammond's Flycatchers prefer cool, shady areas beneath dense forest canopies, while Duskys almost always inhabit warmer, open forests or forest edges.

Hammond's Flycatchers are small, seemingly nervous flycatchers usually found in tall, shaded conifer groves. Some consider them to be white and red fir specialists, but they also breed in pine forests, giant sequoia groves, and shaded riparian areas dominated by cottonwoods and dogwoods. Among the dark foliage they exhibit typical Empid behavior, making short flights from low or high perches to capture insects in flight or to pluck them from leaves, frequently flicking their tails while perched.

Female Hammond's Flycatchers build cup nests of mosses and plant fibers and secure them with spider webs to high, horizontal branches of tall conifers, sometimes over 100 feet above the ground. Due to their lofty heights, nests are extremely difficult to find or examine, and comparatively little is known about the nesting habits of this species. Forest management practices that remove tall trees from mature forests are generally harmful to this species.

STATUS AND DISTRIBUTION Hammond's Flycatchers breed from northern Alaska through the southern Sierra and Rockies, and they winter in mountainous regions of central Mexico and Central America. Though found in southern California in the first week of April, they start arriving in the Sierra foothills from mid- to late April, and they avoid higher elevations for another month. They remain in the Sierra until late August or early September before departing south. Unlike Dusky Flycatchers that wait until arriving on the wintering grounds to molt, Hammond's Flycatchers molt into fresh plumage in the Sierra before departing, with this molt often beginning while the birds are still nesting. This can provide a helpful clue to identification late in the season.

West Side. Fairly common summer residents and breeders in mature conifer forests from 4,500 to 8,500 feet; uncommon postbreeding dispersal above 9,000 feet; fairly common in the Foothill zone during spring migration.

East Side. Uncommon spring and rare fall migrants; no documented breeding records but birds present through summer west of Lake Tahoe.

Gray Flycatcher • *Empidonax wrightii*

ORIGIN OF NAMES "Gray" for the species' overall coloration; Charles Wright (1811–1895) was a self-taught botanist who accompanied the survey team for the cross-country Pacific Railroad in the mid-1800s; Wright collected the first specimen of this species (see Hammond's Flycatcher account for more on this specimen), and it was named in his honor by Spencer Baird (1823–1887), secretary of the Smithsonian Institution, who directed the railroad surveys.

NATURAL HISTORY Gray Flycatchers, thanks to the original observations of Alan Philips (1944), may be the most easily identified of our *Empidonax* flycatchers due to one characteristic and habitual behavior. While most members of this genus will flick their tails upward when perched, Gray Flycatchers dip their tails slowly downward. This may be preceded by a quick upward tick, but the deliberate downward dip is unique among Empids. While Gray Flycatchers were first recognized as a distinct species in the late 19th century, they were often confused with the very similar Dusky Flycatchers in most scientific studies and museum collections. The discovery of their tail-dipping behavior shed new light on the bird's taxonomic status, life history, and distribution, though much remains to be learned.

Gray Flycatchers usually nest in Jeffrey pines, western junipers, pinyon pines, or large clumps of sagebrush or bitterbrush. Breeding season usually begins in late May to early June, but unlike other *Empidonax* flycatchers, Grays have an extended breeding season that includes a second clutch of eggs laid from late June to early July.

STATUS AND DISTRIBUTION Compared to most other *Empidonax* flycatchers, Gray Flycatchers have relatively short migration routes and mostly winter in the oak and pine forests of central Mexico. Their breeding range includes the southern Rockies and the Great Basin, extending westward to the eastern Sierra, where most arrive by late April and depart by mid-August; a few may remain through mid-September.

West Side. Fairly common nesting species in pinyon-pine woodlands of southeastern Tulare County (e.g., Kennedy Meadows and Chimney Creek Campground) and near Walker Pass (Kern County); no breeding records elsewhere; uncommon to rare spring migrants mostly in chaparral or mixed oak-chaparral habitats, occasionally pulses of spring migrants are encountered in significant numbers; few records of fall migrants.

East Side. Fairly common summer breeders below 8,000 feet in open forests and scrub habitats, though reportedly some pairs breed up to 11,000 feet in dry open areas east of the Sierra crest.

Dusky Flycatcher • *Empidonax oberholseri*

ORIGIN OF NAMES "Dusky" possibly for the slightly darker olive-gray upperparts compared with the similar Gray Flycatchers; previously known as "Wright's" Flycatcher due to a misidentified type specimen (see Hammond's Flycatcher account above); L. *oberholseri*, for Harry Church Oberholser (1870–1963), an American ornithologist and curator of birds at the Cleveland Museum of Natural History who described this flycatcher.

NATURAL HISTORY These sun-loving counterparts to the forest-dwelling Hammond's Flycatchers are at home in clearings with shrubs and scattered trees, including openings in dense forests otherwise occupied by nesting Hammond's Flycatchers. Duskys use any open area from clearings in ponderosa pine forests in the Lower Conifer zone up to the gnarled whitebark pines and mountain hemlocks of the upper Subalpine zone, but they are particularly abundant in mountain chaparral habitats where manzanita and ceanothus grow densely under isolated pines and other conifers. In general, Duskys are more common and widespread than Hammond's Flycatchers. The two species are best distinguished by calls and songs. The Hammond's call is a sharp *peek* while the Dusky gives either a dry *whit* or distinctive *doo-hic* call.

At most elevations Dusky Flycatchers begin to nest in late May or early June, when females build tight cups of rootlets and fibers in small trees or shrubs on branches; most nests are less than 10 feet high. However, since many nest at higher elevations than most other neotropical migrants, Duskys are vulnerable to late spring and early summer snow storms. A recent study of a breeding population near Tioga Pass showed that reproductive schedules varied depending on the annual snow depth, with first egg laying dates ranging from late May in dry years until early July in extremely wet years; clutch sizes and overall reproductive success also were lower in years of heavy snow when nesting started later (Pereyra 2011).

Dusky Flycatchers pursue aerial insects from low branches of shrubs and trees, and occasionally pounce on prey items on the ground or hover while picking insects from twigs and leaves. Males, however, regularly sing from the tops of tall lodgepole pines. When perched, they are usually less active than Hammond's Flycatchers and rarely flick their wings except when calling or just after landing.

STATUS AND DISTRIBUTION Compared to Hammond's Flycatchers, Duskys have a larger range in California and breed in mountainous areas throughout the state. They are particularly common on both sides of the Sierra, unlike Hammond's, which breed exclusively on the West Side and occur only as migrants on the East Side. Most Dusky Flycatchers arrive by mid-May and leave the Sierra by mid-August, generally several weeks earlier than Hammond's, which linger to complete their molt; probably accidental in winter, the very few documented winter records appear to be Hammond's rather than Dusky.

West Side. Common summer nesters in open forests, forest edge, and chaparral habitats mostly from 6,000 feet to the crest and less commonly down to 4,000 feet.

East Side. Timing as on West Side; primarily nests in pine forests and aspen groves up to tree line.

Pacific-slope Flycatcher • *Empidonax difficilis*

ORIGIN OF NAMES "Pacific-slope" for the species' distribution; L. *difficilis*, difficult; for the complexity of separating this species from other *Empidonax* flycatchers, which is odd since this is among the easiest of this group to identify (except to distinguish it from Cordilleran, see below).

NATURAL HISTORY Although drab in comparison to many other birds, Pacific-slope Flycatchers are the most distinctive of the Sierra *Empidonax* complex, with brighter yellow underparts and prominent eyerings that are wider behind the eye and often come to a point. They also inhabit the most luxuriant habitats, streamside forests with dogwoods, alders, and black oaks mixed with a variety of conifers.

Until 1989 these flycatchers were lumped with the almost identical Cordilleran Flycatchers *(E. occidentalis)* as "Western Flycatchers." While they are currently recognized as a separate species based on genetic evidence and differences in their typical call notes, taxonomy of this complex group remains under study and could change again in the future. Nearly all studies of the "Western Flycatcher" complex have been conducted on populations of Pacific-slope Flycatchers, so little is known about the life history and behavior of Cordilleran Flycatchers, especially in the Sierra. In areas where their breeding ranges overlap in northeastern California, the sole means of separating these species are the male call notes. Unfortunately, these sounds can blend confusingly, as some Cordilleran males appear to give calls typical of both species, and calls with features of both species have been recorded in areas of overlap. Some experts remain doubtful about the existence of two species and continue to refer to both as "Western" Flycatchers.

Apparently breeding only on the West Side of the Sierra, Pacific-slope Flycatchers build nests in a remarkable variety of sites, including crotches of small trees; holes in cliffs, banks, or tree trunks; and, occasionally in old buildings and the undersides of bridges. Females build cup nests of mosses, dry grass, and bark strips lined with finer materials, while males spend the days perched nearby and calling incessantly.

STATUS AND DISTRIBUTION As their name suggests, Pacific-slope Flycatchers are only known to breed west of the divide formed by the Sierra and Cascade Range. Usually the first of the Empids to arrive in spring, often in mid-March, most have left for Mexico by late August, though some linger into September.

West Side. Common summer residents in moist ravines from the Foothill zone through the cool, shaded forests of the Lower Conifer zone and just into the Upper Conifer zone.

East Side. Uncommon spring and fall migrants at lower elevations; with the exception of the Tahoe Basin, breeding season observations of "Western-type" Flycatchers are likely Cordillerans, as singing males in the steep canyons west of Mono Lake and into north Inyo County appear to be giving the typical calls in that species' favored habitats.

Black Phoebe • *Sayornis nigricans*

ORIGIN OF NAMES "Black" for the color of the species' head and upperparts; "phoebe," imitative of the calls of the Eastern Phoebe (*Sayornis phoebe*); L. *sayornis*, Say's bird, for Thomas Say (see Say's Phoebe account below), and Gr. *ornis*, bird; L. *nigricans*, to become black.

NATURAL HISTORY Bold and charmingly confiding, Black Phoebes are as comfortable living in close proximity to people as any native Sierra bird. Wedded to water in all seasons, even a small pond, irrigation ditch, water trough, or backyard birdbath may provide sufficient moisture to attract Black Phoebes. Their energetic calls and crisply contrasting black-and-white plumage grace the banks of streams and lakes, and no other Sierra flycatcher is so closely tied to water. These flycatchers prefer areas shaded by trees or canyon walls but use open, sunny areas as well. The walls of houses and backyard fences seem acceptable substitutes for canyons to this adaptable phoebe. They may allow very close approach of humans, but cats and dogs can expect a thorough scolding. Black Phoebes forage from rooftops, garden shrubbery, and emergent vegetation in ponds. They dart out from low perches to catch flying insects and may even snatch small fish or small berries from the water's surface.

Black Phoebes are the only Sierra flycatchers that remain on their breeding grounds year-round, and some even winter above the foothills, when sources of unfrozen water can be found. They often begin to form pairs by mid-January, much earlier than most other flycatchers that migrate to the tropics for winter. Females begin nest construction by late February or early March, and all they need is a source of mud. Much like Cliff Swallows, they collect small pellets of mud and plaster them on vertical surfaces then build them out with mixtures of mud, plant fibers, or hair. While their native nesting substrates were rock walls or earthen cliffs, they readily exploit such human structures as buildings, bridges, or culverts. All nests are located near or over water, and usually have natural or artificial overhangs to protect them from direct sun, rain, and predators. One unusual pair nested on top of a spotlight at the Cedar Grove Amphitheater in Kings Canyon National Park.

STATUS AND DISTRIBUTION Black Phoebes are year-round residents of lowlands throughout California. Their numbers and distribution in the Sierra expanded in the past century in response to the construction of human structures and artificial ponds that added key habitat elements that once limited their breeding range.

West Side. Fairly common breeders in the Foothill zone up to about 3,000 feet at all seasons, uncommon up to about 5,000 feet after the breeding season from July to October with an August observation at over 10,000 feet at Kearsarge Lake (Fresno County) a likely record; easily seen at 4,000 feet in Yosemite Valley in all seasons, including winter.

East Side. Generally uncommon but East Side range apparently expanding and numbers growing; now fairly common and increasing as year-round residents in much of the Owens Valley and Reno area; at least one apparently resident pair in Loyalton (Sierra County) and occasional winter records here and there; otherwise uncommon spring and fall migrants or postbreeding wanderers.

Say's Phoebe · *Sayornis saya*

ORIGIN OF NAMES "Say's" for Thomas Say (1787–1834), a prolific, self-taught naturalist who discovered more than a thousand new species of beetles; Say also provided the first scientific descriptions of the coyote, swift fox, Western Kingbird, Band-tailed Pigeon, Rock Wren, Lesser Goldfinch, Lark Sparrow, Lazuli Bunting, Orange-crowned Warbler, and this phoebe, which was named in his honor, along with the name of its genus.

NATURAL HISTORY While Black Phoebes need water, Say's Phoebes shun it. Unlike their close relatives, Say's Phoebes inhabit dry, open terrain with few if any trees or shrubs. Mud is in short supply in these arid environments, and nesting Say's Phoebes can make do without it. Instead, they use dry crevices on cliffs or steep banks as substrates to build their simple nests of weeds and grasses bound together with spider webs. Like Black Phoebes, they also nest in deserted buildings, under bridges, and tree cavities.

One may hear the Say's Phoebe's plaintive, down-slurred *pheee—ur* calls from sagebrush desert of the East Side to annual grasslands of the western foothills. From perches on fences or rocks, they scan for low-flying insects, sometimes hovering like bluebirds before snatching their prey in midair, off low shrubs, or from the ground. Say's Phoebes have a distinctive flight style that allows them to be identified from a long distance. They are notably buoyant in flight, looking like a bird suspended from a rubber band as they bounce along. Say's Phoebes are able to winter in fairly cold areas, a rare trait among flycatchers. This ability may be due to the species' behavior of hunting almost exclusively close to the ground, where it is warmer and insect prey remain most active.

STATUS AND DISTRIBUTION In the Sierra these dry-country flycatchers nest primarily in the Great Basin and deserts of the East Side, winter in West Side grasslands, and are present year-round in the Owens Valley.

West Side. Fairly common in winter in grassland habitats of the Foothill zone; uncommon breeders in the South Fork Kern River Valley and desert mountain ranges and slopes to the south (i.e., Scodie and Piute Mountains, Kern County); occasional summer records in the foothills from Calaveras County south may represent rare breeding attempts.

East Side. Fairly common to uncommon breeders in open country at lower elevations from the Reno area south; in winter, fairly common in the Owens Valley but rare elsewhere.

Female

Male

Vermilion Flycatcher • *Pyrocephalus rubinus*

ORIGIN OF NAMES "Vermilion" for the brilliant, red-orange plumage of males; *pyrocephalus*, a fire head, from Gr. *pyros*, fire, and *cephale*, head; L. *rubinus*, ruby.

NATURAL HISTORY Like dazzling scarlet butterflies, male Vermilion Flycatchers adorn desert riparian groves of cottonwoods and willows in desert oases in the extreme southern Sierra. Flying up to 100 feet above the forest canopy, males perform elaborate courtship displays by fluttering with exaggerated wingbeats while singing a series of buzzy twitters to attract the attention of females. Then, as if to show themselves off even further, males perch prominently on open perches, turning their bodies completely around, bobbing their heads back and forth, and fanning their tails. They watch for passing insects, mostly butterflies, grasshoppers, and dragonflies, then dart out to catch them in midair, but often they forage on the ground for beetles, termites, and spiders. During the breeding season, Vermilion Flycatchers frequent streamside woodlands in desert areas and landscaped parks, where females build flattish, well-made cups on horizontal tree branches less than 20 feet high but occasionally up to 30 feet.

STATUS AND DISTRIBUTION Vermilion Flycatchers breed from South America north to the desert regions of southern California and into the southern tip of the Sierra.

West Side. Uncommon but regular breeding (at least formerly) in the South Fork Kern River Valley, where up to 10 pairs nested annually at the Kern River Preserve; both proof of nesting and observations have declined dramatically in recent years as part of a population fluctuation cycle or a permanent condition; otherwise accidental.

East Side. Casual; a few spring and fall migrants are observed every few years from the Mono Basin south to the Owens Valley.

TRENDS AND CONSERVATION STATUS Formerly considered fairly common in riparian habitats along the lower Colorado River and in desert oases of southern California, Vermilion Flycatchers have declined dramatically in response to the clearing, degradation, and fragmentation of riparian habitats. Parasitism by Brown-headed Cowbirds may also be a factor in their decline, but these flycatchers are uncommon hosts. Due to continuing habitat losses and declining populations, Vermilion Flycatchers were added to the list of California Bird Species of Special Concern in 2008.

Ash-throated Flycatcher • *Myiarchus cinerascens*

ORIGIN OF NAMES "Ash-throated" for the species' coloration, however, most other members of this genus also have grayish throats and breasts; *myiarchus* "ruler of flies," from Gr. *myia*, a fly, and *archon*, ruler or chief; L. *cinerascens*, becoming ashy.

NATURAL HISTORY The subtle elegance of the Ash-throated Flycatcher's plumage is usually hidden away as they sit motionless on a low branch within the canopy of an oak, waiting to venture out to catch a passing bee or to pluck an unsuspecting caterpillar from the underside of a leaf. Luckily, they are frequently vocal with typical calls resembling the rolling warble of a softly blown referee's whistle.

Many flycatchers return to the same perch after pursuing an aerial insect, but Ash-throats often move on to a new one, feeding over wide areas. They frequent areas with scattered trees or patches of chaparral. They are well-adapted to tolerate dry, hot conditions, able to obtain all the water they need from their insect prey. Their breeding season diet is unusually varied, including fruits, small reptiles, and even the occasional mouse. Ash-throated Flycatchers, like other members of their genus, nest in cavities. If natural cavities or nest boxes are lacking, Ash-throats will use abandoned woodpecker holes in telephone poles, walls of buildings, and fence posts as well as hollow metal poles, metal boxes, old cars, vacuum cleaners, and other discarded, human artifacts. Females gather nesting material alone and construct nests by filling cavities with bits of cow dung, dried grass stems, and hairs.

STATUS AND DISTRIBUTION   Ash-throated Flycatchers winter in western Mexico and Central America and begin to arrive in the Foothill zone by late April. Most have left the Sierra by mid-August.

West Side. Common nesters in open or semi-open habitats below 3,000 feet; uncommon up to the Upper Conifer zone after breeding and during fall migration.

East Side. Uncommon breeders in Inyo County, rare breeders further north; uncommon spring (May) and rare fall (August) migrants; some postbreeding wanderers may appear rarely at higher elevations.

Brown-crested Flycatcher • *Myiarchus tyrannulus*

ORIGIN OF NAMES   "Brown-crested" for the species' most distinctive field mark; L. *tyrannulus*, little tyrant.

NATURAL HISTORY   Throughout desert riparian woodlands of the Southwest and Mexico, loudly bickering Brown-crested Flycatchers take up residence for a brief spell each summer. Despite giving conspicuous *huit* calls that run into a series when they're excited, Brown-crests are relatively shy birds that tend to remain hidden among the foliage of tall cottonwoods and willows. From perches high in the canopy, these acrobatic flycatchers sally forth to snatch cicadas, grasshoppers, and beetles from canopy-level foliage, or more rarely pursue flying insects in midair. Like other flycatchers, they specialize on insects in the breeding season but also mix in a few small fruits during migration and in winter.

They usually nest in natural tree cavities or woodpecker holes excavated in large cottonwood or willow trees or large cacti but will also nest in most of the same human-made nest sites occupied by Ash-throated Flycatchers (see account above). Because they arrive after most other species have already begun nesting, they may have to compete aggressively with other species for nest sites. Where they share habitats with Ash-throated Flycatchers, they use larger nest holes than their smaller cousins. Females, possibly with help from the males, add pieces of bark and plant fibers to nesting cavities, finishing off their nests with hairs, feathers, and perhaps a shed snake skin.

STATUS AND DISTRIBUTION While their breeding range extends from the southwestern United States to Central America, Brown-crested Flycatchers reach the northern extent of their range in a few, isolated scattered desert oases and riparian areas in southern California. They arrive in the southern Sierra in mid-May and depart by early to mid-August.

West Side. Regular breeding occurred formerly in the South Fork Kern River Valley, where 10 to 20 pairs bred annually at the Kern River Preserve and South Fork Wildlife Area; currently only being observed consistently at Kelso Creek Sanctuary located 15 miles south of the Kern River Preserve as the remnant of a population fluctuation cycle or a permanent condition.

East Side. No breeding records; spring and fall migrants are casual visitors to desert oases east of the Sierra region.

Western Kingbird • *Tyrannus verticalis*

ORIGIN OF NAMES "Western" for distribution; "kingbird" for their aggressive behavior toward other birds; L. *tyrannus* (see family account above); L. *verticalis*, relating to the top of the species' head, which has a hidden, orange-red patch.

NATURAL HISTORY "Feisty and fretful" (Floyd et al. 2007) succinctly describes the personality of these aggressive and loquacious birds. Their appearance in spring in open lowlands throughout the Sierra is a much anticipated confirmation that winter is over. These large, yellow-bellied flycatchers attract notice as they perch on roadside fences or engage in twisting, turning flights, pursing bees, wasps, and flies. Western Kingbirds, the largest of Sierra flycatchers, prefer dry, open habitats with scattered trees, fences, and telephone lines and generally avoid landscapes dominated by dense trees and shrubs.

As their name suggests, kingbirds are noted for their bold and aggressive territorial defense, directed against any perceived intruder. Males are the most fearless and will attack predators as large as hawks, owls, ravens, and even humans with loud calls or direct attacks with wings and feet. Their most common vocalizations have been described as sounding like an audio tape being played in fast-forward. However, this mnemonic is probably useless to anyone under 40 years of age.

Both parents spend four to eight days building a large, untidy nest of stems, twigs, and plants fibers lined with finer materials and felted together into a tight cup. Preferred nest sites include the outer branches of large trees, including orchard and shade trees near ranches, but they will also use telephone poles, fence posts, and old buildings.

STATUS AND DISTRIBUTION Western Kingbirds are long-distance migrants that breed in low, open habitats throughout the western United States and winter in southern Mexico and Central America. They arrive in California in March and breed in most open habitats throughout the state, and nearly all have departed the Sierra by late August.

West Side. Common spring and summer residents of annual grasslands and oak savannas of the Foothill zone up to about 2,500 feet in the central Sierra; rare individuals may move upslope to the Subalpine zone after breeding.

East Side. Fairly common breeders throughout in appropriate lowland, open habitats; fairly common to common in spring and fall migration, with most records from May; spring migrants arrive later than on the West Side.

SHRIKES · Family Laniidae

Shrikes could fairly be considered "honorary raptors." Much like other birds of prey, they sit motionless on exposed perches in open habitats like prairies, grasslands, tundra, and deserts, and dart out to catch large insects, small birds, or mammals—most often from the ground. Shrikes often impale their prey on thorns, sharp sticks, or barbed wire. Primarily an Old World family, there are more than 30 species in 3 separate genera in Europe, Asia, and Africa; only 2 species, the Loggerhead and Northern, occur in the Sierra or elsewhere in North America. There are no representatives of this family in South America or Australia, but one species made it to New Guinea. The family name is derived from L. *lanius*, a butcher, for the habit of ripping apart impaled prey.

Loggerhead Shrike · *Lanius ludovicianus*

ORIGIN OF NAMES "Loggerhead" refers to the bird's proportionately large head, literally, a "blockhead"— why the term is applied to this species is unclear since the Northern Shrike's head is larger, even compared with its larger body size; "shrike," related to "shriek," a reference to the bird's shrill calls; L. *lanius* (see family account above); L. *ludovicianus*, a formal name for Louis, a reference to Louisiana, where the first specimen was first collected.

NATURAL HISTORY Observant drivers through open shrublands, grasslands, or agricultural areas of the Sierra may find Loggerhead Shrikes perched on fences, utility lines, and exposed branches. In flight, shrikes can be identified by their characteristic style, beating their wings rapidly a few times then gliding with deep undulations. These handsome solitary hunters are among the few North American "songbirds" that regularly kill vertebrate prey. Equipped with remarkable eyesight, they sit patiently and watch for movement. When large insects, small birds, or mice emerge from hiding, shrikes attack them fiercely with sharp claws and hooked beaks. They kill by severing the neck vertebrae near the skull with several quick bites. When prey items are dead, or subdued, shrikes fly with labored wingbeats to favorite perches to eat. In the Sierra large insects—such as grasshoppers, crickets, beetles, caterpillars, and moths—compose the majority of total prey items in their diet, followed by vertebrates and large spiders.

Typical of this family, Loggerheads use barbed wire, thorns, or sharp twigs to shred their victims for immediate consumption or to impale their remains for later meals. A shrike larder in the Sierra foothills might include the decomposing remains of an American Pipit, Savannah Sparrow, or Horned Lark, along with a deer mouse, western fence lizard, grasshopper, or wolf spider. Despite their small size, shrikes rank as major predators of songbirds, especially nestlings and eggs, and have been known to destroy all the nests of other species within their territories. Although they usually pounce on prey from above, shrikes also hawk flying insects in the air. Similar to hawks, owls, and other birds of prey, they cough up pellets of undigested fur, feathers, and bones.

Shrikes require exposed lookout posts and space themselves widely along power lines, fences, or on shrubs or small trees. They maintain feeding territories all year in most of their range, requiring a few dense-foliaged trees or shrubs for cover and nest placement. In the Sierra foothills, shrikes initiate breeding earlier than most other passerines, and nesting occurs from early February until mid-June. Both sexes participate in building large nest structures from thick twigs, roots, and grasses on sturdy, forked branches; sometimes nests from previous years are simply repaired. Nests range in height from about three to eight feet above the ground, and they are well concealed by thick overhanging branches or hidden deep in dense shrubs. Females usually lay five or six grayish-buff eggs, with splotches on the large ends, and incubate them alone for about 16 days. Both parents feed the nestlings and continue bringing food to fledged young until they gain independence after about 45 days.

STATUS AND DISTRIBUTION Resident throughout much of lowland California, some northern breeders move farther south into the state in winter, but band recovery data suggest that most California birds are sedentary and movements from the north are not substantial.

West Side. Fairly common in the Kern River Valley and nearby areas of the Central Valley, becoming uncommon to rare in the low, sloping terrain of the Foothill zone, where they are mostly seen in open grasslands, irrigated pastures, or very sparse blue oak savanna; in September and October they turn up occasionally in large, open meadows as high as the Upper Conifer zone; winter birds stick mostly to the lowest elevations.

East Side. Uncommon to fairly common from April through August in open areas of sagebrush, agricultural land, and pinyon-juniper stands; generally rare in winter except for Mono County, where uncommon, and the Owens Valley, where good numbers are resident.

TRENDS AND CONSERVATION STATUS Breeding Bird Survey and Christmas Bird Count data confirm that Loggerhead Shrikes have declined in California and across North America. In California, widespread conversion of grasslands and shrublands to unsuitable breeding habitats such as urbanized areas, vineyards, orchards, and row crops is certainly partly to blame, but declines are seen even where habitat persists. Pesticide use in agricultural areas, and possibly collisions with vehicles along highways and country roads, has also contributed to declines in shrike numbers in the past few decades. More recently, West Nile virus also impacted Loggerhead Shrikes in the Central Valley, with some local populations reduced to a fraction of their pre–West Nile virus levels (Pandolfino 2008). Loggerhead Shrikes were included on the list of California Bird Species of Special Concern in 2008.

Northern Shrike • *Lanius exubitor*

ORIGIN OF NAMES "Northern" for the species' breeding range in Canada and Alaska; L. *exubitor*, a watchman, or "one who lies down outside," from L. *ex*, out of, and *cubo*, to lie down, although it is unclear why the name was applied to this species.

NATURAL HISTORY The hunting and food storage habits of Northern Shrikes are similar to their close relatives, Loggerhead Shrikes (see above account), except that Northern Shrikes (about an inch longer that Loggerheads) tend to select a higher proportion of small birds, mammals, and reptiles in their diet and a lower proportion of large insects and spiders. In flight, Northern Shrikes exhibit the same undulating flight as Loggerheads, and the two species appear similar when perched. Juvenile Northerns, the most likely age group of this species to be seen in the Sierra, have brownish plumage as compared to Loggerheads, which are grayish in all plumages. A much thinner "mask" and paler upperparts also help to distinguish Northerns from Loggerheads.

STATUS AND DISTRIBUTION While Loggerhead Shrikes are entirely restricted to North America, the Northern Shrike's breeding range extends well beyond northern Canada and Alaska to include most of Eurasia, northern Africa, most of the Middle East, reaching as far as Pakistan and India. In North America the breeding range of the Northern Shrike is more than a thousand miles north of California, but wintering birds visit northern parts of the state intermittently, especially in the coldest winters when prey are scarce at higher latitudes.

West Side. Casual winter visitors with records from the Kern River Valley and annual grasslands west of Yosemite National Park; accidental in most other areas but casual in nearby areas of the Sacramento Valley.

East Side. Uncommon but regular winter visitors to flat, lowland areas north of Lake Tahoe, casual south; most Northern Shrikes observed are brownish juveniles, but some adults may also present; in midwinter, Honey Lake and Sierra Valley are the most likely places to find these visitors from the far north.

VIREOS • Family Vireonidae

Vireos are superficially similar to wood-warblers (family Parulidae), and the two families were once considered close relatives. More recent studies revealed stronger relationships between vireos, shrikes, and jays, and one can see a resemblance in the vireo's somewhat heavy, slightly hooked bills. Unlike colorful male warblers, all Sierra vireos wear subdued plumages year-round, and males and females look the same. Also unlike warblers, vireos frequently sing in fall, making them easier to detect in fall migration than other silent migrants. Vireos are methodical foragers, moving deliberately through foliage, often cocking their heads to peer downward, and only occasionally flycatching. They eat mostly insects in the breeding season but often seek fruits during migration and in winter.

All vireos have similar courtship and nesting behaviors. Males display by swaying their bodies from side to side, erecting their flank feathers, and giving gentle calls. They may show females potential nest sites where they have already added bits of material. Both sexes generally cooperate in building well-made nests over four to seven days, usually in late May or early June. Nests are small, pendulous cups suspended between the forks of horizontal branches near the periphery of tree crowns. Typically they are adorned externally with bits of lichens, mosses, spiderwebs, and a few leaves. Both sexes usually incubate three to five white eggs that are sparsely marked with dark speckles; eggs hatch in about 14 days. Both parents feed the nestlings for another 14 days and continue feeding fledglings for two to three more weeks.

Brood parasitism by Brown-headed Cowbirds is a serious problem for most vireos, with over 80 percent of nests parasitized in some areas. Local vireo populations declined sharply after cowbirds arrived in the Sierra in the 1930s.

There are about 50 species in this New World family, with 13 breeding in North America. Four species occur regularly in the Sierra, and one former breeder, Bell's Vireo, was likely extirpated but may be making a comeback. The family name is derived from L. *virere*, to be green.

Bell's Vireo • *Vireo bellii*

ORIGIN OF NAMES "Bell's" for John Graham Bell (1812–1899), a taxidermist who accompanied John James Audubon on a tour of the Missouri River in 1843; this vireo was one of four new species he collected and described on an 1849 California trip; the scientific name for the Sage Sparrow was also given in Bell's honor.

NATURAL HISTORY Long gone are the pristine corridors of lush vegetation along most of California's great rivers, and along with them these sprightly little brush-loving vireos have mostly disappeared (see "Trends and Conservation Status" below). There are now only a few isolated, but growing, breeding populations in southern California, and recent years have seen encouraging new breeding attempts in the Central Valley and in the Sierra. These plainly attired birds remain well hidden in low, dense shrubs and rarely ascend more than a few feet above the ground. Their presence would often be overlooked if not for the male's loud, ceaseless songs—a mix of rising and falling, rapidly delivered, rough-sounding phrases: *wretchity-wretchity-wretchity-wuh-ree?* These songs may be heard even in the heat of midday.

STATUS AND DISTRIBUTION According to Grinnell and Miller (1944), "Least" Bell's Vireos *(V. b. pusillus)* were once "common, even locally abundant under favorable conditions of habitat." Such habitats included dense riparian groves of the Central Valley and low foothills on both sides of the Sierra.

West Side. Extirpated from most of their former Sierra range for more than 80 years, but recent records provide a glimmer of hope for the future; a pair was present near Lake Success through the breeding season in 2010, and singing males have been noted in the Kern River Valley several times since the early 1990s; singing males and pairs have also been found at several sites in the Central Valley in recent years.

East Side. Formerly nested along the Owens River and near Bishop (Inyo County); no breeding records for over a century until 2008 and 2009, when successful breeding was confirmed near Big Pine (Inyo County)—in 2009 that pair managed to fledge one Bell's Vireo in addition to one Brown-headed Cowbird; a singing male of the Arizona subspecies (*V. b. arizonae*) was observed along Hot Creek (Mono County) in late May 2007.

TRENDS AND CONSERVATION STATUS With the arrival of Brown-headed Cowbirds (originally native to the Great Plains) in California in the late 1890s, a host of songbirds experienced rapid and dramatic declines. Perhaps no other species were affected more severely than "Least" Bell's Vireos, as they disappeared from most of their former range in just a few decades. Cowbird nest parasitism coupled with the loss of most suitable nesting habitats ensured their statewide demise. This subspecies is listed as Endangered by both the U.S. Fish and Wildlife Service and the California Department of Fish and Game. Careful habitat management and cowbird control in southern California breeding areas has increased local populations. Birds dispersing from those populations may be responsible for the Central Valley and Sierra breeding attempts mentioned earlier.

Plumbeous Vireo • *Vireo plumbeus*

ORIGIN OF NAMES "Plumbeous" from L. *plumbeus*, lead-colored.

NATURAL HISTORY Once lumped together as the "Solitary Vireo complex," Plumbeous, Cassin's, and eastern Blue-headed Vireos *(V. solitarius)* were recognized as separate species by the American Ornithologists' Union in 1997. Although they breed in different geographic areas and habitats, they apparently have very similar natural histories, and both Plumbeous and Cassin's Vireos are covered here in one (Cassin's Vireo) account below.

STATUS AND DISTRIBUTION Plumbeous Vireos breed in open pinyon and Jeffrey pine woodlands of the Rocky Mountains and much of the Great Basin, and their range just reaches the eastern edge of the Sierra. The exact dividing line between the ranges of Cassin's and Plumbeous Vireos in the Sierra is difficult to determine with certainty because of the similarity in appearance and song between the two species.

West Side. Rare, breeding documented in a few locations in southeastern Tulare County and northern Kern County; occasional reports of spring migrants elsewhere are difficult to verify since spring Cassin's Vireos in faded plumage can be nearly indistinguishable from Plumbeous Vireos.

East Side. Uncommon breeders in pinyon-juniper woodlands of Mono and Inyo Counties and possibly Kern County; a handful of breeding season records in eastern Alpine County probably represent the northern extent of Sierra breeding range, although three birds reported in 2009 along a Breeding Bird Survey route in eastern Plumas County were in appropriate Plumbeous Vireo habitat; breeders in the Tahoe Basin are mostly Cassin's Vireos, consistent with the habitat present there; rare spring and fall migrants elsewhere.

Cassin's Vireo • *Vireo cassinii*

ORIGIN OF NAMES "Cassin's" for John Cassin (1813–1869), who attended the same Quaker school in Pennsylvania as John Townsend (see Townsend's Solitaire and Warbler accounts) and Thomas Say (see Say's Phoebe account); a life-long businessman, Cassin was a volunteer curator of birds at the Academy of Natural Sciences in Philadelphia in the mid-1800s, where his prolific career included being the first to describe almost 200 species of birds from around the world; while he never visited the western United States, the common names of five western species were given in his honor, including an auklet, kingbird, sparrow, finch, and this vireo.

NATURAL HISTORY Cassin's Vireos, like many other North American vireos, sing as they forage deliberately through the trees and seem to be carrying on pleasant question-and-answer conversations with themselves as they work. "How are you?" . . . "I'm just fine" . . . "and How are *you*?" . . . "I'm doing fine." The vaguely sing-song nature of the tune. and the fairly long pauses between phrases are characteristic. Cassin's also give grating scolding calls that sound somewhat like a rapid *chee-chee-chee.*

Although Cassin's Vireos sing and feed from near the ground to the treetops, most nests are placed below 20 feet, usually shaded by overhanging branches. They usually breed in wetter, denser woodlands than Plumbeous Vireos. Males of both species are aggressive and fearless nest defenders that chase off intruders with loud, scolding calls. Surprisingly, incubating females tolerate close approaches and have been picked off nests by humans. Although loud and aggressive in defense of their nests, these vireos are still heavily parasitized by Brown-headed Cowbirds, especially near residential areas, campgrounds, and stables that attract these nest parasites. Cassin's and Plumbeous vireos may be unique among vireos because pairs often split-up, each parenting half the brood separately.

STATUS AND DISTRIBUTION In spite of this species' susceptibility to cowbird nest parasitism, Cassin's Vireos appear to be increasing both in the Sierra and throughout their range. Spring migrants arrive in the Foothill zone in early April, and peak breeding occurs from mid-May until early July; most depart for wintering grounds in central Mexico by late August, but some linger through September.

West Side. Common in oak and pine forests of the Lower and Upper Conifer zones from 2,500 to about 7,000 feet in the central Sierra; while often seen in riparian forests, they are about equally common in nonriparian oak and conifer forests.

East Side. Fairly common to uncommon breeders in the Tahoe Basin, including the Carson Range on the east side of Lake Tahoe south through Mono County; elsewhere uncommon migrants from late April to mid-May, and in August.

Hutton's Vireo • *Vireo huttoni*

ORIGIN OF NAMES John Cassin (see Cassin's Vireo account above) reluctantly named this bird after William Hutton (birth and death years unknown), a young and unknown amateur naturalist who collected the first specimen of this species in Monterey County in 1847.

NATURAL HISTORY Look and listen for this bird in stands of live oak because no other Sierra bird is as closely tied to these evergreen trees as this vireo. Because live oaks retain leaves through winter, Hutton's Vireos can search for insects in their foliage year- round and are the only Sierra members of their family that do not migrate south for winter. While they can sometimes be found in willows, blue oaks, and ponderosa pines, they never stray far from oak groves.

The drably attired males might be overlooked if not for their songs, distinctive only because they are the most monotonous of any Sierra songbird. Although there is some variety both between and within individuals, each bout of song is relentlessly repetitive. Most songs are distinctly two-noted and usually repeated many times in succession—in one case, 781 times in just over 11 minutes (Bent 1950)! The song is typically represented as *zu-wee, zu-wee, zu-wee* or *vee-ur, vee-ur, vee-ur*. On rare occasions one may encounter a true "virtuoso" Hutton's alternating upslurred and downslurred phrases to sound vaguely like a Cassin's Vireo in training. They may sing any time of year and also give a characteristic peevish, chattering scold that leaves the target of this vocalization with no doubt that they are being reprimanded. Shortly after they begin singing in late winter, Hutton's Vireos establish pairs, although some may remain paired up year-round. They nest in almost any kind of tree or large shrub and usually build their nests at the ends of forked branches.

Moving slowly through the oak foliage mostly staying under overhanging branches, Hutton's Vireos glean a diversity of scale insects, bugs, moths, butterflies, and spiders; in fall and winter, they also pluck the fruits of elderberries, poison oak, and coffeeberries. In winter, they join flocks of other small birds like kinglets, chickadees, titmice, Bushtits, and nuthatches moving noisily through the foothill forests. Hutton's Vireos are often confused with Ruby-crowned Kinglets, but the vireos have larger heads and heavier bills, move more slowly, and flit their wings less often than the more abundant kinglets.

STATUS AND DISTRIBUTION Hutton's Vireos occur in live oak woodlands of the Sierra foothills and the Coast Range year-round, and their global range extends south to Guatemala. They are usually relatively sedentary except for downslope movements in winter, and fall and spring records from the East Side show that they can occasionally wander more widely.

West Side. Fairly common residents of the Foothill zone from about 2,500 to about 4,500 feet; uncommon in black oak forests of the Upper Conifer zone after breeding from early August until mid-October.

East Side. Casual fall visitors, with September records at Lake Tahoe, Mono County Park, and Butterbredt Spring; spring records at Butterbredt Spring, Jawbone Canyon (Kern County), and the Owens Valley.

Warbling Vireo • *Vireo gilvus*

ORIGIN OF NAMES "Warbling" for the species' melodic songs; L. *gilvus*, yellowish.

NATURAL HISTORY This vireo's name was well chosen, for the males sing prolifically throughout the day in beautiful warbling voices, described by William Dawson (1923) as "fresh as apples and as sweet as apple blossoms comes that dear, homely song from the willows." Surprisingly, they even sing from the nest. While at first hearing the songs sound varied and improvised, repeated listening

shows that individuals sing the same complex song over and over again. Plain-colored and snuggled down in deep, woven nest cups with only bills and tails protruding, they are difficult to see, but singing must surely make it easier for predators to find them—how this behavior survived natural selection is a mystery.

Warbling Vireos thrive amid such deciduous trees as cottonwoods, alders, and willows, especially beside streams or moist meadows where they forage among dense canopy leaves. They also favor areas with deciduous black oaks and aspens within conifer stands. Their preference for riparian areas at lower elevations may relate to a need for humidity in these warmer areas, as the species much more readily occupies pure conifer forests at cooler, higher elevations. This vireo's foraging style makes it one of the easier ones to get into clear view. They tend to work a single tree intensively, rather than flit from tree to tree, and they will sing intermittently all the while. They also spend much of their time near the ends of branches harvesting caterpillars, their staple food. Late in the season they may feed heavily on berries before leaving for winter.

STATUS AND DISTRIBUTION Warbling Vireos arrive from wintering grounds in Mexico and Central America by late April (usually a week or two later than Cassin's Vireos), and peak nesting occurs during June. Most have departed the Sierra by mid-August but a few may remain until early September, especially at higher elevations.

West Side. Common to abundant in riparian forests in spring migration (up to 1,400 have been observed in a single mid-May morning at Butterbredt Spring); common breeders in riparian and moist conifer forests of the Lower and Upper Conifer zones from about 3,000 to 7,000 feet in the central Sierra; uncommon to rare up to the Subalpine zone after breeding.

East Side. Common breeders, mostly in riparian areas and aspen groves.

JAYS AND RELATIVES • Family Corvidae

These raucous "songbirds" are often disparaged for camp robbing, eating garbage and carrion, and scolding human trespassers and pets with loud, harsh squawks and calls. However, their superior learning abilities, keen memories, and complex vocal repertoires fascinate many bird students. A few species in laboratory settings have shown they can count and demonstrate self-awareness by repeatedly looking in mirrors, solving complex, multistep problems, and even making their own tools.

They have strong legs, long, stout bills, and rictal bristles covering their nostrils, which serve an unknown function. As omnivorous as humans, Corvid food is anything that can be caught and swallowed, including large insects, small vertebrates, birds, carrion, seeds, fruits, and berries. Most members of this family cache food for later consumption during lean times. They quickly exploit new food sources and have adapted readily to taking advantage of human-made opportunities like roadsides with freshly killed animals. Males and females of most species wear identical plumages and only molt once per year.

Most female Corvids lay three to six greenish or bluish, splotched eggs and incubate them alone (or with some help from their mates) for 16 to 20 days. The young remain in the nests, where they are fed and tended by both adults for about three weeks before they gain independence; larger species (such as magpies, crows, and ravens) care for their young for up to five or six weeks. This cosmopolitan family contains about 115 species residing almost everywhere except in Antarctica, New Zealand, and some oceanic islands—9 of these occur in the Sierra. The family name is derived from L. *corvus*, a raven.

Pinyon Jay · *Gymnorhinus cyanocephalus*

ORIGIN OF NAMES "Pinyon" for the species' preferred habitat; *gymnorhinus*, bare-nosed, combined from Gr. *gymnos*, naked, and *rhinos*, nose—a reference to the species' unfeathered nostrils, an unusual trait among members of this family; L. *cyanocephalus*, blue-headed.

NATURAL HISTORY Due to their unique behavioral and physical characteristics, Pinyon Jays are classified in their own genus, and their closest relatives are thought to be Clark's Nutcrackers. By far the most gregarious members of their family in the Sierra, they spend most of their lives in flocks numbering 250 birds or more. Extended family groups of this highly communal species remain together year-round, including during the breeding season, and many birds never leave their natal flocks. When foraging, some adults, usually males, serve as sentries and watch for predators while their companions feed.

As their name suggests, Pinyon Jays are strongly associated with pinyon pines throughout the year, and the two species have evolved a mutually beneficial relationship. The jays depend on pine seeds as a major staple of their diet, and the pines depend on the jays for long-range dispersal of their wingless seeds. Pinyon Jays harvest the green cones of pinyon pines in fall, hack them apart with their strong bills, and cache the seeds underground for consumption in the winter months. Individual jays store thousands of seeds every year, and have an uncanny ability to find them again months after they were cached, even under snow cover. In years when pinyon pines produce an overabundance of seeds, the jays cache more than they can consume and the remaining seeds can germinate.

Pinyon Jays also take a diversity of other foods, such as acorns, juniper berries, large insects, nestling birds, reptiles, and small mammals. They kill vertebrate prey with sharp blows to the heads and necks of their victims. However, their lives are primarily committed to harvesting, transporting, caching, and retrieval of pinyon pine seeds. Their long, strong bills aid these activities, as do an expandable esophagus for storing hundreds of seeds at a time and long wings to power flights of hundreds of miles in search of food in years when the pinyon crop is poor.

Pinyon Jays are monogamous, and mated pairs stay together for years until one of them dies. Pairs appear to coordinate their seed caching to ensure they both know the general locations. They nest communally in large groups but rarely, if ever, nest in the same trees. Mated pairs begin courtship and nest building by mid-February, much earlier than most Sierra birds, in areas where the flock cached seeds the previous fall. Tree branches, either near the trunk or hidden in dense sprays of foliage near the tips, provide substrates for the nests. Pairs cooperate to build bulky open-cup nests lined with grasses, fine plant fibers, feathers, horsehairs, or shredded juniper bark. Females incubate the eggs, while males return frequently with food.

After the young hatch, both parents feed them; young from previous years may serve as "helpers at the nest" delivering food to the nestlings, similar to the behavior of Acorn Woodpecker extended family groups. Adults and family helpers normally feed only related young, but once they reach near-fledging size, they sometimes receive meals from unrelated adults. Fledgling birds often gather into communal "teenaged groups," or crèches, until they mature and can forage on their own.

STATUS AND DISTRIBUTION Widespread residents of the intermountain west and the eastern Sierra, Pinyon Jays disperse over large distances in the nonbreeding season, especially in years when the local cone crops of pinyon pines fail; one of the more dramatic dispersals was in fall 2003, when birds were found in several locations near the Pacific Coast in California.

West Side. Fairly common almost year-round in pinyon pine woodlands on the Kern Plateau (Tulare County), and on both sides of Walker Pass, Scodie Mountains, and Piute Mountains (Kern County); otherwise, rare and irregular visitors over the Sierra crest from late August to early November, usually above 7,000 feet and north of the Yosemite region.

East Side. Locally common residents and breeders from February through August, but exact locations shift from year to year in response to food; fairly reliable locations include north of Mono Lake along the road to Bodie and near Bridgeport (Mono County), near Indian Creek Reservoir (Alpine County). They wander great distances in fall and winter, as noted earlier, making them one of the most truly irruptive species in North America.

Steller's Jay • *Cyanocitta stelleri*

ORIGIN OF NAMES "Steller" (not "Stellar") for Georg (not "George") Wilhelm Steller (1703–1746), a German naturalist on the Bering Expedition to the Russian Far East and western Alaska (1740–42). He was the first to collect and describe many plants and several animals that now bear his name, including a sea lion, a sea cow, an eider, a sea-eagle, and this jay; *cyanocitta*, a blue jay, combined from Gr. *kyanos*, blue, and *kitta*, a jay.

NATURAL HISTORY Noisy, bold, and inquisitive, these striking black and blue jays are among the most familiar Sierra birds. Steller's Jays are the western counterparts of the similarly crested eastern Blue Jays *(C. cristata)*, extremely rare visitors to California. Male and female Steller's Jays wear identical plumage and sport conspicuous crests with grayish-black heads and throats; the remainder of their plumage is a deep blue, and the wings and tail are especially vibrant.

Steller's Jays employ a complex vocabulary when interacting socially and aggressively or when warning against or mobbing predators. Their diverse calls include a harsh *schaak-schaak-schaak* or *chook-chook-chook* rattle calls and a nearly perfect imitation of the Red-tailed Hawk's shrill cry. On rare occasions, birds give a very soft, surprisingly musical vocalization that can go on for some time. The exact function is unknown, but it is usually given from dense cover—sometimes with another bird close by, sometimes not. They are superb mimics, but the benefit to the jays of such vocal virtuosity is unknown. It is speculated that the hawk imitation may induce small birds to leave their nests to mob the predator, opening an opportunity for the jay to harvest the nest contents.

Steller's Jays' impressive intelligence is demonstrated by their rapid exploitation of new food resources and by their ability to find food from treetops to the ground. In the absence of humans, they subsist largely on seeds, insects, and fruits but will feed eagerly on small mammals, birds' eggs and young, and carrion. Around houses, farms, and campgrounds, they soon learn to seek garbage and human handouts, and they often nest under the eaves of summer cabins and other buildings with low levels of human use. The thriving populations of Steller's Jays around human settlements in the Sierra no doubt reduce the numbers of other songbirds since jays will raid the nests of many species.

Despite their otherwise conspicuous behavior, Steller's Jays are among the more secretive nesters in the Sierra. They hide their large, bowl-shaped twig nests in the dense foliage of conifers, often below 20 feet. Parents keep extremely quiet around the nests and vary their routes to and from feeding areas, to confuse potential predators. Mated pairs may stay together for several years. Courtship and nest building begins in the Sierra by mid-April. Most seek nesting territories with a source of water since mud is required for nest construction. Females do most of the work building sturdy cups of mud and twigs, usually lined with dead leaves, mosses, or pine needles. Males bring food to females during the incubation period. Young birds become independent about a month after leaving the nests, usually by mid-July in the Sierra.

STATUS AND DISTRIBUTION From southern Alaska to Nicaragua, 15 subspecies of Steller's Jays have been described. "Blue-fronted" Steller's Jays *(C. s. frontalis)*, distinguished by the blue vertical stripes above their eyes, reside in the Sierra year-round.

West Side. Locally abundant and conspicuous residents where human-supplied foods are provided; common in wildland settings from the upper edge of the Foothill zone through the Upper Conifer zone; continuous stands of lodgepole pine are avoided, even when they occur within other forest types such as red fir; common in higher elevation Subalpine forests above 8,000 feet only near campgrounds, stables, and other developed areas; some remain at high elevations through the winter, but most retreat to lower elevations, where food is easier to find, with small numbers occasionally showing up on the Central Valley floor.

East Side. Fairly common in most forests including ponderosa and Jeffrey pines, pinyon pines, junipers, mountain mahogany, as well as aspen-dominated riparian woodlands up to about 8,000 feet; most numerous near springs, creeks, and other sources of water; uncommon and irregular during the coldest months from December through February, except where human-supplied foods can be found.

Western Scrub-Jay • *Aphelocoma coerulescens*

ORIGIN OF NAMES "Western" for distribution and to distinguish it from other North American scrub-jays; "scrub" for typical habitat of the species; *aphelocoma*, smooth hair, from Gr. *apheles*, smooth, and *kome*, hair; L. *coerulescens*, bluish.

NATURAL HISTORY The lives of Western Scrub-Jays, like those of the Acorn Woodpeckers, Oak Titmice, and the Western gray squirrel, are intimately linked to oaks. These jays pick acorns directly from trees but, unlike Steller's Jays, which often forage high in trees, they take most other foods on the ground. A central feature of the Western Scrub-Jay's annual cycle is the storage of acorns, a staple food they harvest in fall and hide singly in small holes dug in the ground with their bills.

A. c. superciliosa

A. c. woodhouseii

Some forgotten acorns germinate into seedlings, thus benefiting the subsequent generations of both oaks and jays. It is likely that the majority of California's majestic valley and blue oaks were planted by these jays. Pine nuts, almonds, and other large seeds are cached in a similar fashion. These stores form the bulk of their diet through the winter, but insects become important during the nesting season. Like other members of their family, Western Scrub-Jays are opportunistic omnivores. While seldom found far from oak trees in the Sierra, they also thrive in such arid habitats as foothill chaparral and pinyon-juniper woodlands. They reach their highest densities in residential areas where fruiting trees, bird feeders, and human garbage provide excellent foraging opportunities.

Mated pairs cooperate in building the nests of twigs, mosses, and dry grasses low in trees or bushes, usually less than 20 feet above the ground and almost always near water. The young forage independently after about two months but may associate with their parents through the fall. As with Steller's Jays, they stay extremely quiet around nests to avoid attracting predators. Elsewhere, however, these crestless jays habitually perch in plain view and may approach human visitors to investigate or look for food. Both sexes make a great variety of strident calls, among the most familiar and characteristic sounds of the foothills, and both occasionally sing soft, musical "whisper" songs, not unlike Steller's Jays.

STATUS AND DISTRIBUTION At least 15 subspecies of Western Scrub-Jays have been described from Washington to southern Mexico. Representatives of two distinct groups of subspecies occur in the Sierra: "California" Scrub-Jays, represented by the western subspecies *(A. c. superciliosa)*, and "Woodhouse's" Scrub-Jays *(A. c. woodhouseii)*, which may be recognized as a new species in the future.

West Side. Common to locally abundant year-round residents of the Foothill zone, especially near human habitations; also common in oak woodlands, chaparral, and riparian forests; uncommon residents up to the Lower Conifer zone; occasionally wander above their breeding elevations (e.g., to well over 4,000 feet in the northern Sierra); birds that sometimes show up west of the Sierra crest from Alpine County south in late summer and fall are likely "Woodhouse's" Scrub-Jays from the East Side; "Woodhouse's" Scrub-Jays are also regularly found on the Kern Plateau in southeastern Tulare County well above 5,000 feet.

East Side. Fairly common year-round residents where they mostly breed in pinyon-juniper wood-lands; in fall and winter, they may also forage in sagebrush, riparian woodlands, and around human habitations; "California" Scrub-Jays are the resident subspecies north of Alpine County while "Woodhouse's" Scrub-Jays occur from Alpine County south. Interestingly, Scrub-Jays found just east of the Sierra in Nevada's Pine Nut Mountains look like birds of the "California" group but are genetically closer to the "Woodhouse's" group.

Clark's Nutcracker • *Nucifraga columbiana*

ORIGIN OF NAMES "Clark's" for Wil-liam Clark (1770–1838), coleader of the Lewis and Clark Expedition (1803–1806); "nutcracker" for habit of crack-ing pine nuts; *nucifraga*, a nutcracker, from L. *nux*, a nut, and *frangere*, to crack or fracture; L. *columbiana*, of Colum-bia—a reference to the river where the species was observed during the Lewis and Clark Expedition.

NATURAL HISTORY In the Subalpine zone Clark's Nutcrackers replace Stell-er's Jays as the noisiest and most con-spicuous birds. Pine nuts are the staff of life for these handsome Corvids of the high Sierra. In late summer and fall, a single Clark's Nutcracker may harvest and store more than 30,000 pine seeds, its main food supply until the following summer. Nutcrackers studied at Tioga Pass and Mammoth Lakes (Mono County) had two widely separated cache areas, one in a Subalpine forest and other in a lower-elevation Jeffrey or pinyon pine forest, more than eight miles downslope (Tomback and Kramer 1980).

Nutcrackers harvest pine seeds by prying them from cones hanging on trees or lying on the ground. They seek out large-seeded pines, such as whitebarks, pinyons, and Jeffreys, bypassing the smaller seeds of lodgepoles and western white pines. Some seeds are eaten immediately, but as many as 150 are stored at a time in a "sublingual pouch" at the base of the tongue. Like a chipmunk's, these pouches facilitate the transport of seeds to storage areas. Seeds are cached secretively in hundreds or thousands of small holes dug with the bill, a few seeds per hole, and are subsequently refound through feats of memory possibly unequaled in the animal world. Nutcrack-ers have been seen digging through four feet of snow to recover hidden seed caches with near-perfect precision! Of course, some buried seeds are never used and germinate to spread pine to new areas. Research in the Sierra and in the Rocky Mountains concludes that nutcrackers are the main dispersal agents of whitebark pines and possibly limber pines (Hutchins and Lanner 1982).

In the fall, most descend to lower-elevation areas east of the Sierra crest, where they continue to harvest and cache large pine seeds. Breeding begins as early as mid-February, but a deep snow-pack can delay this until late spring. Both adults gather material to build large stick nests lined with shredded bark, soil, and other fine materials for insulation. Nests sites can be in the forks of branches or in sprays of dense foliage near branch tips, but they are almost always placed on the leeward sides of trees to protect them from high winds. While females take on most of the incubation duties, both parents are equally involved in feeding the nestlings stored seeds and insects until they leave and begin to forage on their own. Both adults feed and protect their young for at least a month, and fledglings follow their parents around for several months to learn the complexities of successful seed storage and retrieval.

STATUS AND DISTRIBUTION Clark's Nutcrackers can be found in two distinct types of habitat in the Sierra region, both characterized by dry, open pine forests. In summer and fall, they primarily inhabit Subalpine forests near tree line on both sides of the crest, but in winter through the breeding season they are mostly confined to lower-elevation pine forests.

West Side. By late May, after most young birds have fledged, family groups begin to move west over the Sierra crest to forage in Subalpine forests, and some descend regularly to the Upper Conifer zone to feast on ripening pine seeds; in exceptional years, probably due to a shortage of pine nuts at high elevations, some descend to the Lower Conifer zone, and a few desert the Sierra region to search for food in the surrounding lowlands; possible nesters on the Kern Plateau (e.g., Troy Meadow and Bald Mountain) as individuals are detected every summer; a few other historical records exist of nutcrackers nesting on the West Side.

East Side. Fairly common but localized breeders and winter residents from November through April; most nesting in large, open stands of Jeffrey and ponderosa pine but also frequent pinyon-juniper habitats; fairly common in open Subalpine forests from May through early November.

Black-billed Magpie • *Pica hudsonia*

ORIGIN OF NAMES "Black-billed" to distinguish it from the Yellow-billed Magpie (see account below); "magpie," a reference to the species' overall black-and-white coloration; L. *pica*, a magpie; L. *hudsonia*, refers to the northernmost boundary of the species' historical range—Hudson Bay in northern Canada.

NATURAL HISTORY Exotically attired compared to most other members of their family, with blue, violet, and green iridescence, and long, graceful tails, Black-billed Magpies look nearly identical to European Magpies *(Pica pica)*. Both were considered members of the same species until recently, but new genetic evidence revealed that they are distinct species. Like most Corvids, Black-billed Magpies consume almost any foods they encounter, including large insects, rotting carcasses of deer and cattle, small rodents, bird eggs and young, berries, seeds and nuts, as well as human garbage and the remains in pet food bowls. Magpies usually forage on the ground by walking or hopping, and they move quickly to catch unsuspecting prey.

Magpies also cache food by poking holes in soil or snow with their bills and regurgitating partially digested food items and covering them with leaves, chunks of wood, or grasses. Such caches are used for temporary storage only and are usually relocated in a few days by sight and/or smell before they rot to the point that even magpies will not touch them. Another notable trait is their habit of perching on large mammals such as deer, pronghorns, and cattle and picking ticks and other ectoparasites from their backs. This mutualistic relationship provides magpies with a source of protein and rids the mammals of the bloodsuckers. Lewis and Clark first encountered these magpies as the birds followed bison herds across the Great Plains or congregated around Native American camps.

Magpies nest in trees or tall shrubs and, since mated pairs only defend the immediate areas around their nests, several pairs may nest in close proximity, forming loose, unstructured breeding colonies. By mid-June both adults participate in building the dome-shaped nests consisting of a rough assemblage of branches twigs forming the platform and a cup lined with finer materials such as shredded juniper bark, dried grasses, and small roots, all cemented together with mud. The entire process can take more than 40 days and is an important element in maintaining pair bonds. Nests from previous years are often rebuilt if not co-opted by hawks or owls. Rebuilt nests grow in size as more material is added and can eventually become more than 40 inches in diameter and nearly 4 feet deep.

Black-billed Magpies form loose flocks outside the breeding season, and dominant, older birds steal food from younger and less experienced birds. Similar to American Crows and Pinyon Jays, magpies form communal roosts in dense-foliaged trees in the cold months to protect them from fierce winds and to communicate the locations of productive feeding areas to each other. At night, they often regurgitate pellets consisting of fur, bones, and other undigested materials from the previous day's feeding.

STATUS AND DISTRIBUTION Widespread in shrub-steppe habitats and adjacent agricultural areas of western North America, the Black-billed Magpie's range extends to the eastern base of the Sierra. Most nesting and roosting areas are in riparian groves near creeks or springs that provide a source of drinking water and mud for nest construction. Magpies disperse from these centralized locations to forage almost anywhere they can find food, including sagebrush flats, irrigated pastures, lakeshores, and ranches.

West Side. Residents just west of the crest in Westwood (Lassen County) and Indian and American Valleys (Plumas County); otherwise casual, with only incidental occurrence of one or a few birds wandering upslope to just over the Sierra crest; a few records from the foothills or lower are most likely escapees, as magpies are sometimes kept as pets.

East Side. Fairly common residents but exact locations vary seasonally and annually depending on where they find reliable food; observed in appropriate habitats from Honey Lake to the Owens Valley.

Yellow-billed Magpie • *Pica nuttalli*

ORIGIN OF NAMES "Yellow-billed" to distinguish it from the former species; *nuttalli* for Thomas Nuttall (see Common Poorwill account).

NATURAL HISTORY These common inhabitants of the Sacramento and northern San Joaquin Valleys range up into the lower foothills in parts of the central and southern Sierra, where grasslands are interspersed with scattered oaks. By day they disperse from their roosting trees to feed in

small flocks on the ground, searching pasturelands, cultivated agricultural fields, and barnyards for grasshoppers, other large insects, and other foods. Similar to their close relatives, Black-billed Magpies, they also feed on carrion, small animals, acorns, and fruit. Unlike Black-bills, however, Yellow-bills are highly sedentary, and individual birds rarely wander more than a few miles from their nesting territories.

Highly social, Yellow-billed Magpies forage and roost together in large groups and are communal nesters—usually in groups of a few up to 30. They prefer to nest in the highest branches of tall trees, especially valley, live, and blue oaks near rivers or ponds in open areas. Less commonly, they nest in orchards and in suburban areas and other open areas with only a few isolated trees. Nesting pairs build dome-shaped nests, about three feet across, using sticks cemented with mud. Nest construction starts in late December, and the entire building process may not be completed until early March. Both parents continue to feed and guard the young for about two months after they fledge. Other species, such as American Kestrels, Western Screech-Owls, and Long-eared Owls, may take over abandoned magpie nests. Yellow-billed and Black-billed Magpies undoubtedly have a shared ancestry, but they were probably geographically isolated during the Pleistocene glacial periods when they evolved into two distinct species.

STATUS AND DISTRIBUTION Yellow-billed Magpies are endemic to the Sacramento and northern San Joaquin valleys, south Coastal Ranges, and the foothills of the western Sierra. Along with Island Scrub-Jays *(Aphelocoma insularis)*, endemic to Santa Cruz Island, they are the only North American species that has never been recorded outside its native state.

West Side. Fairly common residents in scattered locations at lower elevations of the Foothill zone up to about 2,000 feet in the central Sierra, where primarily found in oak savanna as well as developed parks and some landscaped suburban areas; unlike most Sierra birds, Yellow-bills are never seen above the foothills and are absent from many areas of suitable habitat within the proper elevation range; species population declined alarmingly following the 2005 outbreak of West Nile virus (see "Trends and Conservation Status" below).

East Side. No records.

TRENDS AND CONSERVATION STATUS Having survived early 20th-century efforts to exterminate them as agricultural pests, Yellow-billed Magpies now face an even more menacing threat. Their populations have declined significantly since the first widespread outbreak of West Nile virus in

California in 2005 (Crosbie et al. 2008). Following the summer of 2005, their population was reduced by approximately half in the Central Valley, and the species nearly disappeared from some locations where it was formerly abundant. While there has been some evidence of local recovery, numbers are still well below historical levels, and studies suggest that few of the survivors are resistant to this virus. As long as good habitat remains, one can hope that the species will eventually recover, but the long-term effects of this disease on magpies, other Corvids, and migratory birds in general are unknown.

American Crow • *Corvus brachyrhynchos*

ORIGIN OF NAMES "Crow" refers to the species' calls; L. *corvus* (see family account above); *brachyrhynchos*, short-billed, from Gr. *brachus*, short, and *rhynchos*, a bill or beak.

NATURAL HISTORY Among North American Corvids, American Crows are probably most similar in life style to Pinyon Jays since they roost, breed, and forage communally. One big difference is that crows have adapted enthusiastically to life among humans, while Pinyon Jays only rarely associate with our species.

American Crows are intelligent birds that learn quickly to avoid predators, human or otherwise. They are among the few bird species that have been observed modifying and using "tools" to acquire food. Considerably smaller than Common Ravens, American Crows make higher-pitched cawing calls and have rounded rather than wedge-shaped tails. Crows range over large areas in search of open grasslands, cultivated fields, parks, and athletic fields, where they stop to feed on a wide assortment of plant and animal materials including carrion and garbage at landfills. They require trees for roosting and, in winter, gather to spend the night in large flocks, sometimes numbering in the hundreds or more, in dense groves of trees in urban and suburban settings as well as near parks, ranches, and farms. They have a distinctive flight pattern and use regular flapping strokes and rarely glide; they are almost never seen alone.

American Crows are monogamous, cooperative breeders. Mated pairs form large family groups of up to 15 individuals from several breeding seasons that remain together for many years. Offspring from previous nesting seasons remain with the family to assist in rearing new siblings as "helpers-at-the-nest"; family members also preen each other. Nesting starts as early as February and lasts through mid-June. A mated pair may choose any of a large variety of trees for their bulky stick nest but sometimes build in large bushes and, very rarely, on the ground. The young are usually fledged by 35 days after hatching. Unlike most other Corvids, but similar to Common Ravens, American Crows do not attain breeding age for at least two years, and some individuals may not breed until they are four or five. Highly vocal, American Crows employ a complex set of sounds including repeated harsh *caws* and shorter *kus* and *kows* given when flocks are alarmed by predators or humans. Other repetitive sounds are used for defending breeding territories and for communicating with young and other family members.

STATUS AND DISTRIBUTION These lowland birds are most common in the Central Valley but frequent the Sierra foothills on both slopes, especially along roads and in developed areas.

West Side. Common to locally abundant up to at least 3,000 feet in the Foothill zone of the central Sierra, where human-supplied foods are found but uncommon away from human-modified landscapes; uncommon visitors up to the Lower Conifer zone, and rare above the Upper Conifer zone.

East Side. Numbers have increased in the past 20 years, and they are now year-round residents near ranches and towns in the Honey Lake area, in Reno and the Carson Valley, and in parts of the Owens Valley; mainly fall-winter visitors elsewhere but nearly always near human settlements.

Common Raven • *Corvus corax*

ORIGIN OF NAMES "Common" for the species' overall abundance in the Northern Hemisphere; "raven" from OE., a reference to the species' calls; *corax* from Gr. *korakias*, a raven.

NATURAL HISTORY Common Ravens are a featured species in many human stories, myths, and legends. They are sometimes considered bad omens due to their black plumage, habit of eating dead animals, and for being human pests. They typically live about 10 to 15 years in the wild, easily twice as long as most other passerine birds—a few captives have lived for more than 80 years. Largest of the world's "songbirds," ravens might be mistaken for raptors from a distance because of their large size and flight behavior. With wingspans of four feet or more, they approach the size of Red-tailed Hawks and often soar above the treetops. Compared to crows, ravens are considerably larger overall, with more massive heads and bills, and longer, wedge-shaped tails terminating in a rounded point.

Ravens give deep, croaking calls, lower-pitched than crows. Their vocal repertoire is complex, with at least 30 different sounds used for courtship, aggression, and alarm—they also communicate by bill snapping and producing whistling sounds through their wings. When not actively persecuted, ravens are confident, inquisitive birds that strut around or occasionally bound forward with light, two-footed hops. In flight they are buoyant and graceful, interspersing soaring, gliding, and slow flaps. Courting pairs make breathtaking dives and spectacular, tumbling falls in perfect formation. Solitary birds also perform such stunts, seemingly just for the fun of it.

Ravens have large brains relative to their size, and their ability to solve complex, multistep problems in experimental settings leads some to consider them the geniuses of the bird world. Of all the earth's animals, perhaps only humans and ravens have learned to exploit such a wide variety of habitats. Ravens are often the only bird seen midday in the hottest desert or midwinter inside the Arctic Circle, and their presence in both settings is aided by human alterations to those habitats. Ravens in the Sierra are not as social as crows, and they are mostly seen alone or in small groups. Most often, ravens range widely, searching for carrion and other foods.

Although they lack the strong, sharp talons of hawks and owls, ravens are voracious predators that consume small birds and prey on eggs and nestlings. For this reason, their expanding populations in the Sierra and elsewhere pose a significant threat to native songbird populations. Ravens also pursue small mammals, reptiles, and amphibians and peck them to death. In the foothills east of Fresno, ravens were seen plucking White-throated Swifts off the face of Pine Flat Dam and devouring them on top of the structure. One raven attacked a male Gadwall and decapitated him before flying off with the head, leaving the duck running and spurting blood until it collapsed. Ravens also relish scorpions, dragonflies, and other large insects as well as grains, fruits, and garbage. Like many other Corvids, ravens hide food, especially items with a high fat content, to help them survive winters in severe climates when food is hard to find.

Favored sites for their nests of sticks, mud, and animal hair are ledges on rocky cliffs, sheltered by overhangs or trees. Peregrine or Prairie Falcons occasionally reuse these cliff nests. Sometimes ravens use tall trees, typically the largest ones in the vicinity with sufficient foliage at the top to obscure nests from the ground. They may nest in deserted barns or other human-made structures like utility towers. Sierra nesting usually begins in early March and is complete by mid-June. Juveniles begin to court in their first year, but most do not mate successfully until their second or third year. Pairs are believed to stay together for life, but, as with many other bird species, some promiscuity (in the form of extra pair copulations) occurs. Males bring food and help to shelter and defend nestlings. Young birds leave the nest at about 35 to 40 days but stay with their parents for at least another six months.

STATUS AND DISTRIBUTION Among the world's most widely distributed birds, Common Ravens are found across the Northern Hemisphere. There are eight known subspecies with little variation in appearance or behavior.

West Side. Common residents of the Foothill zone in both deciduous and coniferous forests, especially near roads and established campgrounds and hunting camps; increasingly common to abundant in some foothill towns and cities; uncommon to fairly common up to tree line.

East Side. Common residents below about 8,000 feet; uncommon in the Subalpine and Alpine zones, persisting into winter, where most often observed in aspen groves near large meadow systems.

TRENDS AND CONSERVATION STATUS In the Sierra, as well as in most of California, the Common Raven's range has expanded and numbers have increased dramatically since about 1950, as they have followed roads and expanding human presence to the summits in search of road-killed animals and human leavings. Even more recently, they have become common to abundant in low-elevation areas on both sides of the Sierra where they were previously absent to rare. All 25 of the Sierra Christmas Bird Count circles now record double-digit totals, and nearly every one shows increasing numbers in the past 10 to 15 years. Up to the mid-1980s, ravens were reported on only a third of Sierra Breeding Bird Survey routes but now are consistently present on over 80 percent of them.

LARKS · Family Alaudidae

This family comprises a group of more than 90 ground-dwelling species that primarily occur in arid, barren landscapes of the Old World; almost 70 of those are restricted to Africa, and none occur regularly on oceanic islands. Only one species, the Horned Lark, occurs in the Sierra. Larks are distantly related to swallows; they run or walk rather than hop, aided by their unusually long and straight hind claws. Their wings are long and straight, permitting flight under extremely windy conditions. The family name is from L. *alauda*, a lark.

Horned Lark · *Eremophila alpestris*

ORIGIN OF NAMES "Horned" for the species' distinctive, black "horn" feathers; *eremophila*, a lover of desolate places, from Gr. *eremos*, solitary, desolate; *phileo*, to love; L. *alpestris*, of the Alps or high mountains.

NATURAL HISTORY Despite their bold face and breast markings, Horned Larks can be overlooked as large flocks walk or run across open terrain, and the sandy browns of their backs blend with sand, dirt clods, and tussocks of grass. Few birds are as closely wedded to open areas as Horned Larks. They generally shun any landscape with trees and often show a preference for areas of bare, or nearly bare, ground such as plowed fields, burned or heavily grazed grasslands, or dry lakebeds. They pick up insects, weed seeds, and other foods with their short, slender bills. These skittish birds take flight in unison when approached too closely, giving high-pitched *zu-weet* calls when flushed, and they deliver a series of sweet, tinkling notes as they depart in low, undulating flight before alighting on the ground not far away.

Except when breeding, Horned Larks spend the year in large flocks (sometimes numbering in the hundreds) that cooperate in finding food and watching for predators. In the Sierra, elevation appears to be irrelevant as long as the country is open and dry. They breed as early as March in the foothills when males sing in flight, often at great heights, until late June or July at higher elevations. Females excavate simple, slightly protected scrapes on dry ground for their nests, where they usually lay four, greenish and heavily speckled eggs and incubate them alone for about 14 days. Nesting in such exposed locations, females are creative in finding ways to provide some shade, including stacking pebbles on the edge of the nest or placing it next to a pile of cow dung or in old tire tracks. Both parents feed and tend the young until they are capable of flight after about three weeks. Though little studied in the Sierra, it is likely that pairs raise more than one brood per year at lower elevations.

STATUS AND DISTRIBUTION More than 20 subspecies of Horned Larks exist in North America, and at least 4 or 5 of them occur in the Sierra. Differences between the subspecies are subtle and variation within populations further blurs distinctions. In general, the shade of brown of the upperparts is correlated with the type of breeding terrain, with paler subspecies nesting in sandy areas and darker ones in more heavily vegetated habitats. Multiple subspecies may mix together in large winter flocks. In the Sierra they are drawn to heavily grazed, sparse grasslands and plowed fields as well as to Alpine meadows and fell-fields, alkali flats, and the muddy shorelines of lakes and reservoirs east of the crest.

West Side. Fairly common to locally common year-round residents of the Foothill zone, where numbers swell with migrants in winter; uncommon nesters in the Alpine zone, mostly in the central and southern Sierra, where a few remain through fall and winter, but most move downslope to snow-free areas; generally absent from mid-elevation habitats.

East Side. Common year-round residents at low elevations; fairly common up to the Alpine zone during the breeding season.

SWALLOWS · Family Hirundinidae

Swallows and martins spend much of the day diving and turning at exceptional speeds in dexterous pursuit of flying insects, taking them with gaping mouths. Though not as thoroughly aerial as swifts, swallows are similarly well adapted for hunting on the wing with slender, streamlined bodies and long, pointed wings enabling great maneuverability and graceful flight, punctuated by periods of gliding. Their small, weak feet are not suited to walking but do allow them to perch securely on branches, rooftops, and telephone wires, unlike swifts, which are unable to perch.

Most swallows nest in mud structures plastered to vertical walls or in some kind of tree or earthen cavity, either found or of their own making. Some colonial species, like Cliff and Bank Swallows, nest together by the thousands. Females usually lay four to six all-white, slightly elongated eggs (those of Cliff and Barn Swallows have variable amounts of reddish-brown spotting) and incubate them alone, or with some help from the males, for 12 to 14 days. Both parents feed and tend the young for about 20 to 25 days, until the young leave the nests and can fly and forage on their own.

Members of this family occur on all continents except for Antarctica as well as many, large oceanic islands. While swallows and martins are among the best studied of all North American birds, there is considerable uncertainty about the taxonomy of this family. More than 80 species are currently recognized in the world, with 7 species regularly occurring in the Sierra. The family name is derived from L. *hirundo*, a swallow.

Purple Martin · *Progne subis*

ORIGIN OF NAMES "Purple" for the iridescent color of the adult male; "martin," possibly from Fr. *marinet*, swift; L. *progne*, the daughter of King Pandion (see Osprey account), who was turned into a swallow by the gods in Greek mythology; L. *subis*, a reference to a bird that breaks eagle's eggs for unknown reasons, because martins do not deliberately break eggs.

NATURAL HISTORY Purple Martins are the largest North American swallows, and the shimmering purple adult males are the only ones with all-dark bellies. Most people who have lived or traveled in the eastern United States are familiar with the ubiquitous "martin houses" mounted on posts in many neighborhoods. Purple Martins are common in that part of the country, where they nest almost exclusively in these human-made structures. The abundance of occupied martin houses in the eastern United States has spawned many published papers and much anecdotal information on Purple Martins, so most of what is known about their habits is based on those populations. Far less is known about western martins.

Most martins in California nest in abandoned woodpecker holes in trees or in other natural

Male

Female

cavities, but about 10 percent nest in bridges, elevated freeways, and overpasses. An incipient nest box program at Shelter Cove in southern Humboldt County is showing substantial success, but most human migrants from the East who optimistically erect "martin houses" in California find that they remain empty or are used instead by introduced European Starlings or House Sparrows. Western martins also nested historically in abandoned buildings (until the arrival and buildup of European Starling populations in the 1970s) and have persisted, at least very locally in small numbers in the Sacramento area, by making extensive use of drainage "weep holes" under elevated freeway overpasses. Very recently, a few pairs have been found nesting in covered structures on utility towers at the western edge of the Sierra in Tulare County.

Requirements for successful martin nesting colonies include a concentrated source of insect food, often with a prevailing wind source (such as at ridgetops) to enhance foraging efficiency, a supply of nest holes, with open air space for foraging in the immediate vicinity, and a low starling population. More like swifts than most swallows, Purple Martins fly at high altitudes in search of larger swarming insects so most people never see them in swirling, acrobatic flight. A common vocalization is a warbling chortle that is difficult to describe but unmistakable once learned.

STATUS AND DISTRIBUTION Purple Martins are generally uncommon and highly localized in the Sierra, and most observers will not encounter them except by chance or by going to known nesting locations. In the Sierra they need large trees with sizable cavities (often created by Northern Flickers or Acorn Woodpeckers) and prefer open hillside locations (often in recent or older burns) with expansive views. They begin to arrive from South American wintering grounds by late-March, making them one of the last swallows to arrive. They nest from early May through early August, and most have departed the state by late August.

West Side. Uncommon to rare breeders and very sparsely distributed at local colony sites from the Foothill zone up to the Lower Conifer zone, often in recently burned areas such as the "49er Burn" overlooking the South Yuba River, Nevada County, as well as locally in other foothill counties from time to time.

East Side. Casual, most records of individual migrants in flights over lower elevations, making specific locations almost impossible to predict; most observations in May or August.

TRENDS AND CONSERVATION STATUS Grinnell and Miller (1944) considered Purple Martins to be "fairly common," adding that "there is some indication of spreading to occupy certain districts occupied by people in recent years." Unfortunately these authors did not predict the California arrival and rapid increase of European Starlings in the 1970s that outcompeted Purple Martins for nest sites and almost certainly is responsible for statewide population declines. Timber harvest practices such as clear-cuts and postfire salvage logging remove most standing snags from forest stands and reduce the opportunities for the establishment of nesting colonies, although isolated snags in clear-cuts are widely used in southern Oregon. Retaining significant numbers of large snags in clearings, steep slopes, and ridgelines is especially critical for martins in managed, higher-elevation forests. Competition for cavity nest sites with European Starlings continues to be the major limiting factor for Purple Martin populations, especially in developed areas of the Sierra. Purple Martins were included on the list of California Bird Species of Special Concern in 2008.

Tree Swallow • *Tachycineta bicolor*

Male

Female

ORIGIN OF NAMES "Tree" for the species' preferred nesting and resting substrate; *tachycineta*, rapid flight, from Gr. *tachys*, swift, and *kineter*, to move; L. *bicolor*, two colored, for the contrasting white and dark, greenish-blue plumage of breeding adults.

NATURAL HISTORY These sleek, iridescent, aerial acrobats are usually the first swallows to arrive in spring, although the exact timing of arrival and departure is uncertain because this is the one swallow that regularly overwinters in small numbers in the lowlands of central California, including the western foothills of the Sierra. This hardiness is almost certainly due to their ability to subsist on berries and seeds if insects are in short supply, when other insect-dependent swallows might starve.

Tree Swallows make sweet, musical twittering sounds as they forage over open terrain such as lakes, rivers, marshes, and wet meadows. When not in flight, they alight on nearby branches of riparian trees to roost and preen. They usually nest in woodpecker-excavated or natural holes in such deciduous trees as aspens, cottonwoods, and dogwoods, usually near water where they often seek out holes in partially submerged stumps or other flooded trees. They are also quick to accept artificial nest boxes when placed in moist, insect-productive areas but never nest on rocks or cliffs like Violet-green Swallows (see account below). They have benefited from the active nest box programs mainly intended to attract bluebirds. In the lower foothills of the Sierra, Tree Swallows must compete with other cavity-nesting birds such as European Starlings and Western Bluebirds, and they pugnaciously defend limited, potential nests sites with aggressive flights and occasionally physical conflicts that may result in the death of either combatant. Despite their aggressive tendencies, Tree Swallows often nest in close proximity to each other.

STATUS AND DISTRIBUTION Tree Swallows arrive on their breeding territories as early as February, where they remain at least until September. They tend to be most common during spring and fall migrations on both sides of the Sierra. Migratory roosts containing hundreds of thousands of birds were recently detected with Doppler radar and subsequently documented on the Central Valley floor in mid-October (Cousens et al. 2010).

West Side. Fairly common near lakes, rivers, and other wetland habitats in the Foothill zone up to the Lower Conifer zone; uncommon breeders in the Upper Conifer zone and uncommon up to the Subalpine zone after breeding, where most often observed over mountain meadows and ridgetops; uncommon in winter from the floor of the Central Valley up to the lower Foothill zone, where they mostly forage along river corridors and over reservoirs.

East Side. Fairly common breeders in aspen and cottonwood riparian groves south at least to Mono County; uncommon farther south.

Male

Female

Violet-green Swallow · *Tachycineta thalassina*

ORIGIN OF NAMES "Violet-green" for the upperparts of breeding adults; *thalassina*, sea-green, from Gr. *thalassinos*, of the sea.

NATURAL HISTORY Breathtaking probably comes closest to describing the effect of finally getting that perfect look as the sun ignites the rich green and deep violet of one of these beautiful swallows. William Dawson (1923), after briefly lamenting having heaped so many superlatives on the Tree Swallow, proceeds to pull out all the stops as he describes just such an encounter with a Violet-green Swallow: "This child of heaven should be seen on a typical California day, burning bright, when the livid green of back and crown may reflect the ardent glances of the sun with a delicate golden sheen. The violet of the upper tail coverts and rump comes to view only in changing flashes; but one catches such visions as a beggar flung coins."

Darting through the mists above waterfalls or cruising in wide arcs above mountain lakes and meadows, Violet-green Swallows are the most abundant and colorful Sierra swallows. Despite their vivid appearance, large population and breeding range, little is known of their habits compared to most other swallows. Violet-greens pursue aerial insects above cliffs, mountain meadows, chaparral, and open oak woodlands and conifer forests but usually avoid densely forested areas. Yosemite Valley is an especially easy and spectacular place to see the aerial acrobatics of Violet-greens, where they cruise above the water falls along with White-throated and Black Swifts while giving vaguely swiftlike, chittering calls.

Similar to their close relatives, Tree Swallows, Violet-greens nest solitarily or in colonies of more than 20 pairs. They often nest in abandoned woodpecker holes in riparian trees, but they are not always associated with water. They use ledges on cliffs and bridges, similar to Cliff and Barn Swallows, in addition to abandoned buildings and artificial nest boxes. At Mono Lake they have adapted to using crevices on tufa towers exposed by historically falling lake levels. They are attracted to recently burned forests both for the abundant insects and the potential nest holes that woodpeckers create in hollow trees. Their adaptability in choosing nest sites compared to other swallows may help explain why they are so widespread and common in the Sierra. Like Tree Swallows, Violet-greens fiercely defend occupied nests against other swallows as well as other cavity-nesting species such as Mountain Chickadees, Western Bluebirds, and especially the introduced European Starlings.

STATUS AND DISTRIBUTION Violet-green Swallows usually begin to arrive in California by late February, and most depart for wintering grounds in southern Mexico and Central America by late-August, with the last few stragglers remaining until the end of September.

West Side. Common during the breeding season in open habitats with suitable nesting sites from the Foothill zone up to the Subalpine zone; abundant over many wet, mountain meadows and riparian woodlands of the Lower and Upper conifer zones; rare in winter along foothill rivers and reservoirs.

East Side. Common to abundant breeders in Mono Basin, and fairly common in cottonwood and aspen groves with open foraging areas up to about 8,000 feet; in contrast to most other parts of the Sierra, nesting Violet-greens are generally less common than Tree Swallows at higher-elevation aspen groves and other riparian groves where the Tree Swallows probably outcompete them for nest sites.

Northern Rough-winged Swallow •
Stelgidopteryx serripennis

ORIGIN OF NAMES "Northern" for the species' distribution compared to the closely related Southern Rough-winged Swallow *(S. ruficollis)*; "rough-winged" for the microscopic barbs or "hooks" on the species' outer primary feather giving it a rough texture; *stelgidopteryx*, wing-scraper, from Gr. *stelgis*, scraper and *pteryx*, wing; *serripennis*, a saw wing, from L. *serra*, a saw, and *pennis*, wings.

NATURAL HISTORY Compared to the rich rusts and metallic greens and blues of most other Sierra swallows, the modest browns and grays of Northern Rough-winged Swallows are drab, indeed. While other swallows surpass them in beauty, Rough-wings are equally adept in flight. They fly low over water or marshy areas, skillfully darting, almost batlike, around shrubs and vine tangles, rarely mounting high in the air like other swallows. Lacking many distinguishing field marks, they are often confused with Bank Swallows or immatures of other species, but in swirling flocks, Rough-wings have noticeably longer wings. Their typical call is as unmusical as their appearance is plain—a rough *trrriiit* like a wet finger being forced along the surface of a dinner plate.

Preferred nest sites are holes in steep banks of sand, gravel, or clay, often beside lakes or languid streams. Most often they use abandoned rodent or Belted Kingfisher burrows, with some minor excavations by the swallows themselves. However, they will also nest in almost any kind of hole on a vertical surface, including abandoned buildings, bridges, culverts, drainpipes, or gutters. Pairs may nest alone, but more often they join in small colonies of 10 to 25 pairs. Probably because their nesting holes are difficult for humans to access or examine, little is known of their breeding biology, especially in the Sierra. They do not require water at their nest sites, only within easy cruising range.

STATUS AND DISTRIBUTION Northern Rough-winged Swallows usually arrive in California by late March, where they are mostly seen near low-elevation rivers, lakes, and reservoirs. Often the first swallow to disappear in fall, most have departed by late August for wintering grounds in Mexico and Central America.

West Side. Fairly common flying over open habitats from the lowest elevations of the Foothill zone up to about 5,000 feet in the southern Sierra; generally more localized and less common than most other Sierra swallows (except for Purple Martins and Bank Swallows), possibly due to a relative lack of suitable nesting sites; in recent decades small colonies have been observed on steep, gravel banks of historical hydraulic mines; uncommon up to the Alpine zone after the breeding season from late July until late August.

East Side. Fairly common during spring (mid-April until mid-May) and fall (late July through August); uncommon and highly localized during the breeding season (mid-May through July) up to about 7,000 feet.

Bank Swallow • *Riparia riparia*

ORIGIN OF NAMES "Bank" for the species' preferred nesting substrate; L. *riparia,* associated with riverbanks.

NATURAL HISTORY Except for Barn Swallows, Bank Swallows have the largest global distribution of any Sierra swallow, as members of the same species breed throughout the Old World, where they are known as "Sand Martins." Bank Swallows usually forage in large flocks in the company of other swallows seeking insect swarms. Compared with other species, Bank Swallows forage at low heights, usually within about 30 feet of the ground, often skimming inches above water or over open fields or marshes and rarely over densely forested areas.

Throughout their enormous range, they nest by the hundreds or thousands on vertical, erosive banks of sand, silt, or gravel. Bank Swallow colonies are generally larger than Rough-winged Swallow colonies, which often nest nearby on steep, earthen banks. However, unlike Rough-wings, male Bank Swallows excavate their own nests using their conical-shaped bills, feet, and wings to create nesting burrows, usually more than a foot deep. Bank Swallows may select artificial "banks" (such as road-cuts and sand and gravel quarries) for nesting, sites where they are very susceptible to disturbance. Similarly, unvegetated riverbanks are subject to frequent erosion, meaning their nesting sites are ephemeral with exact locations often changing from year to year. Their rather inflexible requirements for breeding sites and lack of dependable nesting locations is probably the main reason Bank Swallows are highly localized and relatively uncommon in the Sierra. Birders looking for this species away from known colonies should beware of young Tree Swallows, which have a less distinct grayish wash across the upper breast and are often mistaken for Banks.

STATUS AND DISTRIBUTION Grinnell and Miller (1944) described the status and distribution of Bank Swallows in California as follows: "Because of colonial nesting [this species] has been considered fairly common or common locally. But colony sites are few and in aggregate numbers this species is the least numerous of all the species of swallows in the state." Bank Swallow colonies continue to be few and far between, and the largest existing California colonies are west of the

Sierra along the Sacramento and Feather Rivers and the Yolo Bypass in the Sacramento Valley. Bank Swallows usually return from wintering grounds in South America by early April and depart by early September.

West Side. Rare in spring and fall migration; only known nesting colonies from Plumas County in Indian, Genesee, and American Valleys.

East Side. Common to abundant only near active nesting colonies such as those near Honey Lake, in Nevada along the Highway 395 corridor south of Reno, Mono County, and in northern the Owens Valley; largest colonies near Crowley Lake (historically up to 2,000 pairs) and Bridgeport Reservoir; otherwise rare or absent and most often found in migration in May and from late July through late August.

TRENDS AND CONSERVATION STATUS Despite being one of the most widely distributed songbirds in the world, Bank Swallows reach the southwest edge of their North American breeding range in California, and there are probably fewer than a hundred active breeding colonies in the state. Much of their formerly suitable nesting habitat has been flooded by dams or destroyed by flood control projects when stream banks are modified or covered by rip-rap to prevent the erosion that produces ideal cut-bank nesting habitat. Due to their declining status and limited suitable nesting habitats, Bank Swallows were listed as Threatened in California in 1989. The 1992 recovery plan for Bank Swallows recommended the installation of artificial, earthen banks and enhanced natural banks near flood control projects to promote the species' recovery.

Cliff Swallow • *Petrochelidon pyrrhonota*

ORIGIN OF NAMES "Cliff" for the species' preferred nesting habitat; *petrochelidon,* a rock swallow, from Gr. *petros,* rock, and *chelidon,* swallow; *pyrrhonota,* red-backed, from Gr. *pyrrhos,* flame-colored, and *notos,* back, for the species' orange-red rumps.

NATURAL HISTORY Cliff Swallows are among the few native birds that have benefited from expansive dams spanning major Sierra creeks and rivers, where they swarm together in pursuit of flying insects and sources of mud to build their nests. The cement faces of dams, undersides of bridges, eaves of buildings, and walls of deserted stone quarries substitute for their natural nesting sites on vertical canyon walls. Indeed, the Sierra is one of the few areas where Cliff Swallows can still be commonly seen nesting on cliffs. They forage within easy flight distance of their nesting colonies over open habitats such as lakes and reservoirs, marshes, grasslands, and chaparral.

Among the most social of North American birds (see Tricolored Blackbird account), a single Cliff Swallow colony may contain thousands of gourd-shaped nests crammed next to or nearly on top of each other. Each Cliff Swallow nest is an impressive structure, a testament to the persistence and architectural skills of these birds. They are roughly the size of elongated cantaloupes and, like Barn Swallow nests, are constructed from mud. Cliff Swallows search for just the right puddle edge, usually the closest one to the colony and form mud pellets with their bills. The entire nest is built bit by bit, one tiny billfull of mud at a time. They fly endlessly back and forth between nest sites and puddles, swarming and fluttering frantically as if the mud were in short supply. They build nests synchronously, and neighbors crowd in upon each other, often squabbling and trying to steal mud pellets. A particularly well-placed mud puddle can seem a living organism, covered by a hundred or more Cliff Swallows, all fluttering their wings at once. Both parents work together for one or two weeks to build their nests using bits of grass and hair to hold them together.

While Cliff Swallows no doubt derive some protection from predators by building their nests on high, vertical walls, they are also vulnerable to prolonged storms that batter their exposed homes and by humans that wash their mud creations off buildings and bridges. These attempts to discourage the swallows are futile, wasteful, and illegal. A simple, effective, and humane solution is to attach a thin strip of clear plastic under the eaves, creating a smooth surface to which the mud cannot adhere.

STATUS AND DISTRIBUTION Historically Cliff Swallows were confined to nesting on steep canyon walls of the Sierra, Cascades, and Rocky Mountains. However, in the past 150 years or so they have expanded their range and now nest under bridges, highway culverts, and sides of buildings across most of North America. They first arrive from wintering grounds in South America in mid-March and remain until early September with a few stragglers present until early October.

West Side. Common to abundant near breeding colonies except during migration, when flocks might be seen almost anywhere; rare and intermittent at higher elevations above about 5,000 feet in the Lower Conifer zone, occurring in greatest concentrations on dams that block Sierra streams or under bridges spanning major rivers; small, high-elevation nesting colonies found occasionally from 7,000 to 8,500 feet in Nevada, Placer, and Amador Counties and probably elsewhere.

East Side. Similar to the West Side, common only near active breeding colonies; populations have increased in recent decades, especially near towns, and they can colonize high-elevation sites some years when the winter snowpack melts early.

Barn Swallow • *Hirundo rustica*

ORIGIN OF NAMES "Barn" for the species' preferred nesting sites on the sides of barns and other buildings; L. *hirundo*, a swallow; *rustica* from L. *rusticus*, rustic or rural.

NATURAL HISTORY Like Cliff Swallows, Barn Swallows are expert masons. They form their cup-shaped nests by painstakingly alternating layers of grass, straw, and mud into pellets that they plaster onto vertical sections of bridges, beams, or barn rafters. Their choice of nest sites and even the nests themselves are much like those of another species well-adapted to living with humans, Black Phoebes. Barn Swallows nest in much smaller colonies than Cliff Swallows, and the two species can often be seen foraging together over open country including marshes, wet meadows, and still waters of ponds and slow-moving streams, usually only 20 or 30 feet above the ground. Due to their global distribution and preference for nesting on human habitations, Barn Swallows have been studied more intensively than any other of the world's swallows.

Barn Swallows are the most widespread and abundant swallows in the world and breed on all continents except for Antarctica. In prehistoric times they are thought to have nested primarily in caves and probably were rare visitors to the Sierra. When Europeans colonized North America, Barn Swallows began a major behavioral shift and started nesting, on the walls and under the eaves of buildings and under bridges. While their transition to nesting on human-created structures was mostly complete in North America by the mid-1900s, Barn Swallows have continued to expand their range upslope in the Sierra and are now fairly common at mid-elevations where they were absent prior to about 1950. Picking out Barn Swallows in a swarm of whirling swallows is easy, even if colors are not visible, because their tails are unique. They are deeply forked when fanned out, but long and needle-shaped when closed. Indeed, of all our swallows, only Barns truly have "swallow-tails." When foraging, Barn Swallows sing a high-pitched warbling chatter with a finale of deeper, guttural notes.

STATUS AND DISTRIBUTION Barn Swallows begin to arrive in California by late March or early April and some remain through October, much later than most Sierra swallows. In general, they are among the latest swallows to arrive in spring and the last to leave in fall (excepting Tree Swallows, which winter in small numbers). They are widespread and familiar near ranches and farms, especially in lowland areas where aerial insects are abundant.

West Side. Common throughout the Foothill zone and fairly common around houses and ranches up to about 5,000 feet in the Lower Conifer zone; increasingly found at higher elevations such as at 6,700 feet at Lake Van Norden near Donner Pass (where they nest along with Cliff Swallows) and as high as 6,000 feet in Tehama, Butte, and Plumas Counties; in migration they might be seen over any open habitat and, like most swallows, avoid densely forested areas; casual to accidental in winter.

East Side. Fairly common only near developed areas at lower elevations, up to about 7,000 feet; can be seen in numbers during fall migration when up to several hundred might pass over from late August through September; casual or accidental in winter.

CHICKADEES AND TITMICE • Family Paridae

These small, lively, and inquisitive birds flock to backyard feeders and seem surprisingly unafraid of humans. Most members of this family flock together during the nonbreeding season, often along with kinglets, nuthatches, or woodpeckers. The sexes are similar and some species show boldly patterned plumages. Sierra species all reside in the region year-round, but some move short distances up or downslope in response to local weather. They vocalize almost continuously and have surprisingly complex and distinctive repertoires, with each serving a specific purpose such as flock cohesion, attracting mates, or territorial defense. Their loud calls and overall curiosity make them effective sentries for predators, and they are frequently the first birds to sound the alarm after one is spotted. Some species give distinct alarm calls depending on the type of predator detected.

Chickadees and titmice are common wherever trees and shrubs predominate, where short, rounded wings and long tails are well suited for flying among branches and leaves. Their short, sharply pointed bills are adept at grasping small prey, tiny insect eggs and larvae, and at cracking open small seeds. While they glean insects and spiders from the tops of foliage and branches, they often swing below a twig or cone and hang upside down, in search of any prey overlooked from above—a skill that yields prey most other birds cannot access. When food is abundant, they gather and hide food items for later consumption.

Chickadees and titmice nest in natural tree cavities or in abandoned woodpecker holes and will also use artificial nest boxes. Females build cup nests inside of these cavities using mosses, shredded bark, leaves, and bits of hair, where they usually lay six to eight white eggs (lightly covered by small reddish-brown spots) and incubate them alone for 12 to 14 days. Both parents provision the young in the nest for about 15 to 20 days, caring for them for three or four weeks until they can fend for themselves.

Members of this family occur across the Northern Hemisphere and Africa, where they are among the world's best-studied and most familiar birds, even to the most casual observers. However, recent genetic studies continue to change their taxonomic relationships, with several changes made in recent years. There are about 60 species of this family in the world, and 4 species reside in the Sierra. The family name was derived from L. *parus*, a titmouse.

Mountain Chickadee • *Poecile gambeli*

ORIGIN OF NAMES "Mountain" for the species' high-elevation distribution; "chickadee" is imitative of their calls; *poecile* from Gr. *poikilos*, dappled or pied—a reference to the black and white plumage of many species in this genus; L. *gambeli*, for William Gambel (1819–1849), who collected several new species of birds while exploring the southwestern states including California; he died of typhoid on the Feather River after an arduous crossing of the Sierra.

NATURAL HISTORY It is tempting to apply the adjective "cute" to these engaging little birds, but the delicate appearance of Mountain Chickadees is deceptive, for they endure the freezing, snowy winters of the Sierra's middle and higher elevations in greater numbers than any other bird. Winter survival presents an enormous challenge to birds so small, but their soft, thick plumage and protected nesting holes enable them to survive subfreezing temperatures on winter nights. While not known to roost communally, paired chickadees, and possibly recently fledged young, share body warmth together in nesting cavities and natural crevices. Under such conditions, they may enter a state of torpor, or temporary dormancy, to conserve energy and body heat. This state depresses bodily functions, with reduced breathing, body temperature, and metabolism. Torpor, an unusual and poorly studied phenomenon for most wintering birds, is known to occur for only a few other Sierra species.

Mountain Chickadees must forage almost continuously to stay warm, and some birds die during prolonged winter storms. During clear, daylight hours they search for eggs, larvae, and adult insects and spiders hidden in bark crevices or under foliage. They also consume large quantities of conifer seeds, which they cache communally. When these stores begin to be depleted, dominant adults guard the remaining supplies for themselves, forcing juveniles to move downslope to milder climates to find food. During past outbreaks of the lodgepole needle miner near Tuolumne Meadows in Yosemite National Park, Mountain Chickadees were the primary predators of the larvae of these destructive moths.

Mountain Chickadees usually forage in loose flocks, often with Red-breasted Nuthatches,

Golden-crowned Kinglets, and other forest birds. They maintain contact by uttering lisping *tsee-tsee-tsee* notes or their more familiar *chick-a-dee-dee* calls. Their songs, given only during the nesting season, are sweet, whistled *dee-do-do* with the identical second and third notes lower than the first. These extremely curious birds often drop down to low branches to inspect passing humans or other animals and may give sharp *tsik-a tsik-a* calls to harass the object of their fears. In late winter, monogamous pairs begin establishing breeding territories by leaving winter groups, and nesting begins in late April or early May. Most young are fledged by the end of June except at the highest elevations. Some pairs may attempt to raise second broods, especially if their first nesting attempt fails due to inclement weather.

STATUS AND DISTRIBUTION Mountain Chickadees are common residents of the mountainous regions of western North America. These engaging mountaineers live year-round in nearly all types of conifer forest as well as in aspens and other deciduous trees on both sides of the Sierra.

West Side. Common to locally abundant from the Lower Conifer zone up to the highest reaches of the Subalpine zone; ponderosa, Jeffrey, and lodgepole pines provide preferred foraging and nesting habitats, but they also forage on firs, junipers, giant sequoias, and black oaks but mostly avoid gray pines; in winter and spring, especially during the harshest winters, a few (mostly juveniles) descend down to the Foothill zone and very rarely the Central Valley.

East Side. Common in pine forests as well as in aspens, cottonwoods, and other deciduous trees; uncommon in pinyon-juniper forests; one bird seen in August 2003 at over 13,000 feet on Mount Tydall (Inyo County) might hold the high-altitude record for this species in the Sierra.

Chestnut-backed Chickadee • *Poecile rufescens*

ORIGIN OF NAMES "Chestnut-backed" for the species' distinctive field mark; *rufescens*, from L. *rufus*, rusty-red.

NATURAL HISTORY Most Californians may think of this species as the chickadee of coastal areas, but since the mid-1900s, Chestnut-backed Chickadees have shared the Sierra with their "mountain" cousins. Similar to their haunts in the Coast Range, birds in the Sierra primarily frequent moist, older forests dominated by mature Douglas-firs, but they also use ponderosa pines and incense cedars, mixed with deciduous trees such as madrones, black oaks, alders, dogwoods, and bigleaf maples. This habitat preference for wetter drainages probably accounts for the patchy distribution of Chestnut-backed Chickadees in the Sierra.

The two chickadee species occasionally share overlapping breeding territories in mid-elevation forests and often displace each other at seed and suet feeders in winter. In the nonbreeding season, these highly social birds also flock together with other forest birds, such as Red-breasted Nuthatches and Ruby-crowned Kinglets. Similar to Mountains, Chestnut-backed Chickadees glean small insects and spiders, conifer seeds, and fruits from terminal limbs, often by dangling or hovering to inspect surfaces hidden from above. Their calls resemble those of Mountain Chickadees but have a more nasal, buzzy quality, and they do not produce a whistled song.

Unlike Mountains, which rely on old woodpecker nests or natural cavities for nest sites, Chestnut-backed Chickadees often excavate their own nest holes in large decaying limbs and stumps at heights varying from near ground level up to about 100 feet; large Douglas-fir snags are preferred.

Both species are also quick to use nest boxes when placed in their nesting territories. At lower elevations of the Sierra, Chestnut-backs begin courtship and breeding by mid-March but not until early May at higher elevations; most young are fledged by late June or early July.

STATUS AND DISTRIBUTION Chestnut-backed Chickadees are year-round residents from southern Alaska south to central California. Grinnell and Miller (1944) described a specimen collected by R. L. Rudd near Sterling City along the North Fork Feather River (Butte County) in 1939 as "an exceptional vagrant, far eastward" of the species' known range. Before that time, Chestnut-backs had not been observed east of the McCloud River (Shasta County) in the Cascade Range. This range expansion is discussed in the "Recent Trends in Sierra Bird Populations and Ranges" chapter.

West Side. Common to fairly common but highly localized in moist, mature forests, within an elevation band from approximately 2,500 to 4,000 feet in the Foothill and Lower Conifer zones south into Madera County (approximately coincident with the southern limit of Douglas-firs in the Sierra); uncommon up to about 5,500 feet in the Upper Conifer zone; rare in severe winters down to the floor of the Central Valley.

East Side. One record at Lake Tahoe, November 1966.

Oak Titmouse • *Baelophus inornatus*

ORIGIN OF NAMES "Oak" for the species' preferred habitat; titmouse from OI. *titr*, something small, and AS. *mase*, a small bird; *baelophus*, short-crested, from Gr. *baios*, short, and *lophos*, a crest; L. *inornatus*, plain.

NATURAL HISTORY "Spring comes to the brown hillsides of California as soon as the first rains break the long autumn drought . . . and the Plain Tit begins his lively if monotonous refrain from the live oaks." As Hoffman (1927) implied, Oak Titmice are wedded to oaks throughout the year, similar to Acorn Woodpeckers, Western Scrub-Jays, and Yellow-billed Magpies. In oak and pine-oak woodlands of the western foothills, the clear melodies of the Oak Titmouse are, more than any other sound, the voice of the oaks. In a rare case of taxonomic justice, the American Ornithologists' Union relieved them of the demeaning name "Plain" Titmouse when they split the species into the Oak Titmouse and Juniper Titmouse in 1997 (see account below). The two species have a similar appearance but their voices are different, and their Sierra distributions do not overlap.

The song of this titmouse is highly variable but nearly always consists of paired notes given in a regular rhythm: *too-wheet, too-wheet, too-wheet*. A complex repertoire of calls includes frequent, wheezy *tcick-a-dee-dee* calls, similar to those of chickadees. While not as likely to join mixed-species flocks as the chickadees, foothill flocks in winter can sometimes include both titmice and chickadees, so care should be taken when identifying these birds only by call. Like other members of this family, titmice are quick to scold humans and other intruders who venture into their territories.

Unusual among songbirds, Oak Titmice form monogamous pairs and may remain together for several years. In the Sierra, Oak Titmice nest from early March until about mid-July. Highly sedentary, they defend small territories (usually four to six acres) at all times of year and rarely stray beyond those boundaries. Woodpecker nests and natural cavities in oaks are their preferred nesting sites, and females do most of the work refurbishing or further excavating nest holes while the males feed them. Titmice eat a varied fare, preferring insects while breeding and seeds and fruits in winter. They pluck insects from bark, foliage, and flowers, and, like other members of their family, they also dangle from twigs, probe into crevices, or drop to the ground to feed. Titmice also pry off bark or pull apart lichens and leaf galls to expose insects, and they often hammer open acorns and hard fruits with their stout bills.

STATUS AND DISTRIBUTION Other than localized breeding populations in southern Oregon and northern Baja California, Oak Titmice are endemic to oak woodlands of California. Vast conifer forests and rocky terrain at the Sierra crest divide Oak Titmice of the West Side from their congeners, Juniper Titmice, to the east. Although both titmice are considered mainly sedentary, patterns of occurrence clearly show some tendency for occasional dispersal, particularly in fall or winter.

West Side. Common residents of the Foothill zone up to about 3,000 feet, where they favor oak woodlands as well as riparian forests; fairly common in black oak woodlands of the Lower Conifer zone up to about 4,500 feet; after fledging, a few juveniles may wander up to about 6,000 feet in the Upper Conifer zone in the central Sierra and rarely up to 8,000 feet in the Subalpine zone; in the extreme southern Sierra, breeds up to about 7,000 feet in the Piute Mountains.

East Side. Uncommon, small population in oaks along Sage Flat Road (Inyo County) and the breeding population in the Kern River Valley extends over Walker Pass for a few miles along Walker Creek (Inyo County) into areas that entirely lack oaks; otherwise no confirmed records.

Juniper Titmouse • *Baelophus ridgwayi*

ORIGIN OF NAMES "Juniper" for the species' preferred Utah juniper habitat; L. *griseus*, gray; L. *ridgwayi* for Robert Ridgway (1850–1929), curator of birds at the U.S. National Museum and an original author of the American Ornithologists' Union's *Check-list of North American Birds*.

NATURAL HISTORY Like their close relatives, Oak Titmice, Juniper Titmice favor sunny, open habitats but, instead of oak woodlands, they dwell among pinyon pines and Utah junipers in canyons and flats of the Great Basin. Aside from a different geographical range and habitat, the natural history of this recently described

species has not been studied in detail but is thought to be similar to the Oak Titmouse. Compared to that species, the Juniper Titmouse is slightly larger and lighter gray overall. Song and calls are usually the best means of distinguishing the two titmice, but the highly variable nature of all their vocalizations must be considered. In general, the paired notes in the Juniper's song are closer in pitch to each other, lending more of a single tone rattle to the song, compared with Oak's bouncy song. Juniper Titmice also tend to give a rapid series of call notes and a somewhat harsher *tschick-a-dee* call.

STATUS AND DISTRIBUTION Juniper Titmice inhabit arid, open pinyon-juniper habitats of the Great Basin east through Colorado and New Mexico to western Oklahoma. They reach the extreme western edge of their distribution in the eastern Sierra and to the north in Modoc County, where their range overlaps slightly with Oak Titmice. The exact status of Juniper Titmice in the Sierra, as we have defined it, is still uncertain.

West Side. No records.

East Side. Fairly common just east of the region in Mono County and east of the Owens Valley, and uncommon in southeastern Lassen County, the only regular breeding season observations in the Sierra are in the northern Mono Basin; other records in that vicinity are mostly from fall or winter, suggesting that these are visitors from nearby breeding range; the Nevada Breeding Bird Atlas noted a couple of "possible" breeding records just east of Lake Tahoe; titmice found at both the north and south end of Lake Tahoe in fall 2003 could have been Oak or Juniper Titmice, though fires in nearby "Juniper Titmouse habitat" in Nevada in preceding months could have caused some dispersal of that species westward; aside from Mono Basin, rare to accidental from eastern Alpine County to as far south as Bishop.

VERDIN • Family Remizidae

Verdins are the only New World representatives of this family that includes the "penduline tits." Closely related to "true tits" (family Paridae), most members of this family (including Verdins, which were formerly classified as members of the Paridae) make elaborate, hanging bag nests. Members of this family forage in a fashion similar to other tits, hanging upside down from small branches to capture insects, their primary prey. In addition to Verdins (the only members of this genus), 12 other species reside in Eurasia and Africa. The family name was derived from OF. *remizi*, a covered carriage—a possible reference to their enclosed nests.

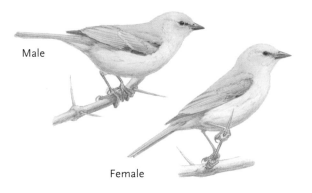

Male

Female

Verdin • *Auriparus flaviceps*

ORIGIN OF NAMES "Verdin" may have been derived from Fr. *verdon*, a bunting; *auriparus*, a golden titmouse, from L. *aureum*, golden, and *parus*, a titmouse; *flaviceps*, yellow-headed, from L. *flavus*, yellow, and Gr. *kephale*, head.

NATURAL HISTORY Called "yellow-headed Bushtits" by some birders, the two species are in different families but share many of the same behaviors (see account below). Occasional reports of Verdins outside their known range are usually misidentifications of "yellow-headed" Bushtits, with their heads covered by yellow pollen. In addition to having different head colors, Verdins also have noticeably shorter tails than Bushtits.

Desert-dwelling Verdins forage solitarily, in pairs, or in small groups by actively hopping from twig to twig while sounding their surprisingly loud *tseet* calls almost continuously. They primarily search for scale insects, caterpillars, or beetles, but they also consume berries, small fruits, and the pods of legumes. They rarely forage on the ground. Like other members of their family, Verdins are well known for constructing overly large nests in thorny plants, often in desert washes. Males build "display nests" to attract females before they start on "breeding nests" with help from their mates. When completed, these prickly, spherical structures may exceed eight inches in diameter and contain over 2,000 twigs. In the southern Sierra, the Verdin's breeding efforts mainly occur from March through May. Females usually lay 4 light bluish or greenish, lightly speckled eggs and incubate them alone for about 14 days. Both parents feed and tend the young for about three weeks until they fledge. After they gain independence, the young may continue to use their natal nests as night roosts.

STATUS AND DISTRIBUTION Verdins reside year-round in arid parts of the southwestern United States and northern Mexico. Their range has apparently expanded in recent years to include the Desert zone of the southeastern Sierra.

West Side. No records.

East Side. Uncommon localized residents in desert canyons west of Highways 14 and 395 from Kern County into southernmost Inyo County.

BUSHTIT • Family Aegithalidae

All 13 members of this mostly Old World family live in forest and shrub habitats, including the one North American representative, the Bushtit (the only member of the genus *Psaltriparus*). Except for the Long-tailed Tit *(Aegithalos caudatus),* which is widely distributed across Europe and Asia, all the other species live in the Himalayas and/or other mountainous regions of Eurasia and China. Collectively known as "long-tailed tits," members of this family are sedentary and live in small social groups that defend common territories. The family name is derived from Gr. *aigithalos*, a titmouse.

Bushtit • *Psaltriparus minimus*

Female

Male

ORIGIN OF NAMES "Bushtit" for the species' habitat in bushes or thickets; "tit" from OI., anything small; *psaltriparus*, a "harp-playing titmouse," from Gr. *psaltria*, a harp player, and L. *parus*, a titmouse, although it's unclear why this name was applied to this species as its songs are not at all harp-like or in any way musical; L. *minimus*, the smallest.

NATURAL HISTORY "Bushtits are bird children that never grow up. . . . To see a flock of these merry mites trouping over the bushes, you would think they were playing an endless game of tag." As William Dawson (1923) implied, Bushtits are among the smallest, most social, and active songbirds in North America. These long-tailed, grayish mites, scarcely larger than most humming-birds, always socialize during the nonbreeding season. Loose flocks of 10 to more than 30 birds, all in continuous motion, float like slowly advancing waves through the foliage of trees and shrubs and may surround the observer as they appear so intent on their work that they are oblivious to humans. Warblers, chickadees, and kinglets often join their foraging flocks. Scattered Bushtits maintain contact with the rest of their group with constant, high-pitched *tsee-tsee-tsee* calls. Much like chickadees, Bushtits forage for small insects and spiders by hanging effortlessly to scrutinize the undersides of leaves and twigs. Occasionally they eat seeds or nectar, but arthropods are the mainstay of their diet.

Winter flocks (possibly composed of related individuals) break up by late February or early March as they begin to pair off for breeding. While undistinguished in plumage or song, Bushtits create exquisite and remarkable abodes: long, soft pouches woven from lichens, mosses, willow down, spider webs, or feathers, decorated with flower blossoms or moth's wings. Construction of these splendid nests may take up to six weeks. These nests, suspended from twigs, seem overly large for their five to seven tiny, white eggs. Both parents incubate eggs for about 12 days and feed and tend the young until they leave the nests after about 14 days; the parents continue feeding the young for about 3 more weeks.

STATUS AND DISTRIBUTION Bushtits are year-round residents of scrub and oak-pine forests from British Columbia to the highlands of South America. Two similar but different subspecies occur in the Sierra: the brown-capped and grayish-faced, "coastal" birds *(P. m. minimus)* of the West Side, and the gray-capped and brownish-faced "interior" birds *(P. m. plumbeous)* of the East Side. Members of both subspecies have been seen in the Tahoe Basin and Mono Basin.

West Side. Common to locally abundant residents in blue and live oak woodlands, foothill chaparral, and riparian habitats of the Foothill zone; fairly common in black and canyon live oaks of the Lower Conifer zone; uncommon in similar habitats in the Upper Conifer zone; the "coastal" subspecies occurs in the Kern River Valley and across Walker Pass.

East Side. Fairly common residents in riparian woodlands; most conspicuous from early October through March, when large, boisterous flocks are frequently encountered; though the species is largely sedentary, winter numbers may be augmented by migrants from elsewhere; most records from desert canyons in southern Inyo and northeastern Kern Counties are probably wintering birds; however, the first breeding in eastern Kern County was confirmed at Butterbredt Spring in 2005.

NUTHATCHES · Family Sittidae

Early British birdwatchers observed small, tree-climbing birds wedging nuts into tree crevices and hacking them open with their stout, upturned bills. First called "Nuthacks," they later became known as "Nuthatches." These agile, little climbers use their sharp claws to travel up, down, or sideways along bark surfaces. In this fashion they can examine crevices for food overlooked by woodpeckers or Brown Creepers, which, unlike nuthatches, almost always travel upward using their tails as props. Nuthatches fly with relatively slow wingbeats that are clearly audible when they flutter between trees.

Most nuthatches excavate their own nesting cavities in living trees or snags but will also use abandoned woodpecker holes and artificial nest boxes. Females do most of the excavation work, with occasional help from the males or other unrelated adults (possibly offspring from previous years). They build cozy, cup nests of soft grasses, feathers, and shredded bark, where they usually lay 4 to 8 white eggs, liberally covered with tiny reddish spots, and incubate them alone for 12 to 14 days. Both parents feed and tend the young in the nest for 14 to 18 days until they fledge, and for about 2 or three 3 until they can forage on their own.

About 24 species of nuthatches (all in the same genus) are found across Eurasia and North America. Three species are year-round residents of the Sierra but generally reside in separate ecological zones, favor different kinds of trees, and differ in plumage and voice. The family name was derived from Gr. *sitte*, a nuthatch.

Red-breasted Nuthatch · *Sitta canadensis*

ORIGIN OF NAMES "Red-breasted" for the coloration of adults, brighter reddish for males; *sitta* (see family account above); L. *canadensis*, of Canada, where the species is common in boreal forests.

NATURAL HISTORY Once learned, the nasal *ank-ank-ank* calls of Red-breasted Nuthatches are seldom forgotten. These hornlike sounds are surprisingly loud and penetrating to have emanated from such small birds. They chatter socially in the tops of conifers but often drop down to lower branches to forage or inspect intruders to their woods. Among the most common of Sierra birds, Red-breasted Nuthatches favor mature, mid-elevation conifer stands where they gather seeds and acorns into winter food caches that, along with insect larvae and spider eggs, sustain them through the long, freezing winters. According to David Gaines (1992), rock climbers "have observed nuthatch mountaineers stuffing nuts into cracks and behind flakes on granite walls hundreds of vertical feet above the valley floor. Undoubtedly they have planted most of the pines that cling tenaciously to Yosemite's walls."

Even on the coldest days, Red-breasted Nuthatches actively feed with other winter residents, such as Mountain Chickadees, Brown Creepers, and Golden-crowned Kinglets. Every other year or so, many Red-breasted Nuthatches abandon coniferous forests at higher latitudes and elevations in response to poor cone crops and move into the foothills or even down to the Central Valley. In the Sierra, Red-breasted Nuthatches begin nesting in mid-April at lower elevations (i.e., 3,000 to 4,000 feet) but not until mid- to late May in the heavy snow zone higher up (i.e., above 5,000 feet). They usually excavate their own nests in snags or large, decaying limbs, or find abandoned natural cavities; firs and aspens are preferred nesting trees. Perhaps to protect their nests from ants, predators, and competitors, Red-breasts smear pitch around the entrances of their nest holes, requiring them to fly directly inside rather than landing first, then climbing in like other nuthatches or woodpeckers. While they usually nest at about 10 or 15 feet, nest heights vary from just above the ground to more than 40 feet.

STATUS AND DISTRIBUTION Red-breasted Nuthatches are common, widespread residents of fir- and spruce-dominated forests from southern Alaska to southern California and Arizona. In the Sierra they are particularly common in mature stands of Douglas-firs, white firs, and red firs but, to a lesser extent, they also work over many other species of conifers and deciduous trees.

West Side. Common to locally abundant year-round residents of from the Lower Conifer zone into the Subalpine zone in the central Sierra; uncommon to rare in pure lodgepole pine forests of the Subalpine zone; fairly common in some winters in the Foothill zone, where conifers dominate but generally avoid pure stands of gray pine; postbreeding birds may move up- or downslope in response to local food supplies and weather and might be found at almost any elevation.

East Side. Fairly common residents of ponderosa and Jeffrey pine forests and aspen-lined streams up to the Subalpine zone; uncommon in pinyon-juniper forests; depending on local cone crops may be fairly common in some winters, absent in others.

White-breasted Nuthatch · *Sitta carolinensis*

ORIGIN OF NAMES "White-breasted," for the coloration of adults; L. *carolinensis*, of Carolina, where the type specimen was collected.

NATURAL HISTORY Unlike their Red-breasted cousins, White-breasted Nuthatches avoid dense stands of conifers at mid-elevations of the western Sierra. Instead, two different subspecies occupy distinct ranges and habitats above and below this ecological zone (see "Status and Distribution" below). In addition to occupying different parts of the Sierra, these subspecies can be distinguished by their voices: foothill birds of the West Side usually give one to four, nasal *wheer . . . wheer* notes, while those living on the East Side and in forests near the Sierra crest on the West Side give a much longer series of rapidly paced, higher-pitched *chur-chur-chur-chur* calls, with a somewhat rougher quality to the notes. These two subspecies are not consistently distinguishable in the field by appearance alone. However, based on vocalizations and supported by recent genetic evidence, these subspecies could someday be recognized as separate species, so birders should learn to distinguish their distinctive calls.

Largest of the Sierra nuthatches, White-breasted Nuthatches are equally adept at scampering up, down, or sideways along vertical bark surfaces. At lower West Side elevations, White-breasted Nuthatches are strongly associated with oak savanna and woodlands. Oak trees provide acorns, a staple they cache in crevices for winter use. They do not depend solely on oaks, however, for when acorns are in short supply they often gather the seeds of gray and ponderosa pines in the Sierra foothills; they also depend on abundant insects and spiders throughout the year. At high elevations of the Sierra they harvest and cache the seeds of lodgepole pines and mountain hemlocks for winter use. Cross-country skiers occasionally surprise these hardy mountaineers at their food caches when traveling through the highest forests. Whether in oaks and pines of the foothills or Subalpine forests near the Sierra crest, they prefer open stands with large, well-spaced trees.

White-breasted Nuthatches are more sedentary than other members of their family and generally reside in breeding territories of about 25 to 30 acres year-round. Once established, pairs are apparently monogamous and stay together for years until one member dies or disappears. Pairs occasionally join foraging flocks of chickadees, kinglets, and vireos but rarely stray far from their home turf. Unlike most nuthatches, White-breasts do not excavate their own nest holes. Instead, they use abandoned woodpecker holes and natural cavities mostly in broad-leaved trees, conifers, and occasionally orchard trees. Nesting usually begins by early March in the western foothills but not until late June or early July at the highest elevations. They often nest near houses and other buildings and are quick to colonize artificial nest boxes. Mostly working alone, females create cup nests of shredded bark and wood chips and line them with lichens, hair, and feathers.

STATUS AND DISTRIBUTION Several subspecies of White-breasted Nuthatches have been described across North America. Based on distinct vocalizations, these subspecies fall into three groups: "Eastern," "Interior West" (Rocky Mountains and Great Basin), and "Pacific." Two subspecies, one of the "Pacific" group and one of the "Interior West" group, occupy distinct, nonoverlapping habitats and ranges in the Sierra. "Slender-billed" White-breasted Nuthatches *(S. c. aculeata)* are found in the western foothills, and "Inyo" White-breasted Nuthatches *(S. c. tenuissima)* in high-elevation forests on both sides of the Sierra. Based only on specimens, both Grinnell and Miller (1944) and Orr and Moffitt (1971) concluded that the range of "Inyo" nuthatches did not reach the Sierra.

However, more recent work (Gaines 1992, Floyd et al. 2007, Dunn and Garrett, in prep.) indicates that this taxon's range extends well into the Sierra. Although much more study is needed to confirm exact ranges, a broad band of dense mixed conifer and red fir habitats of the Lower and Upper Conifer zones divides the ranges of these subspecies on the West Side, and they apparently only come in contact in southeastern Tulare County in the vicinity of Chimney Creek Campground. Both subspecies are year-round residents and generally do not make seasonal movements like other nuthatches.

West Side. "Slender-billeds" are common to abundant in oak savannas and woodlands as well as broad-leaved trees in riparian areas of the Foothill zone; fairly common in black oak woodlands of the Lower Conifer zones; rare to absent in dense forests of the Lower and Upper Conifer zones; "Inyo" nuthatches are uncommon from the Upper Conifer zone to the crest in wooded habitats such as around Lake Almanor and down to approximately the 7,000-foot level in the central Sierra and a bit lower in the southern Sierra; in northern Kern County, "Slender-billeds" are found in the Greenhorn and Piute Mountains, even in habitats lacking oaks, while "Inyo" nuthatches are found further east along the north fork of the Kern River into southern Tulare County.

East Side. "Inyo" nuthatches are fairly common to uncommon residents in stands of Jeffrey pines up to the Subalpine zone, where nesting has been observed at least up to 10,000 feet in Mono County; uncommon residents in pinyon-juniper woodlands as well as in groves of aspens and cottonwoods around wet meadows and along streams; rare in fall and winter below breeding range.

Pygmy Nuthatch • *Sitta pygmaea*

ORIGIN OF NAMES "Pygmy" for their small size compared with other nuthatches; L. *pygmae*, a pigmy or dwarf.

NATURAL HISTORY Birders use the term "warbler neck" to describe the pain ones gets from staring straight up trying to see warblers, but Pygmy Nuthatches can induce a similar condition as ceaselessly "peeping" flocks refuse to descend from the highest treetops. Pygmies are the smallest and most social of Sierra nuthatches. For most of the year they flock together in groups of 10 to 15 birds combing the bark and fluttering back and forth between large ponderosa and Jeffrey pine trees. Unlike other nuthatches, they often venture out to the outermost twigs, hanging upside down like Mountain Chickadees. They are most easily detected by their high-pitched *teedee-teedee-teedee-teedee* calls.

Pygmy Nuthatches mostly eat pine seeds except during the breeding season, usually from early May through July, when adults and young feed primarily on insects and spiders. For breeding, they prefer large snags in open pine stands where adults excavate nests or expand existing cavities in rotten wood. Females line the nests with feathers, bark, and soft grasses. Unusual among North American songbirds, Pygmy Nuthatches employ "helpers at the nest," when up to three males (probably progeny from former years) defend the nests and feed the young. They also roost communally on cold winter nights, and up to 150 birds have been found snuggling together in one cavity.

STATUS AND DISTRIBUTION Pygmy Nuthatches are year-round residents of long-needled pine forests throughout western North America, especially in open, mature stands that retain numbers of large snags.

West Side. Highly localized residents in near-pure stands of ponderosa and Jeffrey pine near Lake Almanor and other areas of Plumas (e.g., Indian Valley) and Sierra Counties just west of the crest, the Yosemite region, and portions of Madera, Fresno, Tulare, and Kern Counties above about 6,000 feet, where stands of these pines predominate; occasional incursions below these breeding areas as low as 3,000 feet in Madera County suggest some pre- and postbreeding wandering.

East Side. Common to fairly common residents of similar habitats up to the Subalpine zone, where they also breed in aspen groves.

CREEPERS • Family Certhiidae

Members of this mostly Old World family are sometimes called "tree-creepers" since, much like woodpeckers, they spend their time creeping up large trees. Only one species, the Brown Creeper, occurs in North America, but it looks remarkably similar to eight other species from Europe and Asia. Members of this family also resemble "woodcreepers" (subfamily Dendrocolaptinae) of the Neotropics in appearance and behavior. However, they are not closely related and their similarities are due to convergent evolution, resulting from their common requirements to forage on and blend with tree bark. All members of the family Certhiidae have short legs and never walk on the ground. They make short, weak flights between tree trunks on stubby, rounded wings. The family name comes from Gr. *kerthios*, a small tree-dwelling bird.

Brown Creeper • *Certhia americana*

ORIGIN OF NAMES "Brown" for the back color; "creeper" for the species' trunk-creeping habits; L. *certhius*, a creeper; *americana*, of America.

NATURAL HISTORY Brown Creepers can appear as bits of tree bark come to life with streaky-brown heads, backs, and wings blending so well against conifer bark that they can be difficult to find. However, those familiar with their whistled *tsee-see-tsee-tseedle-tseedle-see* songs, have little trouble locating them. Distinguishing their simple but very high-pitched calls from those of Golden-crowned Kinglets can be a challenge, but the Brown Creeper's calls are more forceful and clearer-toned. Unlike nuthatches, which often head downward and sideways, creepers almost always forage upward on the trunks or large branches of big conifers. After spiraling to the top of one tree, they flit to the bottom of another and resume hitching their way up again, using stiff, pointed tail feathers as props, much like woodpeckers. Creepers use their down-curved tweezerlike bills to extract beetles, scale insects, and spiders (and their larvae or eggs) from the deeply furrowed bark of mature trees.

In the Sierra, Brown Creepers usually nest from mid-April through late June or early July. Groves of incense cedars, giant sequoias, and Douglas-firs provide ideal breeding habitat. Loose planks of bark provide ideal sites for their hammock-like nests of mosses, lichens, grasses, and feathers. Females do most of the nest construction and usually lay 4 to 6 whitish eggs covered with small, brownish or reddish spots and incubate them alone for 14 to 16 days; both parents feed and tend the young. Solitary for most of the year, Brown Creepers sometimes join wintering flocks of chickadees, kinglets, or nuthatches, spending most of their time in search of food to provide energy to survive the freezing nights. They sometimes roost communally (possibly in family groups) under slabs of bark and tree crevices. Up to five individuals have been seen huddled together under the protected eaves of a mountain cabin in Nevada County.

STATUS AND DISTRIBUTION Brown Creepers are widespread and fairly common in conifer-dominated forests from southern Alaska to Central America. Studies in the Sierra confirm that they primarily forage and nest in mature conifer forests, with sugar pines and incense cedars used most intensely (Adams and Morrison 1993). However, they also forage in stands with a mix of broad-leaved trees such as black oaks, aspens, or cottonwoods.

West Side. Common to fairly common year-round residents of the Lower and Upper Conifer zones from about 4,000 to 8,500 feet in the central Sierra; uncommon to rare in the Subalpine zone (on both sides of the Sierra), and avoid uniform stands of lodgepole pines; uncommon in the Foothill zone in some winters, where they might be seen on gray pines, live or blue oaks; small numbers winter on the Central Valley floor, where they use a variety of native and non-native trees.

East Side. Fairly common year-round residents of ponderosa and Jeffrey pine forests and aspen/cottonwood groves; rare or absent in pinyon-juniper woodlands, where trees have small stature and do not provide suitable bark surfaces for foraging; some movement to lower elevations in winter, where they can be fairly common some years and absent in others.

WRENS • Family Troglodytidae

Small and cocky, wrens are a dynamic group of birds whose outsize voices match their spunky and curious nature. Although many aspects of their life histories are highly variable, most wrens can be immediately recognized by their short, cocked-up tails, long curved bills, and lively songs. Other traits, not so easily observed, are the moderately flattened skulls that allow wrens to reach their heads farther into crevices in search of insects and spiders. This feature is accentuated in Canyon Wrens, whose spinal cord attaches at the back of the skull rather than to the underside like other wrens. Being almost exclusively insectivorous, wrens are especially susceptible to extreme weather conditions and must either migrate or move below the snow zone in winter.

Female wrens usually lay 4 to 8 white, buff, or brownish eggs (with varying amounts of red or brown spots) and incubate them alone for 14 to 18 days. Males typically turn their attention to building additional nests while females incubate. Both parents tend the nestlings for 15 to 20 days and provide food for 3 or 4 weeks after fledging. There are approximately 80 species of "true" wrens (in 20 different genera), and 7 species occur in the Sierra. The family name comes from Gr. *troglodytes*, a cave dweller, since members of this family usually nest in domed, earthen structures or rock or tree cavities.

Cactus Wren • *Campylorhynchus brunneicapillus*

ORIGIN OF NAMES "Cactus" for the species' preferred nest sites; *campylorhynchus*, curve-beaked, from Gr. *kampylos*, curved, and *rhynchos*, beak; *brunneicapillus*, brown-haired, from L. *brunneus*, brown, and *capillus*, hair on the head.

NATURAL HISTORY Midmorning in the desert and it is quiet but for the crackling sounds of insects in the heat . . . and then: "Soft and low it comes, a rich yodeling alto of uniform tone—uniform, that is, save for the light crescendo with which the series opens, and the fading murmur of its closing note." Here William Dawson (1923) captures perfectly a typical encounter with a Cactus Wren. Their song is background for so much of what happens in the desert canyons at the south-

eastern edge of the Sierra and also the backdrop for many western action films, including areas far beyond their actual range.

By far the largest of the Sierra wrens, Cactus Wrens seek larger insects such as grasshoppers, moths, and wasps, as well as large spiders, and will readily eat fruit when available. They are not ideal neighbors as they frequently seek out and destroy the nests and eggs of other birds attempting to breed within their territories—a trait they share with some other members of their family. Pairs remain together and defend territories year-round. Nests are large, globular affairs, nearly always built in a cactus or spiny shrub that presents an insurmountable challenge to most prospective nest-robbers.

STATUS AND DISTRIBUTION Widely distributed in desert regions of North America, the range of Cactus Wrens barely extends into the southern and southeastern fringes of the Sierra, where they are restricted to areas with thorny trees, shrubs, and cacti.

West Side. Locally common in scattered locations in Joshua tree woodlands in the watershed of the South Fork Kern River; found along the lower portions of Chimney Peak Road and along Kelso Creek; also occurs in Walker Basin and the Piute Mountains.

East Side. Locally fairly common in desert canyons and washes from Nine Mile Canyon (Inyo County) south.

Rock Wren • *Salpinctes obsoletus*

ORIGIN OF NAMES "Rock" for the species' preferred habitat; "wren" from AS. *wraenna*; *salpinctes* from Gr. *salpinktes*, a trumpeter; L. *obsoletus*, shabby, indistinct.

NATURAL HISTORY Never was a bird better named than the Rock Wren. They are firmly wedded to rocky habitats for foraging and nesting and even match their plumage to the subtle grays and browns of the rocks. Once detected by their distinctive *tsik-keeer* calls or variable but repetitive songs, Rock Wrens can usually be spotted bouncing from rock to rock, investigating every crevice, and bobbing incessantly. Almost nothing is known of their nesting biology except that they tuck their nests into narrow openings in the rocks and often create a "sidewalk" of pebbles at the entrance, with no known purpose. They hunt for a variety of insect prey as they work deep into crevices and forage in grasses and shrubs within and at the edges of their preferred habitat.

It is a mystery why such a widespread and relatively approachable bird remains almost completely unstudied, but an opportunity exists for anyone with the time and inclination to become the world's expert on Rock Wrens. Nesting in the lowest elevations of the Sierra begins as early as March and, based on data from other parts of the range, second broods are possible anywhere below the Subalpine zone.

STATUS AND DISTRIBUTION Anywhere in western North America where one finds large, rocky areas, Rock Wrens may be present. In the Sierra they might be seen on the boulder-strewn faces of dams as well as along exposed, barren shorelines of low-elevation reservoirs, rocky areas exposed by wildfire, and all the way up to the highest Alpine talus slopes. Few other birds occupy such a broad altitude range. While mostly found in open habitats, rock outcrops in oak savanna or in open patches surrounded by woodlands may also harbor these birds. There is some downslope movement in winter but always to areas with a rocky component.

West Side. Uncommon to fairly common breeder in appropriate habitats in the Foothill zone and uncommon from the Subalpine into the Alpine zones; present here and there within the conifer zones, always on rock outcrops; fairly common in winter at lower elevations, when residents are likely joined by downslope migrants.

East Side. Common to fairly common breeders; widespread but patchily distributed breeders throughout; uncommon to rare in winter at lower elevations in appropriate habitats, the Owens Valley and Reno area Christmas Bird Counts record them almost every year.

Canyon Wren • *Catherpes mexicanus*

ORIGIN OF NAMES "Canyon" for one of the species' preferred habitats; *catherpes,* a down creeper, from Gr. *kat,* downward, and *herpes,* creeper; L. *mexicanus,* of Mexico where the type was collected.

NATURAL HISTORY The steep-walled canyons for which these wrens are so aptly named make perfect amphitheaters to capture and enhance one of nature's most sublime songs. The series of sweet phrases finds an emotional connection in almost every heart as the notes drop in pitch as though tumbling down the canyon walls themselves. Finding the singer is another matter, but the visual reward is nearly as satisfying as the aural one. As with some other wrens, males may burst into song even on a midwinter's day.

Perfectly adapted to their vertical world, Canyon Wrens, like sideways-creeping crabs, scale sheer rocky cliffs at least as effortlessly as the Yosemite rock climbers who sometimes encounter these remarkable birds thousands of feet above the valley. Our knowledge of their life history is nearly as poor as for Rock Wrens, though this lack of information is understandable given the lofty and inaccessible nesting locations Canyon Wrens prefer. They spend much of their time in the nooks and crannies, whether searching for insects and spiders or constructing their surprisingly lush, soft nests tucked under ledges or into cracks in the rock face. They may share similar habitats with Rock Wrens, but Canyon Wrens are rarely found far from water and canyon walls with a stream below. Based on observations outside the Sierra, nest building probably begins between mid-April to late May, depending on elevation.

Widely distributed across western North America, Canyon Wrens are mainly found in steep-sided river canyons. In the Sierra they occur from the lowest foothill canyons up to the river gorges of middle altitudes. Winter movements are poorly understood, but some downslope migration is shown by Christmas Bird Count data and other records outside of breeding areas.

West Side. Uncommon to fairly common residents from the Foothill zone to the lower elevations of the Upper Conifer zone; possibly absent from higher elevations in winter; unlike Rock Wrens, they are rare in the Subalpine and Alpine zones.

East Side. Uncommon residents up to about 8,000 feet; occasional observations at higher elevations (on both sides of the crest) probably represent postbreeding wanderers.

Bewick's Wren • *Thryomanes bewickii*

ORIGIN OF NAMES John James Audubon first collected this bird in Louisiana and named it in honor of his friend Thomas Bewick (1754–1828; pronounced like the car "Buick"), an English wood carver who made plates for several significant natural history books he authored, including *Birds of Great Britain* in the late 1700s; *thyromanes*, reed cup (for its nest), from Gr. *thryon*, reed, and *manes*, cup.

NATURAL HISTORY A conversation in the Sierra foothills between a new student of bird song and their teacher might go something like this:

STUDENT: What's that sound?

TEACHER: It's a Bewick's Wren.

STUDENT: And what's that one?

TEACHER: It's another Bewick's Wren.

STUDENT: Okay, now *that* is something different.

TEACHER: Yes, it is, it's yet another Bewick's Wren song.

A single male Bewick's Wren may have up to two dozen different songs in his repertoire. Add in variation in song types from site to site throughout the species' range, and one can see why even experienced birders occasionally chase after that "rare bird" only to discover another Bewick's Wren. Still, the songs can almost always be recognized as those of this species by noting characteristic tone and structure. Nearly all variations consist of a trill and buzz somewhere in the phrases—with practice, even this excessively talented songster can be identified correctly.

Bewick's Wrens are generally sedentary, and males on neighboring territories learn and match each other's songs. Their year-round maintenance of territories may help to explain why they sing throughout the year in most of the West, unlike their more migratory relatives, House Wrens. Bewick's Wrens prefer a variety of brushy habitats where they actively glean insects and spiders from leaves, branches, trunks, and the ground, mostly at or below eye level. They are opportunistic cavity nesters and will use anything from a natural tree cavity to an abandoned automobile. Nest building may begin as early as late February in the Sierra, and fledglings may begin to appear in early April. It is not known if Sierra birds regularly produce second broods as dry, hot conditions (especially on the West Side) may make it difficult to successfully fledge young in summer. Both parents are actively engaged in nest building and feeding young, but apparently only the female incubates eggs and broods nestlings while being fed by the male.

STATUS AND DISTRIBUTION Across western North America, Bewick's Wrens are found in a wide variety of habitats from chaparral to oak woodlands to riparian areas but always with a significant shrubby understory. In the Sierra they are largely year-round residents and sing to maintain territories through the winter.

West Side. Common residents below 2,500 feet, becoming fairly common to uncommon up to just over 5,500 feet, particularly in extensive areas of postfire chaparral; some evidence for postbreeding wanderers to higher elevations.

East Side. Fairly common residents at lower elevations throughout, especially in riparian thickets; uncommon in pinyon-juniper woodlands.

House Wren • *Troglodytes aedon*

ORIGIN OF NAMES "House" may be an allusion to the species' careful nesting habits or their tendency to nest near human structures; Gr. *troglodytes*, one who creeps into caves; Gr. *aedon*, a nightingale.

NATURAL HISTORY Few other bird songs communicate such a spirit of pure exuberance as the complex jumble of notes from House Wrens. The song rises in pitch and volume, then quickly descends with a bubbly quality that always inspires a smile. Females also sing, but theirs is usually a shorter, simpler version. Although eastern populations of House Wrens often nest around old houses, barns, and buildings, our western wrens tend to nest in woodlands and forest edges, usually far from human habitation but occasionally near rural houses and ranches. Their elevational range overlaps significantly with the similar Bewick's Wren (see earlier account), but the two species avoid competition by breeding in slightly different habitats. Indeed, studies in the Sierra foothills where breeding areas overlap found no evidence for competition between these two wrens. Both species will use nest boxes and a wide variety of cavities, but House Wrens prefer more open settings offering clear views in the immediate vicinity of their nest sites.

House Wrens have a well-earned reputation for being intolerant of other birds nesting within their territories and will diligently seek out and destroy nests, eggs, and nestlings of any bird with the temerity to invade their domain. Humans wandering by will also receive a thorough scolding with harsh chatters delivered in classic "wren pose"—tail cocked defiantly skyward. Males arrive in mid- to late March and select one or more prospective nest sites, but the later-arriving females make the final choice and do most of the nest building, beginning in April at lower elevations.

STATUS AND DISTRIBUTION House Wrens are common breeders in riparian thickets, open woodlands, and meadow edges across most of North America. They vacate all but the very lowest elevations of the Sierra in winter for the Central Valley floor or more southerly areas.

West Side. Most males arrive by late March with females following in April; common nesters from the Foothill zone into the Upper Conifer zone; while small numbers winter at the lowest elevations, most have left by the end of September; juveniles have been found nearly up to tree line in late summer.

East Side. Fairly common nesters in aspen and riparian groves up to 8,000 feet and sometimes higher; in desert areas found only in the wettest riparian zones; most arrive in early April and leave by mid-September; accidental or absent in winter.

Pacific Wren • *Troglodytes pacificus*

ORIGIN OF NAMES "Pacific" to distinguish the distribution of this recently described species from the similar eastern Winter Wren *(T. hiemalis)* as well as the Eurasian Wren *(T. troglodytes)*; all were formerly lumped together as "Winter" Wrens; L. *pacificus*, of the Pacific.

NATURAL HISTORY Surely Pacific Wrens can lay rightful claim to having the largest ratio of song to body mass of any bird on the planet. When these tiny birds sing, their entire bodies vibrate, and a deliriously rapid and stunningly long and loud cascade of notes bursts forth. These songs can extend for several seconds while dancing up and down the scale at a nearly frantic pace, and usually emanate from cool, moist ravines where large trees dominate. The conditions found in giant sequoia groves and other old growth forests are ideal for this bird, with large snags and the upturned roots of some fallen forest giant offering ideal nest sites and an open, shady understory perfect for foraging along the ground. Intruders who chance upon the territory of a pair of Pacific Wrens will be confronted energetically by a protesting male. With his minute stump of a tail cocked at a jaunty angle, he scolds with emphatic doubled *tick* notes.

Males build several nests with the females choosing among them and adding a lining. Based on studies in Idaho, this nest-building activity peaks in May. The males then have little to do but defend the territory until the nestlings are fairly large, when he begins to participate actively in feeding them. Fledglings may remain in the male's territory for a couple weeks while being fed by both parents. If weather conditions are good, a second brood may be attempted.

STATUS AND DISTRIBUTION Pacific Wrens are fairly common residents of the Pacific Northwest, Coast Range, and Sierra. They are most often found in moist, cool drainages of the Lower Conifer zone, particularly where older forest conditions persist. It is believed that most birds remain through the winter, though some downslope movements occur and small numbers winter on the Central Valley floor.

West Side. Fairly common residents up to about 5,000 feet and uncommon locally to over 7,000 feet; densities may be highest in giant sequoia groves; postbreeding wanderers may move upslope in late summer and many move downslope for winter, though exact winter status is not well understood at higher elevations.

East Side. Previously thought to be absent as breeders, nesting has been confirmed in recent years along Pine Creek in northern Inyo County and at Lake Tahoe; rare to casual in winter at lower elevations in the Reno area south to northern Owens Valley; rare in Alpine and Mono Counties in fall and winter.

Marsh Wren • *Cistothorus palustris*

ORIGIN OF NAMES "Marsh" for the species' preferred habitat; *cistothorus*, a shrub leaper, from Gr. *kistos*, a type of shrub, and *thourus*, leaping; *palustrus*, a marsh dweller, from L. *palus*, marshland.

NATURAL HISTORY The word "song" may not be the first one that comes to mind when listening to the clicks, sputters, and chattery trills rising from a wetland edge, but Marsh Wrens have a legitimate claim to the largest and most complex repertoire of songs of any Sierra bird. Male Marsh Wrens may sing up to 200 distinct song types, each slightly different, some quite short, and generally strung together into a stereotyped sequence. The songs and sequence vary from place to place, but males in a given locale tend to match each other's song types and sequence to produce extended and truly impressive bouts of matched counter-singing.

The prodigious number of song types is probably driven by the highly competitive nature of this bird's lifestyle. Marsh Wrens are polygynous, with each male attempting to attract and mate with as many females as possible. He competes with his fellow males both by displaying his musical virtuosity and his architectural ambition. Males construct a dozen or more nests, each an elaborate structure of cattail strips and grasses, woven together to produce a full domed cavity and usually incorporating lashed-together cattail stalks. When a prospective mate arrives, the male takes her on a tour of his creations. It is unclear if the female generally uses one of the male's nests (after adding a soft interior lining of soft plant fibers and feathers) or more often builds one of her own within his territory. It is also unclear exactly what purpose this abundance of nests serves. It may demonstrate his fitness to the female, but these nests may also serve as places for fledglings to roost and hide, or they may simply make it much harder for predators to find the active nest.

Marsh Wrens share with other members of this family a determination to prevent other birds of any species from nesting within their territories. They will destroy the eggs of other Marsh Wrens or any other species found nearby. Indeed, some have claimed that the communal nature of Tricolored Blackbird nesting colonies was partially driven by the need to overwhelm the capacity of these tiny nest raiders to destroy their eggs. Turnabout being fair play, various species of blackbirds have been documented destroying Marsh Wren nests. However, with so many "dummy" nests in place, they probably have less than a 1-in-10 chance of happening upon an active nest.

STATUS AND DISTRIBUTION The harsh, complex songs of Marsh Wrens resonate from large and small marshes across most of North America. In the Sierra cattail and tule edges of stock ponds and reservoirs have probably allowed populations from the Central Valley (subspecies *C. p. aestaurinus*) to colonize some foothill areas on the West Side that lacked breeding season habitat prior to the arrival of Europeans. On the East Side seasonal wetlands were present long before people arrived, though Grinnell and Miller (1944) noted breeding Marsh Wrens only in the Honey Lake and south Lake Tahoe areas. Today Marsh Wrens, most likely of the subspecies *C. p. pulverius* (using the taxonomy of Phillips 1986), are resident throughout the East Side in lower-elevation marshes.

West Side. Though common residents in nearby Central Valley wetlands, Marsh Wrens are uncommon and patchily distributed residents in the Foothill zone; migrants or postbreeding wanderers are sometimes found in wetland habitats at higher elevations in late summer or fall.

East Side. Locally common breeders in appropriate habitats below 6,500 feet; some remain year-round except in the coldest winters and many East Side Christmas Bird Counts record them most years; most common and widespread during fall migration from late September through October.

GNATCATCHERS • Family Polioptilidae

These dainty birds are most similar to Old World warblers (family Sylviidae) and wrens (family Troglodytidae) in their anatomical structure and habits. Like those related species, gnatcatchers move restlessly through the foliage plucking insects with long, slender bills. Similarly, their short wings and long, fanned tails facilitate agile flights through the dense vegetation where they feed and nest. The gnatcatcher family is comprised of about 20 species that mostly reside year-round in tropical and subtropical areas of North and South America, and only the Blue-gray Gnatcatcher occurs in the Sierra. The family name means "gray-winged" from Gr. *polios*, gray, and *ptilon*, a feather or wing.

Male

Female

Blue-gray Gnatcatcher • *Polioptila caerulea*

ORIGIN OF NAMES "Blue-gray" for the species' overall coloration; "gnatcatcher" for their preference for flying insects; *polioptila* (see family account); L. *caerulea*, a shade of dark blue.

NATURAL HISTORY Soft, nasal calls usually precede the emergence of these sprightly, inquisitive birds from thick tangles of shrubs or oak foliage. Blue-gray Gnatcatchers forage much as kinglets do, picking small insects from foliage or flying out to seize them in mid-air. Similar to wrens, their tails are cocked in a jaunty, uptilted fashion and twitch constantly as they flutter between perches. Flashing the white outer feathers of their long tails, they look and behave somewhat like miniature Northern Mockingbirds and share that species' proclivity to mimic songs of other birds. In the Sierra, gnatcatchers only warble their wheezy, jumbled songs during the peak of breeding season, usually from late April until early July.

Arid hillsides of scattered oaks and dense shrubbery serve as prime nesting habitat for Blue-gray Gnatcatchers. They also use riparian thickets, usually those in close proximity to foothill chaparral or scrubby oak woodlands. By early May both parents begin building intricate nests, a laborious process that can take up to two weeks. They create deep cup nests from small twigs and line them with plant down, mosses, lichens, and shredded bits of wood, and suspend them from horizontal limbs of small trees or shrubs where they are protected from direct sun by overhanging branches. Females usually lay four to six pale blue eggs covered with small reddish spots. Both parents incubate the eggs for 14 to 16 days and care for the nestlings for another 2 or 3 weeks.

STATUS AND DISTRIBUTION Unlike most of their sedentary relatives that spend the entire year in the neotropics, Blue-gray Gnatcatchers migrate to northern latitudes to breed. They also have the largest breeding range of any member of their family, spanning across the entire United States, while other North American relatives are confined to much smaller geographic areas of the arid

West. Sierra breeders usually return from wintering grounds in Mexico and Central America in April and migrate south again by September. Significant numbers winter in riparian areas on the Central Valley floor, but it is unknown if these come from the Sierra breeding population.

West Side. Fairly common but localized in hot, dry chaparral (chamise is a favored species) and oak woodlands as well as dense riparian areas of the Foothill zone up to the Lower Conifer zone; uncommon in mountain chaparral habitats of the Upper Conifer zone in late summer up to about 5,000 feet but seldom much higher; in winter, rare in northern foothills; becomes uncommon south of Mariposa County but can be found in juniper habitat on the west side of Walker Pass, along Chimney Peak Road in both Kern and Tulare Counties, and at Chimney Creek Campground (Tulare County).

East Side. Fairly common to uncommon in lower-elevation scrub habitats (mostly below 7,000 feet) situated within open stands of ponderosa, Jeffrey pines, or pinyon pines with a scattered mix of dense shrubs such as mountain mahogany, sagebrush, bitterbrush, or desert peach; also nests in riparian thickets with scattered aspens or cottonwoods and a thick cover of willows or other shrubs; casual in winter, with nearly all records from Mono and Inyo Counties.

DIPPERS • Family Cinclidae

American Dippers are found from the Aleutian Islands and northeastern Alaska, south through the mountainous regions of western North America and the highlands of Mexico to western Panama. There are five members of this family, and all have similar habits and form and live in swift mountain streams on every continent except Africa and Antarctica. The family name is derived from Gr. *kinklos*, a bird mentioned by Aristotle for its bobbing behavior.

American Dipper • *Cinclus mexicanus*

ORIGIN OF NAMES "Dipper" for the familiar dipping and bobbing behavior; *cinclus* (see family account above); L. *mexicanus*, of Mexico, where the bird was first collected.

NATURAL HISTORY The American Dipper was John Muir's favorite bird. He called it by its old name, "Water Ouzel," which he described as "mountain streams' own darling, the hummingbird of blooming waters, loving rocky ripple-slopes and sheets of foam as a bee loves flowers . . . among all the mountain birds, none has cheered me so much in my lonely wanderings" (Muir 1878). Dippers are a familiar sight in the Sierra along most perennial rivers and creeks. Their sharp chatters rise above the noise and mists of swift-flowing water as they flit nervously from rock to rock. When flying, they skim low over the water, following the stream course and forsaking shortcuts over land. These plump, sooty-gray birds bob incessantly up and down on bended knees while flashing their white eyelids. When feeding, they cling to precipitous rock faces, dive beneath foaming waters, and reappear some distance downstream.

Their powerful legs, long toes, and streamlined bodies allow them to walk with ease over mossy boulders or beneath cascading rapids. They do not have webbed feet but occasionally swim across still pools like tiny ducks. Dippers have evolved many specialized adaptations to suit their aquatic lifestyles, such as enlarged oil glands for waterproofing their outer feathers, a thick undercoat of down, and specialized flaps over their nostrils and eyes (nictitating membranes) to protect them from the pounding spray. On deep dives they "fly" under water using their short, rounded wings as flippers and flattened, stubby tails as rudders.

Dippers remain in mountain streams of the Sierra year-round. The surrounding landscape

can be buried by snow and nearly devoid of other bird life, but Dippers still frolic in the icy waters. Only when the highest streams freeze do Dippers venture downslope. According to studies in Colorado, winter is the critical survival period for Dippers (Price and Brock 1983); some may migrate downstream to avoid the heavy snow zone, and some of those that brave the winter in the high country die due to limited food and a lack of protected roosting sites at a time when energy costs to stay warm are high. Although solitary most of the year, males and females pair for the breeding season, from early March to late June. Males sing long, wren-like courtship calls in early spring. These songs are unfamiliar to most Sierra visitors because they usually cease before the spring snowmelt and are hard for human ears to hear above the roar of the water. Females, which look just like the males, build bulky nests of mosses and grasses that are about one foot in diameter—usually cradled on inaccessible rock ledges above streams or behind waterfalls, where mist keeps mosses green and provides good camouflage. Occasionally, dipper nests are found on log jams, beneath dense vegetation along rushing streams, or under bridges. Females lay 4 to 5 white eggs and incubate them alone for 15 to 17 days. Both parents feed their spot-breasted young for about 25 days until they leave their nests. Dippers prefer fresh, unpolluted waters that support abundant supplies of their preferred foods—caddisflies, stone flies, mayflies, midge larvae, snails, and small fish.

STATUS AND DISTRIBUTION Dippers occur the length of the Sierra, where they are found on steep-gradient streams and sometimes at lakeshores at higher altitudes.

West Side. Fairly common residents along creeks and rivers from about 2,000 to 9,000 feet in the central Sierra; uncommon above tree line; most individuals winter below the heavy snow zone but some may remain at elevations above 6,000 feet year-round; uncommon in the Foothill zone, rarely down to the Central Valley in some winters.

East Side. Fairly common up to about 8,000 feet in spring and summer; winter status similar to the West Side.

KINGLETS • Family Regulidae

Delicate, active birds that flit fairylike among twigs and leaves, members of this family have slender, pointed bills for capturing insects. They breed in boreal or high mountain forests, where most species remain year-round. In all seasons they flick their wings continuously while flitting among dense branches and leaves in search of small, soft-bodied insects and spiders. Kinglets and their kin specialize on eating prey items hidden at the tips of slender branches, where their light weight and ability to hover give them access to food unavailable to larger, less agile birds. Their unusually fluffy, thick plumages keep them warm on long winter nights. Members of this family sing complex songs during the breeding season and call frequently the rest of the year, when they are often seen in flocks of chickadees, nuthatches, and other forest birds.

In the Sierra, kinglet nests are usually well hidden high up amid dense sprays of conifer needles and rarely observed. The sexes, quite similar but not identical, both participate in building deep cup nests by gathering needles and grasses, binding them together with spider webs, decorating them with bits of mosses and lichens, and lining them with feathers and hair. Males bring food to the females while they incubate 8 to 10 (larger clutch sizes than other similar-sized songbirds) brown-speckled yellowish eggs for 14 to 16 days. Both sexes then tend the young on the nest for two to three weeks.

Of the six members of this family (and genus), four occur in Europe, Asia, and/or Africa, and the other two are restricted to North America and reside in the Sierra. The family name was derived from L. *regulus*, a little king, a reference to their brightly colored, regal crowns.

Golden-crowned Kinglet • *Regulus satrapa*

Male

Female

ORIGIN OF NAMES "Golden-crowned" for their bright red, gold, and black crowns; "kinglet," a diminutive reference to crowns; L. *regulus*, a little king; *satrapa* from Gr. *satrapes*, a ruler's crown.

NATURAL HISTORY Though common and widespread in the Sierra, Golden-crowned Kinglets could be missed entirely but for their high-pitched calls emanating from the highest treetops. These calls are a rapid series of *see-see-see* notes, much like the Brown Creeper's, which are longer, clearer, and more forceful. During their breeding season, kinglets also produce a song that begins with the similar high-pitched notes and descends into a jumble of chattery notes. Kinglets are among the smallest of Sierra birds, only slightly larger than Bushtits and hummingbirds. They flit quickly between thick sprays of conifer foliage or flutter up to pick off insects, caterpillars, and other delicacies from terminal needles. A birder with luck and persistence may finally catch a glimpse of the kinglet's "jeweled" crown if the birds decide to drop to lower foliage.

Golden-crowned Kinglets build cup nests high in tall conifers where they are seldom seen. In the Sierra their poorly documented breeding season is thought to extend from late May through late July at higher elevations, with most pairs producing two broods per year. When snow blankets their forest haunts in fall and winter, a few drop down to more hospitable climates of the foothills or valley floor, where they mix with Ruby-crown Kinglets, chickadees, or nuthatches in oak or riparian woodlands and urban non-native conifers. Most Golden-crowns, however, remain in the highest, coldest forests all winter, where they forage almost continuously for eggs and larvae of insects or spiders concealed in snow-covered foliage or under flakes of bark. How these tiny birds, which are not known to roost in holes or go into torpor, survive subfreezing temperatures is still a mystery. One recent clue came in the midst of a Maine winter evening when four kinglets were discovered tightly huddled together, head first into a fluffed-up ball of feathers (Heinrich 2003).

STATUS AND DISTRIBUTION Golden-crowned Kinglets are one of the (if not *the*) most abundant birds of the Sierra's dense conifer forests. Most remain at high altitudes all year, with some dropping to lower elevations in some winters.

West Side. Common to locally abundant year-round residents of the Lower and Upper Conifer zones from about 4,500 to 9,000 feet in the central Sierra, especially in dense, mature white fir, red fir, and Douglas-fir forests; uncommon nesters in Subalpine forests dominated by lodgepole pines and mountain hemlocks; uncommon to locally fairly common down to the Foothill zone in winter.

East Side. Fairly common year-round residents of ponderosa and Jeffrey pines; uncommon in drier forests south of Mono County.

TRENDS AND CONSERVATION STATUS Golden-crowned Kinglets expanded their range southward over the past century. In their 1914 to 1920 Yosemite region surveys, Grinnell and Storer (1924) described them as only "moderately common, at least during the summer season in the Canadian [Upper Conifer] zone . . . sparingly in the Transition [Lower Conifer] zone on the west slope." Things have changed dramatically since then, as they now outnumber all other bird species in these forests. Fire suppression has allowed shade-tolerant white and red firs to fill in the gaps of formerly open stands of ponderosa and Jeffrey pine and incense cedar, making them attractive to these tiny insectivores. Management burns and uncontrolled wildfires may eventually reverse this trend but, in the meantime, Golden-crowns are thriving.

Ruby-crowned Kinglet • *Regulus calendula*

ORIGIN OF NAMES "Ruby-crowned" for the species' best but often concealed field mark; L. *calendula*, a little lark.

NATURAL HISTORY When agitated or otherwise aroused, even in winter, male Ruby-crowned Kinglets erect their crown feathers to unveil stunning, scarlet patches that are hidden most of the time. From fall into early spring, Ruby-crowned Kinglets abound in the foothills of the western Sierra, frequenting all habitats with dense growths of trees and shrubs, including riparian and oak woodlands, foothill chaparral, and suburban gardens. They tend to be solitary in winter but sometimes join small flocks of other foraging birds, flitting about picking insects from twigs and leaves or fluttering out to catch them in midair.

Ruby-crowns sound harsh chatters most of the year, but males preparing to depart for northern or higher-elevation breeding grounds begin to sing territorial songs. These melodies start with a few faint whistles and end with a rising crescendo of loud trills and sweet warbles, described by Grinnell and Storer (1924) as *see, see, see, oh, oh, oh, property, property, property* and *si-si-si, o, oh-oh, cheerily, cheerily, cheerily.* They continue to sing throughout the summer on their breeding grounds, and song can be heard in fall when they first return to wintering areas. Despite their affinity for broad-leaved trees and shrubs in winter, they always nest in conifers. In the Sierra they usually breed from mid-May through July and possibly later at the highest elevations. Similar to Golden-crowns, they build deep cup nests, where they are difficult to monitor or even see, and little is known of their nesting habits in the Sierra. The breeding population may not be the same ones that winter in the foothills; rather, they may migrate farther south.

STATUS AND DISTRIBUTION Widespread breeders in boreal forests across North America, Ruby-crowned Kinglets in the Sierra are near the southern edge of their current breeding range. While still common and widespread as winter visitors, their status as breeders has declined dramatically in the past century (see the "Recent Trends in Sierra Bird Populations and Ranges" chapter for details).

West Side. Common to locally abundant winterers in the Foothill zone from late September through mid-April; uncommon and now highly localized breeders south into Tulare County in the Subalpine zone in stands of lodgepole pine and mountain hemlock up to about 9,000 feet, especially near creeks or meadow edges; uncommon breeders in the Lower and Upper Conifer zones.

East Side. Fairly common spring (April) and fall (September and October) migrants in aspen groves and other riparian thickets up to the Subalpine zone; uncommon and patchily distributed as breeders above 6,000 feet from south of Sierra Valley into northern Inyo County; generally rare in winter at lower elevations except near Reno and in the Owens Valley where common.

WRENTIT • Family Sylviidae

Wrentits have been the equivalent of a taxonomic "pinball," bouncing around from family to family with each new analysis of their origins. Since the mid-1800s, Wrentits have been classified as members of many different families by many different authors. This confusion arose because they seemed to have a mixture of characteristics of true wrens (family Troglodytidae) and tits (family Paridae). More recently they were considered to be the only members of the family Chamaeidae. Modern genetic studies suggested that they were the only North American representatives of the Babblers (family Timaliidae), and in 2010 they were moved again to become the only North American members of the Old World warblers (family Sylviidae). This family historically included hundreds of species but more recently was trimmed to fewer than 60 species found in Asia, Africa, and Europe. The current family name is derived from L. *sylvia*, a wood or forest.

Wrentit • *Chamaea fasciata*

ORIGIN OF NAMES "Wrentit" for their physical and behavioral resemblance to both wrens and tits; *chamaea* from Gr. *chamai*, a ground dweller; L. *fasciata*, striped or barred—a reference to the species' faint breast streaks.

NATURAL HISTORY More than any other sounds, the ringing, staccato notes of Wrentits express the spirit of the chaparral. Both males and females echo these "bouncing-ball" calls, but the latter do not trill at the end. Wrentits call in all seasons and are more often heard than seen. They also frequently give a short, repeated *churr* call, which, if imitated, may bring one into view. Mostly they seem reluctant to expose themselves to view, always staying near the

ground in dense shrubbery on the driest slopes as well as (to the surprise of many) in moist, wooded riparian areas with lush understories. A West Coast endemic species, their skulking nature leaves them in the "heard-only" category for many visiting birders. When feeding, they flit between twigs and leaves, picking insects, spiders, and small berries from foliage. Males and females look identical, with grayish-brown plumage, cream-colored eyes, and wrenlike, uptilted tails.

Wrentits mate for life and spend much of their time in close company with each other, foraging, preening, and roosting. Among the most sedentary of North American birds, Wrentits as mated pairs can remain together on the same breeding territories for up to 12 years, moving less than a quarter of a mile in their lifetimes. They may seem less common than they really are because they space themselves out uniformly. In truth, they are among the more common of foothill chaparral and riparian species in many areas.

Both parents share in building cup nests concealed in low shrubs. Females usually lay 4 bluish-green eggs that both parents incubate for about 15 days. Young are tended in the nests for about two weeks before they fledge and begin foraging on their own. After the spring breeding season, a few individuals (mostly immatures) move up to forage amid the snowberries, manzanitas, and huckleberry oaks as high as the Upper Conifer zone, where they may remain until fall.

STATUS AND DISTRIBUTION Wrentits occupy a relatively small geographic range extending from the Columbia River southward through the Cascades, Sierra, and Coast Ranges to the deserts of Baja California.

West Side. Common year-round residents of the Foothill zone down to the Central Valley; fairly common up to the Lower Conifer zone; a fairly isolated breeding population in the Feather River drainage in the vicinity of Quincy (Plumas County) approaches the eastern edge of the West Side; uncommon in mountain chaparral of the Upper Conifer zone above 5,000 feet in late summer and fall; present during the breeding season up to 5,000 feet in southern Tulare County.

East Side. A pair presumably bred in Sand Canyon at the southwestern corner of Inyo County in 1992, and birds were detected at two locations near Lone Pine in summer of 2001. However, it is unclear if these records represent a range extension into the East Side.

THRUSHES AND RELATIVES • Family Turdidae

With moderately long and slender bills, thrushes take insects, worms, and spiders from the ground and low plants or snatch them from midair. They also dine on fruits and berries from late summer through the winter. The young of all seven Sierra species leave their nests with heavily spotted breasts, but only two, Hermit and Swainson's Thrushes, retain their breast spots as adults. Many of the thrushes display soft browns and grays, and the vivid hues of bluebirds offer a striking contrast to those muted plumages. For many birders it is the music that makes this their favorite family, and many thrush songs have an ethereal, flute-like quality unsurpassed in the bird world.

Thrushes build the familiar cup nests in trees or on the ground, but bluebirds use holes in trees as well as artificial nest boxes. Females of most species lay 4 to 6 bluish or light greenish, unmarked eggs and incubate them alone for 10 to 12 days. Townsend's Solitaire eggs are white to pink and heavily scrawled with reddish-brown scrawls and splotches. Both parents tend and feed nestlings for 12 to 14 days until they fledge, and for 20 to 30 days until they can forage successfully on their own. The taxonomy of this family has been revised in recent years, and it now includes about 175 species in more than 20 genera that occur on all continents except Australia and Antarctica. The family name is derived from L. *turdus*, a thrush.

Western Bluebird • *Sialia mexicana*

Male

Female

ORIGIN OF NAMES "Western" for distribution compared to the similar Eastern Bluebird *(Sialia sialis)*; "bluebird" for overall coloration; *sialia* from Gr. *sialis*, a kind of bird; L. *mexicana*, of Mexico, where the species was first collected.

NATURAL HISTORY A flash of brilliant blue darting across a country road in the foothills is often the first sign a male Western Bluebird is near. When foraging, Western Bluebirds perch low on branches or fences in open grassy areas and dart out to capture insects on the ground, on low plants, or in midair. They often hover a few feet above the ground before dropping to seize their prey, mostly large insects, including grasshoppers, beetles, caterpillars, and bees, as well as large spiders. While primarily insectivorous in spring and summer, they switch to fruits and berries from late summer through winter. Consequently, their nonbreeding distribution varies from year-to-year, depending on where they find plentiful sources of berries from natural sources like mistletoe, juniper, elderberry, and wild grape or from sources planted by people. In winter, Western Bluebirds sometimes mingle with other fruit-loving birds like Yellow-rumped Warblers, Cedar Waxwings, or American Robins at good clumps of berries in oak woodlands and foothill chaparral.

Western Bluebirds are a classic edge species, typically inhabiting the borders between two markedly different habitats such as between woodlands and open grasslands or meadows. Such edges provide abundant food and perching sites not available in either habitat alone, and few species are as tied to habitat edges as Western Bluebirds. Unlike Eastern and Mountain Bluebirds, Westerns generally avoid large meadows and treeless grasslands, preferring open oak woodlands and deciduous forests, wooded riparian areas, suburban gardens, farmlands, and recently burned forests. They will use large, open areas, however, if nest sites are present in wooden fence posts or artificial nest boxes.

Bluebirds nest in cavities. In the foothills they compete for nest sites with introduced European Starlings and House Sparrows. This competition for cavities can result in physical aggression for this limited resource, and bluebirds often come out on the losing end. In higher-elevation meadows, native Tree and Violet-green Swallows are the main competitors for nest holes. Woodpeckers invade recently burned forests in large numbers and bluebirds follow, taking advantage of the newly created nest cavities, abundant insects, and edge habitats. Bluebirds are also known to use abandoned mud nests of Barn and Cliff Swallows.

In the Sierra, Western Bluebirds begin to form pairs in early March and their nesting season extends through early August, depending on the elevation. While pairs remain together during the breeding season, females are promiscuous, and recent genetic studies suggest that more than half the eggs in any given clutch are fertilized by males other than the female's mate (Germaine and Germaine 2006). Males continue to feed and guard the young for several weeks, while females attempt to renest, often with different males. Sometimes yearling male Western Bluebirds help adults, presumably their parents, with feeding and guarding the young.

STATUS AND DISTRIBUTION Western Bluebirds occur in western North America from southern Canada to southern Mexico. Their breeding range includes most of California outside the deserts. They reside year-round on both sides of the Sierra.

West Side. Common residents and breeders in open oak woodlands and savannas of the Foothill zone, and locally abundant in winter where berries are plentiful; uncommon breeders up to the Upper Conifer zone; in late summer and fall, some transient Westerns venture up to the Subalpine zone in search of ripening juniper berries and other foods.

East Side. Uncommon but regular breeders in locations such as Sierra Valley, Tahoe Basin, and Carson Valley from early April through late October, where usually associated with stands of juniper; rare and highly localized spring and summer visitors as far south as the Owens Valley; in winter, casual north of Inyo County and locally uncommon from the Owens Valley south.

TRENDS AND CONSERVATION STATUS Artificial nest boxes are readily used by Western Bluebirds, and concerted programs to provide boxes that exclude starlings may be responsible for recent increases in this species on Sierra Christmas Bird Counts. A recent study showed that bluebirds have higher reproductive success in nest boxes than in natural cavities or woodpecker holes.

Female

Male

Mountain Bluebird • *Sialia currucoides*

ORIGIN OF NAMES "Mountain" for high-elevation distribution compared to other bluebirds; *currucoides*, a possible hybrid of L. *carruca*, an old name for warblers, and Gr. *oides*, similar to.

NATURAL HISTORY In the visual wealth of high Sierra landscapes, the stunning, sky-blue plumage of the male Mountain Bluebird stands out as a special gem, and no photograph or painting can prepare one for the full effect of this particular shade of blue. The grayish females and young also show flashes of this intense blue in their wings, rumps, and tails. Unlike Western Bluebirds, which favor habitat edges between forests and grasslands, Mountains typically occupy large, open meadows or shrublands, dry grassy openings, and rocky fellfields from the Subalpine zone to above tree line. In these open landscapes they scan for insects from low shrubs or rocks and fly out to seize them on the ground or in midair. Often they hover before dropping on their prey, usually grasshoppers, butterflies, ants, beetles, and spiders. Mountain Bluebirds also consume small fruits and berries such as currants, juniper berries, and wild grapes.

Less dependent on trees than Western Bluebirds, they sometimes nest in rock crevices or human structures, though they prefer abandoned woodpecker holes when available and will use artificial nest boxes. Competition for holes is keen, and Mountain Bluebirds have been seen chasing chickadees, nuthatches, and swallows away from potential nest sites. Rarely Western Bluebirds nest as high as the Subalpine zone, and fierce competition for nest holes has been observed where the two bluebirds co-occur, such as in a recently burned area in Yosemite National Park.

In the Sierra, Mountain Bluebirds initiate nesting in mid-April in most localities, and their breeding season extends into mid-August, especially if second broods are attempted. Females select the nest sites and fill them with loose cups constructed from grasses, stems, and small twigs and lined with plant down. After young Mountain Bluebirds leave the nest, their parents continue to feed them for about a month, and often lead them to nearby stands of dense vegetation, presumably to hide them from predators until they gain independence.

STATUS AND DISTRIBUTION Mountain Bluebirds breed from southern Alaska to northern Mexico. They remain in the Sierra from early March through late October in most years before leaving for warmer locations farther south or lower elevations.

West Side. Fairly common breeders in the Subalpine and Alpine zones, especially in open terrain and recently burned forests; uncommon to rare in migration and winter at lower-elevation meadows and grasslands with the exception of the Kern River Valley, where they range from uncommon to abundant in winter.

East Side. Uncommon breeders from the highest Alpine fell-fields down to the sagebrush flats of the Great Basin; some winters fairly common in places such as Honey Lake, Carson Valley, Owens Valley, and southern desert areas.

Townsend's Solitaire • *Myadestes townsendi*

ORIGIN OF NAMES "Townsend's" for John Kirk Townsend (1809–1851), who collected a number of new bird species in the Rocky Mountains and Oregon, including this solitaire and a warbler and sent them to John James Audubon, who named them in Townsend's honor; "solitaire," a suggestion of a shy or solitary bird (see below); *myadestes*, a fly-eater, from Gr. *myas*, a fly, and *edestes*, eater; L. *townsendi*, for Townsend.

NATURAL HISTORY Although no more solitary than many other forest birds, Townsend's Solitaires wear subtle plumage, and their quiet nature during the nesting season makes them seem less common than they really are. They fly out from perches to hawk insects in midair or pick them off bark, foliage, or ground. Some forage conspicuously over montane chaparral, using tall, scattered conifers as perches and song posts, while others stick to open understories of shaded forests. There their presence is betrayed only by the flash of their white tail edges, buff wing patches, and the snapping of their bills as they capture flying insects.

In exuberant contrast is the male's song, often given in flight, which is rightly extolled for its virtuosity. He often selects a perch near the top of a tall conifer and delivers a complex series of musical warbles and trills that may continue for half a minute or more. The song has the clear flute-like quality of other thrushes, while their rapid, warbling patterns are reminiscent of Cassin's Finches. Solitaires also make short, mellow single whistles that sound much like the squeak of a rusty gate. Solitaires sing throughout the fall and winter, and both males and females aggressively defend nonbreeding territories centered on patches of juniper trees that sustain them through the winter. Occupants of the best berry-rich areas have higher winter survival rates. At lower elevations they prefer mistletoe berries but also eat the fruits of toyon, manzanita, and elderberry.

Solitaires typically nest on forested ridges or canyon slopes from early May to about mid-August. Their simple nests are built on or near the ground, in well-drained spots under protecting overhangs such as exposed tree roots, root wads of fallen trees, crevices in earthen banks, rock walls, or road cuts. They also nest under the eaves of carports and other structures in some areas. Young are fed and tended by both parents for about two weeks until they leave the nests. Parents divide broods into two groups and care for them separately through mid-September.

STATUS AND DISTRIBUTION Townsend's Solitaires breed in boreal and other coniferous forests from southern Alaska to the Sierra Madre of central Mexico. After breeding in the Sierra, Solitaires disperse up- and downslope in search of ripening juniper berries, where they remain until these food supplies are exhausted. As a result, their winter distribution varies considerably from year-to-year.

West Side. Fairly common in the Upper Conifer zone and into the Subalpine zone from April through October; uncommon up to the Alpine zone from late August through early November; uncommon to rare in winter down to the Foothill zone, where they compete with Cedar Waxwings, American Robins, and Phainopeplas for mistletoe berries.

East Side. Fairly common summer residents from April through August up to about 10,000 feet; can be locally common as numbers increase from September through March, when West Side breeders cross the crest to winter in juniper-dominated habitats.

Swainson's Thrush • *Catharus ustulatus*

ORIGIN OF NAMES See Swainson's Hawk account; "thrush" from OE. *thrusche*, possibly derived from L. *turdus* (see family account above); *catharus*, from Gr. *katharos*, pure—a possible reference to the bird's clear and beautiful songs; L. *ustulatus*, singed or burned, for the species' somewhat russet-colored plumage.

NATURAL HISTORY Swainson's Thrushes look a good deal like their close relatives, Hermit Thrushes. Swainson's are best distinguished by the relative lack of contrast between their plain brown backs and tails. Swainson's have a buffy wash beneath their breast spots, and their conspicuous buffy eye rings extend to their bills, forming spectacles. These marks can be missed under poor viewing conditions. A good behavioral clue is that Swainson's are usually calm and demure, unlike Hermits with their flicking wings and pumping tails. Secretive by nature, Swainson's Thrushes are more often heard than seen. They deliver an ascending series of short, clear-toned phrases, among the finest of Sierra birdsongs, as well as distinctive "water-drop" call notes. Their song lacks the long, drawn-out introductory note characteristic of the Hermit Thrush song. Swainson's Thrushes tend to forage in shrubs and trees, and less often on the ground, compared with Hermits. Their diet consists of beetles, ants, bees, caterpillars, moths, crane flies, and especially mosquitoes taken in flight during short bursts of flycatching. They also consume berries from a variety of low-growing shrubs.

In the Sierra, Swainson's Thrushes nest in dense thickets of willows, alders, or brushy forest undergrowth near streams, lakes, and wet meadows. They breed from late May until mid-August, with peak activities in June. Females build compact cups from grasses, mosses, and lichens and hide them under dense foliage less than eight feet above the ground. Both parents bring food to the young, but females do most of the brooding and nest maintenance. Young leave the nests after about two weeks and become independent a few weeks later.

STATUS AND DISTRIBUTION The four subspecies of Swainson's Thrushes breeding across of North America can be divided into the "Russet-backed" and "Olive-backed" groups. The subspecies found on the West Side (*C. u. oedicus*, one of the "Russet-backed" group) breeds in low- to mid-elevation riparian habitats and winters mainly in western Mexico and Central America. An "Olive-backed" subspecies (identified historically as *C. u. almae*, but now considered part of *C. u. swainsoni*) is believed to be the primary breeding species on the East Side and winters much farther south in northern portions of South America.

West Side. Uncommon breeders from late May through mid-August; historically widespread and common (see "Trends and Conservation Status" below), now present in widely scattered locations; in the northern Sierra, Buck's Lake is among the most reliable spots, also found at Blodgett Forest (El Dorado County) and the Clark Fork of the Stanislaus River (Tuolumne County); documented nesting sites (at least up until the early 1980s) in Yosemite National Park near Tuolumne Grove and White Wolf, and east of Harden Lake; in Sequoia National Park, likely breeding populations may still exist at Zumwalt Meadows, Grant Grove, and Mineral King Valley; northward migrants are fairly common in dense riparian forests along the South Fork Kern River during the last two weeks of May, where up to 200 individuals have been recorded in a single morning's survey.

East Side. Rare, irregular breeding records in riparian habitats near Johnsville, Plumas County, western Tahoe Basin (Placer and El Dorado Counties), and possibly in the Mono Basin.

TRENDS AND CONSERVATION STATUS Grinnell and Miller (1944) described Swainson's Thrushes as "common to abundant" throughout the West Side. As dramatically confirmed by recent resurveys of many Sierra sites visited by Grinnell and his associates in Plumas County and Yosemite National Park (Moritz 2007), this species has disappeared from most of its historic Sierra range since the mid-20th century. This decline is one of the unsolved ornithological mysteries of the Sierra. In most historical breeding sites, the habitat is still intact and nearly all other species are still present. Explanations for their decline are elusive since obvious culprits like nest-parasitizing Brown-headed Cowbirds are apparently not suspects. California studies revealed less than 3 percent parasitism rates, and most nests contained no cowbird eggs at all. Impacts on the wintering grounds may be to blame, but other populations of this species that have overlapping wintering areas have not shown such significant declines.

Hermit Thrush • *Catharus guttatus*

ORIGIN OF NAMES "Hermit," a reference to the species' solitary, reclusive habits; L. *guttatus*, spotted, for their distinctive breast spots.

NATURAL HISTORY "He who has never heard the evening requiem of the Hermit has missed the choicest thing which Nature in California has to offer." This is a bold statement from William Dawson (1923) but certainly defensible. The haunting song begins with a long, fluted note, followed

by echolike, harmonious phrases, then a pause, then another long note that may be considerably higher or lower in pitch than the first, followed by more matching harmony. These elusive thrushes move inconspicuously under shady canopies, where their brownish plumage makes them easy to overlook unless they vocalize. One reward of learning bird songs is being able to locate and identify secretive species like Hermit Thrushes without ever glimpsing them. Even the nonsinging winter visitors to lower elevations can be detected by their characteristic *chup* calls or by a Spotted Towhee–like *wheeze*.

In the spring and summer, Hermit Thrushes probe for beetles, caterpillars, and spiders in grassy areas or forest litter, hopping then halting to look and listen, much like American Robins, though seldom in the open like robins. In winter, they mostly consume fruits and berries, especially those produced by toyon, madrone, elderberry, and Himalayan blackberry. At mid-elevations Hermit Thrushes nest in moderately dense stands of ponderosa pines, incense cedars, and Douglas-firs, but higher up they frequent open stands of red fir and lodgepole pine and Alpine habitats up to tree line. Their choice of dense forests at lower elevations may reflect a need for cool temperatures while breeding. They usually nest lower than 10 feet, sometimes on the ground, but more often in clumps of saplings with clear views of the nearby surroundings. Females build sturdy cup nests from twigs, mosses, and strips of bark, and plaster them together with mud. Fine plant materials and pine needles line the nest. Both parents feed their young in the nest and for about two weeks after fledging until the young are able to start foraging on their own.

STATUS AND DISTRIBUTION Hermit Thrushes are the only members of their genus that winter in North America. Members of the Sierra breeding subspecies *(C. g. sequoiensis)* depart for wintering grounds in northern Mexico by late August, and a smaller contingent of breeders from Alaska and Canada representing different subspecies *(C. g. guttata* and *C. g. nanus)* move into low-elevation locations in September and October and remain through early April.

West Side. Fairly common breeders from about 4,500 feet in the central Sierra into the Alpine zone; fairly common in winter from below the snow-covered areas down to the Central Valley.

East Side. Fairly common nesters in pine forests as well as in riparian woodlands and aspen groves up to tree line; rare winter visitors below about 7,000 feet.

American Robin • *Turdus migratorius*

ORIGIN OF NAMES "American" to distinguish them from a very distantly related Eurasian species with a red breast, also called a robin; L. *turdus* (see family account above); L. *migratorius*, for their migratory habits.

NATURAL HISTORY Robins are among the most familiar of North American birds. They occur almost everywhere and are one of the few thrush species that regularly forage in the open, far from shade and cover. Visual hunters, they dash across meadows or turf, stop short and cock their heads to listen and peer at the ground ahead. After a momentary freeze, they jab their bills at unsuspecting earthworms or large insects such as beetles, caterpillars, or grasshoppers. Robins also forage in riparian woodlands and moist forest understories, and in the fall and winter, they consume fruits of toyon, dogwood, mistletoe, and elderberry as well as those produced by ornamental trees and shrubs. Along with Cedar Waxwings, they flock to the fermenting berries of *Pyracantha* and wild grapes, and they become drunk if too many are eaten, causing them to stagger on foot or fly erratically.

Male robins proclaim their nesting territories with simple sing-song tunes, among the first to be heard at dawn. Their songs are similar to those of Black-headed Grosbeaks and Western Tana-

Female   Male

gers, species that share many forest habitats with robins. Robin songs lack the rough, burry quality of tanager songs and are less musical and variable than a grosbeak's. They nest throughout the Sierra, preferring moist, open coniferous forests near open habitats. Human developments such as irrigated lawns and pastures benefit robins and have allowed them to extend their breeding range to every nondesert part of California.

In the Sierra, courtship begins in mid-April, and nesting activities can extend through late July, especially for pairs that attempt two or three nests per season; a new nest is built for each successive brood. They are among the earliest nesters in many areas, and the young from first clutches are out of the nests before most other birds have finished incubating. Female robins build nests in forked branches of small trees or shrubs, usually less than 15 feet above the ground. Occasionally they nest under decks or on top of roof supports in farmlands and suburban areas. Nests are constructed from twigs, coarse grasses, lichens, and mosses cemented together with mud and lined with feathers, fine grasses, and other soft materials. Young robins beg for food conspicuously for almost a month until they can forage for themselves. On leaving the nest, the young resemble their parents but for their speckled rather than reddish breasts. This juvenile plumage is lost by early fall.

STATUS AND DISTRIBUTION No other Sierra birds occupy as many elevations and habitats as American Robins. They are among the most common birds almost everywhere, and they thrive from the lowest oak woodlands to Alpine fell-fields as well as in most forest types, chaparral habitats, landscaped gardens, and golf courses. They reach their greatest abundance, however, in areas where trees and shrubs are interspersed with meadows, grasslands, and other open habitats. In winter, robins invade the Sierra foothills, sometimes by the thousands. A flock seen in December 1984 was spread out over a three-mile section of the Tuolumne River canyon and contained at least 100,000 birds. Doubtless many wintering robins in the Sierra come from colder climes to the north, while some or all of the breeding birds migrate south for the winter, rather than downslope below the heavy snow zone.

West Side. Common from the Foothill zone up to the Alpine zone; locally abundant in winter below about 5,000 feet in the central Sierra, where they find productive sources of berry-producing trees and shrubs on irrigated lawns and suburban areas where fruiting trees and shrubs are planted.

East Side. Common spring and summer residents in most habitats below tree line; fairly common but irregular fall visitors; often fairly common in winter at lower elevations where snow cover is minimal or when juniper berries are abundant.

Varied Thrush • *Ixoreus naevius*

ORIGIN OF NAMES "Varied" for the species' patterned plumage compared to other thrushes; Gr. *ixoreus*, for mistletoe berries, a preferred food; *naevius*, from L. *naevia*, spotted or varied colors.

NATURAL HISTORY These flashy visitors from the north are among the few songbirds that visit the Sierra in winter but do not remain to breed. Despite their fancy plumage, Varied Thrushes are easily overlooked due to their quiet nature and affinity for shaded, moist forests. Often, the first clues to their presence are their vibrant and musical calls that resemble someone trying to whistle and hum at the same time. A simple but plaintively lovely sound, typically delivered on one pitch, then switching to a different pitch for the next call. They can also *chup*, very like a Hermit Thrush. Besides eating toyon, manzanita, mistletoe, and dogwood berries, they search for acorns and invertebrates in the damp forest litter. They may be solitary, form sizable flocks, or mingle with American Robins, where they can go unnoticed due to similar shape, size, and behavior.

STATUS AND DISTRIBUTION Varied Thrushes breed in coniferous forests from northern Alaska south to the Coast Range in northern California. They do not breed in the Sierra; instead, they occur as an irruptive species visiting intermittently in winter. From October to early April, Varied Thrushes may be locally common in some years but rare or absent in the same localities in other years. When they visit the Sierra in numbers, Varied Thrushes concentrate where berries are plentiful, generally remaining below the heavy snows. This species, similar to other irruptives like Bohemian Waxwings, venture south when northern food supplies are lean.

West Side. Uncommon to fairly common from mid-October through mid-April below about 5,000 feet in some years but rare or absent in others; usually occur in shaded conifer or riparian habitats unlike American Robins that usually forage in more open areas; in fall and spring, migration can occur almost anywhere.

East Side. Rare visitors to pine forests, juniper woodlands, riparian, and aspen groves from mid-October through mid-December with irregular occurrences on Christmas Bird Counts from Honey Lake, South Lake Tahoe, Mono Basin, and the Owens Valley; rare in spring migration.

MOCKINGBIRDS AND THRASHERS • Family Mimidae

As the name of this family suggests, mockingbirds, thrashers, and their relatives are inspired singers and accomplished mimics. Most display subtle patterns of gray or brown, identical for both sexes. These slender, long-tailed birds use strong, down-curved bills to forage on the ground or in dense brush cover.

Both parents build large, bulky cup nests with twigs, stems, and grasses, usually less than 10 feet above the ground and always hidden by dense, overhanging branches or foliage. Usually heavily marked with reddish-brown or maroon spots and splotches, 2 to 4 bluish or whitish-blue eggs are incubated by both sexes (except for Northern Mockingbirds, where only females incubate) for 14 to 18 days. Both parents feed and care for the young for 12 to 14 days until they fledge, and for about 2 to 3 weeks afterward. While young birds are in the nests, and shortly thereafter, their parents feed them protein-rich animal foods such as grasshoppers, beetles or spiders. During the rest of the year, however, members of this family depend largely on fruits and berries gleaned from plants or picked off the ground. Mimids are restricted to the New World, and more than 30 species of mockingbirds, thrashers, catbirds, and tremblers occupy large portions of North, Central, and South America. The family name is derived from L. *mimus,* to mimic or imitate.

Northern Mockingbird · *Mimus polyglottos*

ORIGIN OF NAMES "Northern" to distinguish this species from about 15 other species of subtropical and tropical mockingbirds; "mockingbird" for the species' ability to mock or imitate the sounds of other species and various outdoor noises; *mimus* (see family account); *polyglottos,* many tongues, from Gr. *polus,* many, and *glotta,* a tongue.

NATURAL HISTORY As Ralph Hoffman (1927) aptly expressed: "A Mockingbird at the height of the breeding season pours forth one brilliant phrase after another, modulating skillfully from one to the next, and when song alone fails to express his ardor he flies up a few feet from his perch and flutters down again with wings and tail spread." Both male and female Northern Mockingbirds sing, and some learn more than 150 different songs during their lifetimes. Even more impressive, the song repertoires used by males in spring are almost completely distinct from those sung in fall. These enchantingly varied songs can be heard in the Sierra foothills at all times of year, save perhaps in late summer and early winter, when they may be silent for a few weeks. They sing melodies composed of many motifs, each usually repeated three or more times before switching to a different tune. They imitate the sounds of many lowland birds (such as Mallards, Killdeer, Western Kingbirds, Western Scrub-Jays, Acorn Woodpeckers, Oak Titmice, and Bullock's Orioles) as well as toads, frogs, crickets, cats, dogs, squeaky gates, lawn mowers, or almost anything that strikes their fancy. At least 10 percent of their song is mimicry and the remainder, pure mockingbird. Unattached males frequently sing long into the night, especially on warm, moonlit evenings: often starting well before dawn. If one is keeping you awake, your only hope might be to find him a girlfriend.

Mockingbirds are primarily monogamous, and mated pairs defend nesting territories vigorously against all intruders. With white wing patches and outer tail feathers flashing, they fly aggressively at other birds, cats, dogs, and even humans. The wing patches are often flashed in ritualized manner as they walk along the ground, but the true purpose of this common behavior remains unknown. They nest from February until September, a very long breeding period compared to most Sierra birds.

STATUS AND DISTRIBUTION Among the most widespread, familiar, and loved birds of North America, Northern Mockingbirds are recognized as the "state bird" of five states. Prior to the arrival of Europeans in California, they were probably confined to low-elevation sagebrush, chaparral, and deserts from the San Joaquin Valley south. As settlers moved into the state and planted more fruiting trees and shrubs, mockingbirds gradually spread northward and can now be found in all but the highest and driest areas of the state.

West Side. Fairly common year-round residents of the Foothill zone, where they strongly prefer farmlands, suburbs, and country gardens and avoid heavily wooded areas; uncommon up to the Lower Conifer zone and rarely occur much higher.

East Side. Formerly rare throughout but now fairly common breeders and uncommon residents in the Owens Valley; fairly common to uncommon residents from Reno south through the Carson Valley; still rare but becoming more regular in Honey Lake, Sierra Valley, and Mono Basin areas with most observations from September through November; casual in winter except in the Owens Valley and Reno-Carson Valley areas; habitats as on the West Side.

Sage Thrasher • *Oreoscoptes montanus*

ORIGIN OF NAMES "Sage" for the species' preferred habitat; "thrasher," probably a variation of OE. *thrusher*, a thrush; *oreoscoptes*, a mountain mimic, from Gr. *ores*, mountain, and *scoptes*, a mimic; L. *montanus*, of the mountains, as they were once called "Mountain Mockingbirds."

NATURAL HISTORY Smallest of the thrashers, Sage Thrashers have subtle, streaky plumage that blends with the shrubs they inhabit, making them hard to find. They are usually detected by their songs, as they are the loudest and most prolific singers in sagebrush flats east of the Sierra escarpment and often sing from atop large shrubs. Their songs, somewhat reminiscent of Northern Mockingbirds, include long melodies that imitate the songs of other birds as well as nonbird sounds in their local environments. Unlike mockingbirds, Sage Thrashers generally do not repeat the same phrases in succession.

Sage Thrashers are almost entirely tied to large, dense stands of big sagebrush on flat or rolling terrain, and they generally are absent from steep hills and canyons. Little is known of this species' nesting habits in the Sierra, but in other parts of their range their nesting season extends from mid-April through mid-July. Both parents participate in building large cup nests from sticks, plant stems, and shredded bark in low (two or three feet off the ground) forks of big sagebrush plants. Less often, they nest in greasewood but avoid using smaller and more delicate shrubs such as rabbitbrush and bitterbrush as nest sites.

STATUS AND DISTRIBUTION Confined to western North America, Sage Thrashers breed throughout the Great Basin to the eastern edge of the Sierra; they winter in deserts of the Southwest and

Mexico. They begin to arrive on their nesting grounds in late February and early March, where they remain until late September or early October.

West Side. Rare to casual visitors to the Foothill zone fall through early spring with most records in fall.

East Side. Common to fairly common in appropriate habitats throughout the breeding season; rare to casual in the nonbreeding season.

California Thrasher • *Toxostoma redivivum*

ORIGIN OF NAMES "California," where the type specimen was collected; *toxostoma*, bow-mouthed, from Gr. *taxon*, bow, and *stoma*, a mouth—a reference to the decurved bills typical of species in this genus; *redivivum*, from L. *redivisus*, revived—a reference to William Gambel's (see Mountain Chickadee account) collection and description this species after it had been "lost" in museum drawers for an unknown period of time.

NATURAL HISTORY Hidden deep beneath the cover of protective chaparral, California Thrashers rake their long, curved bills sideways through leaf litter and probe the soil for beetles, ants, wasps, or spiders. These long-tailed, brownish birds also relish berries and seeds taken from bushes or from the ground. Males and females both sing exuberant, repetitious songs from the tops of shrubs or low trees. Like their relatives, the mockingbirds, their songs are richly varied. However, their repertoires are not as extensive, and they mimic less freely—thrashers do not repeat themselves as often, and utter each phrase only once or twice.

Wearing drab attire, California Thrashers are the largest members of their genus. They avoid heavily forested areas and instead reside in dense growths of chamise, buckbrush, or manzanita, where their songs mingle with those of Wrentits and the similarly plumaged California Towhees. Shrubby habitats interspersed with oaks are also favored. Less commonly, thrashers occur in brushy streamside thickets under a canopy of riparian trees and in rustic suburban gardens. Their protracted nesting season may extend from December until August, and mated pairs often raise two broods of young per year. Deep within dense foliage of bushes or small trees, both parents build a nest coarser than that of the Northern Mockingbird.

STATUS AND DISTRIBUTION California Thrashers are sedentary, year-round residents of coastal and foothill regions of California, and their range extends south to central Baja California.

West Side. Fairly common on brushy slopes and dense riparian habitats below about 3,500 feet in the Foothill zone; a small population occupies the chaparral-covered, western slope of Walker Pass up to about 5,000 feet in Kern County; otherwise, rare up to the Lower Conifer zone and never wanders into the higher mountains.

East Side. Locally uncommon resident only in desert canyons of southern Inyo and northern Kern Counties, such as Sand Canyon and Butterbredt Spring.

Le Conte's Thrasher · *Toxostoma lecontei*

ORIGIN OF NAMES "Le Conte's" for John Lawrence Le Conte (1825–1883), an independently wealthy physician (who never practiced medicine); he specialized in collecting insects and was the president of the American Association for the Advancement of Science; he also collected the first specimen of this species; *L. lecontei*, a Latinized form of his name.

NATURAL HISTORY Dressed in a subdued shade of grayish-brown, Le Conte's Thrashers blend with the background sands of their desert environments. In most of their range it may rain only a few days per year, and these thrashers rarely drink. Instead, they gain their moisture from eating large, juicy arthropods (such as beetles, grasshoppers, scorpions, and large spiders) and some seeds. With long tails cocked upward, they use their long, sturdy legs to run after prey, somewhat like miniature roadrunners. They are most easily detected by their long, complex songs that may include several phrases repeated hundreds of times. Distinguishing among thrasher songs can be a challenge, but Le Conte's tends to sing high-pitched, squeaky notes more than other thrashers with overlapping ranges.

In the southern Sierra, Le Conte's Thrashers reside in open, arid habitats with scattered shrubs such as creosote, sagebrush, or in Joshua tree woodlands in gently rolling terrain such as on alluvial fans. Both parents build large, bulky nests of twigs, grasses, leaves, and feathers in thick, thorny plants such as cacti, acacia, or mesquite. They nest earlier than most Sierra birds, usually from mid-February to mid-May.

STATUS AND DISTRIBUTION Le Conte's Thrashers are year-round residents of desert habitats of the southwestern United States and northern Mexico. Their range barely extends into the extreme southern Sierra.

West Side. The few recent records are from Canebrake Creek and Kelso Creek wash habitats in the Kern River watershed.

East Side. Fairly common localized residents in desert canyons of northern Kern County with some observations in similar canyons of southernmost Inyo County; most reliable in Butterbredt and Jawbone Canyons (Kern County) and west of Highway 14 along the Los Angeles Aqueduct; a few Owens Valley floor occurrences outside of their known breeding range in the vicinity of Fish Slough.

TRENDS AND CONSERVATION STATUS Throughout their limited range in California, Le Conte's Thrashers have been adversely affected by agricultural and oil development. Some formerly occupied areas of desert scrub in the southern San Joaquin Valley have been converted to cotton fields, residential uses, or oil development. High-intensity, summerlong grazing by cattle and other domestic livestock in occupied areas is another ongoing threat. For these reasons Le Conte's Thrashers were included on the list of California Bird Species of Special Concern in 2008.

STARLINGS • Family Sturnidae

Members of this family, including starlings and mynahs, are native to Europe, Asia, and Africa as well as to northern Australia and many Pacific islands. None are native to North America. All species have strong feet, rapid and direct flight, and exhibit gregarious social behaviors. Their preferred habitats are open landscapes with a few scattered trees, shrubs, or other perching posts. Several species, including European Starlings, are highly dependent on humans in their native locales as well as in areas where they have been introduced. The nearly 120 species in this family include some of the most brightly colored and distinctive birds in Eurasia and northern Africa. The family name is derived from L. *sturnus*, a kind of bird.

European Starling • *Sturnus vulgaris*

ORIGIN OF NAMES "European" for the species' native distribution; "starling," a modern spelling of ME. *sterlyng*; L. *sturnus*, see above; L. *vulgaris*, vulgar or common.

NATURAL HISTORY Starlings have plagued North America since their introduction from Europe more than a hundred years ago, reportedly by a group of misguided but well-intentioned Shakespearean enthusiasts. Their goal was to introduce every bird mentioned in his plays and sonnets into North America. The first successful colony of about 100 European Starlings was established in New York City's Central Park in the early 1890s, and by 1942 they had spread with phenomenal speed and success across the continent to California. Their current North American population has been estimated at over 200 million birds, but data from both Christmas Bird Counts and Breeding Bird Surveys suggest that numbers have stabilized or may even be decreasing. Surprisingly, starling

Breeding

Nonbreeding

populations in Great Britain and elsewhere in western Europe have declined dramatically in recent decades, possibly a result of declining pasturelands with urbanization and intensification of agriculture; residents there who miss their starlings are welcome to take all of ours back!

Starlings flourished in North America because of their varied diet and ability to adapt to new food sources supplied either directly or indirectly by humans. They mostly feed on the ground and readily devour many types of insects, fruits, cultivated grains, and weed seeds. They are superb mimics and seem particularly fond of producing exact replicas of Killdeer, California Quail, and Western Meadowlark calls.

In the eyes of many, the starling's worst habit is usurping the homes of native, cavity-nesting birds such as Acorn Woodpeckers, Ash-throated Flycatchers, Purple Martins, and Western Bluebirds. The starling also takes over mud nests created by Cliff and Barn Swallow colonies. They evict residents of artificial nest boxes placed for bluebirds, wrens, chickadees, titmice, and other native birds. The cumulative effects of starlings on native birds have not been measured accurately, but there is no doubt that many Sierra birds (especially larger species that cannot use holes small enough to exclude starlings) have suffered from competition with starlings since they first arrived in California.

Resident populations are especially abundant near towns, farms, and livestock feeding areas where human-supplied foods are plentiful and dead trees and woodpeckers provide an ample supply of cavities. In California most female starlings produce two broods each year, with the second one being started about 10 days after they finish caring for the first brood. In the Sierra foothills first resident pairs get breeding activities under way by early March, but migrants breed slightly later in late March or early April.

As part of their courtship behavior, male starlings deposit grasses and small twigs in nest cavities to attract prospective females that usually lay five or six light bluish eggs in their mate's chosen domain. Both sexes incubate the eggs for about 15 days until the young starlings hatch and feed them for another 20 days until they leave the nest holes. Young birds are independent of their parents after only four or five days, giving the adults plenty of time to start a second brood.

Putting aside prejudices, starlings are attractive birds with their spangled breasts in fall and iridescent glossy sheen in spring and summer. When breeding, their bills also turn bright yellow. Starlings might be confused with blackbirds but are stockier, with shorter tails, and can be distinguished in flight by these features as well as by their faster wingbeats.

STATUS AND DISTRIBUTION Sadly, introduced European Starlings are the most abundant and widely distributed birds in North America and common in most agricultural, urban, and suburban areas throughout the developed lowlands of California. In fall, millions of northern breeders join the state's large, resident population to roam the Central Valley and Sierra foothills in vast, highly synchronized flocks searching for productive feeding areas. Starlings also form huge winter roosts in marshes and agricultural fields in association with Red-winged, Tricolored, and Brewer's Blackbirds.

West Side. Common residents, and locally abundant in winter, primarily in the Foothill zone, where populations are established in most foothill towns, cities, and ranches; uncommon up to the Lower Conifer zone such as in Yosemite Valley; casual at higher elevations.

East Side. Common residents in developed areas, mostly below about 7,000 feet; locally abundant in winter near towns and ranches where human-supplied foods are plentiful.

PIPITS • Family Motacillidae

Pipits belong to a large family of fairly small, ground-dwelling birds, all with long, wagging tails. In addition to pipits, this family includes wagtails and longclaws, species with primarily Eurasian and African distributions. Worldwide, this family includes about 65 species in 6 different genera but only 1, the American Pipit, occurs in the Sierra. The family name is derived from L. *motacilla*, a wagtail.

American Pipit • *Anthus rubescens*

ORIGIN OF NAMES "American" for the primary range of this recently recognized species, formerly lumped with a closely related Eurasian species *(Anthus spinoletta)* that were both called "Water Pipits," for their preferred wetland habitats; "pipit" from L. *pipio*, to peep or chirp—a reference to the characteristic flight calls; *anthus* from Gr. *anthos*, a kind of bird from a Greek myth where a boy turned into a bird after being trampled to death by his father's horses; *rubescens*, reddish, for the breast coloration of breeding adults.

NATURAL HISTORY This species could just as well have been named American "Wagtail" since they incessantly bob and wag their tails as they hunt for insects along the ground. Loosely associated feeding groups of a few or hundreds of birds spread out near standing water or moist soils, and unlike the highly gregarious Horned Larks, which often share the same winter foraging habitats, individuals appear to pay little attention to each other and may fly off together or separately. Just as walking pipits seem unable to stop bobbing their tails, flying pipits seem compelled to give their characteristic *spipit* calls.

American Pipits choose some of coldest and most barren habitats in North America to breed, either in high Arctic tundra or on the highest mountain peaks. In the Sierra, females create nests by digging small scrapes in Alpine turf, almost always near standing water or wet areas and positioned near rocks or tussocks that provide shade. Nests are lined with dry grasses, mosses, and sedges before the females lay 4 or 5 grayish-white eggs and incubate them alone for about 14 days. Both adults feed the young for about 15 days until they leave the nests and are independent after about 30 days.

STATUS AND DISTRIBUTION In winter, "Western" American Pipits *(A. s. pacificus)* migrate from their breeding grounds in Alaska and western North America to visit the grasslands and open agricultural habitats of the Central Valley, with smaller numbers found in similar habitats of the lower Sierra foothills and East Side valleys and basins. A different subspecies, the "Rocky Mountain" American Pipit *(A. s. alticola)*, appears to have recently colonized the Sierra as a breeder, with nesting first confirmed in 1975 at the Hall Natural Area, just east of Yosemite National Park (Gaines 1992). Since then, nesting has occurred at several other Sierra locations suggesting that American Pipits breed in the Alpine zone along the Sierra crest from at least Fork Lake south of Mount

Whitney (Inyo County), north to the Tahoe region (Mount Rose), with most records between about 10,000 and 12,000 feet.

West Side. Rare and highly localized breeders in Alpine habitats; fairly common to common from fall through spring in low-elevation, open grassy areas.

East Side. Breeding status as on the West Side; fairly common from October through April in open wet meadows with short grasses, open agricultural fields as well as around the shores of most large lakes and reservoirs; can be absent from areas in winter when snow cover is deep and can also be locally abundant in some areas such as Mono Lake from October through November.

WAXWINGS • Family Bombycillidae

Unlike most "songbirds," or oscine passerines, waxwings have no real songs and their relationships to other passerine families are uncertain; their closest relatives may be extinct. Many authorities consider them most closely related to pipits (family Motacillidae) and starlings (family Sturnidae), but waxwings also show similarities to silky-flycatchers (family Ptilogonatidae), including Phainopeplas. Of the three species of waxwings, two (the Cedar and Bohemian) occur across North America and both visit the Sierra. Bohemians are rare irruptive visitors (see Appendices 1 and 2). The third species in this family is found in Japan and northeastern Asia. The family name is combined from Gr. *bombykos*, silky, and L. *cilla*, an erroneous definition of this word, thought to mean "a tail" by some early taxonomists.

Cedar Waxwing • *Bombycilla cedrorum*

ORIGIN OF NAMES "Cedar," a reference to the species' preference for the berrylike cones of eastern red cedars; *cedrorum*, L. of the cedars.

NATURAL HISTORY Compact flocks of Cedar Waxwings perched in the tops of leafless trees fill the air with excited, high-pitched calls. A whistled *zee-zee-zee* is their only adult vocalization. They continue to call as they burst forth in unison and fly in compact flocks. Waxwings are well known for their flocking behavior, soft crests, velvety, tailored-plumage, and waxy spots on their wings that resemble red sealing wax from which their name derives. The exact function of these wing ornaments is unknown, and a few adults lack them.

Waxwings feed in flocks ranging in size from a dozen to several hundred birds. While they consume large insects by flycatching and gleaning leaves, they mostly rove about looking for abundant growths of wild berries, fruits, flowers, and buds, often gorging themselves on such delicacies. Pairs or even groups sometimes pass ripe fruits from bill to bill, perhaps to strengthen social bonds. Waxwings and berry bushes are closely linked; flocks gather to devour one crop and then, almost mysteriously, disappear to find the next.

These gregarious birds travel nomadically through the Sierra foothills, and their numbers and locations vary from year to year. Although they might be seen in any month (rare in July), they appear primarily during fall, winter, and spring. Their fondness for sweet fruits and open wooded habitats draws them to suburban parks, gardens, rural farms, and orchards, where they feast on berry-producing trees and shrubs, including those of ornamental species like cherries, crabapples, and pepper trees. Preferred native berry crops in the Sierra are those produced by toyon, mistletoe, mountain ash, wild grape, and madrone. Similar to American Robins and other berry-eaters, Cedar Waxwings sometimes get "drunk" from eating fermenting fruits. Contrary to some popular opinions, however, fruits do not pass through the digestive tracts of waxwings faster than most other birds; this notion may have come from those unfortunates who unsuspectingly parked their cars under "waxwing trees" with hundreds of birds feeding, producing abundant feces and messy windshields. Waxwings play an important ecological role in dispersing seeds of fruiting trees.

STATUS AND DISTRIBUTION Cedar Waxwings breed north of the Sierra, but flocks linger in the region late into May or June before departing for northern breeding grounds—much later than most wintering songbirds. Some adults with dependent young begin returning to the Sierra by early August. A nesting pair was found at Buck's Lake (5,150 feet), and other possible breeding pairs have been reported in Lassen and Mono Counties.

West Side. Fairly common but irregular and localized in the Foothill zone; only common where fruiting trees and shrubs are plentiful; some waxwing flocks also visit ponderosa pine and black oak forests of the Lower Conifer zone periodically to feast on berries of mistletoe and madrone; casual at higher elevations.

East Side. Uncommon and local; regular fall-spring in the Reno area and much of the Owens Valley, occasional at other locations mostly associated with junipers, oaks, and riparian areas or near towns and ranches with fruiting trees; casual above the East Side Conifer zone.

TRENDS AND CONSERVATION STATUS Sierra Christmas Bird Counts show a significant positive trend in recent decades with birds occurring more frequently and in larger numbers throughout the region, probably due to landscaping with berry-producing plants in residential areas.

SILKY FLYCATCHERS • Family Ptilogonatidae

Silky Flycatchers are a small family represented by only four species in three separate genera, thought by some experts to be related to the waxwings (family Bombycilidae). All are beautiful, long-tailed birds, and three, including the Phainopepla, sport crests on their heads. Phainopeplas, the only family member represented in the Sierra, are the northernmost representatives of this primarily Central American family that inhabits woodlands from the southwestern United States south to the mountains of western Panama. The family name is derived from Gr. *ptilogonys*, a tapered feather, in reference to the pointed tails of all members of this family.

Phainopepla • *Phainopepla nitens*

ORIGIN OF NAMES "Phainopepla," "shining robe," from Gr. *phainos*, shining, and *peplos*, a robe, in reference to the adult male's glossy black plumage; *nitens* from L. *nitere*, to shine.

NATURAL HISTORY A clear view of a male Phainopepla shining in the sunlight is a fine treat for the eyes. The key is to look for oaks infested with mistletoe and listen for their characteristic, upslurred *bwoop?* calls (sounding very like a large water drop falling from a faucet). Phainopeplas fly with slow, unsteady wingbeats, flashing their white wing patches. In the breeding season, they often sing in the air or from a high perch, making soft, low whistles and wheezy warbles. Expert flycatchers, they dart from perches and hover, snatching insects one after another before alighting. However, their primary staple foods are ripening fruits of mistletoe and elderberries, plucking these from outer clusters while fluttering in midair.

Female

Male

In the Sierra, Phainopeplas are primarily found in oak woodlands, foothill chaparral, and desert riparian habitats, especially those with heavy growths of mistletoe or other productive berry crops. Breeding pairs in desert habitats are territorial but in oak woodlands and chaparral, Phainopeplas tend to breed in loose colonies. Males perform most of the nest construction, a part of their courtship activities to attract mates. Females add a nest lining of soft plant down and spider webs and usually lay two or three pale, whitish-gray eggs liberally marked with fine, black or purplish spots. Both parents share the incubation duties about equally for 14 days until the young hatch. Young birds are fed and tended by the adults for 18 or 19 days until they leave the nest.

STATUS AND DISTRIBUTION Inhabitants of deserts and arid woodlands of the southwestern United States and Mexico, Phainopeplas reach the northern edge of their breeding range in California. Their status in the Sierra is very complex. While some are present year-round in the western foothills, others apparently migrate to southern deserts to breed from February through April. Some presumably return to the region to breed a second time, though this has not been directly confirmed. They are most common and widespread in the Sierra from fall through winter, when resident birds are joined by migrants from southern deserts. Three competing hypotheses have been proposed to explain the observed patterns of seasonal distribution: (1) individual birds breed once in the desert and again later in the year in oak woodlands; (2) two distinct populations exist, one breeding early in the desert and the other later in oaks; and (3) oak woodland nesters are birds that failed to breed successfully in the deserts.

West Side. Fairly common from fall through late winter in the Foothill zone from Yuba County south, especially in areas dominated by live oaks with infestations of mistletoe; fairly common year round in the Kern River Valley area; fairly common nesters May through June in the south-central foothills (Amador-Mariposa Counties), but uncommon nesters elsewhere; rare above about 3,000 feet.

East Side. Uncommon early spring nesters in Joshua tree woodlands and desert riparian areas from the Owens Valley south; accidental to casual farther north, with the most northerly record from Alpine County.

WOOD-WARBLERS • Family Parulidae

The small, slender-billed wood-warblers enliven the Sierra with bright colors and lively songs—some call them the "butterflies of the bird world." At least some yellow is visible in the plumages of most Sierra species as they flit busily through the foliage. None of their songs are actually "warbles" nor are they particularly musical, but they are key to finding and identifying the various species. Warblers live mainly on insects gleaned from branches and leaves or captured in midair, but some species also eat fruits, allowing them to remain farther north in winter. They occupy a wide range of habitats, including lush streamside thickets, the lofty boughs of the tallest conifers, and dry chaparral. Their complex songs may be heard throughout the nesting season but less frequently by midsummer. By late summer and fall, many low-elevation nesters and migrants from the north appear in the higher mountains. Here they often travel in large, mixed-species flocks that may include six or more species congregating around forest and meadow edges or in moist patches of willows, alders, and dogwoods along Sierra streams.

Most western warblers winter in montane pine and oak forests of Mexico and Central America. This large family, unrelated to the Old World warblers (family Sylviidae), is represented by about 115 species, and all are restricted to the Western Hemisphere. Eleven species breed regularly in the Sierra, and the Townsend's Warbler is a fairly common spring and fall migrant in the Sierra but has not been confirmed to breed in California. Sierra warblers nest in trees and shrubs or on the ground. Females do most of the nest building and usually lay four to six whitish or greenish eggs (variably covered by reddish or purplish spots or splotches). Females incubate the eggs alone for 12 to 14 days. Both parents feed the young for 10 to 12 days until they leave the nests and for about 2 weeks after—until the young can fly and forage on their own.

The family name means "a little tit" and was derived from L. *parus*, a tit or little bird, and *ula*, small; in 1758 Linnaeus misidentified the Northern Parula (until recently known as *Parula americana*) as a close relative of the chickadees and titmice (family Paridae) rather than as a warbler, and the family name stuck. Recent changes in taxonomy in this family have drastically altered historical relationships. A number of genera were eliminated entirely and one of those, *Dendroica*, previously included the largest number of North American warblers (*Dendroica* warblers were all moved to *Setophaga*, which formerly included only the American Redstart [*Setophaga ruticilla*]).

Orange-crowned Warbler • *Oreothlypis celata*

ORIGIN OF NAMES The name of this warbler is misleading, as the orange crown rarely shows, even on breeding males; "warbler" from "warble," meaning to sing or trill; *oreothlypis*, mountain bird, from Gr. *oros*, mountain, and *thlypis*, a bird; *celata*, from L. *celatus*, hidden—a reference to the species' usually concealed orange crowns.

NATURAL HISTORY Among the first signs of spring in the Sierra foothills are the songs of Orange-crowned Warblers—rapid, musical trills that slow and drop in pitch at the end. Compared to most other wood-warblers, Orange-crowns move slowly and deliberately through foliage and lack distinctive field marks such as bright colors, eyerings, or wingbars.

Orange-crowns begin establishing nesting territories in March in open or semi-open habitats dominated by shrubs. Oak woodlands, riparian thickets, and steep canyons clothed in chaparral are especially attractive. They subsist largely on insects and spiders captured within the protective cover of dense trees brush, usually at low and intermediate heights in the foliage and occasionally on the ground. They also consume sap oozing from "wells" drilled by sapsuckers on conifer and orchard trees and will eat fruits, especially in winter. Orange-crowns often sing from the highest branches of oaks, cottonwoods, and sycamores, but build their nests on dry ground under the shelter of shrubs or other foliage. Females build the nests alone using coarse grasses, strips of bark, and small leaves, and line them with fine grasses, feathers, or even deer hair.

STATUS AND DISTRIBUTION Orange-crowned Warblers breed across North America, and three subspecies might be seen in the Sierra, including the "Gray-headed" *(O. c. celata)*, "Rocky Mountain" *(O. c. orestera),* and "Lutescent" *(O. c. lutescens).* "Lutescent" Orange-crowns begin to arrive in numbers in the Sierra foothills by mid-March, and by late May some finish nesting and drift upslope to higher elevations to forage. On the East Side in Mono County, "Rocky Mountain" Orange-crowns breed and are also seen in spring and fall migration throughout the region, though rarely on the West Side. "Gray-headed" Orange-crowns are rare to casual in spring and fall migration on the West Side, but they migrate later in both spring and fall than any other subspecies. Birders should note that "Rocky Mountain" Orange-crowns can also have gray heads, especially first fall birds.

West Side. Fairly common nesters ("Lutescent") in riparian woodlands and chaparral up to the Lower Conifer zone; they do not breed at higher elevations but may be common along meadow edges or patches of shrubs within meadows and along streams up through the Subalpine zone through mid-September; by mid-October most have left, but a few remain through the winter in the Sierra foothills and riparian forests and other lowland areas of California.

East Side. Common in the summer at Yuba Pass, Sierra Valley, and Tahoe Basin ("Lutescent") as postbreeding upslope wanderers; uncommon breeders in East Side canyons of the Mono Basin ("Rocky Mountain"); common in fall migration and fairly common in spring migration in most riparian and scrub-dominated areas up to about 8,000 feet; in winter, uncommon in the Reno area and the Owens Valley and casual to accidental elsewhere.

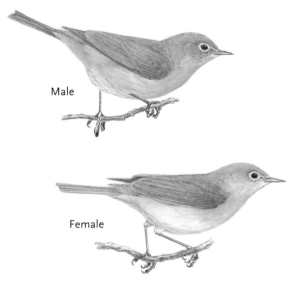

Male

Female

Nashville Warbler • *Oreothlypis ruficapilla*

ORIGIN OF NAMES For Nashville, Tennessee, where the species occurs only in migration but where it was first collected by Alexander Wilson (see Wilson's Snipe account); L. *ruficapilla,* rusty-haired or crowned, from L. *rufus,* red, and *capillus,* hair, for the species' reddish crown that shows clearly on breeding males, unlike closely related Orange-crowned Warblers.

NATURAL HISTORY Energetic songs proclaim the presence of Nashville Warblers, though birds may be difficult to spot as they flit through dense foliage between singing perches. They search for adult insects and caterpillars in all kinds of vegetation, seldom straying far from broad-leaved trees or dense brush cover. The song usually has two distinct parts, a bouncing first part, *chip-a, chip-a, chip-a,* followed by a fairly slow trill.

They establish nesting territories in riparian habitats or in stands of firs, pines, and cedars mixed with black oaks, often with well-developed understories of manzanita, ceanothus, or other shrubs. They form monogamous pairs by late April, and their nesting season extends through late June, with peak activities from mid-May to early June depending on the elevation. Despite their fondness for trees and shrubs as feeding areas and song posts, Nashville Warblers hide their nests on the ground beneath stumps or in rocky crevices, usually sheltered by overhanging grasses, mosses, ferns, or shrubs. Females build open cups made of leaves, ferns, and bark strips, lined with grass, hair, and needles, often with a rim of moss.

STATUS AND DISTRIBUTION Two subspecies of Nashville Warblers, with entirely separate distributions, occur in North America. The western subspecies *(O. r. ridgwayi),* once recognized as a full species, "Calaveras Warbler," has somewhat brighter plumage than the eastern subspecies *(O.r. ruficapilla),* which mostly breeds from central Canada eastward through New England. The two subspecies differ in song and behavior and may someday be resplit into two species. In the Sierra, Nashvilles arrive in numbers by early April and depart by late August with a few lingering until late September.

West Side. Common summer breeders in the Lower and Upper Conifer zones, primarily in forest stands where black oaks predominate; fairly common in spring and fall migration in streamside thickets and open forests through the Subalpine zone.

East Side. Fairly common breeders in riparian habitats south to at least the Tahoe Basin; uncommon nesters as far south as Mono County; fairly common in spring and fall migration but somewhat more regular in fall; most high-elevation records (usually July and August) are probably birds that drifted up from the West Side or that nested farther north.

Virginia's Warbler • *Oreothlypis virginiae*

ORIGIN OF NAMES Named by Spencer F. Baird (1823–1887), a prominent zoologist at the Smithsonian Institution, for Mary Virginia Anderson, the wife of William Wallace Anderson (1824–1911), a Confederate Army surgeon who collected this warbler in New Mexico and sent it to Baird.

NATURAL HISTORY Virginia's Warblers are less colorful versions of their close relatives, Nashville Warblers. Both share the habit of bobbing their tails up and down, but Virginia's seem to do it in a more energetic and pronounced fashion. Even the songs of these two species are similar, with the Virginia's just a bit lower pitched than Nashville's. At least one hybrid be-

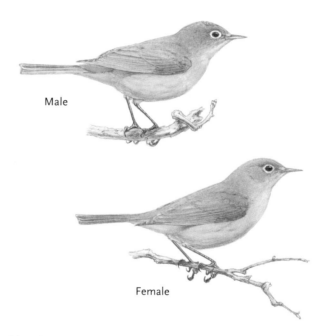

Male

Female

tween these similar species was captured and banded near Tioga Pass, and observers should watch for others in the eastern Sierra where their breeding ranges overlap.

Virginia's are mostly confined to dry pinyon-juniper woodlands and scrub riparian habitats during the breeding season. Typical shrubs found on their breeding territories in the Sierra include mountain mahogany, serviceberry, manzanita, currant, and snowberry, where they glean adult and larval insects from leaves. They also catch insects in aerial flights and probe for them in flowers and buds. Migrants tend to avoid tall trees and prefer to forage in low brushy or weedy habitats, especially in fall. Similar to Nashville's, female Virginia's build open cup nests on the ground or in rock crevices from mosses, grasses, roots, and strips of juniper bark. Virginia's Warblers, along with many other songbirds, are susceptible to nest parasitism by Brown-headed Cowbirds throughout the arid West.

STATUS AND DISTRIBUTION Widely distributed across the Great Basin and southern Rocky Mountains, the western limit of the Virginia's Warbler's breeding range barely extends into eastern California. They leave their wintering grounds in southern Mexico by mid-April and usually begin nesting in the eastern Sierra by mid-May.

West Side. Accidental, late July to early September observations from the Sierra crest and the Yosemite region, and a late August record near Donner Summit (Placer County); no documented breeding records west of the crest.

East Side. Rare and irregular breeders in pinyon-juniper habitats of Mono and Inyo Counties, with a handful of probable breeding records in aspen groves near Monitor Pass (Alpine County); most occupied breeding habitats are between 7,000 and 9,000 feet; more common breeders just east of the region at Glass Mountain (Mono County) and the White Mountains (Inyo County); spring migrant records from the Tahoe Basin but no documented records farther north in the Sierra.

Male

Female

MacGillivray's Warbler • *Geothlypis tolmiei*

ORIGIN OF NAMES First named "Tolmie's War-bler" *(Sylvia tolmiei)* by John Kirk Townsend (see Townsend's Solitaire account) in honor of his friend William Tolmie (1812–1886), who collected birds in British Columbia when employed by Hudson's Bay Company; the common name was later changed by John James Audubon to credit William MacGillivray (1796–1852), a Scottish naturalist who helped him edit *Birds of America*; *geothlypis*, an earth or ground bird, from Gr. *geo*, earth, and *thlypis*, a finch.

NATURAL HISTORY A loud, metallic "chip," similar to a junco's, is often the first clue that a MacGillivray's Warbler is near. They usually utter these notes while skulking through dense foliage, staying hidden or offering only brief glimpses. During nesting season, however, males occasionally provide clear views as they ascend the tallest bushes and break into their rich, burry song. Even then, these are shy birds, much more likely to be seen well by a solitary birder than by a large group. They are among the most responsive of all Sierra warblers to birder's "pishing" sounds.

Unlike most other western wood-warblers, MacGillivray's Warblers generally avoid areas with dense tree cover and keep to low vegetation, where they seek out moist, low tangles of vines, dogwoods, willows, and alders along streams or wet meadows. However, they will use such habitats when they occur well under a high canopy of very large, mature conifers. They also nest in drier mountain chaparral habitats of huckleberry oak, snowbrush, or manzanita, as long as they are near water. Females build small compact cup nests from grasses, weed stems, and rootlets and fasten them to forked branches of low shrubs, usually less than five feet above the ground. In addition to nesting low, MacGillivray's Warblers also generally feed less than three feet above the ground, gleaning insects from foliage and probing in leaf litter for grubs and spiders.

STATUS AND DISTRIBUTION The western population of the MacGillivray's Warbler breeds from southern Alaska to the southern Sierra and south Coastal California, and winters in southern Mexico. They arrive in the Sierra in early May and remain until early September.

West Side. Fairly common in scrub riparian and mountain chaparral habitats of the Lower and Upper Conifer zones; rare nesters up to the Subalpine zone; after nesting, small numbers move through meadow edges and riparian corridors up to the Alpine zone, where they mingle with Orange-crowned, Nashville, and Wilson's Warblers.

East Side. Similar to West Side; recent breeders up to 10,000 feet at the Hall Natural Area, just east of Yosemite National Park; fairly common migrants in riparian habitats and pine forests; up to the Subalpine zone from early August until mid-September.

Common Yellowthroat · *Geothlypis trichas*

ORIGIN OF NAMES "Yellowthroat" for the distinctive field mark; Gr. *trichas*, a thrush—a name incorrectly applied to this species by Linnaeus.

NATURAL HISTORY "From the tules bordering a pond or from the bushes at the edge of low wet ground comes a 'tschek' of protest" (Hoffmann 1927). Common Yellowthroats are denizens of marshes, lakeshores, and meandering streams and rarely stray far from wetlands. Aquatic habitats need not be extensive, as damp, brushy margins of sloughs and drainage ditches, patches of tules and cattails around farm ponds, or small stands of willows near streams may support a nesting pair. Yellowthroats hide from view, like Marsh Wrens, in thick vegetation and only their frequent, flat calls reveal their presence, as noted earlier. Sometimes males ascend the tallest reeds and burst into loud, energetic *wichety-wichety-wichety* songs. Yellowthroats pick dragonflies, grasshoppers, butterflies, and caterpillars off marshland plants.

Male

Female

Loss of natural wetlands to urbanization and agriculture has greatly reduced Yellowthroat populations throughout California. However, they can still be found where suitable habitats remain in the Central Valley and the Sierra and are likely benefiting from better protection of wetlands. Females build large, bulky nests from grasses, leaves, and plant stems and secure them to low plants near water.

STATUS AND DISTRIBUTION Among the most widespread wood-warblers, at least 13 subspecies have been recognized in North America. While some subspecies are migratory and spend the winter in Mexico and Central America, the "Western" Common Yellowthroat *(G. t. occidentalis)* is a short-distance migrant that resides in the Central Valley year-round and breeds on both slopes of the Sierra.

West Side. Common breeders in the Kern River Valley; elsewhere uncommon in wetlands of the Foothill zone from May through September, such as at Lake Success and below Terminus Dam (Lake Kaweah); rare at higher elevations up to 7,000 feet in wet meadows of Yosemite National Park, such as Crane Flat and Peregoy Meadow; causal as postbreeding wanderers up to the Subalpine zone from late July to mid-October; rare in winter.

East Side. Fairly common spring and fall migrants; uncommon localized breeders south to the southern Owens Valley, mainly at lower elevations, but singing (possibly unmated) males observed in abandoned gravel pits at about 6,000 feet near Truckee (Nevada County) and at the Hall Natural Area above 10,000 feet just east of Yosemite National Park; fairly common in spring and fall migration; in winter, casual in the Owens Valley, accidental elsewhere.

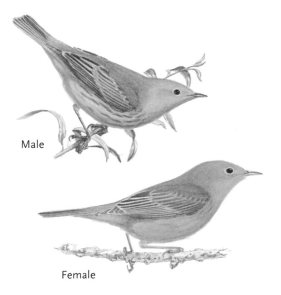

Male

Female

Yellow Warbler • *Setophaga petechia*

ORIGIN OF NAMES "Yellow" for the species' bright overall coloration; *setophaga,* an insect eater, from Gr. *seto,* an insect, and *phaga,* to eat; L. *petechia,* a rash—a reference to the chestnut-red breast stripes of breeding males.

NATURAL HISTORY Like golden flecks of sunlight, Yellow Warblers flit among the lush green foliage of Sierra riparian forests and meadows. Most western wood-warblers have bright yellow patches somewhere in their plumage, but none can match the uniformly brilliant hues of the Yellow Warblers. At a distance males appear to be solid bright yellow above and below, but a closer view reveals that their breasts are decorated with narrow, chestnut-red streaks.

During spring migration, the rapid, cheerful *sweet-sweet-I'm-so-sweet* songs of Yellow Warblers issue from a variety of habitats, but while nesting, most pairs prefer deciduous streamside forests of alders, willows, or cottonwoods with thick understories of tangled shrubs. In the Sierra some breed in wet shrubby meadows, dry montane chaparral dominated by huckleberry oak, manzanita, ceanothus, and rarely open conifer woods. In moist habitats Yellow Warblers are frequent hosts to Brown-headed Cowbirds that parasitize nests by laying their own eggs. The larger more aggressive cowbird nestlings outcompete the host nestlings for food, and the warblers are very rarely able to rear any of their own offspring. When host birds, including Yellow Warblers, discover cowbird eggs in their nests, they sometimes build a new layer on top of the original nest or abandon it altogether.

Nest predation by Steller's Jays, Western Scrub-Jays, American Crows, garter snakes, western gray squirrels, and Douglas tree squirrels has been documented elsewhere in the Yellow Warbler's range, and these species almost certainly prey on their eggs and nestlings in the Sierra as well. Females build deep cup nests of grasses and bark on forks of trees and shrubs and line them with fine plant fibers. In California, Yellow Warblers often make several breeding attempts each season but usually only produce one successful brood.

STATUS AND DISTRIBUTION Among the most widespread and taxonomically diverse of the wood-warblers, more than 40 subspecies of Yellow Warblers have been recognized. The two subspecies that breed in the Sierra (*S. p. brewsteri* on the West Side and *S. p. morcomi* on the East Side) are lumped together by some taxonomists. West Side males are generally more yellow than the East Side males, but distinguishing them consistently in the field is extremely difficult. Yellow Warblers begin to arrive in the southern Sierra by early April and remain until early September, with a few staying into October.

West Side. Abundant only in the South Fork Kern River Valley, where more than 300 individuals have been tallied in a single day in mid-June; elsewhere common in riparian and mountain chaparral habitats from the Foothill zone to the Upper Conifer zone; during spring migration, one occasionally encounters "pulses" of large numbers of Yellow Warblers moving through in loosely associated groups; postbreeding dispersal to higher-elevation mountain chaparral from late July to mid-September; rare up to the Subalpine zone.

East Side. Common nesters in riparian forests up to about 7,000 feet, postnesting birds travel widely through mid-elevation pine forests, where they usually forage in broad-leaved trees; some depart for wintering areas soon after the young are fledged (mid-July); good numbers are seen in fall migration, somewhat more in spring.

TRENDS AND CONSERVATION STATUS Clearing of riparian habitats for agriculture, inundation by dams, and diverting water from rivers and creeks have destroyed extensive areas of the Yellow Warbler's preferred breeding habitat in the Sierra. Brood parasitism by Brown-headed Cowbirds has also nearly extirpated Yellow Warblers as a breeding species in the Central Valley, where they were formerly common. These factors likely have greatly reduced their breeding population in the Sierra compared with historical numbers. The only dense Sierra nesting population remaining is in the Kern River Valley, where huge tracts of riparian forest have been restored and cowbirds are trapped and destroyed. Due to their statewide decline, Yellow Warblers were included as a California Bird Species of Special Concern in 2008.

Yellow-rumped Warbler • *Setophaga coronata*

Male

Female

ORIGIN OF NAMES "Yellow-rumped" for the species' characteristic field mark; L. *coronata*, a crown, for the yellow spot on the crown.

NATURAL HISTORY At any time of year Yellow-rumped Warblers can be found in almost every available habitat. This remarkable flexibility was celebrated by William Dawson (1923) with his typical enthusiasm: "And so this tiny warbler . . . threads mazes no surveyor has ever chained, scales heights which no aneroid has ever measured, sleeps in a hundred and forty different beds, from the lowly weed to the fir tree's loftiest pinnacle, lunches at ten thousand counters, and comes back to us winter by winter."

Somewhat larger and hardier than most of their relatives, Yellow-rumps are the only wood-warblers that are commonly found year-round in the Sierra. In winter they frequent oak woodlands and lower-elevation conifer and riparian forests. They mostly glean insects from foliage but often flutter out from tree crowns to snap up flying insects in midair. During the colder months they also take seeds, berries, or wild grapes, showing a flexibility of diet (including feeding on sap and insects at sapsucker-drilled holes) that allows them to thrive after most other warblers have headed south. They are equally indiscriminate about winter foraging habitats and will search for food from the ground to the treetops, in open areas, forests, backyards, vineyards, and orchards.

Arriving in the higher elevations with snowdrifts still blanketing the ground, Yellow-rumps sing variable and loosely structured songs that signal the beginnings of spring. They inhabit all types of coniferous and deciduous forests from the lowest oak woodlands up to tree line but nest most commonly in the Lower and Upper Conifer zones. No other warblers occur across such a wide range of elevations or occupy as many habitats while breeding, and only the Wilson's shares their nesting haunts in the Subalpine zone. Yellow-rumps tolerate a wide range of forest conditions, including dense and overgrown "dog haired" forests, but they seem to prefer open stands, forest-meadow edges, lakesides, and other habitats providing ample air space for hawking insects.

Females build nests from twigs, moss, lichens, and strips of bark, lined by finer materials such as grasses, animal hairs, and feathers and place them both low in shrubs or high in tall trees, almost always hidden by thick sprays of foliage. Postbreeding Yellow-rumps that nested in the Sierra, or migrants from farther north, remain in mid- and high-elevation forests through the end of October. During the fall, flocks of hundreds stream through tree canopies, flashing yellow rumps and white tail spots, and uttering their characteristic and loud *chip* notes. These flocks may also include a few Orange-crowned, Hermit, and Townsend's Warblers, Mountain Chickadees, Red-breasted Nuthatches, and Golden-crowned Kinglets—providing more viewing opportunities than most birders can assimilate.

STATUS AND DISTRIBUTION In 1973 the American Ornithologists' Union lumped two formerly distinct species, the "Audubon's" *(S. c. auduboni)* and "Myrtle" *(S. c. coronata)* Warblers together as "Yellow-rumped Warblers" because the two subspecies sometimes interbreed in the southern Canadian Rockies. All of the nesting, and most of the wintering, Yellow-rumps in the Sierra are "Audubon's." "Myrtle" Warblers winter uncommonly in riparian habitats of the Sierra foothills and are also seen in spring and fall (mainly late September into October) migration. In March and April, Yellow-rump populations wintering in the Sierra foothills and elsewhere in California's lowlands head for northern breeding grounds and about the same time, birds that will nest in the Sierra arrive from their wintering areas farther south. Yellow-rumps nest from mid-May through late July with a peak in mid-June in all but the highest elevations.

West Side. Common breeders in most conifer forest types from the Lower Conifer zone up to the Subalpine zone, especially near forest openings, meadows, and creeks; large flocks can be locally abundant in the Upper Conifer and Subalpine zones from mid-August through the end of October or until driven downslope by snow; locally common in the Lower Conifer and Foothill zones from November through March.

East Side. Common summer residents in the pine forests up to tree line; abundant breeders in riparian habitats; locally common to abundant in spring and fall migration, and rare winter residents except in some lower-elevation areas like Reno, Carson Valley, and Owens Valley, where they can be uncommon to common in winter.

Male

Female

Black-throated Gray Warbler •
Setophaga nigrescens

ORIGIN OF NAMES "Black-throated" for color of chin and throat; "gray" for the color of the back; L. *nigrescens*, blackish, for the species' black throat, head, and flank markings.

NATURAL HISTORY Like their "black-throated" cousins, Hermit and Townsend's Warblers, male Black-throated Gray Warblers sing buzzy songs, but none are as conspicuously or as consistently buzzy as theirs. More than any other Sierra wood-warblers, nesting Black-throated Grays are associated with oaks, preferring dry, open forests with thick shrub cover. At lower elevations they seek extensive stands of canyon live oaks on the steep slopes of river canyons. At higher elevations they frequent stands of black oaks, including small oak patches within conifer-dominated forests of the Upper Conifer zone. They primarily glean small insects, especially the green caterpillars that parasitize oaks, and seldom flycatch. Compared with most other western wood-warblers, these strikingly handsome birds seem fairly tame and tolerate close approach by humans. Females build nests alone in low brush or occasionally in broad-leaved trees near streams; males remain nearby singing and scolding but apparently do not help with construction.

STATUS AND DISTRIBUTION Black-throated Gray Warblers have an entirely western distribution, breeding from southern British Columbia to northern Baja California; most winter in the oak-pine forests of western Mexico. Black-throated Grays arrive on their Sierra breeding grounds in numbers by mid-April, and most pairs initiate nesting by early May, with peak breeding activities in early June. While most depart for the south by mid-August, a few remain in the Sierra until early October.

West Side. Fairly common breeders in the Foothill zone and uncommon in oak-conifer forests to the Upper Conifer zone; rare in the Subalpine zone in August and September; casual in winter.

East Side. Uncommon to rare in the breeding season in juniper-dominated habitats; rare from early August until early October up to the tree line, where they are primarily found in lodgepole pine–dominated forests; accidental in winter.

Townsend's Warbler • *Setophaga townsendi*

ORIGIN OF NAMES See Townsend's Solitaire account.

NATURAL HISTORY Townsend's Warblers are golden counterparts of Black-throated Grays. The two species closely resemble each other in overall pattern and appearance, save the Townsend's substitution of brilliant yellow for white on the face, breast, and flanks. Such similarities of plumage suggest that these warblers share a common ancestry, along with Hermit Warblers. As noted earlier, the wheezy songs of these three species also resemble each other. Townsend's and Hermit Warblers hybridize often where their breeding ranges overlap in Oregon and Washington, and the progeny of such mixed-matings should be watched for in migration. Townsend's Warblers primarily glean insects from foliage including caterpillars, moths, beetles, and spiders, but on their wintering grounds "honeydew" produced by scale insects comprise a major component of their diet.

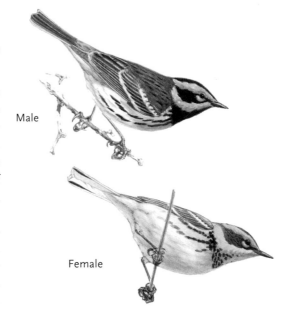

Male

Female

STATUS AND DISTRIBUTION Townsend's Warblers have never been recorded nesting in California but Sierra records well into June and in late July are intriguing, and it is possible that breeders may someday be found here. These warblers pass through while migrating between their breeding grounds amid the towering spruces and Douglas-firs of the Pacific Northwest and their wintering grounds in Mexico and Central America, where they are among the most common birds in mixed foraging flocks. Townsend's Warblers are fairly common winter residents of coastal California but are rare in the Sierra foothills during the same period. In spring, Townsend's Warblers migrate uncommonly through low-elevation oak Sierra habitats from April to June. By the end of July, they begin to reappear in the Sierra on their way south, where they might be spotted in almost any type of forest. In late August and September their numbers increase as they join mixed flocks of other warblers such as Yellow-rumps, Hermits, Nashvilles, and Orange-crowns moving through conifer forests, meadow edges, and riparian thickets.

West Side. Uncommon spring (mostly May) and fall (early August to early October) migrants mainly in the Foothill zone, but uncommon up to the Subalpine zone from late July until early September; locally uncommon to rare in mixed-species flocks in the Lower Conifer zone in winter.

East Side. Similar to West Side.

Hermit Warbler • *Setophaga occidentalis*

Male

Female

ORIGIN OF NAMES "Hermit" for the species' rarity and secretive habits presumed by early eastern ornithologists; L. *occidentalis*, western, for the species' range.

NATURAL HISTORY Hermit Warblers were considered rare, solitary, and reclusive by early eastern ornithologists. When the species was first described, few "birdwatchers" lived in the western half of the continent, binoculars had not been invented, and serious ornithologists primarily observed birds over a gun barrel. Therefore, their real status was not discovered until much later, when Hermit Warblers were found to be among the most numerous wood-warblers in mid-elevation coniferous forests of the Cascades and Sierra.

Hermit Warblers forage and sing in the tallest pines, firs, and other conifers and often hawk flying insects from treetop perches, like Yellow-rumps. Unlike Yellow-rumps, they rarely leave the treetops and are probably responsible for more cases of "warbler-neck" (from spending too much time looking straight up) among birders than any other warbler. They will occasionally descend to hunt for prey near the ground in patches of low shrubs or conifer seedlings. Favored foods include flies, moths, bugs, wasps, and small spiders. Their rapid, energetic songs are highly variable and may be confused with those of the closely related Black-throated Gray and Yellow-rumped Warblers, where their breeding distributions overlap in Sierra conifer and oak forests. Once learned, however, the male Hermit's "diagnostic" song stands out clearly from the others and fills the woods with its haunting quality. The key is the final phrase, a very high-pitched, thin buzzy note that generally rises and then sometimes falls in pitch. More problematic is distinguishing Hermit songs from those of Townsend's Warblers, which will sing as they migrate through in spring. Hermit Warblers start singing as soon as they arrive on their breeding grounds, usually by mid- to late April, and continue to serenade their mates through the nesting cycle.

Despite the abundance of Hermit Warblers in Sierra forests, details of their breeding behavior are mostly unknown. Careful observations of a few nests in Nevada County suggest that females build nests well out on the branches of tall conifers always well concealed by overhanging foliage—making them difficult to study even with a high-powered spotting scope and a lot of patience. While some nests have been reported at heights of more than 150 feet above the ground, many others are located much lower, and at least one nest was found at ground level at the base of a conifer. At least for the Nevada County nests, males aggressively defend incubating females and successfully fended off Brown-headed Cowbirds trying to parasitize their nests. Peak nesting activities occur from mid-May until mid-June, with most nesting efforts completed by mid-July. Females build cup nests from small twigs, lichens, plant fibers, and strips of cedar bark bound together with spider webs.

STATUS AND DISTRIBUTION Hermit Warblers breed in coniferous forests from Washington to the southern Sierra. They migrate in small numbers through the foothill woodlands in April, but they confine nesting to the Lower and Upper Conifer zones, where tall pines, firs, and cedars provide suitable nesting sites. After nesting in late July, they move upslope as high as the Subalpine zone and join mixed foraging flocks with Yellow-rumped, Orange-crowned, and Townsend's Warblers. Most Hermit Warblers have left the Sierra for their wintering grounds in the highlands of central Mexico by mid-September.

West Side. Common in Lower and Upper Conifer forests with large trees from mid-May until late July; southward migrants are fairly common in the Upper Conifer and Subalpine zones from late July until mid-September; accidental in winter at all elevations.

East Side. Uncommon localized breeders only in Tahoe Basin; rare to casual in fall migration from late July until early September; one record of a singing male in June in Mono County; one of the rarest East Side warblers with fewer overall records than some eastern "vagrants" such as the Northern Parula, Black-and-white Warbler, Northern Waterthrush, and American Redstart.

Wilson's Warbler • *Cardellina pusilla*

ORIGIN OF NAMES First collected and named (for himself) by Alexander Wilson (see Wilson's Snipe account); *cardellina*, a small finch, from L. *carduelis*, a finch; and *pusilla*, small.

NATURAL HISTORY Signaling the arrival of spring in the high country, Wilson's Warblers adorn the naked, snow-laden branches of willows and alders like animated golden ornaments. They favor shrub-bordered streams, lakes, and wet meadows from the Upper Conifer zone to above tree line. At mid-elevations on the West Side, they sometimes share nesting haunts with MacGillivray's Warblers, but Yellow-rumps are the only other warblers that share their breeding habitats in or above the Subalpine zone. Restricted to moist, shrubby habitats while breeding, Wilson's Warblers venture into arid terrain during migration. In spring many appear in foothill woodlands and chaparral, and in fall some travel through coniferous forests. Wilson's do some flycatching but usually

Male

Female

forage for insects and sing their chattery songs within the protective cover of low foliage. Theirs is a loud, insistent song, usually two-parted with the second part lower in pitch and usually faster than the first: *Che . . . che . . . che, che-che-che-che.*

In contrast to most songbirds, male Wilson's Warblers in the Sierra sometimes secure several mates. A long-term study near Tioga Pass in Yosemite National Park showed that almost one-third of the territorial males mated with up to three different females (Stewart et al. 1977). Surprisingly, Wilson's Warblers breeding in coastal California are thought to be entirely monogamous.

Selecting moist glades with luxuriant growths of willows, alders, ferns, dogwoods, chokecherries, and other low vegetation, females Wilson's Warblers build bulky cup nests from shreds of bark, fine grasses, mosses, and animal hair and hide them on the ground under shrubs, roots, or logs. Because males often secure several mates, females in the same "harem" may nest in close proximity. Many of these highest-elevation populations do not finish nesting until August due to delayed snowmelt in Alpine meadows.

STATUS AND DISTRIBUTION Wilson's Warblers nest at high elevations and high latitudes across most of North America and winter from central Mexico through Central America, where they are the only migrant warblers that winter in tropical high plains. Migrants begin to arrive in the Sierra foothills in April, move to high-elevation breeding grounds by mid-May, and remain in the region through early October.

West Side. Fairly common nesters from about 4,000 feet up to the Subalpine zone; common in the Foothill zone in spring migration and fairly common in the highest-elevation willow patches in fall migration from mid-August through September, where they often flock with Yellow-rumped, Orange-crowned, and Nashville Warblers and occasionally venture into conifer forests.

East Side. Fairly common breeders and common spring migrants in streamside habitats in East Side riparian forests; also widespread breeders and fall migrants in the Subalpine zone.

TRENDS AND CONSERVATION STATUS Data from Sierra Breeding Bird Surveys showed significant declines in Wilson's Warblers in the past few decades, consistent with continentwide negative trends. Research on a coastal California population suggests that declines are due to issues on the breeding grounds rather than in the winter range, but similar studies have not been conducted in the Sierra. Habitat degradation from grazing may have impacted this species. Any future climate trends toward drier conditions are likely to affect the moist habitats needed by Wilson's Warblers.

Yellow-breasted Chat · *Icteria virens*

ORIGIN OF NAMES "Yellow-breasted" for the species' distinctive field mark; "chat" for chattering vocalizations; Gr. *icteria*, yellow or jaundice—a reference to the breast color; L. *virens*, green, for the species' back color.

NATURAL HISTORY It is impossible to suppress a smile when listening to a Yellow-breasted Chat's bizarre series of churs, buzzes, whistles, and chatters coming from a patch of dense riparian habitat.

Chats offer songs as variable (though not as repetitive) as mockingbirds. Like mockers, they will incorporate sounds of other birds into their repertoire, but they are satisfied with delivering each phrase once and moving on to the next after a short pause for effect.

Historically grouped with the wood-warblers by taxonomists, Yellow-breasted Chats are the only members of their genus. They look and behave very differently from most other warblers, prompting some to question their taxonomic position. Much larger than other wood-warblers, they have thick, heavy bills, and a dazzling array of songs and calls unlike any other warbler. Considerable controversy surrounds their placement with the wood-warblers, and some anatomical studies suggest that they may be more closely related to tropical tanagers (family Thraupidae) or possibly some other tropical family. However, recent genetic evidence indicates that they are most properly placed with the wood-warblers.

Yellow-breasted Chats in the Sierra skulk through broad stands of moist, lowland riparian forest, where they dwell in the jungle-like tangles of willows, wild grape, berries, and other vines. In such dense habitats they are much more likely to be heard than seen. While foraging, they usually stay low in vegetation, searching for insect swarms and occasionally fruits and seeds. Females build nests from grass stems, weeds, and leaves from near ground level up to about five feet, always hidden in dense shrub growths near water. Some pairs, especially if their first nesting attempt was unsuccessful, will try to nest again in June or July.

STATUS AND DISTRIBUTION   Highly migratory, Yellow-breasted Chats winter in southern Mexico; they are the last of our warblers to arrive in spring and generally the first to leave in fall. They are rare before May and most are gone by mid-August. There are no documented winter records for the Sierra.

West Side. Uncommon breeders in foothill riparian habitats up to about 2,500 feet; the largest population is in the South Fork Kern River Valley (30 to 45 pairs); late summer and fall records up to Yosemite Valley.

East Side. Rare breeder in northern Inyo County; recent July records in Mono and eastern Alpine Counties may indicate an expansion of breeding range into recovering riparian habitats; uncommon spring migrant and rare fall migrant.

TRENDS AND CONSERVATION STATUS   Yellow-breasted Chats were once common and widespread in the Central Valley and along lower foothill streams of the western Sierra, but dams, water diversions, and direct removal of riparian habitats have greatly reduced their populations, and they are now uncommon and restricted to the remaining large, wide stands along lowland Sierra creeks and rivers. Brood parasitism by Brown-headed Cowbirds is probably also a factor in the chat's decline. Data from Sierra Breeding Bird Survey routes show continuing declines in the past 30 years. Due to statewide decline, Yellow-breasted Chats were included as a California Bird Species of Special Concern in 2008.

SPARROWS AND RELATIVES • Family Emberizidae

While many members of this family are superficially similar (mostly brownish and streaky), Sierra sparrows are fascinating birds and reward the patient observer by revealing subtle distinctions in appearance and behavior. Partly because many sparrow species are similar in appearance, occupy similar niches, and eat similar foods, they have evolved a complex variety of songs, behaviors, diets, and morphological characteristics unique to each species. Try watching, for example, how sparrows in a mixed-species flock differ in the ways they flush and take cover when startled, with longer-winged, agile species like Vesper Sparrows making long flights and landing on exposed perches, and shorter-winged fliers like Song Sparrows diving immediately for cover into dense thickets. Sparrow songs, which can range from the delightfully musical tunes of Song and Lark Sparrows to the thin buzzes of Grasshopper Sparrows, are one of the best ways to find and differentiate these birds. Some species such as White-crowned and Golden-crowned Sparrows enliven winter by continuing to sing throughout the year.

Sparrows are grassland or brush-loving birds with stout, conical bills adapted for crushing and husking seeds. Larger-billed species tackle large seeds from shrubs, and smaller-billed species focus on small, thin-walled seeds like those of grasses. Most of a sparrow's life is spent foraging and nesting either on the ground or in low vegetation, with the exception that males of some species ascend onto high, conspicuous perches while singing (and some grassland species sing in flight).

Females work alone to build nests, incubate eggs (typically four white to pale blue or pale green eggs marked with reddish-brown spots and squiggles), and brood the nestlings. Males often make significant contributions to the feeding of young birds, which generally leave the nest at 9 to 12 days of age. Unlike some of the closely related finches, which subsist on seeds year-round, sparrows are notable for switching to a diet of insects during the breeding season. There are few Sierra-specific studies of breeding biology of sparrows, so most of what we know is based on research done elsewhere.

It is thought that sparrows evolved in the New World, with 85 percent of the world's 320 species found in the Americas. Sixty species have occurred north of Mexico, and seventeen species occur regularly in the Sierra Nevada. Taxonomy in this family is complex, and recent DNA studies have inspired several changes at the genus level and some taxa, such as Sage and Fox Sparrows, may be split into several new species in the near future. The name "sparrow" is a bit of a misnomer erroneously used by early European zoologists and now in wide acceptance, even though the members of this family are actually buntings; the "true" sparrows are all members of the Old World family (Passeridae). The family name is derived from OG., *embritz*, a bunting. Sparrow comes from AS. *spearwa*, to flutter.

Green-tailed Towhee • *Pipilo chlorurus*

ORIGIN OF NAMES "Green-tailed" for the species' distinctive field mark; "towhee" is a name of unknown origin but may be imitative of the loud calls of most species; L. *pipo*, to chirp; *chlorurus*, green-tailed, from Gr. *chloros*, green, and *oura*, tail.

NATURAL HISTORY There is nothing drab, brownish, or streaky about these gorgeous sparrows of the East Side shrublands and high-elevation West Side chaparral. The trick is to find and get a good look at these skulkers. Distinctive catlike "mewing" and a lively, bouncing song (similar to the Fox Sparrow's with which it shares some of this habitat) are the best ways to find this towhee and, with patience, a male may perch up in the morning light, throw back his head to display his stunningly white throat and rusty crown, and sing.

Green-tailed Towhees use a wide variety of scrub habitats but prefer areas with a rich variety of shrubs and often with some scattered trees. Postfire landscapes that have had a few years to produce a good mix of shrubs can be ideal. They feed using the classic two-legged "towhee hop," like their other close relatives. This consists of a quick hop forward with both legs, followed immediately by a vigorous backward thrust of both legs, thus scattering leaf litter as the bird searches for small seeds and insects hidden below. It is an efficient but noisy method that can often give away the bird's presence.

They leave the Sierra for wintering grounds in the southwesternmost parts of the United States and northern Mexico, where they occupy similar, though more arid, habitats. However, not being as well adapted to desert life as some other sparrows, they tend to stay fairly close to a source of water in winter. They are found occasionally in winter in the Central Valley, where they can show up in most any unforested landscape, including suburban backyards.

We know relatively little about the breeding habits of these secretive sparrows. Nests are large, sturdy constructions and nearly always very well concealed within a shrub. Like most other sparrows, it seems the male's only contribution is to occasionally feed the female while she incubates the eggs or broods the young, and to help feed the young birds. Nesting may begin in May or June, depending on elevation and snow cover, and young birds are usually independent by early August.

STATUS AND DISTRIBUTION Green-tailed Towhees breed in the western United States and most winter in Mexico. They begin to appear in April at low elevations on both sides of the Sierra on their way up to higher breeding grounds. Though more common and widespread in the breeding season in East Side habitats, good numbers can be found in mountain chaparral on the West Side, usually above 6,000 feet, though nesting has occurred below 3,000 feet. They move regularly up to the Subalpine zone in early August, but nearly all have left by mid-September.

West Side. Locally fairly common in higher-elevation mountain chaparral; found rarely in winter at foothill locations and down to the Central Valley floor.

East Side. Common in a variety of shrubby habitats dominated by sagebrush, bitterbrush, or mountain mahogany across the elevational gradient except for the highest rocky areas and the driest lowlands.

Spotted Towhee • *Pipilo maculatus*

ORIGIN OF NAMES "Spotted" to distinguish the species from the similar Eastern Towhee (*P. erythrophthalmus*), which lack white spots on their backs; L. *maculatus*, spotted.

NATURAL HISTORY Easily the most strikingly colorful of our sparrows, Spotted Towhees are a good bit easier to bring into view than their green-tailed cousins and will usually respond to a bit of "pishing" by briefly perching up in clear view. They use shrubby habitats but tend toward mid-lower-elevation sites with manzanita or ceanothus chaparral and riparian thickets, where broadleaf shrubs dominate, especially non-native Himalayan blackberry patches. They tolerate the presence of a considerable tree canopy as long as there is well-developed shrub understory below. They also use the "towhee hop" when foraging (see Green-tailed Towhee account above), and they read-

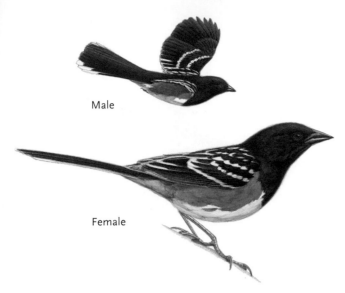

Male

Female

ily come to backyard feeders, often using the same foraging technique while standing on a bed of seeds as though they need to uncover them from nonexistent leaf litter.

Spotted Towhees use drier and more open habitats than their close relatives, the Eastern Towhees, which may explain why our towhee is spotted, a pattern offering better camouflage in dappled light. There are many subspecies recognized with the Sierra-breeding subspecies *(P. m. falcinellus)* found mainly on the West Side, while *P. m. curtatas* and *P. m. montanus* occur on the East Side. Females of all Sierra subspecies are somewhat paler, with larger white spots than the coastal subspecies.

Breeding biology, though little studied in the Sierra, appears similar to other sparrows, with the female carrying most of the burden. Nests are placed on or near the ground in a wide variety of shrubby locations. Nesting can begin in early April in lower-elevation locations or later in spring higher up. Some breeders, and all those using higher elevations, move to lower elevations in winter.

STATUS AND DISTRIBUTION Spotted Towhees breed across most of western North America. In Sierra they are generally year-round residents, but some altitudinal movements occur with low elevations hosting much larger populations in winter.

West Side. Common year-round in the Foothill and Lower Conifer zones, becoming less common and more seasonal above 4,000 feet; occasional upslope movement in fall before the snows begin.

East Side. Common and widespread year-round in appropriate habitats with some altitudinal shifts similar to the West Side.

TRENDS AND CONSERVATION STATUS Sierra Breeding Bird Survey and Christmas Bird Count data show highly significant positive population trends for Spotted Towhees in the past 40 years. Data from elsewhere in California and across the species' range show stable populations, so it is unclear why Sierra numbers are increasing. Christmas Bird Count tallies in the Sierra may be up due to birds taking advantage of backyard feeders to remain at higher elevation in winter. Increases in breeding birds may also be influenced by chaparral habitats that have expanded in response to more frequent stand-clearing fires and to the expansion of introduced Himalayan blackberries.

Rufous-crowned Sparrow • *Aimophila ruficeps*

ORIGIN OF NAMES L. *aimophila*, blood loving, from Gr. *aima*, blood, and *philos*, loving—why the name was applied to this genus is uncertain; *ruficeps*, red-headed, from L. *rufus*, red, and *ceps*, head.

NATURAL HISTORY Distinctive *dear-dear-dear* calls coming from sparse, scrubby vegetation on a steep hillside may be your only clue to the presence of these shy sparrows. Their preference for dry, rocky slopes with thinly distributed shrubs links them with recently burned areas. They often occupy such areas in the first several years following a fire but disappear after the vegetation matures. While often found in treeless habitats in much of their range, Rufous-crowned Sparrows in the Sierra frequently use areas with scattered small oaks. In such areas males can be found singing from within and atop the oaks. Their song has a complex musical pattern with a stumbling cadence, often similar to that of the Lazuli Bunting, which may occupy the same habitats.

The nesting season begins as early as late February in warmer areas and is generally complete by late May. Nests are built on the ground under shrubs or in thick grass. Mated pairs are thought to maintain their territories year-round.

STATUS AND DISTRIBUTION "California" Rufous-crowned Sparrows *(A. r. ruficeps)* are geographically isolated from other subspecies and are endemic to California and northern Baja California. They are mainly sedentary in the Sierra and restricted to a narrow band of the lower Foothill zone along the entire length of the West Side.

West Side. Locally fairly common residents on relatively steep rocky slopes with a scattering of shrubs and sometimes oaks, from just above the Central Valley floor up to about 2,500 feet, and very rarely slightly higher; some postbreeding wandering to somewhat higher elevations and a few winter observations on the Central Valley floor.

East Side. No documented records; uncommon residents in Kern County nearly up to Walker Pass, so some may possibly slip over to the East Side somewhere in that vicinity.

California Towhee • *Melozone crissalis*

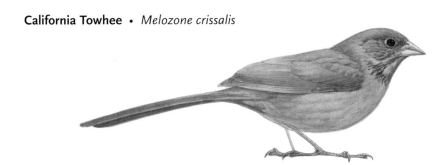

ORIGIN OF NAMES "California" since the species is restricted to that state as well as Baja California; *melozone* from Gr. *melos*, a song, and *zone*, a band; L. *crissalis*, crissum—a reference to the cinnamon undertail converts, one of species' few distinctive field marks.

NATURAL HISTORY If any sparrows really deserve to be called "L.B.J.'s" ("little brown jobs"), California Towhees are perhaps the best candidates. However, what they lack in fancy plumage they make up for in a confident and confiding nature, making them one of the more easily observed of our sparrows. William Dawson (1923), who could wax lyrically about nearly any bird, could only say of this one: "There is, honestly, no particular reason why we should be fond of this prosy creature, save that he is always around."

Even when hidden in the wide variety of brushy habitats, California Towhees reveal their presence by ringing, metallic *chink* notes, not to be confused with the much louder alarm call of the California ground squirrel. Its song, heard almost exclusively in spring, is simply a set of these call notes strung together into a slow trill. Recent analyses of DNA data led to California Towhees and a few of their similarly drab, brownish relatives being plucked from the genus *Pipilo* and placed in the genus *Melozone* along with several other similar species that mostly occur in Mexico.

Very nearly endemic to California/Baja California, California Towhees are at home in almost any brushy habitat interspersed with open areas (where most foraging occurs), including foothill chaparral, streamside thickets, and suburban backyards through all but the wettest, driest, coldest, or highest parts of the state. Mostly sedentary, these towhees remain on the same territories year-round and engage in frequent disputes with neighbors that can be quite violent. Nesting begins early (February–March) for California Towhees, and they may raise two broods in a good year. Nests are loosely constructed mostly of grasses and placed in a bush usually a few feet above the ground. Parents tolerate their offspring in the nest vicinity for as long as six weeks unless a second brood is started.

STATUS AND DISTRIBUTION California Towhees are year-round residents from southern Oregon to southern Baja California. In the Sierra they reside in open areas with scattered shrubs throughout the Foothill zone.

West Side. Common and widespread in appropriate habitats, becoming more sparsely distributed at elevations above 2,000 feet; some significant upslope wandering in the fall with observations as high as 8,000 feet in Yosemite National Park in September.

East Side. While mostly restricted to the West Side, there are resident birds in brushy canyons on the western edge of the southern Owens Valley from near Olancha and south into northern Kern County; there are also nonbreeding records farther north in the Owens Valley.

Chipping Sparrow · *Spizella passerina*

ORIGIN OF NAMES "Chipping" for the species' familiar calls; L. *spizella*, a finch; L. *passerina*, of or like a sparrow.

NATURAL HISTORY These attractive little sparrows are among the most widespread of any member of this family, with a breeding range that covers nearly all of Canada and the United States, encompasses most of the life zones of North America, and includes nearly all of California except the Central Valley and deserts. Across this expansive range, they can be found in a stunning variety of coniferous or deciduous woodlands but always prefer areas with scattered trees and shrubby understories close to open grassy areas. In the Sierra they are equally catholic in their choice of breeding habitat using oak savanna and open pine woodlands in lower elevations and a variety of open conifer habitats nearly up to tree line. Few other Sierra birds breed across such a broad altitudinal range.

Their songs—dry, flat trills on a single pitch—are unimpressive to the ear, but catch one in your binoculars in midsong and watch how their entire bodies vibrate. Some variations of the Chipping Sparrow song are very difficult to differentiate from the variations of the Dark-eyed Junco's song, and the two species overlap in many habitats. Though not difficult to see when flocks gather during migration, or when males are singing early in the breeding season, they become more secretive once breeding begins and confirming breeding can be difficult. Males are usually on territory and singing by early April at lower elevations. Nests are usually built in trees about 10 feet above the ground and often near the outer edge of branches. Since two broods per year are not unusual elsewhere in its range, the same may be true in our region, at least at lower-elevation sites.

STATUS AND DISTRIBUTION Chipping Sparrows are patchily distributed in the breeding season across a wide range of elevations, and most leave in September to winter in southern California or Mexico.

West Side. Uncommon to locally fairly common in the breeding season in oak savanna and woodland in the Foothill zone; most breed in the Lower and Upper Conifer zones, and few can be found into the Subalpine zone.

East Side. Fairly common and widespread in breeding season above the valley floors nearly to tree line north of Inyo County; rare to absent further south except in migration.

Brewer's Sparrow • *Spizella breweri*

ORIGIN OF NAMES Thomas Mayo Brewer (1814–1880), a Boston physician, politician, and publisher who coauthored *History of North American Birds* and wrote *North American Oölogy* in the mid-1800s.

NATURAL HISTORY For anyone who appreciates the muted browns and grays that characterize much of the Great Basin landscape, these are the sparrows for you. Drab to some, subtly handsome to others, no one can deny this species its place as the singer with the greatest endurance. A crisp spring morning in this habitat is usually punctuated with the Brewer's Sparrow's long song—a string of trills, wheezes, and bubbly rapid warbles linked together in an unbroken series that leaves the listener wondering when the bird is going to stop to breathe. The song is usually given while perched atop a bush, making this the best time to get a good look at this bird, which otherwise forages on the ground under shrubs or flits from bush to bush mostly staying out of view.

Brewer's Sparrows are found mostly in continuous stands of sagebrush, bitterbrush, or other typical desert shrubs but most closely associated with Great Basin sagebrush. They will use habitats with pinyon pines, junipers, or other conifers if there is a well-developed brushy understory and the trees are widely scattered. They usually choose the largest shrubs available for their tightly constructed nests and often line the nest with hair. Their breeding season runs from May through July and, based on studies elsewhere, it is likely these sparrows produce second broods when weather permits.

STATUS AND DISTRIBUTION Brewer's Sparrows breed mainly where sagebrush dominates, but migrants can be found in April and again in September along the California coast and, less frequently, in the Central Valley or foothills of the West Side; the species winters from southern California, east into western Texas, and south into Mexico.

West Side. Common breeders on Kern Plateau and Kennedy Meadows in southeastern Tulare County and highly localized breeders in other open, brushy habitats, including the Piute Mountains and Butte Meadow (Butte County); otherwise rare migrants; postbreeding, upslope movements probably account for summer observations west of the Sierra crest.

East Side. Common breeders in appropriate scrub habitats along the entire length of the Sierra.

Black-chinned Sparrow • *Spizella atrogularis*

ORIGIN OF NAMES "Black-chinned" for this distinctive field mark of breeding males; *atrogularis*, black-throated, from L. *ater*, black, and *gula*, throat.

NATURAL HISTORY These reclusive sparrows may be the least studied of any regularly occurring Sierra songbird. The bird's presence is most often betrayed when it gives its unmistakable song, best described by Ralph Hoffmann (1927): "One hears in summer a sweet, plaintive song, beginning with several liquid notes in a minor key followed by a long run which descends the chromatic scale." This song may be given from the top of a shrub in the cooler morning and evening hours, providing brief but excellent views of this lovely sparrow.

These are birds of arid, rocky shrubland, often associated with stands of chamise. Generally sparsely distributed even within areas of ideal habitat, they seem to prefer areas with hilly topography and a few, large scattered shrubs. In many places they favor conditions found a few years after a fire and avoid such areas when the chaparral becomes overmature after a long fire interval. Little is known of this species' breeding biology or foraging habits. Birds arrive in late April and are hard to find by mid-August. Nests are found above ground in dense shrubs.

STATUS AND DISTRIBUTION Black-chinned Sparrows are localized breeders in dry, rocky shrublands mostly in the southern half of California; their range (at least of intermittent breeding) has apparently extended north in the past 50 years into chaparral habitats along the West Side of the Sierra and in the Coast Range. In those years when the species is discovered in these northern areas, they tend to be found simultaneously in widespread locations, and those years roughly correlate with major drought periods (1977–76, 1987–92, 2007–8), when their southern habitats may have become unproductive. Birds arrive in late April and depart in late summer for wintering grounds (mainly in Mexico); rarely found in migration.

West Side. Uncommon and highly localized breeders; increasingly rare and irregular as one moves north of the Yosemite region.

East Side. Rare and localized in a narrow strip west of the Owens Valley south of Independence, the subspecies found here (*S. a. evura*) is the same one found from Arizona to Texas and distinct from the West Side breeders (*S. a. cana*); these subspecies are essentially indistinguishable in the field; scattering of records farther north with one near Mount Rose (Washoe County, Nevada) in August 2001 being most notable.

Vesper Sparrow • *Pooecetes gramineus*

ORIGIN OF NAMES "Vesper," an evening prayer, a reference to the species' habit of singing into twilight; *pooecetes*, grass-dweller, from Gr. *poe*, grass, and *oiketes*, dweller; L. *gramineus*, of the grasses.

NATURAL HISTORY In the Sierra, Vesper Sparrows choose breeding areas with scattered clumps of sagebrush interspersed with grassy patches, the same basic shrubby-grassy structure preferred throughout their range. Though named for their habit of singing in the evening, these sparrows are just as likely to sing in the morning or any other time of day in the breeding season. It is a sweet, pleasant song, reminiscent of some versions of the Song Sparrow's tune, usually with two introductory notes on the same pitch, followed by a trill or more complex series of phrases with a great deal of geographic variation.

These are very much "ground" sparrows doing most of their foraging for insects and seeds on the ground, usually in patches of grasses. The nest is constructed almost entirely of grass and is placed on the ground, often in a shallow hollow, and usually at the base of a shrub or tall patch of grass. Unlike most sparrow fathers, the males are known to help with incubation of eggs and brooding of young. They may also tend a batch of fledglings while the female incubates a second clutch of eggs, a common sparrow behavior.

STATUS AND DISTRIBUTION Vesper Sparrows occupy much of North America, breeding across most of Canada and the northern United States and wintering in the southern third of the United States and much of Mexico. Their seasonal status in the Sierra is complex, with two different subspecies present in winter. "Great Basin" Vesper Sparrows *(P. g. confinis)* are fairly common breeders in sagebrush interspersed with grassy patches mainly on the East Side from April to September, and a few apparently remain to winter in the Owens Valley. Small numbers winter on the West Side from Stanislaus County southward, but most winter further south in the southwestern United States and into Mexico. "Oregon" Vesper Sparrows *(P. g. affinis)* are somewhat darker and shorter-tailed than "Great Basin" Vesper Sparrows and migrate from breeding areas in the Pacific Northwest to winter in grasslands within the Foothill zone along the western edge of the Sierra and into the Central Valley in places.

West Side. "Oregon" Vesper Sparrows are uncommon to locally common in winter in grasslands throughout the Foothill zone and favor areas with some structure (bushes, small trees, even fences or corrals). "Great Basin" Vesper Sparrows are highly localized breeders in big sagebrush habitats near Mountain Meadows Reservoir and Lake Almanor; rare winter visitor in the Foothill zone.

East Side. "Great Basin" Vesper Sparrows are common breeders in appropriate habitats such as Martis Valley (Placer and Nevada Counties) and at Troy Meadow (Tulare County); Sierra Valley and the vicinity of Mono and Crowley Lakes are also good places to find this species; wintering birds in the Owens Valley may come from nearby breeders or from birds breeding farther north.

TRENDS AND CONSERVATION STATUS "Oregon" Vesper Sparrows were included on the list of California Species of Special Concern in 2008, mainly due to their restricted breeding and wintering ranges and threats to their grassland habitats on both breeding and wintering grounds. Vesper Sparrows are not well surveyed by California Christmas Bird Counts, so no significant trends are evident. However, Breeding Bird Surveys from British Columbia and Oregon show significant declines in recent decades. In California the grassland habitats of the Sierra foothills and the Central Valley edge on which this species depends are being rapidly fragmented or lost to orchards, vineyards, and development.

Lark Sparrow • *Chondestes grammacus*

ORIGIN OF NAMES Named for the species' larklike songs, or possibly for their bold head patterns and white outer tail feathers; *chondestes* from Gr. *khondros*, grain, and *edestes*, eater; Gr. *grammacus*, a painted line, in reference to the species' distinctive head patterns.

NATURAL HISTORY Lark Sparrows are the only member of their genus and are probably most closely related to Vesper Sparrows. If you are going to introduce a beginner to sparrows, this is

where you should start. Arguably the handsomest of sparrows, these birds are easy to see as they forage for seeds on the ground, perch on fence wires, or sing from trees or bushes. Larger than most other sparrows, they possess an abundance of field marks from the strikingly patterned faces to the bold white edges to their long, rounded tails.

These sparrows are blessed not only with good looks but with one of the most complex and enthusiastic songs of any Sierra bird. It is a rapid jumble of trills, sweet notes, and musical warbles, at times bordering on frantic. Perhaps the best way to recognize this song is to listen for the occasional harsh *raspberry* notes scattered here and there, so out of place in an otherwise very musical song. They are one of the sparrows that treat us to song even in winter, when their song is more subdued and fragmented but sweet nonetheless.

In the Sierra they are found primarily in oak savanna habitats on the West Side. Lark Sparrows spend much of their time foraging on the ground for seeds or leaping to catch grasshoppers (their favorite prey), and their diet tends toward seeds in winter and insects in summer. Gregarious outside the breeding season, sizable flocks are encountered along roadsides and groups will gather at backyard feeders.

It seems everything about these sparrows is extraordinary, including their sex life. Unlike most songbirds, and certainly unique to the sparrows, Lark Sparrows engage in prolonged and elaborate courtship behaviors. Males prance and display to their prospective mates with showy flashes of those white tail spots, drooping and dragging their wings along the ground and gently passing a twig to the female during copulation. The female then takes the twig to the nest site, usually on bare ground with some grassy or shrubby cover or sometimes low in a tree or bush. Also unusual is the tendency to reuse old nests of Lark Sparrows and of other birds.

STATUS AND DISTRIBUTION Throughout their range in western North America, Lark Sparrows use grassland edge habitats where either shrublands or scattered trees abut grassland expanses. They are residents in oak savanna and grassland-woodland edge habitats all along the western edge of the Sierra and sparsely distributed breeders in shrubland-grassland habitats on the East Side.

West Side. Common residents in appropriate habitat at lower elevations throughout; flocks seen in spring and fall may be supplemented by migrants from farther north.

East Side. Uncommon, breeds mostly at low densities and in scattered locations, including Sierra Valley, the Carson Valley, and the Owens Valley; unknown if these East Side breeders winter on the West Side or move farther south; casual in fall and winter.

TRENDS AND CONSERVATION STATUS Lark Sparrows are one of a large group of common grassland-associated species (e.g., Western Meadowlarks, Killdeer, Loggerhead Shrikes, American Kestrels) showing long-term declines in both breeding and wintering populations all across North America. While Sierra-specific data are insufficient to show significant trends for this species, Breeding Bird Survey and Christmas Bird Count data from California and the United States show significant declines over the past 30 years.

Black-throated Sparrow · *Amphispiza bilineata*

ORIGIN OF NAMES "Black-throated" for the striking black triangles on the throats and breasts of adults; a*mphispiza* from Gr. *amme,* sand or soil, and *spiza,* a finch; *bilineata* from L. *bi,* two, and *linearis,* lined—a reference to the two white stripes on the head.

NATURAL HISTORY Crisply attired in tasteful black, white, and gray, these sparrows are perfectly adapted to some of the driest conditions endured by any North American bird. Field and laboratory studies have shown that Black-throated Sparrows can get by without drinking water at all when some green vegetation or insects or even dry seeds were available, provided temperatures are not too high. Their ability to retain dietary water and maintain full activities under the driest of circumstances may be unequaled among North American songbirds. They can even subsist on highly saline water and maintain normal activity levels. Black-throated Sparrows even shun standing pools of surface water, as long as they can extract enough water from their food. Put succinctly in the *Atlas of Breeding Birds of Nevada* (Floyd et al. 2007): "It is hard to imagine a shrubland too hot and dry for the beautiful Black-throated Sparrow."

Throughout their range, Black-throated Sparrows select habitats based on structure rather than any particular plant species. Relatively open areas of scattered shrubs are used, and in the Sierra they mostly occur on rocky hillsides as well as pinyon-juniper and Joshua tree woodlands with adequate brushy understories. Birds forage actively both on the ground between shrubs and within the shrubs themselves, harvesting mainly seeds in late summer and gleaning insects in the spring.

Males defend large territories in spring, establishing and maintaining them with frequent singing from the tops of the larger shrubs. Their songs are relatively simple and always include a distinctive trill with a rich ring to it, which can be heard at remarkably long distances. The nest is tightly constructed, often includes some animal hair, and is placed above the ground in a bush. In areas of its range where late summer monsoons are common, this species may attempt a second nesting.

STATUS AND DISTRIBUTION Black-throated Sparrows breed from California east to Colorado and winter in the southwestern United States and Mexico. In the Sierra they are localized breeders (April through August), but often unpredictable in some locales, present some years, and absent in others. They are rare but regular wanderers in fall and winter to the California coast and occasionally the Central Valley.

West Side. Regular breeders and residents in Kern County west of Walker Pass, the Kern River Valley, and along Kelso Valley Road; breeding farther north highly localized and inconsistent in patches of dry chaparral, often a few years after a large fire; status north of Kern County is complex and historically uncertain; the species may have expanded its breeding range into scattered locations in the past 50 years, or these isolated, intermittent populations may have simply gone undiscovered; status in these areas may be analogous to Black-chinned Sparrow (see account above), with birds tending to show up at multiple sites in a given year and occupation of these sites possibly linked to unusually severe drought conditions in southern California deserts; examples of some of these locations include near Johnsondale and west of Springville (Tulare County), Cherry Lake Road (Tuolumne County), Icehouse Road (El Dorado County), Foresthill Bridge (Placer County), and Cohasset Ridge (Butte County).

East Side. Locally common and reliable breeders in desert canyons in Kern and southern Inyo Counties; becoming progressively more localized and less predictable as one moves north; hillsides west of Highway 395 from Alpine County, and north through Douglas, Carson City, and Washoe Counties (in Nevada) harbor some breeders most years; rare farther north.

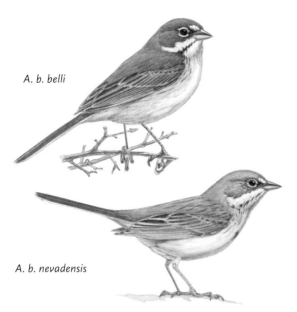

A. b. belli

A. b. nevadensis

Sage Sparrow • *Amphispiza belli*

ORIGIN OF NAMES "Sage" for the preferred habitat of the most widespread subspecies (see Bell's Vireo account).

NATURAL HISTORY There is much controversy about whether this species should be split into two or possibly three species, each of which breeds in the Sierra (discussed in more detail below in "Status and Distribution"). Sage Sparrows on the East Side and in the southern Sierra have much paler grayish upperparts, with much less distinct malar stripes than the darker, more brownish birds breeding to the west. The songs are also somewhat different, with the western Sage Sparrows singing a more flowing, musical, tinkling tune, and the eastern ones a thinner, buzzier one.

Differences in habitat selection by different subspecies are related to geography. Eastern and southern Sierra subspecies use sparser, drier, flatter shrublands, and western subspecies favor denser chaparral, often on hillsides. What little we know of the habits of this species comes mostly from studies in the Great Basin and may or may not relate to all subspecies. This is most certainly a sparrow of the ground, foraging almost entirely there for seeds and insects. Even when disturbed, this bird is much more likely to run along the ground from cover to cover rather than fly. This behavior, combined with its habit of singing while still concealed in a bush, make it difficult to see.

Sage Sparrows start breeding earlier than most other sparrows. Birds are on territory in most of the range by late March, and eggs may be laid in March on the West Side or in the southernmost parts of the range, and early April on the East Side. It seems that resident pairs on the West Side may remain together throughout the year, while migratory birds on the East Side arrive already paired, both unusual occurrences among sparrows. Nests are usually built within a shrub but sometimes on the ground below. Like nearly all sparrows, females do all the nest-building, egg incubation, and brooding of the nestlings, and males only help feed the young.

STATUS AND DISTRIBUTION Migratory across most of their range in the Great Basin, three sub-species of Sage Sparrows reside year-round in the Sierra. To keep the Latin to a minimum, we refer to them as "Great Basin" Sage Sparrows *(A. b. nevadensis)*, "Bell's" Sage Sparrows *(A. b. belli)*, and "canescens" Sage Sparrows *(A. b. canescens)*. "Great Basin" Sage Sparrows are strongly linked with sagebrush throughout their range, seldom found in areas without at least some sagebrush, usually Great Basin sagebrush. They reach the eastern edge of the Sierra as breeders as far south as the northern edge of Inyo County; "canescens" Sage Sparrows are residents on the East Side, from northern Inyo County (where some intergrades with the "Great Basin" Sage Sparrows have been documented), south into Kern County, crossing over onto the West Side in northern Kern County;

"Bell's" Sage Sparrows are residents of foothill chaparral on the West Side in the central Sierra. Due to lack of consistently distinct characteristics, some authors do not consider "canescens" Sage Sparrow a separate subspecies, with some lumping it with "Bell's" and others lumping it with "Great Basin."

West Side. "Bell's" Sage Sparrows are highly localized residents of chaparral (usually including chamise) from Placer County south into Mariposa County; locations include Foresthill Bridge area just northeast of Auburn (Placer County), Pine Hill Preserve (El Dorado County), and Hunter's Valley Mountain (Mariposa County); "canescens" Sage Sparrows are fairly common breeders in northern Kern County (Kern River Valley, Greenhorn and Piute Mountains).

East Side. "Great Basin" Sage Sparrows are uncommon and highly localized breeders north of Mono County, becoming common in Mono County and just into northern Inyo County; "canescens" Sage Sparrows are fairly common residents from northern Inyo County south into Kern County.

Savannah Sparrow • *Passerculus sandwichensis*

ORIGIN OF NAMES Named by Alexander Wilson (see Wilson's Snipe account), who collected a specimen in Savannah, Georgia; L. *passerculus*, little sparrow; *sandwichensis* based on the original English name, "Sandwich Bunting," assigned by John Latham to birds collected at Sandwich Bay, Alaska.

NATURAL HISTORY Savannah Sparrows are truly open country birds, found mainly in grasslands, wet meadows, hay and alfalfa fields—landscapes often devoid of trees or even shrubs. Like some other members of this family, taxonomy is complex and at least three subspecies can be found in the Sierra, but only two in significant numbers (discussed in more detail below in "Status and Distribution"). With plumage variations due to subspecies, age, molt, wear, and individual differences, it is best to simply be aware that these sparrows, perhaps more than any other, may not look like "the one in your field guide." Consistently distinguishing these subspecies in the field ranges from difficult to impossible.

These are generally easy birds to see most of the year, readily perching up on fences or any higher patch of vegetation. The song, heard only in breeding season, is thin and buzzy but with definite structure, usually finishing with a set of notes on different pitches. The long trill sometimes rises gently in pitch, followed by a short trill at a lower pitch. Even when not singing, this sparrow can be located by listening for a very high, thin *tseep* call note. However, be aware that this species also gives a short, buzzy *bzzzeet* call and, more rarely, a flat *tick*, very like a Dark-eyed Junco.

Savannah Sparrows choose a variety of Sierra habitats for breeding (mainly on the East Side), with wet meadows a favorite. However, they may also breed in grasslands interspersed with shrubs and agricultural fields of hay or alfalfa. In any case, patches of taller vegetation are used for singing. Nests are placed on the ground and require concealment by some sort of herbaceous or shrubby cover. Wintering birds on the West Side choose open grasslands and frequently forage in fairly large flocks.

STATUS AND DISTRIBUTION The 28 named subspecies of Savannah Sparrows can be roughly divided into 5 "groups," some of which may someday be recognized as separate species. All Savannah Sparrows found in the Sierra are of the "Sandwichensis" group. "Nevada" Savannah Sparrows

(P. s. nevadensis), the only subspecies confirmed to breed in the Sierra, are locally common from April to September mainly on the East Side and winter from the Central Valley to Baja California and into central Mexico; *"anthinus"* Savannah Sparrows *(P. s. anthinus)* migrate from breeding range in Alaska and northern Canada to winter on the West Side (note that due to a complex and controversial nomenclature change, Grinnell and Miller [1944] cited this subspecies as *P. s. alaudinis*, the name now given to the subspecies resident along the northern and central California coast).

West Side. "Anthinus" Savannah Sparrows and smaller numbers of "Nevada" Savannah Sparrows common in foothill grasslands in winter; some upslope movements in fall and birds of both subspecies have been found above 8,000 feet; "Nevada" Savannah Sparrows resident in the Kern River Valley; Aleutian-breeding subspecies (*P. s. sandwichensis*) may also winter in small numbers on the West Side and, to further complicate the picture, all three taxa occur as migrants, chiefly in March and September.

East Side. "Nevada" Savannah Sparrows common breeders in appropriate habitats throughout up to nearly 7,500 feet and occasionally higher; uncommon to rare in winter north of Inyo County, uncommon to fairly common from Inyo County south.

Grasshopper Sparrow • *Ammodramus savannarum*

ORIGIN OF NAMES Named for grasshopper-like song; *ammodramus*, sand runner, from Gr. *ammos*, sand, and *dramein*, to run; *savannarum* is a reference to the species' preferred savanna habitat (note: the Savannah Sparrow has a different spelling because it was named for Savannah, Georgia—not for a type of grassland).

NATURAL HISTORY Like most members of their genus, these sparrows are champion skulkers, able to elude detection even in the sparsest habitats. Nearly all observations are of singing males and even the song—a thin, buzzy, insectlike trill—can be hard to hear and does not carry particularly well. This song is somewhat like a simpler version of the Savannah Sparrow's song, consisting of a trill on a steady pitch, usually preceded by a short *hiccup*. Females may also give a very similar song, but only males use a second song type—a rapid thin warble, sometimes given in fight. This second song is more rarely heard and may surprise the unprepared listener with its musicality. If you are lucky, the bird will be perched on a tall grass stem or weed; if not, it will sing hidden from view. If flushed, Grasshopper Sparrows usually fly a short distance close to the ground, then disappear into the grass and run along the ground in a mouselike fashion, making them almost impossible to locate.

It may be easier, and more important, to report what we do not know about the natural history of these birds. Almost no research has been done on Grasshopper Sparrows in California, and extrapolating from studies done elsewhere can be misleading. This is a species of open, usually expansive grasslands in California. Although some studies from the Great Plains suggest that the species avoids grazed habitats, this does not seem to be the case in California, where grasslands are usually dominated by non-native annuals and abundant grazers. Indeed, one of the very few California studies found that Grasshopper Sparrows used grazed plots exclusively. Links to historical native plants are likewise unclear. While a positive connection to native bunch grasses has been suggested in San Diego County, this does not necessarily apply elsewhere. An association with the presence of some native plants has been shown, but most native plants were found preferentially in grazed grasslands. This is likely because most grasslands of the Central Valley and Coast Range tend to become dominated by weedy invasive plants when not grazed. Our best guess about the preference of Grasshopper Sparrows is sort of a "goldilocks" theory: they like grasslands that are grazed enough to keep out the invasive weeds, but not so heavily grazed to eliminate good cover.

Grasshopper Sparrows place their nests directly on the ground, usually constructing a cover of grasses pulled over the nest to allow an entrance from the side. Two broods are common elsewhere, so that seems likely to occur in California as well. At least one study in Nebraska observed non-parental and unrelated birds helping to feed nestlings, a very unusual behavior among sparrows.

STATUS AND DISTRIBUTION Twelve subspecies of Grasshopper Sparrows are currently recognized in North America, and the "Western" subspecies *(A. s. perpallidus)* breeds from Oregon to northern Baja California. Their exact status in the Sierra is uncertain, since the species is so difficult to detect, but they appear to be uncommon and highly localized breeders in grasslands of the West Side. They are rarely encountered in winter, but that may have more to do with difficulty of detection rather than lack of presence. While not considered a distinct subspecies, it is of interest that California's Grasshopper Sparrows are geographically isolated with the nearest breeding populations in northern Oregon.

West Side. Regular, though highly localized and irregular from March through September, with nearly all observations of singing birds in foothill grasslands; apparently becoming more common from El Dorado County south; recent observations along White Rock Road (Mariposa County); some summer records as high as 4,500 feet; winter status uncertain but some remain through winter.

East Side. Very rare to casual migrants (mostly in fall) and at least one winter record in the Owens Valley; at least one historical summer record in Lassen County, just north of the region, and occasional credible summer reports in Sierra Valley may represent breeding attempts.

TRENDS AND CONSERVATION STATUS Grasshopper Sparrows were included as a California Bird Species of Special Concern in 2008 based on negative long-term population trends in California, the species' limited and uncertain range, and threats to habitat from urbanization and conversion of grasslands to more intense agriculture like orchards and vineyards. Data from Breeding Bird Surveys are too sparse to draw any conclusions about trends in the Sierra.

Fox Sparrow • *Passerella iliaca*

ORIGIN OF NAMES "Fox" for the reddish, foxlike plumage of eastern subspecies; L. *passer,* a sparrow; *ella,* a diminutive form; *iliaca* from L. *ilia,* side of body—a reference to the species' streaked flanks.

NATURAL HISTORY Depending on your point of view, taxonomy of Fox Sparrows can be fascinating or bewildering, with at least 18 subspecies recognized and a wealth of studies on their evolutionary relationships available. Because the Sierra hosts representatives of four distinct and field-identifiable types, and since some authors advocate species status for one or more of these groups, we will address each in the "Status and Distribution" section below. With regard to natural history, we concentrate only on the Sierra-breeding *megarhyncha* group ("Thick-billed" Fox Sparrows).

Stands of montane chaparral in summer are rarely without these bold, highly vocal sparrows. Males frequently perch on the highest point on a large shrub or even in a tree to sing their cheerful, bouncing, highly variable song. The song can easily be confused with that of the Green-tailed

P. i. megarhyncha

P. i. unalaschcensis

Towhee, and both species share similar habitats on the East Side and higher elevations on the West Side. Both may start with a similar flourish, but the Fox Sparrow's song is sweeter and usually lacks the characteristic *burrs* or trills included in the towhee's song. Birders who only know the flat *tick* call of Fox Sparrows from the lowlands in winter, will be surprised to hear this subspecies give a metallic *chink*, very like a California Towhee.

While most abundant in manzanita-dominated chaparral, these sparrows can be found in all types of shrub habitats, even when infiltrated with a significant canopy of conifers. They take good advantage of habitat created by fires, usually occupying these areas several years postfire and persisting even when the habitat becomes quite overgrown and mature. Like nearly all our sparrows, Fox Sparrows forage on the ground and in leaf litter, often using their version of the two-legged "towhee" hop to clear the surface looking for seeds or insects. Nests may be low in bushes or trees or even on the ground but are always very well concealed. Apparently Sierra birds do not produce more than one brood per season. Fox Sparrows follow the standard sparrow parenting model with the males' role limited to singing, copulation, and sharing in feeding the young birds.

STATUS AND DISTRIBUTION Genetic, morphological, and behavioral data all suggest that the myriad named Fox Sparrow subspecies can be organized into four groups and, thankfully for birders, most individuals can be identified to group in the field with good views and careful study. Members of the *iliaca* group ("Red" Fox Sparrows) breed in boreal forests across North America and are rare to casual winter visitors to the Sierra. Those of the *unalaschensis* group ("Sooty" Fox Sparrows) breed along the Pacific Coast from Alaska south into Washington and are the predominant wintering Fox Sparrows in the Sierra. Representatives of the *schistacea* group ("Slate-colored" Fox Sparrows) breed inland from British Columbia south into Utah and Nevada and are fairly rare fall and winter visitors to the Sierra. Those of the *megarhyncha* group ("Thick-billed" Fox Sparrows) breed in much of the Sierra and winter in small numbers at lower elevations, with most birds of this group migrating to coastal southern California and northern Baja California.

West Side. "Thick-billed" Fox Sparrows are common breeders in chaparral and mixed chaparral-woodland habitats from the upper Foothill zone (above 3,000 feet) to the Subalpine zone and rare in winter in the Foothill zone, mainly in the south; "Sooty" Fox Sparrows arrive in mid- to late September and are fairly common in riparian and wetland thickets and chaparral mainly in the Foothill zone; "Slate-colored" Fox Sparrows are rare fall migrants and winterers in the Foothill zone; "Red" Fox Sparrows are casual in winter in the Foothill zone.

East Side. "Thick-billed" Fox Sparrows are fairly common breeders in appropriate habitat; more likely to be found in streamside areas on the drier East Side than on the West Side; Fox Sparrows are casual on the East Side lowlands in winter with most records probably of "Sooty" Fox Sparrows, though members of all four groups have been reported.

Song Sparrow • *Melospiza melodia*

ORIGIN OF NAMES "Song" for the species' habits of singing throughout the day in most seasons; *melospiza* from Gr. *melos*, song, and *spiza*, finch; L. *melodia*, a pleasant song.

NATURAL HISTORY The beautiful, ringing songs of Song Sparrows are familiar and welcome sounds in backyards, parks, and wet shrubby habitats throughout much of North America. This song may have inspired more mnemonic descriptions than any other North American bird vocalization, from Henry David Thoreau's "Maids, maids, maids, hang on your tea-kettle-ettle-ettle-ettle-ettle" to William Dawson's "Peace, peace, peace be unto you, my children." The two or three notes on one pitch introduce a more complex mix of trills and twitters. This lovely song can be heard in any season and often brightens the dullest winter day. Since wintering Song Sparrows may include both resident birds and migrants, it would be interesting to know if residents are more or less likely to sing in winter than visitors.

These sparrows are almost never far from water, using streamside thickets, brushy understory in riparian woodlands, or marsh edges year-round. Nesting may begin as early as March at lower elevations. The female builds a sturdy nest of sticks lined with soft grasses and the well-insulated nature of this nest may aid in early nesting. The nest may be on the ground, low in a bush, or in marsh cattails or tules but is never far from water and often built over water. In higher-elevation mountain meadows, Song Sparrows may share the same willow habitats as Lincoln's Sparrows, likely a recent phenomenon (see "Trends and Conservation Status" below). Though unverified in the Sierra, second and perhaps even more broods are likely each season, at least in warmer areas.

STATUS AND DISTRIBUTION A bewildering array of Song Sparrow subspecies reside across North America, and their exact taxonomic status in the Sierra is poorly understood; trying to differentiate among the many subspecies in the field is risky business. They are residents in lower elevations in brushy riparian understory and wetland habitats on both sides of the crest and breed at higher elevations up to well above 8,000 feet. Winter numbers at lower elevations are augmented by migrants from further north and east and by downslope movements of some breeders.

West Side. Common residents into the Foothill zone; fairly common in breeding season in riparian, meadows, and some chaparral habitats to over 7,000 feet; common in winter below 2,000 feet.

East Side. Fairly common to common breeders in meadows and streamside thickets from the lowest elevations throughout the East Side zone and nearly to tree line in places; becoming less common from mid–Inyo County south and rare to absent in the drier habitats from south Inyo County into Kern County; breeding status south of Walker Pass in desert hills uncertain, they generally occur as migrants there but may remain to breed in small pockets of riparian vegetation; a hardy species, they remain fairly common through all but the snowiest winters at lower-elevation locations like Honey Lake, Carson Valley, and Owens Valley, and even in the 6,000 foot Tahoe Basin.

TRENDS AND CONSERVATION STATUS According to Grinnell and Miller (1944), Song Sparrows rarely nested above 200 feet on the West Side (at least in the northern Sierra), but they are now fairly common breeders up to 7,500 feet—and often higher. Their populations on the East Side have apparently increased as well, with new populations being found at higher elevations or in new locations. This is a true range expansion that has been well documented throughout the Sierra, but no data are currently available on which subspecies is (or are) involved.

Lincoln's Sparrow • *Melospiza lincolnii*

ORIGIN OF NAMES "Lincoln's" for Thomas Lincoln (1812–1883), a young naturalist who accompanied John James Audubon on a trip to Labrador, where Lincoln collected this sparrow named by Audubon in his honor.

NATURAL HISTORY Find a lush, green montane meadow with a healthy strand of willows following the meanders of a central stream and fringed with tall conifers. That's where you need to look, or better yet, listen for Lincoln's Sparrows. These sparrows can be hard to see in the breeding season and, even when singing, sometimes remain hidden deep in the willows and corn lilies. Their song is unique in both structure and tone, usually composed of three or four different trills, each on a different pitch and with varying tempos delivered one right after the other. The quavering quality of the notes gives the song a vaguely thrushlike effect. "It is a vigorous, joyous outburst, surprisingly loud for such a shy and diminutive singer" (Hoffmann 1927). Lincoln's Sparrows do not sing in their low-elevation winter haunts, so you need to visit the mountains in summer to fully appreciate this bird.

Lincoln's Sparrows are one of the most poorly studied members of this family, and we have little information on breeding behavior, particularly as it may relate to Sierra-breeding birds. Nests are placed on the ground well-hidden under willows, elevated just enough to stay dry in the wet habitats preferred by this species. Given the high-altitude locations of much of its breeding habitat, snowmelt is likely the determinant of when breeding begins, and there may be scant opportunity for more than one brood per year.

STATUS AND DISTRIBUTION Lincoln's Sparrows breed in mountain and boreal areas across North America. In the Sierra they are highly localized breeders mostly in the Upper Conifer zone but will breed up to the Subalpine zone in appropriate habitats. Most Sierra-breeding birds are thought to winter south of the Sierra, and Lincoln's Sparrows seen in migration or in lower elevations in winter are most likely from subspecies breeding in Alaska and Canada.

West Side. Fairly common breeders in appropriate habitats from 4,000 to 10,000 feet as far south as the Greenhorn Mountains in northern Kern County, representing a recent range extension; in winter uncommon to fairly common mostly in wetland and dense riparian understory in the lower Foothill zone and in the Kern River Valley.

East Side. Fairly common breeders in appropriate habitats in the East Side zone from the north end of the region to northern Inyo County; rare to absent as breeders farther south; in winter generally rare north of the Owens Valley; common some winters from the Owens Valley south.

TRENDS AND CONSERVATION STATUS Across their North American breeding range, Lincoln's Sparrow populations appear stable or possibly in decline. Sierra Breeding Bird Surveys do not include enough of their habitat to detect any trends. However, it is likely that given its very narrow habitat niche, Lincoln's Sparrows have been impacted by factors affecting mountain meadows. Historical grazing not only reduced or eliminated willow habitats from many meadows, but also changed the hydrology such that willows cannot regenerate even when grazing is removed. Brood parasitism by Brown-headed Cowbirds poses a threat to most Sierra songbirds. The expansion of Song Sparrows into these same habitats in the past 60 to 70 years may also have impacted Lincoln's Sparrows, though no studies to demonstrate competition between these closely related species in the Sierra have been conducted.

White-throated Sparrow • *Zonotrichia albicollis*

ORIGIN OF NAMES "White-throated" for the species' most distinctive field mark; *zonotrichia* from Gr. *zone*, a belt, and *trikhas*, a thrush—a reference to the banding on the heads of birds in this genus; *albicollis*, white-necked, from L. *albus*, white, and *collum*, neck.

NATURAL HISTORY White-throated Sparrows breed in the boreal forests of Canada and are familiar winter birds in backyards and parks throughout the eastern United States. In the Sierra and in California in general, they are uncommon but regular winter visitors, usually as single birds found in flocks of White-crowned and Golden-crowned Sparrows in shrubby thickets and roadside bushes. A diligent observer who methodically picks through all the flocks of wintering sparrows is likely to find one or more White-throated Sparrow each winter. Though most often silent, on occasion they sing, *Old . . . Sam, Peabody-Peabody-Peabody*—the first two notes suggestive of the first notes of the Golden-crowned Sparrow's song.

STATUS AND DISTRIBUTION Primarily breeders across Canada and the eastern United States, White-throated Sparrows are uncommon but regular winter visitors in lowland areas and interior valleys of the Sierra from mid-September through April. Fall migrants might be seen from mid-October to late November, with records up to nearly 10,000 feet.

West Side. Uncommon or rare with a dozen or more sightings per winter from Lake Almanor to the Kern River Valley.

East Side. Rare but fairly regular winter visitors and migrants in the Owens Valley, becoming casual farther north.

White-crowned Sparrow • *Zonotrichia leucophrys*

ORIGIN OF NAMES "White-crowned" for the species' most distinctive field mark; *leucophrys* from Gr. *leukos*, white, and *ophrus*, eyebrow.

Z. l. oriantha

Z. l. gambelii

NATURAL HISTORY Except for game birds and some endangered species, White-crowned Sparrows are among the best-studied birds in North America. These lively and confiding sparrows are ideal subjects for casual nature study as well as scientific research because they are common and can be readily observed in the open habitats where they breed and winter. The male's clear, sweet song is easily the most characteristic sound of high mountain meadows in the Sierra, but their songs come in many variations. Ornithologists have even discovered local *oriantha* dialects that vary noticeably from one river drainage to the next.

They breed in willow thickets around high-mountain meadows and lakes near tree line, where they seem to require a combination of wet, grassy areas, surface water, and patches of dense shrubby vegetation. Nests are placed on the ground or in the low branches of dense shrubs, with nesting commencing in May if the snows have melted and extending into July. These high-eleva-tion breeding sites preclude double broods per season and require greater amounts of time devoted to incubating eggs and brooding nestlings, activities that remain entirely up to the females.

STATUS AND DISTRIBUTION The subspecies that breeds in the Sierra, "Mountain" White-crowned Sparrows *(Z. l. oriantha)*, may be recognized by their black lores—the small area between each eye and the upper edge of the bill. This feature can be used to distinguish these birds from "Gambel's" White-crowned Sparrows *(Z. l. gambelii)*, the wintering subspecies in the lowlands, which may mix with "Mountain" White-crowned Sparrows in migration. "Mountain" White-crowned Sparrows breed in wet meadows and brushy thickets in Upper Conifer and Subalpine zones, often above tree line; most winter south of the Sierra. "Gambel's" White-crowned Sparrows are winter visitors (from breeding grounds in Alaska and Canada) to low-elevation locations on both sides of the crest from mid-September through April.

West Side. "Mountain" White-crowned Sparrows fairly common breeders from 6,500 to 10,500 feet, and sometimes higher, becoming rare to absent south of Tulare County; "Gambel's" White-crowned Sparrows common to abundant from mid-September through April below 3,000 feet but can be locally common at higher elevations in fall migration.

East Side. "Mountain" White-crowned Sparrows fairly common breeders in appropriate habitat from 5,000 to over 11,000 feet, rare to absent south of Inyo County; "Gambel's" White-crowned Sparrows common to abundant in winter below 5,000 feet, rare at higher elevations except during fall migration, when they can be common above 10,000 feet.

Golden-crowned Sparrow • *Zonotrichia atricapilla*

ORIGIN OF NAMES "Golden-crowned" for the spe-cies' most distinctive field mark; *atricapilla*, black-capped, from L. *ater*, black, and *capillus*, hair on head.

NATURAL HISTORY One of fall's most anticipated sounds is the plaintive *Oh, dear, me* song of Golden-crowned Sparrows returning for another winter in California. Though they may sing all winter, they are heard more regularly just after arriving in fall and shortly before leaving in spring. They are often found in large flocks of other sparrows, though mostly with their close relatives, White-crowned Sparrows. They forage on the ground for seeds but never far from some shrubby cover or a brush pile. They will readily come to backyard feeders, especially when food is provided on or close to the ground. The young birds making their first visit to California may entirely lack any "gold" in their crowns and can be the plainest of sparrows. They will not acquire that beautiful gold crown set off by contrasting black borders until their second year.

STATUS AND DISTRIBUTION Golden-crowned Sparrows breed in Canada and Alaska, with winter visitors to California arriving mid-September and remaining through April at low elevations on both sides of the Sierra crest.

West Side. Common to abundant in winter below 3,000 feet.

East Side. Fairly common to uncommon in fall through winter below 5,000 feet north of Mono County and in northern Owens Valley; rare to casual in midwinter in Mono Lake Basin and around Lake Tahoe with most records in fall migration; can be common at higher elevations during fall migration (September through October).

Dark-eyed Junco • *Junco hyemalis*

ORIGIN OF NAMES "Junco" from L. *juncus*, a reed, an odd reference since this species prefers woodlands and forests over marshes; *hyemalis* from L. *hiems*, winter.

NATURAL HISTORY With their crisp black hoods and flashy white tail edges, these dapper little sparrows are among the most familiar birds in the Sierra, and few if any other species are more abundant or widespread. Many people know them from backyard feeders, campgrounds, and picnic areas, but juncos are equally common along

Male

Female

forested trails and meadow edges everywhere in the Sierra. From early spring well into summer, they sing a simple, ringing trill on a single pitch. This can sometimes sound fairly "dry" and be confused with Chipping Sparrow's song, where they overlap at mid-altitudes. Their sharp *tisk!* call note is distinctive. Foraging flocks in winter can produce a chorus of softer chips reminiscent of the contact calls of a Bushtit flock.

Nearly everything about junco behavior can be characterized as "variable," and they are common breeding birds in nearly every habitat of the Sierra—from oak woodlands to tree line. There are nesting records from the Central Valley floor up to more than 11,000 feet. Nests are often placed in natural cavities or crevices formed by peeled bark, rock outcrops, or roots of upturned trees and sometimes under tall grasses in mountain meadows. Parental roles follow the standard sparrow pattern with females building the nest, incubating eggs and brooding nestlings with no assistance from her mate aside from feeding the young. Other aspects of breeding behavior show tremendous variability across the species' geographic and altitudinal range. For example, nest construction, nest location, and number of broods per year vary with location and altitude, and it is this plasticity of behavior that may be a key to the success of these bold little birds. They are equally open-minded in winter and visit a range of habitats from oak savanna to riparian forests to orchards to suburban backyards, usually in sizable flocks constantly searching the ground for seeds.

STATUS AND DISTRIBUTION The number of recognized subspecies of Dark-eyed Juncos can be mind-boggling, and several of these may visit the Sierra in winter or during migration. However, the only Sierra breeders are of the "Oregon" Junco *(J. h. thurberi)* group. Wintering birds are probably mostly of this same subspecies, which move downslope in response to snow and cold. However, other "Oregon" subspecies visit from farther north during winter, though these are generally indistinguishable in the field. One rare but regular winter visitor that is instantly recognizable is of the "Slate-colored" Junco *(J. h. cismontanus)* group. Much more rare in winter are "Gray-headed" *(J. h. caniceps)* and "Pink-sided" *(J. h. mearnsi)* Juncos, with the latter almost certainly overreported when confused with female "Oregon" Juncos.

West Side. Common to abundant year-round with most breeders from 3,000 to 10,500 feet but breeding records at lower elevations relatively uncommon; winter numbers abundant in lower elevations and good numbers can persist at higher elevations depending on snow levels and temperatures.

East Side. Status similar to West Side, but absent as breeders in the dry canyons from southern Inyo County into Kern County.

GROSBEAKS AND RELATIVES · Family Cardinalidae

Much like wood-warblers, members of this brightly colored family primarily winter south of the Sierra. Males wear vivid plumages compared with the relatively drab females. The heavy, conical bills of buntings and grosbeaks are designed to crack the hard coats of seeds and nuts, but adults also eat fruits and buds. Insects provide an important source of protein for growing nestlings. Despite the similarities of their common names, Black-headed and Blue Grosbeaks in the family Cardinalidae are only distantly related to Pine and Evening Grosbeaks in the family Fringillidae, which includes a variety of other red and yellow "true finches."

Until recently, Western and Summer Tanagers were included in the family Thraupidae along with about 240 neotropical species of tanagers, euphonias, flower-piercers, honey-creepers—a stunning array of species that rank among the world's most colorful birds. However, recent genetic studies showed that the four migratory tanagers to North America (Western, Summer, Hepatic, and Scarlet; all members of the genus *Piranga*) are actually more closely related to grosbeaks and buntings than they are to most of the spectacular tropical tanagers.

Most female Cardinalids build flimsy, cup nests of sticks and strips of bark, line them with fine grasses or hair, and bind them to supporting branches high in tree canopies. They usually lay three to five light greenish or bluish eggs, covered with fine brownish or purplish spots and streaks. They incubate the eggs alone for about 12 to 14 days. Both parents feed the young for 10 to 12 days until they can fly, and continue to feed them for about 20 to 25 days until the young learn to forage for themselves. This family is represented by more than 40 migratory species, all restricted to the Western Hemisphere. The family is named is from L. *cardinalis*, important, or "one in charge"—in reference to the robes of Catholic cardinals and the red plumage of Northern Cardinals *(Cardinalis cardinalis)*, the "type species" of this family.

Male

Female

Summer Tanager · *Piranga rubra*

ORIGIN OF NAMES "Summer" for the species' time of residence in the United States; "tanager" from *tangara*, "a kind of bird" in Tupi, a South American Indian language; *piranga*, possibly another Tupi name for this bird; L. *rubra*, red.

NATURAL HISTORY Male Summer Tanagers are the only all-red birds that regularly occur in the Sierra. A few other species—such as Vermilion Flycatchers, Red Crossbills, Pine Grosbeaks, *Carpodacus* finches (i.e., House, Purple, and Cassin's)—are mostly red, but unlike male Summer Tanagers, all of them have at least some gray or brown plumage on their backs, wings, or tails.

The Summer Tanager's historical range in California was confined to the vast riparian forests that once lined the lower Colorado River. Changes in water management and clearing of native vegetation along this critical habitat corridor eliminated mature cottonwoods and willows along most of the floodplain. Summer Tanager populations declined dramatically, and the few remaining pairs were forced to explore new breeding habitats to the north and west. They were first detected as a breeding species at the South Fork Kern River in 1977. This extensive, riparian jungle also provides breeding habitat for a host of other declining birds (such as Yellow-billed Cuckoos,

Southwestern Willow Flycatchers, and Brown-crested Flycatchers) that occur regularly nowhere else in the Sierra.

In the western United States, breeding Summer Tanagers require mature willows and Fremont cottonwoods at least 30 feet tall, and continuous stands of about 25 acres are preferred. They search tree canopies in a slow methodical fashion for cicadas, grasshoppers, bees, and other large insects, occasionally sallying out to catch them in midair. Males arrive on the breeding grounds in the South Fork Kern River Valley slightly ahead of females, usually by the last week of April, and set up and defend territories from other competing males with loud frequent songs—somewhat reminiscent of American Robin songs but hoarser.

STATUS AND DISTRIBUTION "Western" Summer Tanagers *(P. r. cooperi)* reach the westernmost extent of their breeding range in the extreme southern Sierra, where they are almost entirely confined to large tracts of riparian forest. They breed from mid-May until mid-July and depart by the end of September (and rarely into the first week of October) for wintering grounds from Central Mexico to Central America.

West Side. Uncommon breeders in the South Fork Kern River Valley, where about 30 to 40 pairs have nested annually since 1985, primarily in tall, extensive cottonwood-willow riparian forests; one male, possibly of the "Eastern" subspecies *(P. r. rubra),* returned to the same location near Colfax, above the North Fork American River, Placer County, for five successive years (2006 to 2010). This bird mated with a female Western Tanager in 2007, and they fledged young, presumably the first documented example of hybridization between these two species (Pandolfino et al. 2010).

East Side. Casual, single pairs recorded breeding as far north as Lone Pine, Inyo County; spring and fall migrants observed north to the Mono Basin, Mono County.

TRENDS AND CONSERVATION STATUS Recent studies suggest that there are probably fewer than 100 Summer Tanager breeding pairs in the entire state. These small, isolated populations have declined due to removal of mature riparian vegetation from former nesting strongholds such as the lower Colorado River. Dams, water diversions, fires, and direct clearing of willows and cottonwoods are the greatest continuing threats to the state's population. Due to the small and dwindling population of this species in the state, California included Summer Tanagers as a California Bird Species of Special Concern in 2008.

Western Tanager • *Piranga ludoviciana*

ORIGIN OF NAMES "Western" for the species' distribution; L. *ludoviciana,* for the Louisiana Purchase, which included most of the western United States.

NATURAL HISTORY Western Tanager is too drab a name for these bursts of sunshine from the tropics. Males in their crimson, gold, and black raiment adorn Sierra forests during spring and summer. There are few Sierra birding experiences that rival the view of a male singing from a treetop in full sunlight. Five or more may flock together during spring migration, occasionally mingling with other long-distance

Male

Female

migrants such as flycatchers, vireos, and warblers. Birders who learn their unique *pit-u-dik* call notes will find them surprisingly common in migration. Their montane breeding areas include tall trees, open forests, and edges of mountain meadows that provide lofty song perches and ample airspace for hawking flying insects. Tanagers also search for large insects by moving slowly and deliberately along sprays of leaves. They usually keep to high, dense foliage, perhaps hiding from *Accipiter* hawks and other predators that might spot their bright yellow and red plumage. Later in the summer, they glean fruits and berries from ripening shrubs.

Female Western Tanagers build cup nests from conifer twigs and mosses on horizontal branches from 5 to 50 feet above the ground. Male tanagers sing during most of the nesting season, sounding something like American Robins, but the tanager's notes are noticeably hoarser, aptly described as sounding like "a robin with a sore throat." Black-headed Grosbeak songs are also similar, but the tanager's is much less musical and varied. After the nesting season, males begin to lose their bright red head feathers, but their wings remain black. By October these glorious birds have departed for warmer tropical latitudes, and their vibrant hues are replaced by the muted fall colors of black oaks, aspens, and cottonwoods.

STATUS AND DISTRIBUTION In the Sierra, Western Tanagers prefer mature, open stands of conifer forests or the edges of denser forests, but they also inhabit deciduous trees along streams and lake shores as well as clumps of black oak growing with conifers. They begin to filter into lowland riparian forests, foothill woodlands, and chaparral of the Sierra by late April. By mid-May, Western Tanagers move upslope into coniferous forests and begin nesting, where they remain until mid-September; a few may wander up to the Subalpine zone until early October. While some occasionally winter in southern California, most fly farther south to the oak and pine forests of southern Mexico and east as far as Costa Rica.

West Side. Fairly common in spring and fall migration in riparian woodlands and oaks of the Foothill zone; common breeders in ponderosa pines, incense cedars, and black oaks of the Lower Conifer zone; uncommon breeders amid the red firs of the Upper Conifer zone but avoid dense stands of lodgepole pines; rare up to tree line in late August until mid-September.

East Side. Fairly common breeders in Jeffrey pines, white and red firs, and western white pine, as well as in riparian habitats; generally not observed in stands of pinyon pine and juniper, except in spring and fall migration.

Male

Female

Rose-breasted Grosbeak • *Pheucticus ludovicianus*

ORIGIN OF NAMES "Rose-breasted" for the male's bright breast spot; "grosbeak" from F. *grosbec*, thick billed; *pheucticus*, probably from Gr. *phycticos*, painted with colors, a reference to the male's rosy breast spot and underwing linings; L. *ludoviciana*, of Louisiana, where the species was first collected.

NATURAL HISTORY Aside from the northern Rocky Mountains where their breeding ranges overlap, Rose-breasted Grosbeaks inhabit eastern deciduous forests in the breeding season, while

Black-headed Grosbeaks (see account below) mostly occur in oak woodlands, conifer forests, and riparian habitats of the western states. While the males of these two species differ dramatically in appearance, females are remarkably similar. Both species share many aspects of their natural history, but Rose-breasts do not breed in the Sierra.

STATUS AND DISTRIBUTION Rose-breasted Grosbeaks breed across Canada and the eastern United States and are uncommon spring and fall migrants in the western states, with multiple records every year in the Sierra. Similar to Black-headed Grosbeaks, Rose-breasts are attracted to feeders where many observations are made. Records are well distributed throughout lower elevations on both sides of the Sierra, and hybridization with Black-headed Grosbeaks has occurred.

West Side. Uncommon, recorded at Butterbredt Spring almost annually in spring and fall migration; otherwise, rare migrants with late May and June records for most counties; most migrant records from below about 3,000 feet in the central Sierra.

East Side. Rare spring migrants with most records in late May or early June from riparian habitats or bird feeders; casual in fall.

Black-headed Grosbeak • *Pheucticus melanocephalus*

ORIGIN OF NAMES "Black-headed" to describe breeding males; *melanocephalus*, black-headed, from Gr. *melas*, black, and *kaphle*, head.

NATURAL HISTORY Among the earliest sounds of spring mornings in the Sierra are the loud, beautifully complex and cheerful warbles of Black-headed Grosbeaks. These rapid songs resemble those of American Robins but have a richer and more varied quality, full of clear trills and chirps—described by some as sounding like "a robin that has taken singing lessons." Male grosbeaks often burst into song while fluttering out from high

Male

Female

perches, displaying their vivid plumage and lemon-colored underwings. Black-headed Grosbeaks sing from prominent perches usually within 100 feet of their nests. Both males and females sing but have different songs and are known to sing from the nest while incubating. Though rarely described, these grosbeaks sometimes engage in mimicry and can produce a remarkable variety of sounds.

These conspicuous birds favor broad-leaved trees such as oaks, cottonwoods, and willows, especially near water, but also live in dry, open conifer forests with at least some shrubby undergrowth. They thrive near human habitations and commonly forage at bird feeders and in orchards and campgrounds, where they snatch crumbs from picnic areas along with other scavengers such as Steller's Jays. These grosbeaks also relish fruits, berries, buds, insects, and seeds, which they crack easily with their massive bills. When courting females, males fly with their wings and tails spread. Females build nests among the dense foliage on outer branches of tall broad-leaved trees or shrubs, up to 35 feet above ground, usually near streams. They occasionally build in dense shrubs near the ground and may use blackberry thickets for nest sites.

STATUS AND DISTRIBUTION Black-headed Grosbeaks return to lower-elevation oak and riparian forests of the western Sierra from wintering grounds in southern Mexico by early to mid-April, where they often frequent bird feeders. Males arrive before the females to establish their nesting territories, and most nesting occurs from mid-May until late July in oak woodlands of the Foothill zone up to the firs, pines, and black oaks of the Upper Conifer zone. They rarely venture upslope above their normal nesting range, and by early September they have departed for tropical wintering grounds in southern Mexico and Central America.

West Side. Common breeders in most riparian, oak, and conifer woodlands of the Foothill and Lower Conifer zones; uncommon in the Upper Conifer zone and rare up to the Subalpine zone.

East Side. Uncommon breeders mostly in aspen groves and riparian areas, avoiding pinyon-juniper woodlands; rare up to tree line through late August.

Male

Female

Blue Grosbeak • *Passerina caerulea*

ORIGIN OF NAMES "Blue" for color of breeding males; L. *passerina*, a sparrow; L. *caerulea*; cerulean, a shade of blue.

NATURAL HISTORY Once classified as the only members of the genus *Guiraca*, Blue Grosbeaks were recently lumped into the genus *Passerina* along with Lazuli and Indigo Buntings based on similarities of genetics, songs, plumages, and behavior. Blue Grosbeaks inhabit open riparian edge habitats with dense, low shrubs and grasslands near sloughs, irrigation ditches, roadsides, and small streams. They forage for insects, spiders, seeds, and wild fruits in low brambles and trees but also feed on the open ground. Males proclaim nesting territories by singing from the highest branches, often from dead branches or leafless twigs. Females construct compact cup nests from twigs, rootlets, dead leaves, and almost anything else they find near their nest sites including bits of plastic or cellophane. Preferred nest sites include naturally armored plants such as thistles, nettles, or Himalayan blackberries, but they will also use low tree branches and shrubs.

STATUS AND DISTRIBUTION Although Blue Grosbeaks are fairly common breeders in much of the Central Valley, they very rarely breed in the West Side foothills and are uncommon, localized breeders in just a few areas of the Sierra. They arrive on their breeding territories in early May and begin fall migration by the end of August. Most Sierra breeders probably winter in western Mexico.

West Side. Uncommon but regular, annual breeders only in the South Fork Kern River Valley; rare to casual above 500 feet in the Foothill zone with only a couple of probable breeding records (e.g., Red Hills, Tuolumne County; Hensley Lake, Madera County).

East Side. Uncommon, localized breeders in scrubby riparian habitats along the upper Owens River in Inyo County, north to southern edge of Mono County; accidental farther north.

Lazuli Bunting • *Passerina amoena*

Male

Female

ORIGIN OF NAMES "Lazuli" from L. *azulus,* blue stones, for the male's brilliant turquoise plumage; "bunting" derived from the names of several British birds, now applied to a variety of sparrow- and finchlike North American birds; L. *amoena,* dressy or pretty.

NATURAL HISTORY The exquisite turquoise plumage of the male Lazuli Bunting equals in splendor the semiprecious stone lapis lazuli. It is a shade of blue possessed by no other breeding bird north of Mexico and must be seen in the field to be fully appreciated. Male buntings sing an intricate, complex combination of syllables that is difficult to describe. While individual birds tend to sing just one type of song, there is tremendous variation between individuals. The rapid mix of slurred, musical notes is often difficult to differentiate from Rufous-crowned Sparrow song, a bird that sometimes shares lower-elevation breeding habitats with this bunting. Perhaps the best handle is that the notes of the bunting tend to be paired and the song lacks the stumbling cadence of the sparrow's song. Lazuli Bunting males will sing persistently throughout the day from the tops of bushes and trees while the dull, brownish females skulk secretively amid the tangled undergrowth below. In spite of the male's bright colors and tendency to sing from exposed perches, they can be frustratingly hard to see, but the reward is well worth the effort.

Throughout their range in the Sierra, these buntings favor shrubby riparian growths, usually with at least a few taller cottonwoods, sycamores, or oaks to serve as song posts. They also occur in foothill chaparral including chamise as long as some source of moisture is available. Even in the high country, they frequent low, dense riparian shrubs such as alders and willows. Areas of forest or chaparral that have recently burned are particularly favored habitats and the sight of this turquoise bunting singing from a blackened branch is a memorable image.

In habitats especially rich in insects or seeds, males holding choice territories may attract several mates. Unsuccessful males, often first-year birds, lurk in less productive peripheral areas. Females build cup nests from fine materials like twigs, grasses, and rootlets and mold them together with spider webs. Nests are built in low brush, grasses, or vines above dry ground but usually near water.

STATUS AND DISTRIBUTION Lazuli Buntings arrive in the Sierra foothills by mid-April but do not invade the higher mountains until mid-May, and most have left for southern wintering grounds in western and central Mexico by mid-September. Lazuli Buntings are sometimes paired with Indigo Buntings, and hybrids between the two are becoming more frequent and widespread (see Indigo Bunting account below).

West Side. Fairly common in the foothill chaparral and riparian areas up to about 4,000 feet in the central Sierra; uncommon in riparian areas, shrublands, recently burned and second growth forests, and meadow edges to the Upper Conifer zone; rare up to Subalpine zone meadow edges after breeding.

East Side. Fairly common summer residents and breeders in riparian areas, aspen groves, and moist chaparral up to about 8,000 feet; uncommon up to tree line after breeding.

Male

Female

Indigo Bunting • *Passerina cyanea*

ORIGIN OF NAMES "Indigo" from L. *indicum,* India, the country where dark blue dyes originated; *cyanea,* L. blue.

NATURAL HISTORY See Lazuli Bunting account above. Breeding behavior and habitat selection is similar to Lazuli Bunting, except that Indigos may be less wedded to open habitats. Indigo songs are nearly identical to those of Lazuli Buntings, but the notes tend to be even more distinctly paired. In areas where breeding ranges overlap these two species readily hybridize and, with subsequent backcrosses, the lovely turquoise and chestnut shades of the Lazuli lineage generally disappear.

STATUS AND DISTRIBUTION Indigo Buntings, the Lazuli's all-blue cousins, historically nested almost entirely in eastern North America. Possibly as early as 1940, they began pushing westward and their numbers have increased in recent decades.

West Side. Uncommon breeders in the Kern River Valley, where a few nesting pairs are found annually at the South Fork Wildlife Area (especially near Patterson Lane) and Kern River Preserve (e.g., along the canal just south of the headquarters); Indigo × Lazuli hybrids also have been observed there and many other places in California and the Sierra; so far, this vast riparian area is their only regular breeding outpost in the Sierra; rare to casual, mainly in spring, elsewhere on the West Side.

East Side. Rare or casual spring and fall migrants, primarily observed in aspens, cottonwoods, and other deciduous trees with most records from May, June, and September; Indigos have paired with Lazulis on several occasions and, as elsewhere in the West, such observations are increasing.

BLACKBIRDS AND RELATIVES • Family Icteridae

Members of this family generally have long, pointed bills, and males often flaunt bright orange, yellow, or iridescent black plumage; females are typically drab in color. Many species are highly gregarious, forming large flocks after the breeding season and through the winter. Mixed flocks of blackbirds and cowbirds often travel and feed together. Sierra members of this family are most common in the foothills and are mostly uncommon or rare above the middle elevations.

Female Icterids build bulky cup nests or deep pendant bags suspended from trees, shrubs, or wetland plants, where they usually lay three to five whitish, greenish-blue, or buff eggs, scrawled with dark maroon or blackish markings. The females alone incubate the eggs for 12 to 14 days. Both parents provision the young in the nest for about 10 to 12 days and tend them for about 2 weeks until they can fend for themselves. Brown-headed Cowbirds (see account below) are brood parasites and lay all of their eggs in other bird's nests. This family is confined to the New World and includes about 100 species of blackbirds, orioles, meadowlarks, grackles, cowbirds, oropendolas, and caciques; the majority living in subtropical or tropical regions of Mexico, Central, and South America. The family name was derived from Gr. *ikteros,* jaundiced or yellow.

Red-winged Blackbird · *Agelaius phoeniceus*

Bicolored
Male

Female

ORIGIN OF NAMES "Red-winged" for the conspicuous red "epaulets" of breeding males; "black-bird" for the overall coloration of males; *agelaius*, gregarious or flocking, from Gr. *agele*, a herd; L. *phoeniceus*, deep red.

NATURAL HISTORY In marshes choked with tules and cattails, male Red-winged Blackbirds voice their throaty *oak-a-lee* songs. Perched on swaying stems, they puff out their glossy black wings, prominently displaying scarlet shoulder patches. These bright "epaulets" and the frequently repeated, gurgly songs signal neighboring Red-wings and other intruders that nesting territories will be defended vigorously. Unlike their "sister" species, Tricolored Blackbirds (see account below), which nest in dense colonies, male Red-wings attempt to stake out territories of a half-acre or so. Through sheer force of numbers, settling Tricolor colonies will force Red-wings from established territories to the periphery of marshes or to seek alternate breeding sites. The two species undoubt-edly have a shared ancestry, but the Tricolors were geographically isolated during the Pleistocene glacial periods when they evolved into two distinct species.

Adult males arrive first on the nesting grounds, often by mid-March, and begin to defend small territories with songs and chases. A few days later when the streaky females arrive, most bound-ary disputes will have been resolved and the entire marsh divided into individual male territories. Females inspect the displaying males, and the quality of their real estate, before choosing their nest sites. Successful males possessing particularly productive territories may attract harems of up to five females. Females, sometimes assisted by males, construct cup nests and fasten them to erect marsh plants, low tufts of grass, dense shrubs, or small trees, usually near water and often directly above it. Some pairs may attempt second broods if time permits, but most have completed nesting by mid-July, even in the high country.

In winter, huge Red-winged Blackbird flocks roam the Central Valley and Sierra foothills, often in association with European Starlings, Brown-headed Cowbirds, and Brewer's and Tricolored Blackbirds. They prefer to forage in moist habitats such as flooded fields and marshlands, where they probe the soil and glean low vegetation for weed seeds, waste grain, and insects.

STATUS AND DISTRIBUTION Among the most abundant and widespread birds in North America, approximately 26 subspecies of Red-winged Blackbirds breed from southern Alaska to Central America as well as on some Caribbean Islands. Northern breeders, including many different subspecies, migrate south for winter to join the resident Red-wings at lower latitudes. "Nevada" Red-wings *(A. p. nevadensis)* and most other subspecies have yellow margins to their red epaulets. They breed at middle and high elevations of the Sierra and out into the Great Basin and mostly winter in the Central Valley and Sierra foothills. "Bicolored" Red-wings *(A. p. californicus)* have entirely red epaulets and reside year-round in the Central Valley and lower foothills of the Sierra. Grinnell and Miller (1944) described the distribution of the highly localized "Kern" Red-wings *(A. p. aciculatus)*: "In nesting season, restricted area in the mountain valleys of east-central Kern County," including the Kern River Valley, where they also spend the winter.

West Side. Common to locally abundant in marshy habitats around ponds, lakes, streams, roadside ditches, and agricultural fields of the Foothill zone from March through August, with smaller numbers remaining through winter; small colonies breed in moist mountain meadows and other wetlands up to the Subalpine zone, such as Tuolumne Meadows in Yosemite National Park, but retreat to the lowlands in winter.

East Side. Common to locally abundant in habitats similar to those used on the West Side during the breeding season from early April through July; winter status varies from common near Honey Lake, Reno south into Carson Valley, and the Owens Valley, to uncommon or rare elsewhere.

Male

Female

Tricolored Blackbird • *Agelaius tricolor*

ORIGIN OF NAMES "Tricolored" for the black, red, and white plumage of breeding males; L. *tricolor*, three colors.

NATURAL HISTORY With the extinction of the Passenger Pigeon *(Ectopistes migratorius)* in 1914, Tricolored Blackbirds have the distinction of forming the largest breeding colonies of any North American landbird. As described by Ralph Hoffman (1927), teeming marshes "issue a medley of droning and braying sounds and endless lines of blackbirds fly out in all directions to the neighboring fields and fly back with food for their young."

Standing near, or within, an active colony the *keer-aaah* and *keer-a-rooow* calls of thousands of breeding males can be almost deafening. Tricolors nest in compact colonies rather than spreading out like their relatives, the Red-wings. Individual nests might be spaced within a foot of each other, stacked from above and below. A nesting marsh must be large enough to support the minimum colony size of about 50 pairs; some colonies are immense, numbering tens or hundreds of thousands (at least historically). For example, pioneering Tricolored Blackbird biologist Johnson Neff (1937) described a breeding colony in Glenn County in 1934 "that covered about 60 acres . . . even at one to ten square feet (per nest), the nests in this marsh would number about 260,000." Assuming about 1.5 females per male, this colony was comprised of almost 400,000 breeding birds!

Colonial nesting provides protection from predators through mutual warning and defense. Their basic requirements for selecting breeding sites are open accessible water, a protected nesting substrate (either flooded cattails or tules, Himalayan blackberries, or other thorny vegetation), and a suitable foraging space providing abundant insect prey within a few miles of the nesting colony. Breeding usually extends from mid-March through early August; autumnal breeding (September through November) has been documented at several sites in the Central Valley and Marin County, but most of those colonies were not successful.

These gregarious birds live in flocks all year long, and wintering Tricolored Blackbirds often congregate in huge, mixed-species blackbird flocks that forage in grasslands and agricultural fields and at dairies and feedlots. In February, however, they segregate into pure Tricolored Blackbird flocks that roam across the landscape in search of suitable nesting habitats with an abundant insect source nearby. Instead of feeding near their nests like most other blackbirds, Tricolors search for insects and seeds in open farmlands, flooded fields, and open pond margins, sometimes several miles away.

STATUS AND DISTRIBUTION Except for small nesting colonies found locally in Oregon, Washington, Nevada, and northern-coastal Baja California, Tricolored Blackbirds are endemic to California. Within their restricted range, they make extensive migrations and movements, both in the breeding season and in winter. During the breeding season, Tricolors often exhibit "itinerant breeding," when individuals move north after first nesting efforts (March to April) in the San Joaquin Valley to new breeding locations in the Sacramento Valley, low Sierra foothills, and northeastern California (Hamilton 1998).

West Side. Locally abundant only near active breeding colonies; until recent years, colonies of several thousand pairs bred most years in the Kern River Valley at the South Fork Wildlife Area (when flooded), Kern River Preserve (often at Prince's Pond), or in Himalayan blackberry clumps at the Canebrake Ecological Reserve; a small colony has nested near Cherokee in recent years, and foraging flocks can often be seen nearby at Table Mountain (Butte County) in April and May; small colonies of a few hundred to 1,500 pairs are found in Calaveras and Mariposa Counties up to about 1,000 feet; otherwise, most known breeding colonies are below 500 feet; uncommon in winter at higher elevations, but foraging flocks have been found in the past several years during the Grass Valley Christmas Bird Count (Nevada County) at about 2,600 feet and a few birds in mixed blackbird flocks in American and Indian Valleys (Plumas County).

East Side. Generally casual; small nesting colonies have been recorded near Honey Lake and in the Carson Valley; otherwise absent from the East Side as a breeder; most East Side reports are probably misidentifications, but there are occasional well-documented observations of single birds or small groups, mostly in April and May.

TRENDS AND CONSERVATION STATUS A statewide survey in 2011 estimated the Tricolored Blackbird population at 260,000 individuals, about 34 percent lower than a similar survey performed in 2008. The majority of the population still breeds in colonies of tens of thousands, but the number of such colonies is now small. The greatest threats to this species are direct habitat loss and degradation from human activities. Many native habitats that once supported nesting and foraging Tricolored Blackbirds have been replaced by urbanization and agricultural croplands unsuited to their needs. Entire colonies (up to tens of thousands of nests) in cereal crops and silage in the San Joaquin Valley have been destroyed by harvesting and plowing of agricultural lands.

Over the last 15 years, public agencies and Audubon California have used public funds to buy silage crops from landowners to delay harvest so that Tricolored Blackbirds could finish nesting. At freshwater marshes entire colonies have been lost to predation by Black-crowned Night-Herons,

Cattle Egrets, and Common Ravens. Incidence of such predation may be higher than historically, as declines in nesting habitat concentrate nesting colonies in fewer, regularly used areas. Established breeding colonies are also destroyed by fluctuating water levels at reservoirs and marshes, either by flooding active nests or by drying up and creating access for coyotes, raccoons, and other mammalian predators. Food shortages have also been implicated in nestling starvation and the abandonment of entire colonies. Due to their declining population and ongoing threats, Tricolored Blackbirds were included as a California Bird Species of Special Concern in 2008.

Western Meadowlark • *Sturnella neglecta*

ORIGIN OF NAMES "Western" for the species' distribution compared to the Eastern Meadowlark *(Sturnella magna)*; "meadowlark," descriptive of the species' habitat; *sturnella*, from L. *sturnus*, a starling; L. *neglecta*, neglected, a name given by John James Audubon since early ornithologists originally lumped this species with the Eastern Meadowlark *(S. magna)* based on museum specimens without having the benefit of hearing both of them sing.

NATURAL HISTORY The preeminent voice of western grasslands, male Western Meadowlarks broadcast their delightful, gurgling, flute-like songs from tall weeds, fences, or treetops. Meadowlarks generally stop singing by late summer but, thankfully, begin again in the crisp days of fall—as though the bird had "caught the music of a May day, just at its prime, in a crystal vase, and was now pouring out the imprisoned song" (Dawson 1923).

In the Sierra foothills they breed from mid-April through at least mid-June. Unlike most female Icterids, which wear subdued plumages for the sake of nest camouflage, the female Western Meadowlark is as vividly dressed as her mate. Her bright golden breast disappears when she sits on the nest, however, and her streaked brown back renders her almost invisible. As a further precaution against predators, females conceal their nests under domes of dried grasses, and approach them circuitously through tunnels of weeds. Meadowlarks' coloration also camouflages them when foraging amid dirt clods, tussocks of grass, and low shrubs. Like most blackbirds, they pick adult and larval insects and waste grains from the ground.

In addition to Western Meadowlarks, several other open country birds—including American Pipits, Horned Larks, and Vesper Sparrows—have white-edged tails that stand out when they flush from the ground, suggesting this may serve as a generalized alarm signal among these flocking, ground-dwellers.

STATUS AND DISTRIBUTION Throughout their range in western North America, Western Meadowlarks are familiar and cherished for their enchanting songs and striking appearance. Six states have recognized them as their "state bird." Only the Northern Mockingbird (another esteemed singer), with five states, is close. They are year-round residents of annual grasslands, oak savannas, and open shrublands of the Sierra foothills, where they prefer sites with relatively tall grasses and exposed song and lookout perches.

West Side. Locally common from fall through early spring in the Foothill zone; uncommon in the breeding season in scattered, open habitats up to the Lower Conifer zone; small flocks of post-breeding birds sometimes travel upslope to mountain meadows, exposed ridgelines, and Alpine fell-fields in late summer and fall; they typically winter below the snow zone, but one bird was flushed from a juniper tree near the North Fork American River (6,000 feet) in January 1978 following one of the heaviest snow storms on record.

East Side. Fairly common to locally common breeders in open stands of big sagebrush, rabbitbrush, and greasewood as well as in pastures and other agricultural areas; uncommon above about 7,000 feet; fairly common in winter in Honey Lake area, elsewhere uncommon to absent depending on snow cover; a pulse of migrants often seen late September through October.

Yellow-headed Blackbird •
Xanthocephalus xanthocephalus

ORIGIN OF NAMES "Yellow-headed" for the striking field mark of males; *xanthocephalus*, yellow-headed, from Gr. *xanthos*, yellow, and *kephale*, a head.

NATURAL HISTORY Called "banana-heads" by some birders, the loud *kuk-kuk-kow-wheeee* songs of male Yellow-headed Blackbirds resound across western marshlands, sounding as if the poor birds are being strangled as they try to sing. Jaramillo and Burke (1999) noted: "Under no circumstances could one call its song beautiful, but it is distinctive." The more secretive females are mainly brown with dull yellow throats and breasts, and they give only muted chatters.

Male

Female

Yellow-headed Blackbirds breed in loose colonies, often sharing the same deep, cattail, and tule marshes with Red-winged Blackbirds. During the breeding season, males are extremely territorial and spend most of their time displaying from the tops of marsh plants or chasing off other territorial males or intruders. Polygamists, they guard harems of up to six females, and breeding territories are grouped together in the most favorable marshes. In the Sierra they usually nest from mid-April through July. Working alone, females attach cup nests to erect cattails, usually over standing water. When breeding, they take aquatic insects from the water's surface and mud or pick them from marsh plants. Adult and larval damselflies and dragonflies are especially preferred. Outside the nesting period, Yellow-headed Blackbirds forage for waste grain in agricultural fields, rolling across the landscape, often in company with other blackbirds.

STATUS AND DISTRIBUTION Yellow-headed Blackbirds are abundant and widespread breeders in deep marshes throughout western North America but are more localized and patchily distributed in the southwestern part of their range, including California. They arrive in the Sierra in mid-April, and most depart for wintering grounds in Mexico by mid-September.

West Side. Rare to uncommon in spring in the Foothill zone; regular breeding only at Lake Almanor and in the Kern River Valley (but not in all years at either location); most Foothill zone observations are of one or a few males in spring, likely "prospecting" for potential breeding sites; casual at higher elevations, with scattered fall records up to the Subalpine zone, as at Tuolumne Meadows in Yosemite National Park; casual in winter.

East Side. Common to locally abundant only at breeding colonies such as at Marble Hot Springs in Sierra Valley, June Lake, Mono Basin (DeChambeau Ponds and Simons Spring), Bridgeport Reservoir, Carson Valley, Owens Lake and other Owens Valley marshes; intermittent breeders at the old gravel ponds near Truckee (Nevada County), south Lake Tahoe, and Honey Lake; casual in winter.

TRENDS AND CONSERVATION STATUS Similar to Tricolored Blackbirds, Yellow-headed Blackbirds have declined with the loss of wetlands to development, flood control, and water diversions for irrigation of agricultural crops. Both species are adversely affected by sudden increases in water levels that flood their nests, and water draw-downs that make it easy for coyotes and other predators to find their nests. Pesticide applications in agricultural areas have also harmed their breeding efforts. The rapid expansion of Great-tailed Grackles (see the "Recent Trends in Sierra Bird Populations and Ranges" chapter) may also add an aggressive competitor for nesting sites to the list of threats to this species. Yellow-headed Blackbirds were included as a California Bird Species of Special Concern in 2008.

Male

Female

Brewer's Blackbird • *Euphagus cyanocephalus*

ORIGIN OF NAMES "Brewer's" for Thomas Mayo Brewer (see Brewer's Sparrow account); *euphagus*, a good eater, from Gr. *eu*, well, and *phago*, to eat; *cyanocepalus*, blue-headed, from Gr. *kyanos*, blue, and *kephale*, a head.

NATURAL HISTORY Even non-birdwatching visitors to the Sierra may notice these glossy, bright-eyed birds strutting about, picking crumbs from parking lots, picnic tables, and campsites. While native to North America, they often associate with introduced European Starlings and House Sparrows and occupy in the same, human-modified habitats: residential areas, shopping centers, cemeteries, golf courses, and mowed urban parks. Brewer's Blackbirds can make a living without the aid of human leavings and can be found in pristine Sierra meadows. However, they tend to concentrate around developed areas and near farms and corrals where livestock are fed. Besides eating waste grain and scraps, they probe the ground with their sharp bills and turn over rocks looking for insect larvae and seeds.

Visitors to urban parks with planted conifers and other densely foliaged ornamental landscaping might be attacked by Brewer's Blackbirds defending their nests and giving harsh *tshupp-shupp* and *chuck-chuck* calls. Pairs may breed alone or in loose colonies of four or five up to dozens of nests. Males are usually monogamous and defend small territories around their nests. Depending on the elevation (see "Status and Distribution" below), they begin nesting in the Sierra foothills as early as February but not until May or early June in the high country; most nesting efforts are completed by mid-July. Females construct cup nests from grasses, marsh plants, or discarded paper on the forks of trees or shrubs, always hidden by dense, overhanging foliage. They most often nest in conifers but unusual nests have been found in rock crevices, on stumps along lake margins, within tussocks of grass on the ground, under culverts, and along railroad grades.

Outside the breeding season, Brewer's Blackbirds patrol urban and suburban areas for food or join mixed-species flocks with other blackbirds roaming marshes and agricultural fields in search of productive feeding areas. In winter, some Sierra breeders may join huge communal roosts of Red-winged and/or Tricolored Blackbirds in the Central Valley and Sacramento/San Joaquin Delta.

STATUS AND DISTRIBUTION Widespread across most of western North America, Brewer's Blackbirds are year-round residents of developed and agricultural areas of California, including both sides of the Sierra.

West Side. Common to locally abundant in cities, towns, and farms of the Foothill zone and fairly common up to the Subalpine zone, with small breeding populations near campgrounds, pack stations, and heavily used horse trails up to Tioga Pass in Yosemite National Park; postbreeding records above 10,000 feet at Rae Lakes and Sixty Lakes Basin in Kings Canyon National Park; uncommon in winter above the Lower Conifer zone except in towns and other developed areas where snow is removed regularly.

East Side. Common to locally abundant in the same habitats as the West Side from April through November; common to fairly common in winter in Honey Lake, Sierra Valley, Owens Valley, and Reno south through the Carson Valley; fairly common to absent elsewhere depending on the severity of the winter.

Great-tailed Grackle • *Quiscalus mexicanus*

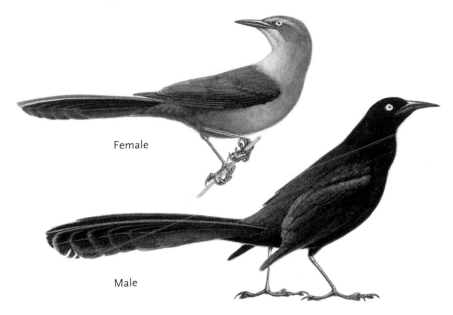

Female

Male

ORIGIN OF NAMES "Great-tailed" for the long, broad tails of males; "grackle" from L. *gracula*, a jackdaw, or possibly imitative of their calls; *quiscalus*, from L. *quiscalis*, a quail; *mexicanus*, of Mexico, where the type specimen was collected.

a cowbird chick while still trying to raise their own offspring. A single female may lay up to 40 eggs per season and be responsible for the loss of countless nests over her lifespan (see accounts for Willow Flycatcher, Bell's Vireo, and Yellow Warbler for examples of cowbird impacts). Cowbirds parasitize the nests of almost any songbirds except the largest species and cavity nesters, and they prefer hosts that feed insects to their young.

Male

Female

Cowbirds often remove a host egg and so fool the unsuspecting foster parents. A female cowbird may return to lay more eggs in the same nest, as over half the parasitized songbird nests reveal more than one egg. Returning parents may faithfully incubate the alien eggs as if nothing were amiss. Some larger species, such as American Robins and Bullock's Orioles, defend against cowbirds by abandoning their parasitized nests, covering a clutch of eggs over with new nest material, or tossing cowbird eggs from nests altogether. For these reasons, more than half the cowbird eggs laid never hatch. Red-winged and Tricolored Blackbirds regularly drive cowbirds away from their nesting areas, but few other birds do this. If the host parents cooperate, cowbird young usually hatch a day or two ahead of their nest mates. Being larger and noisier, they demand, and often receive, a larger share of the food brought by the foster parents. Some or all of the host's own nestlings may miss so much food that they weaken and die.

STATUS AND DISTRIBUTION Although native to North America where they originally followed bison herds across the Great Plains, Brown-headed Cowbirds visited California only rarely in winter prior to 1900. In their early explorations of the Yosemite region, Grinnell and Storer (1924) observed cowbirds only a few times along the lower reaches of the Merced and Tuolumne Rivers on the West Side but reported: "At Mono Lake in the season of 1916 a number of Nevada Cowbirds were obtained." Apparently cowbirds first invaded Yosemite Valley in 1934, and their range in the Sierra has expanded ever since. As their name suggests, cowbirds often associate with cattle, horses, and sheep to feed on waste grain and stirred-up insects on the ground and in manure. The increase in pack stations and livestock grazing has provided food that drew cowbirds into the high country; backyard bird feeders may also be aiding this expansion. However, the most recent Breeding Bird Survey data suggest Sierra populations are now in decline.

West Side. Common in the Foothill zone and fairly common up to the Subalpine zone from April through July, frequenting moist meadow edges and riparian thickets during spring and summer, avoiding unbroken tracts of forest, and commuting miles daily from breeding areas to livestock feeding areas; uncommon in fall through winter in the Foothill zone and rare up to the Lower Conifer zone.

East Side. Fairly common to locally abundant in open, developed habitats and livestock feeding areas and near aspen and cottonwood riparian woodlands up to the Subalpine zone from mid-April through August; uncommon in fall and uncommon to absent in winter with fair numbers sometimes wintering in Honey Lake, Sierra Valley, Carson Valley, and Owens Valley.

Hooded Oriole · *Icterus cucullatus*

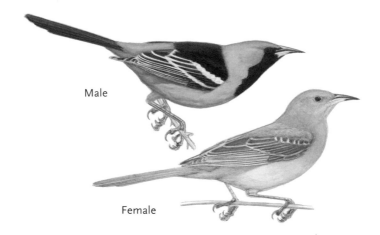

Male

Female

ORIGIN OF NAMES "Hooded" for the bright, orange-yellow "hoods" of males; "oriole" from Fr. *oriol*, a yellow bird; *icterus* (see family account above); L. *cucullatus*, hooded.

NATURAL HISTORY Prior to the arrival of Europeans, Hooded Orioles were restricted to nesting in riparian trees in desert oases of the Southwest and Mexico. However, they have a great fondness for planted palms and have expanded their range in southern and coastal California and more recently into the lower Sierra elevations with the widespread planting of *Washingtonia* and other ornamental palms. Their nests are tightly woven cups suspended from the undersides of a palm frond, leaf, or branch. In the Sierra, breeding occurs from mid-April through mid-June. Although Hooded Orioles in much of their range sing short, variable songs, they rarely sing in northern California and in their small range in the Sierra. A rising *wheet* call or chatter (drier and more wooden-sounding than Bullock's Oriole's chatter) are the vocalizations we most often hear from these birds.

When breeding, Hooded Orioles forage in trees and shrubs for such large insects as beetles, grasshoppers, ants, caterpillars, and spiders. They also supplement their diets with fruit and nectar, and are often attracted to hummingbird feeders. A few may overwinter to take advantage of this free source of food.

STATUS AND DISTRIBUTION Hooded Orioles winter in Baja California and south coastal Mexico and usually arrive in California by mid- to late March and depart by mid-September; in the Sierra they are almost entirely confined to suburban areas and ranches where palms have been planted.

West Side. Uncommon and very localized early April through mid-September in the Foothill zone, where scattered pairs nest up to about 2,500 feet; recent breeding locations include the westernmost edges of Nevada and El Dorado Counties, Auburn (Placer County), and above Lake Don Pedro (Tuolumne County); regular spring migrants and localized breeders in Kern River Valley.

East Side. Uncommon spring (April and May) and rare fall migrants from the Owens Valley south into north Kern County; rare to casual spring migrants further north; may have bred recently near Bishop and possibly elsewhere in the Owens Valley; no winter records on either side of the Sierra.

Bullock's Oriole • *Icterus bullockii*

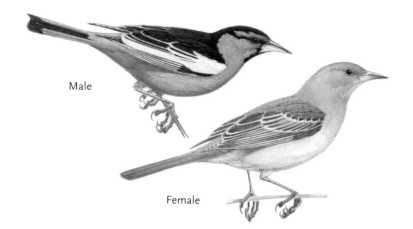

Male

Female

ORIGIN OF NAMES "Bullock's" for William Bullock (1775–1840), a British mine owner and traveler who first collected this species near Mexico City; William Swainson (see Swainson's Hawk account) named it in Bullock's honor; formerly lumped with Baltimore Oriole *(Icterus galbula)* as "Northern" Oriole.

NATURAL HISTORY The vibrant oranges and blacks of male Bullock's Orioles stand out like living embers in the green foliage of Sierra trees. Males voice their loud, melodic songs and utter harsh chatters from the treetops. The yellowish females, plain only by comparison with their flaming mates, filter in several days later. Bullock's Oriole pairs are seasonal monogamists that cooperate in weaving deep, intricate hanging pouches up to 15 inches long. These bags are suspended from high, forking branches of deciduous trees, usually those growing along streams such as sycamores, cottonwoods, willows, and deciduous oaks; less commonly they nest in drier sites and choose live oaks or orchard trees. Nests are woven from plant fibers, primarily bark and fine grass fiber, and animal hair is also commonly used.

During breeding season, they consume mostly larval insects, including caterpillars, as well as fruits (those dark-colored berries being preferred). They also sip nectar from orchard trees and planted Eucalyptus groves and are frequent visitors to hummingbird feeders. Unlike most of their relatives, Bullock's Orioles are not known to flock after breeding.

STATUS AND DISTRIBUTION Widespread breeders in open woodlands of the western United States, male Bullock's Orioles arrive in the Sierra foothills by early April, with females following a week or two later; nearly all have departed for their wintering grounds in coastal Mexico and northern Central America by mid-August.

West Side. Common breeders in open oak woodlands, and riparian forests of the Foothill zone but prefer streamside sites for feeding; uncommon to fairly common in black oak woodlands of the Lower Conifer zone up to about 4,000 feet; uncommon in the Upper Conifer zone up to about 6,000 feet after breeding and during fall migration.

East Side. Fairly common in aspen groves, stands of cottonwoods (or even isolated trees near ranches), and other open, deciduous woodlands; rare up to the Subalpine zone in late summer; accidental in winter on either side.

Male

Female

Scott's Oriole • *Icterus parisorum*

ORIGIN OF NAMES "Scott's" for Winfield Scott (1786–1866), a general in the Mexican-American War who served as emissary for every president from Thomas Jefferson to Abraham Lincoln; one of his lieutenants collected this bird and named it in Scott's honor; while the common name was retained, the scientific name was changed to back to L. *parisorum*, in honor of the Paris brothers (French traders in natural history specimens), for whom Charles Bonaparte (see Bonaparte's Gull account) had previously named this species.

NATURAL HISTORY The brilliant yellow and black plumage of a male Scott's Oriole adds a splash of brightness to the otherwise muted colors of the desert landscapes they inhabit. In spring and summer, they sing clear, melodic phrases persistently from before dawn until dusk. In the southern Sierra, Scott's Orioles breed in a variety of arid habitats including desert canyons, pinyon-juniper forests, and Joshua tree woodlands. In all seasons they feed mostly on insects and insect larvae, and they also eat nectar and fruit, including those of cacti. On their wintering grounds, they are known to feast on dense colonies of roosting Monarch butterflies, which can be toxic to some species.

Scott's Orioles nest in a variety of trees, including yuccas, palms, pinyon pines, and Joshua trees. Nests are hanging pouches made of leaf fibers or grasses and lined with soft plant material, usually placed under the shade and less than 10 feet above the ground. Their nests are oval, somewhat lopsided structures, not as round or pendulous as those of Bullock's Orioles.

STATUS AND DISTRIBUTION Scott's Orioles breed in the arid southwestern United States and northern Mexico and winter in southern Mexico. They arrive in the southern Sierra mid-March to mid-April and depart by late August.

West Side. Fairly common to uncommon breeders in the Desert zone of Kelso Valley, the east end of South Fork Kern River Valley, and west of Walker Pass; may breed some years as far north as southeastern Tulare County; accidental elsewhere with a few records in May and early June.

East Side. Fairly common to uncommon breeders in appropriate habitats from the southern edge of Inyo County (e.g., Nine Mile Canyon) south at least to the Chuckwalla Mountains (Kern County); as on the West Side, accidental elsewhere mainly as a spring vagrant.

FINCHES AND RELATIVES • Family Fringillidae

Bird taxonomists sometimes refer to members of this family as "true finches." These seed-eating birds have coned-shaped bills, with sharp cutting edges angled downward at the base, making them especially adept at holding and cracking seeds. In the Sierra, breeding male Fringillids display bold patterns of red, yellow, or gold. Their natural diets help maintain these bright plumages; captive birds fed artificial foods often molt into drab browns or grays like those of females and juveniles. Finches flock together in undulating flight, often in full song. Unpredictable, they can be abundant in an area one year and rare or absent the next. Many species rove widely in search of bountiful crops of cones, buds, berries, or seeds, and some also eat insects while nesting. Only the Lawrence's Goldfinch leaves the Sierra entirely in winter.

Most Sierra members of this family build nests in trees, shrubs, or hidden by grasses on the ground; House Finches often build nests on or near buildings, on abandoned farm equipment, or other human-associated structures. Goldfinches build neat cup nests, while nests of Pine and Evening Grosbeaks are messy clumps of twigs. Females usually lay four to six pale blue eggs, variably marked with red or purplish spots and splotches. Females alone incubate eggs for 12 to 16 days while the males return with food at frequent intervals. Both parents feed and guard the young until they leave the nest after about two weeks.

More than 200 species are currently recognized in this family, with most representatives residing in the Northern Hemisphere; one subfamily is endemic to the neotropics *(Euphoniinae),* and another includes only the Hawaiian honeycreepers *(Drepanidinae).* Despite having "grosbeak" as part of their common name, Pine and Evening Grosbeaks are only distantly related to Black-headed and Blue Grosbeaks (family Cardinalidae). The family name, Fringillidae, was derived from L. *fringilla,* a finch.

Gray-crowned Rosy-Finch •
Leucosticte tephrocotis

Female
(breeding)

Male
(nonbreeding)

ORIGIN OF NAMES "Gray-crowned" for the species' head color compared to other rosy-finches; "rosy-finch" for the rosy-red wings and underparts of adults; *leucosticte,* variably white, from Gr. *leucos,* white, and *sticte,* variable—a reference to the scaly-looking underparts of most members of this genus; *tephrocotis,* gray-eared, from Gr. *tephros,* gray or ashy, and *otos,* ear.

NATURAL HISTORY Grinnell and Storer (1924) described Gray-crowned Rosy-Finches as "the most typically alpine of all Californian birds. The mountaineer does not meet with it until he reaches the main Sierran crest or at least the loftiest of the outstanding spurs." Windswept, snow-laden peaks are home to Gray-crowned Rosy-Finches. No other Sierra birds venture to such lofty heights with the apparent ease of these Alpine specialists. The chilling winds of fall force most other birds from the mountaintops and leave the hardiest rosy-finches to brave winter alone.

Female Gray-crowned Rosy-Finches secure sturdy nests of woven grasses and mosses beneath boulders or in crevices on rock walls, well out of reach of most ground predators. They incubate clutches of snow-white eggs while their mates return frequently with their squirrel-like cheek pouches filled with food. They feed actively on glacial cirques, snow banks, or in lush patches of grasses, mosses, or heather. In addition to gathering seeds from Alpine plants, they exploit a unique food resource—insect larvae and adults entombed within melting ice or frozen on the surface of snowfields. Most of these insects are blown upslope from more productive habitats below. Rosy-Finches can be quite tame and easily approached but may move nervously and take flight, with buzzy *peent* calls, for no apparent reason. In flight, their wings appear noticeably longer than those of most other birds their size.

STATUS AND DISTRIBUTION Residents of only the highest mountain peaks, usually above 10,000 feet, the Gray-crowned Rosy-Finch's range extends from northern Alaska south to the southern Sierra and Rocky Mountains. The "Sierra Nevada" subspecies *(L. t. dawsoni)* resides above tree line year-round from Sierra County south to the highest peaks of Tulare County. An especially likely place to find them is in the vicinity of Tioga Pass at the eastern edge of Yosemite National Park, along the trails to Gaylor Lakes and Mount Dana, and on the slopes above Saddlebag Lake (Mono County). In the northern Sierra few peaks exceed 9,000 feet, and rosy-finches are rare with isolated populations on some of the higher summits, such as Granite Chief (Placer County) and the Sierra Buttes (Sierra County). Members of the far more migratory "Hepburn" subspecies *(L. t. littoralis)* show more extensive gray on their faces and breed outside the Sierra from Mount Shasta northward, but some winter farther south into the Sierra. The scarcity of winter bird records from the high Sierra means ornithologists do not know for certain where most rosy-finches go during the coldest months.

West Side. Fairly common to uncommon residents of the Alpine zone from Sierra County south to Tulare County, where some probably remain year-round, but during the most severe winter storms some may wander down to the relatively protected Subalpine forests.

East Side. Fairly common but sparsely distributed breeders (and possibly residents) at high-altitude sites from Squaw Valley High Camp (Placer County) south at least to Cottonwood Lakes Basin (Inyo County); winter flocks, sometimes numbering into the hundreds, have been observed in Nevada from Washoe Lake into the Carson Valley, at Conway Summit (Mono County), in Round Valley (Inyo County), and especially near feeders in Aspendell (Inyo County). These winter flocks may include a mix of "Sierra Nevada" and "Hepburn" Gray-crowned Rosy-Finches and even Black Rosy-Finches (see Appendix 2).

Male

Female

Pine Grosbeak • *Pinicola enucleator*

ORIGIN OF NAMES "Pine" for one of the species' preferred foraging and nesting trees; "grosbeak" from OE. *grosbeck*, thick-billed; *pinicola*, pine-dweller, from L. *pinus*, a pine, and *incola*, an inhabitant; L. *enucleator*, a seed-extractor.

NATURAL HISTORY Pine Grosbeak flocks roam widely in search of conifer cones, tender buds, insects, and berries, and their whereabouts from year to year may be hard to predict. They occur in high-elevation forests, sometimes in dense stands, but more often at meadow edges, in forest clearings, or in deciduous trees along mountain streams. Their short, conical bills enable them to easily extract the small seeds from conifer buds, mountain alders, and mountain ashes. In areas where ripe cones abound, they are fairly common at times, yet many experienced birdwatchers have searched such habitats in vain for these elusive birds. Familiarity with their warbly, rolling songs and calls will greatly help to locate them. Once found, Pine Grosbeaks seem quite tame and often allow close approach. Sometimes they feed near mountain cabins and may even venture inside to pick crumbs from floors, tables, and shelves.

In the Sierra, Pine Grosbeaks breed from mid-May through July and only produce one brood per season. They usually select low branches (below 10 feet) in Subalpine trees such as mountain hemlocks, whitebark, and lodgepole pines, and rarely red firs for nest sites. Females build bulky nests from conifer twigs and line them with fine grasses, rootlets, and mosses. Despite their dependence on conifer buds for most of the year, males bring insects to the females and young in the nest, mostly captured on foliage or ground but occasionally by snatching them in midair. Similar to most other Fringillids, Pine Grosbeaks are drawn to mineral springs and salt licks; the areas near Soda Springs in Tuolumne Meadows and the horse corrals there and at White Wolf in Yosemite National Park are especially good places to observe them.

STATUS AND DISTRIBUTION Pine Grosbeaks occur in boreal and subarctic forests across the Northern Hemisphere. Like other boreal forest specialists such as Great Gray Owls and Black-backed Woodpeckers, Pine Grosbeaks reach the southwestern edge of their North American distribution in the Sierra. The "California" Pine Grosbeak subspecies *(P. e. californica)* is endemic to the Sierra and resides in the higher mountains year-round from Plumas County south to Golden Trout Wilderness Area (Tulare County). Unlike some other subspecies, "California" Pine Grosbeaks are not known to be irruptive and may seldom stray far from their breeding territories. However, their winter whereabouts remain largely a mystery since so few birders visit their frozen haunts at that time of year—even when these areas are visited, documented observations of Pine Grosbeaks are exceedingly few. David Gaines (1992) related one example of "hundreds" descending into Yosemite Valley (4,000 feet) in May 1955, but this appears to be an exceptional occurrence.

West Side. Generally uncommon and localized around forested margins of mountain meadows, lakes, and streams of the Subalpine zone and down to red fir forests of the Upper Conifer zone; rare to casual below about 6,000 feet in the central Sierra at any time of year.

East Side. Uncommon and localized from Sierra County south into Mono County; most easily found at the crest near Yuba Pass (Sierra County), along the Castle Peak (Nevada County) and Pacific Crest trails (in Nevada and Placer Counties), near Mount Rose (Washoe County, Nevada), and near Tioga Pass and above Mammoth Lakes (Mono County).

Purple Finch • *Carpodacus purpureus*

ORIGIN OF NAMES "Purple" for the reddish-purple of breeding males; "finch," AS. *finc*, a finch; *carpodacus*, a fruit-biter, from Gr. *karpos*, a fruit, and *dakos*, a biter; L. *purpureus*, purple.

NATURAL HISTORY Clear, liquid, warbly songs of Purple Finches resonate through moist forest glades of the Sierra during spring and summer. Songs of Purple Finches are highly variable and may include copied snippets from vireos, warblers, goldfinches, towhees, or other *Carpodacus* finches, House and Cassin's. First-year male Purple Finches often sing and may even breed while in immature, "female-like"

Male

Female

plumage. First-year males of all three of the *Carpodacus* finches sing while in this same plumage, leading some to assume that females sing. In fact, females of each species do sometimes sing very short, simple songs, but those are distinct from full male songs. The songs of Purple, Cassin's, and House Finches sound similar, but the Purple's is the most consistently melodic of the three and is more rapid than the House and lower pitched than the Cassin's songs. The surest vocal clue to identification is the loud, sharp *pik* or *tick* calls given by Purple Finches either perched or in flight.

Flocks of burgundy-red males and streaked females and juveniles reside year-round in shaded groves, canyon bottoms, and meadow edges. From mid-April through July they nest in open woodlands and along forest-meadow edges. Females build cup nests in dense outer foliage of conifers or deciduous tall trees near a source of water while their mates sing nearby. Even while nesting, these highly social birds tend to feed in flocks. They seek the fresh buds of Sierra junipers, mountain ashes, alders, and willows as well as the fruits of mountain gooseberries, Sierra currants, and California coffeeberries; they are also frequent visitors to orchards and to bird feeders. While breeding, they supplement their diet with adult insects and caterpillars, mostly captured in the foliage of tall trees but occasionally on the ground.

STATUS AND DISTRIBUTION The range of the "California" Purple Finch *(C. p. californicus)* extends from northwestern Canada south to northern Baja California. In the Sierra their range overlaps with that of the House Finch at low altitudes and with Cassin's Finch in higher forests (see various accounts below). Many Purple Finches withdraw from the higher mountains in winter for warmer climates in the foothills and southern California. Grinnell and Miller (1944) reported for Purple Finches: "Altitudinal migration occurs along the slopes of the Sierra Nevada . . . winter visitant populations that scatter over western lowlands from October to April thus are composed of birds from near-by breeding areas mixed with at least a few individuals from far north." Backyard birders unfamiliar with the extreme variation in male House Finches frequently mistake them for Purple Finches when they encounter birds that do not look like the House Finch illustrations in their field guides; care should be taken when reporting low-elevation Purple Finches.

West Side. Common to fairly common in moist forests and edges of mountain meadows of the Lower and Upper Conifer zones but uncommon or rare in the Subalpine zone; fairly common below 3,000 feet in the Foothill zone in some winters where they might be found in forest clearings, chaparral, or oak savanna habitats, but rare or absent in other years when cone crops in the higher mountains provide adequate food supplies.

East Side. Uncommon to rare in winter and fall from Honey Lake south to Alpine County and casual farther south; likely breeding only in the vicinity of Susanville (Lassen County) and in Alpine County near Markleeville and up Highway 4, with most other breeding season reports probably misidentifications.

Cassin's Finch • *Carpodacus cassinii*

ORIGIN OF NAMES "Cassin's" for John Cassin (see Cassin's Vireo account).

NATURAL HISTORY While Cassin's Finches sing and devour conifer buds among the highest branches, they also search for seeds on grassy forest floors, in clearings, and along the edges of mountain meadows. They generally favor somewhat drier sites than Purple Finches, but the two species can sometimes be found breeding in the same mid-elevation forests on the West Side. On the East Side, Cassin's Finches may breed in the same riparian woodlands as House Finches.

Cassin's Finches are usually found in flocks, even while breeding, when they might be seen at mineral deposits along with Red Crossbills or Pine Grosbeaks. In the Sierra, Cassin's Finches breed from May through mid-July. Females build cup nests of small twigs and line them with shredded bark or hair and secure them at the ends of flat sprays of conifer foliage, anywhere from 30 to 60 feet above the ground.

Male

Female

As with Purple Finches, male Cassin's Finches, still wearing streaky brown "female-like" plumage, sing in their first year (see Purple Finch account above), and might be incorrectly assumed to be "singing females." Singing males regularly imitate the songs of other birds demonstrating that this is a learned, not inherited, behavior. Songs of Cassin's resemble those of Purple Finches but are usually longer, higher pitched, and not as melodic; their distinctive *ooh-wee-ooh* calls are diagnostic.

STATUS AND DISTRIBUTION Unlike Purple Finches, which breed across North America, Cassin's Finches are restricted to the western mountains. In the Sierra they occur mostly in higher forests and, like many other Fringillids, their local populations fluctuate widely in abundance from one year to the next. Postbreeding movements of Cassin's Finches in the Sierra are not well understood. Some remain in breeding areas through winter and, depending on weather or food availability, move to lower elevations on both sides of the crest; others may migrate to southern California and Baja California.

West Side. Common breeders in open lodgepole pine forests of the Subalpine zone; fairly common breeders in red fir and mixed conifer forests of the Upper Conifer zone but generally uncommon breeders below about 5,000 feet; uncommon to fairly common winter visitors to the Lower Conifer zone, and rare to uncommon and unpredictable in the Foothill zone, usually in severe winters.

East Side. Common to fairly common breeders in higher-elevation pine forests up to about 10,000 feet and in riparian woodlands down to about 6,000 feet and casual breeders in the Owens Valley; fairly common to uncommon in pinyon-juniper woodlands in winter.

House Finch • *Carpodacus mexicanus*

ORIGIN OF NAMES "House" for the species' fondness for human habitations; *mexicanus*, of Mexico, where the type specimen was collected.

NATURAL HISTORY In the spring and summer, House Finches sing warbling phrases throughout the day from the tops of barnyard trees, suburban houses, and power lines. Although native to western North America, they, like some introduced species, commonly frequent human habitations, urban parks, and farms. They feed on grains, seeds of thistles, dandelions and other weeds fruits, and berries, and they are among the birds that feast on ripening grapes in vineyards. House Finches are often among the most common birds at most backyard seed feeders. Like their other *Carpodacus* cousins, first-year males sing while still in drab immature plumage (see Purple Finch

Male

Female

account above). Compared to Purple and Cassin's Finches, House Finches sing a more variable *sing-song* warble including slurred *shur-ree* phrases and noticeable down-slurred notes at the end.

Their primary requirements are a source of water, shrubs and trees for escape cover, and high song perches. Their vegetarian preference may explain why House Finches do not invade the insect-rich higher mountains in late summer and fall as many insectivorous birds do. The generalist breeding behavior of House Finches probably accounts for their association with humans. They nest almost anywhere, in the eaves and rafters of barns and other buildings, in chimneys, cavities in cliffs, or in trees. They quickly occupy nesting sites abandoned by other species and fortify them with horsehair, straw, or twine. Even while nesting, from April to July, most of them continue to flock, except when incubating eggs or feeding young. Occasionally male House Finches show orange (or even yellow), rather than red, and the intensity of their coloration is due to their diet. Captive birds fed unnatural foods often lose their bright colors after the first molt. The extreme variation in the appearance of male House Finches leads to many misidentifications as birdwatchers try to find a bird in their field guide that matches the one they see.

STATUS AND DISTRIBUTION Native to western North America, a small group of House Finches was released on Long Island in the early 1940s and they have now spread throughout most of the eastern United States. In California they reside mostly in developed, agricultural, and rural landscapes. In the Sierra they are generally the lowland counterparts of the mountain-dwelling Purple and Cassin's Finches, although House Finches are now commonly found at elevations above 6,000 feet in association with human development. Expansion of human settlement and increasing availability of backyard feeders has allowed House Finches to move upslope and remain year-round in areas where they were historically absent or rare.

West Side. Common to locally abundant year-round residents of the Foothill zone; fairly common around towns and ranches up to the Lower Conifer zone; uncommon to rare away from human developments above 3,500 feet at any time of year.

East Side. Common to locally abundant in the same habitats as the West Side from spring through fall; in winter, locally abundant in urban areas like Reno, Carson City, and Bishop, and uncommon to fairly common in other developed areas below 6,000 feet.

Red Crossbill · *Loxia curvirostra*

ORIGIN OF NAMES "Red" for the overall coloration of breeding males; "crossbill" for the species' overlapping mandibles; *loxia* from Gr. *loxos*, crooked; *curvirostra*, curved-billed, from L. *curvus*, curved, and *rostrum*, a bill.

NATURAL HISTORY Few other Sierra birds have adapted so closely to a single food source as Red Crossbills. Their crossed bill tips make perfect tools for extracting seeds from unopened lodgepole pine cones. As they pry cone scales apart, they deftly lift out the seeds with their tongues. In this manner they achieve access to a rich food source that is available only to Clark's Nutcrackers, Douglas squirrels, and a few other animals. In addition to conifer seeds, they feed on fresh buds and insects. Like many Fringillid finches, Red Crossbills rove widely in search of productive food sources and may be abundant in an area one year and rare, or absent, the next.

Male

Female

Throughout their range in the Northern Hemisphere, Red Crossbills have been observed breeding at all times of the year. Their nesting usually synchronizes with the variable availability of conifer cones rather than with season. Lodgepole pine cones, for example, generally ripen in late summer and fall, but some stay closed for years. In the Sierra most crossbill nesting records fall between February and June, before the cone scales open and their seeds disperse. Crossbills typically construct cup nests near the ends of horizontal conifer branches.

Highly gregarious, Red Crossbills often gather in nervous flocks and flit between trees uttering distinctive *kip-kip* calls in flight. Flocks of young and old birds vary greatly in color; adult males are brick red, females are yellow and olive-gray, immature males are intermediate, and young fledglings have streaked breasts.

STATUS AND DISTRIBUTION Red Crossbills are wide-ranging inhabitants of boreal and montane forests from southern Alaska south to the pine forests of Central America. Within this large range they exhibit considerable variation in bill shape and body size, apparently adapted to exploit specific conifer cone types. The taxonomy of Red Crossbills is complex, and currently ten distinct "call type" groups have been described in North America based on distinct calls. Some or all of these groups may represent separate "cryptic" species, which, while nearly identical in appearance, appear to be reproductively isolated from each other. For some call groups, research has shown that even when nesting in close proximity, they do not interbreed. At least four different "call type" groups have been recorded in the Sierra, with Type 2 by far the most common breeder. Type 4 is found with some regularity and also breeds in the Sierra. Call Types 3 and 5 are rarely encountered. An excellent introduction to identification of these calls is available online (Young 2008).

West Side. Nomadic flocks are uncommon to locally common in ponderosa or lodgepole pine–dominated forests of the Subalpine zone and Upper Conifer zones, depending on the annual cone crop; unpredictable but generally rare elsewhere; crossbills reside in the higher mountains year-round but in "irruption" years with poor cone crops at high elevations may descend to lower elevations, where they are usually found in planted conifers.

East Side. Status similar to the West Side, with most observations above 6,000 feet; rare south of Mono County, though wandering birds may appear anywhere, even in desert oases.

Pine Siskin • *Spinus pinus*

ORIGIN OF NAMES "Pine" being one of the species' preferred foraging and nesting trees; possibly from Sw. *siska*, a chirper; *spinus* from Gr. *spinos*, a linnet or finch; L. *pinus*, a pine.

NATURAL HISTORY At rest, Pine Siskins may look somewhat drab and undistinguished, but in flight their wings and tails flash bold golden patches, signaling their close kinship with the brilliant goldfinches. Pine Siskins travel in flocks of a few to more than a hundred birds that often erupt in a chorus of hoarse, wheezy *zwew-eet* calls and long, rising *Zeeeeeee* calls.

Widespread in the western Sierra, Pine Siskins usually live in or near conifer forests. They scour all kinds of trees for fresh buds and seeds, and pick insects from the foliage. They often concentrate at forest clearings, logged areas, mountain meadows, and roadside ditches to look for thistle patches or other sources of seeds. They are also among the most common visitors to thistle-feeders, along with Lesser Goldfinches. Siskin flocks move nomadically through forests, and their presence in any specific area is highly unpredictable. Like many finches, Pine Siskins visit mineral springs and salt licks to supplement their diet.

Gregarious all year, pairs form in winter and usually split off from the main group for nesting in late April or early May, when local food supplies are greatest; nesting is usually completed by mid-July, even in the highest mountains. Even while breeding, Pine Siskins remain in small colonies, spacing their nests about 10 yards apart and concealing them within the densest conifer boughs, near the ground or high above it. Breeding pairs occasionally visit each other's nests and often join communal foraging flocks even while incubating eggs.

STATUS AND DISTRIBUTION Widespread inhabitants of coniferous forests across the northern latitudes of North America, Pine Siskins also reside in pine forests as far south as Guatemala. They remain in the highest forests of the Sierra year-round but also make frequent but unpredictable movements to lower elevations, especially to urban parks and gardens where thistle feeders may attract them in droves.

West Side. Common to locally abundant in mixed conifer and red fir forests of the Upper Conifer zone through the lodgepole pine forests of the Subalpine zone; unpredictable, but locally fairly common in the Lower Conifer zone in the breeding season; locally abundant to absent in the Foothill zone in winter, when they often forage in deciduous trees.

East Side. Habitats similar to the West Side, breed in Subalpine forests up to about 10,000 feet; locally fairly common to uncommon at lower elevations in winter, except locally at feeders where may be abundant.

Lesser Goldfinch • *Spinus psaltria*

ORIGIN OF NAMES "Lesser" for the species' smaller size than other goldfinches (see accounts below); "goldfinch" for the bright yellows of breeding males; Gr. *psaltria*, a lute player, for the species' somewhat musical songs.

NATURAL HISTORY As their name suggests, Lessers are the smallest of the goldfinches. Plaintive *pe-eee—pee-ee* calls of Lesser Goldfinches fill the air as flocks flush from weed patches or fly overhead. Most widespread of the goldfinches in the Sierra, they inhabit oak woodlands, riparian forests, lowland chaparral, grassy hillsides, and residential neighborhoods, and are among the most frequent visitors to thistle feeders. Flocks of these social birds hunt for seeds, and often hang on weed stalks that bend precariously with their weight. They also eat insects or ripening fruit and, like many other finches, seek out mineral springs and salt licks. Songs are complex, rapid jumbles of notes and frequently incorporate phrases from other bird songs as this species is an accomplished mimic.

Male

Female

Lesser Goldfinches nest in a diversity of lowland habitats, from moist riparian forests to arid chaparral. They consume great quantities of water, however, and never stray far from a place to drink. Flocks break up as breeding pairs begin to form in April and nesting is usually completed by early July. Males occupy breeding territories of about 100 feet in diameter while the females actively defend the immediate area around the nest from other intruding pairs. Females weave compact cup nests between well-shaded terminal branches of oaks, conifers, or shrubs, often less than 10 feet above the ground.

STATUS AND DISTRIBUTION Five different subspecies of Lesser Goldfinches inhabit oak woodlands, riparian forests, chaparral, and residential areas from southern Washington south across southwestern North America to Panama. Males of the "Western" subspecies (S. p. hesperophila), including those in the Sierra, have greenish backs, while those of other subspecies have black backs.

West Side. Common to locally abundant year-round residents of the Foothill zone; fairly common in oak-dominated forests of the Lower Conifer zone up to about 5,000 feet in the Sierra; fairly common to uncommon up to the Subalpine zone in late summer and fall, with records in the southern Sierra above 12,000 feet; they usually keep to shrubfields, mountain meadows, or other sparsely forested areas.

East Side. Fairly common to uncommon in the same habitats as the West Side; fairly common to uncommon in winter at most locations below 6,000 feet; rare in Mono Basin.

Lawrence's Goldfinch • *Spinus lawrencei*

ORIGIN OF NAMES "Lawrence's" for George N. Lawrence (1806–1895), an amateur New York ornithologist and taxidermist who worked on specimens from the Pacific Railroad Surveys; John Cassin (see Cassin's Vireo account) named this bird in Lawrence's honor.

NATURAL HISTORY What a great treat to come upon a flock of Lawrence's Goldfinches in the midst of a hot day in the foothills. The most beautiful of Sierra goldfinches, they are also the rarest and least predictable. As Ralph Hoffman (1927) noted: "A valley . . . may be filled with the black-chinned gray-bodied birds one summer and the next year not contain one."

Like many other Fringillid finches, their numbers and distribution vary erratically from year to year. In spring and summer, flocks of these black, gray, and gold birds visit the Sierra foothills,

Male

Female

where they extract seeds from fiddlenecks, thistles, and other weeds. Like their other goldfinch cousins, Lawrence's also flock to thistle feeders. Lawrence's are excellent mimics and their songs may be composed almost entirely of the sounds of other foothill birds ranging from American Kestrels and California Quail to those of flycatchers, wrens, vireos, sparrows, and other goldfinches. Listen for distinctive *tinkling* notes given as flight calls and also incorporated into their songs.

Although they may forage on open, rocky hillsides, Lawrence's Goldfinches usually nest in thick stands of blue or live oaks. They prefer arid sites near a source of water such as a creek or small lake. Their need for water within arid landscapes where sources in spring and summer are unreliable may partially explain the intermittent and unpredictable nature of their presence. Their unusually long breeding season may extend from late March until early August, and a few pairs may attempt second broods, especially if the first one fails. The bright males display golden wings and sing vigorously in defense of their small nesting territories. They breed near each other in small, loose colonies with males guarding small territories only about 50 feet in diameter that may encompass a single tree. Females suspend their nests woven from grasses and lichens on terminal branches 20 feet or more above the ground. They incubate their eggs alone, but their attentive mates bring food at frequent intervals.

STATUS AND DISTRIBUTION Lawrence's Goldfinches are confined to California and northern Baja California during the breeding season, but most winter away from the Sierra in warm, arid portions of southern California, other southwestern states, and northern Mexico.

West Side. Common and reliable breeders in Kern River Valley but otherwise locally fairly common but unpredictable breeders in the Foothill zone; fairly common to uncommon breeders up to about 4,500 feet in the Lower Conifer zone and up to over 6,000 feet on the Kern Plateau in southern Tulare County; after breeding from August to October, flocks may stray up to meadow edges and streamside thickets in the Upper Conifer zone; rare in winter in the lower Foothill zone.

East Side. Regular spring migrants in some desert canyons of Kern County such as Butterbredt Spring and Jawbone Canyon; elsewhere casual migrants in spring and fall; accidental in winter with one Owens Valley record in December 1995; one breeding record in the Owens Valley in 1996; has been recorded on the Kelso Valley Breeding Bird Survey Route and has probably bred at Butterbredt Spring.

American Goldfinch • *Spinus tristis*

ORIGIN OF NAMES "American" to distinguish this species from the European Goldfinch *(Carduelis carduelis)*; L. *tristis*, sad, an allusion to the species' "whiny" calls, which sound mournful to some.

NATURAL HISTORY In their bright yellow breeding raiment, male American Goldfinches are unsurpassed in brilliance. Called "wild canaries" by some, they are a familiar sight to residents of the Sierra foothills. Flocks sing cheerfully on spring mornings as they feast on willow catkins, caterpillars, or grasshoppers. For no apparent reason, they move from one tree to the next, calling *puh-chik-o-ree* or *potato-chip* as they bound through the air in disorganized flocks. Just as suddenly, they drop down to feed on thistle heads, dandelions, or seeds of annual grasses and flowers (especially those in the sunflower family). The almost entirely vegetarian diet of American Goldfinches may explain why most Brown-headed Cowbird eggs laid in their nests fail to produce young, since cowbird chicks require insect food (see Brown-headed Cowbird account).

Male

Female

In April and May, female American Goldfinches construct compact cup nests from plant fibers, willow catkins, and thistle down and anchor them to forked branches of shrubs and trees, usually at heights less than 10 feet. They often nest in riparian woodlands or other moist areas where males defend small territories or none at all. Gregarious in all seasons, American Goldfinches range widely through a great variety of habitats during winter, searching for abundant sources of buds or seeds, often in mixed flocks with Lesser Goldfinches and sometimes with Pine Siskins. In their subdued winter plumage, American Goldfinches may be distinguished from Lessers by their dusky white, rather than yellowish, bellies and rumps.

STATUS AND DISTRIBUTION Common in appropriate habitats across all of temperate North America, American Goldfinches reside year-round in the Sierra foothills. Although they occasionally nest in open grasslands, chaparral, oak woodlands, and orchards, ideally they prefer lush streamside groves of willows, cottonwoods, and alders.

West Side. Locally fairly common breeders in the Foothill zone south to Mariposa County but rare to absent in breeding season farther south; never as widespread or common as Lesser Goldfinches; after breeding, some may wander rarely through riparian corridors and montane chaparral up to about 5,000 feet in the Upper Conifer zone in fall but retreat to the lowlands for winter, where they can be common around feeders and developed areas.

East Side. Rare during the breeding season; migrant flocks may appear in open sagebrush flats in fall; uncommon to locally common in winter at lower elevations from Honey Lake south to northern Inyo County; large numbers may congregate near towns and ranches in winter.

Evening Grosbeak • *Coccothraustes vespertinus*

ORIGIN OF NAMES "Evening" because some early ornithologist thought they only emerged at dusk to sing, but they actually utter loud, unmusical calls in broad daylight; "grosbeak" (see Pine Grosbeak account); *coccothraustes*, a nut-breaker, from Gr. *kokkos*, a kernel or nut, and *thrauo*, to break; *vespertinus* from L. *vespertina*, evening.

NATURAL HISTORY Evening Grosbeaks' huge pale-green bills and bold white, black, and gold plumages make these gaudily dressed birds as striking as any warbler or tanager. Few other Sierra birds

Male

Female

are as certain to produce a chorus of "oohs" and "aahs" from a group of birders who finally get one focused in their binoculars. They seldom travel alone but move in flocks to scour forests for seeds and buds of conifers and hardwoods as well as insects. Continuously giving rapidly paced calls to maintain contact, they fly in undulating waves that vividly display their white wing patches.

Four discrete flight calls "types" have been observed in various Evening Grosbeak populations across North America; however, only two of these (Types 1 and 2) have been recorded in the Sierra (Sewall et al. 2004). Type 2 calls are by far the most common—they first rise in pitch, then fade out in the middle, and descend again at the end of the call, sounding something like *tee-oo*. Type 1 calls are usually encountered in the Pacific Northwest and are rare in the Sierra—they begin at a higher frequency than Type 2 calls, then drop throughout the call, sounding something like *chee-eer*. Discriminating between these flight call types takes significant attention and a great deal of practice.

Evening Grosbeaks are often drawn to roadside ditches, where they might be seen with Purple or Cassin's Finches, Pine Siskins, or Brewer's Blackbirds. Areas chosen for nesting have adequate food, a source of water, and often a mineral spring or salt lick nearby. Evening Grosbeaks forage in a wide variety of Sierra trees, including oaks and cedars, but seem to gather in largest numbers in stands of fir or ponderosa/Jeffrey pines. Females construct flimsy stick nests and position them up to 60 feet high in dense conifer foliage, where most observers never glimpse them. Nest site selection appears to be more influenced by local food availability (e.g., insect outbreaks) than by habitat type or structure. In the Sierra most clutches are started in late May or early June, and young typically fledge by August. In fall and winter, Evening Grosbeaks roam conifer forests of the Sierra, and some years they descend into foothill oak and pine woodlands. Flock sizes tend to swell somewhat during the nonbreeding season, when they sometimes number in the hundreds.

STATUS AND DISTRIBUTION Irruptive residents of montane and boreal forests across the higher latitudes of North America, Evening Grosbeaks require a good supply of ripe cones, insect larvae, or other foods before they settle in any particular place. Unpredictable food supplies in the mountains make these nomadic birds disappear mysteriously from areas where they were once common and not return for several years or more. As described by Grinnell and Storer (1924): "The California Evening Grosbeak is so irregular as to its seasonal behavior in the Yosemite region that no prediction can be made concerning its occurrence in any stated locality at any given time of the year."

West Side. Locally common or absent depending on food conditions but breed in mature conifer forests from about 3,500 feet in the Lower Conifer zone to red fir forests of the Upper Conifer zone, at least up to 8,000 feet; fairly common to uncommon from the Upper Conifer to the Foothill zone in winter, when they often forage in oaks and other broad-leaved trees; some may winter in forests of the Subalpine zone, but winter status in the higher mountains is largely unknown.

East Side. Locally fairly common in summer in pine forests south through Mono County with highest concentrations in the Tahoe Basin; unpredictable but in some years locally fairly common from October through March south into the northern Owens Valley.

TRENDS AND CONSERVATION STATUS Data from both Christmas Bird Counts and Breeding Bird Surveys across North America suggest a decline in numbers of Evening Grosbeaks, and the spectacular winter irruptions historically seen in the eastern United States have not occurred on similar scales in recent decades. Data from the West are inadequate to speculate on population trends. However, occasional winter irruptions of Evening Grosbeaks into coastal and low-elevation locations in Washington, Oregon, and California that were common up to the early 1990s have not been seen since. The largest irruption of this type since the 1990s occurred from fall 2010 into spring 2011. Interestingly, the birds found at low-elevation sites in California during this irruption were from populations breeding much farther north in the Pacific Northwest and the northern Rocky Mountains and not from the Sierra-breeding population (Davis et al. 2011).

OLD WORLD SPARROWS • Family Passeridae

Old World sparrows, distinct from New World sparrows (family Emberizidae), are primarily seed-eaters, though they also consume small insects when nesting and feeding young. Much like crows, gulls, and pigeons, many species scavenge for food around cities, where they mostly consume human leavings. About 50 species are native to Europe, Asia, and Africa, but a few species, including House Sparrows, have been widely introduced outside their native range. The family name is derived from L. *passer*, a sparrow.

House Sparrow • *Passer domesticus*

ORIGIN OF NAMES "House" for their species' preference for nesting in buildings and other human-made structures; *passer* (see family account); L. *domesticus*, house.

NATURAL HISTORY If seen for the first time in the picturesque countryside of their native Europe, House Sparrows might seem handsome with their tailored patterns of black, brown, and gray. In North America, however, these introduced birds are often viewed with disdain. They abound wherever humans or livestock dwell and were even more common before the automobile replaced the horse, and waste grain and manure around stables allowed them to flourish. Since they now like to hang out at fast food restaurants, some birders call them "Burger Kinglets."

Introduction of the House Sparrow to North

Male

Female

America was unfortunate both for humans and native birds. House Sparrows damage grain crops and usurp cavity nesting sites needed by Western Bluebirds and other birds as well as nests built by Cliff and Barn Swallows. The promiscuous, black-bibbed males actively court as many females as possible. Males have no musical songs to attract their mates; rather, they screech *cheep-cheep-cheep* in shrill, tedious tones. Almost any crevice or cranny will do for a nest site, including the eaves or decorative grill work of buildings, abandoned farm equipment, as well as holes in trees. Both sexes gather discarded paper, cloth, or straw and build rounded, dome nests. Females usually lay four grayish-white eggs, covered with variable amounts of black splotching and incubate them alone for 12 to 14 days. Males may occasionally help with incubation and bring food at regular intervals, and both parents feed the young insects for about two weeks until they fledge. House Sparrows typically raise two or three broods a year—a major reason why they have been so successful. Like the equally maligned European Starlings, they are an inescapable part of our birdlife and deserve at least grudging respect for their ability to exploit human-altered habitats.

STATUS AND DISTRIBUTION Indigenous to Europe, Asia, and Africa, House Sparrows have followed humans all over the world, and now occur on every continent except Antarctica. About 100 of these boisterous birds were introduced to Brooklyn in the early 1850s. They followed railroads and livestock westward and reached California by the early 1870s. Their spread into the Sierra came later, as they were not detected in Yosemite Valley until the early 1920s, presumably having followed stock trains into the mountains. Luckily for many native Sierra birds, House Sparrows are rarely found away from areas of human habitation.

West Side. Common to abundant year-round residents of towns, farms, ranches, and livestock stables throughout the Foothill zone; fairly common in similar habitats up to the Lower Conifer zone, but uncommon above about 4,000 feet.

East Side. Similar to the West Side, but occur regularly in developed areas above 6,000.

Checklist of Sierra Birds

Name	Notes

Waterfowl (Anatidae)

☐ Greater White-fronted Goose *(Anser albifrons)*

☐ Snow Goose *(Chen caerulescens)*

☐ Ross's Goose *(Chen rossii)*

☐ Brant *(Branta bernicla)*, RARE

☐ Cackling Goose *(Branta hutchinsii)*

☐ Canada Goose *(Branta canadensis)*

☐ Trumpeter Swan *(Cygnus buccinator)*, CASUAL

☐ Tundra Swan *(Cygnus columbianus)*

☐ Whooper Swan *(Cygnus cygnus)*, ACCIDENTAL

☐ Wood Duck *(Aix sponsa)*

☐ Gadwall *(Anas strepera)*

☐ Falcated Duck *(Anas falcata)*, ACCIDENTAL

☐ Eurasian Wigeon *(Anas penelope)*

☐ American Wigeon *(Anas americana)*

See Appendix 2, "Rare, Casual, and Accidental Birds of the Sierra Nevada," for details of status.

Name	Notes
☐ Mallard *(Anas platyrhynchos)*	
☐ Blue-winged Teal *(Anas discors)*, RARE	
☐ Cinnamon Teal *(Anas cyanoptera)*	
☐ Northern Shoveler *(Anas clypeata)*	
☐ Northern Pintail *(Anas acuta)*	
☐ Baikal Teal *(Anas formosa)*, ACCIDENTAL	
☐ Green-winged Teal *(Anas crecca)*	
☐ Canvasback *(Aythya valisineria)*	
☐ Redhead *(Aythya americana)*	
☐ Ring-necked Duck *(Aythya collaris)*	
☐ Tufted Duck *(Aythya fuligula)*, ACCIDENTAL	
☐ Greater Scaup *(Aythya marila)*, RARE	
☐ Lesser Scaup *(Aythya affinis)*	
☐ Harlequin Duck *(Histrionicus histrionicus)*	
☐ Surf Scoter *(Melanitta perspicillata)*, RARE	
☐ White-winged Scoter *(Melanitta fusca)*, RARE	
☐ Black Scoter *(Melanitta americana)*, ACCIDENTAL	
☐ Long-tailed Duck *(Clangula hyemalis)*, CASUAL	
☐ Bufflehead *(Bucephala albeola)*	
☐ Common Goldeneye *(Bucephala clangula)*	
☐ Barrow's Goldeneye *(Bucephala islandica)*	
☐ Smew *(Mergellus albellus)*, ACCIDENTAL	
☐ Hooded Merganser *(Lophodytes cucullatus)*	
☐ Common Merganser *(Mergus merganser)*	
☐ Red-breasted Merganser *(Mergus serrator)*	
☐ Ruddy Duck *(Oxyura jamaicensis)*	

Quail (Odontophoridae)

☐ Mountain Quail *(Oreortyx pictus)*	
☐ California Quail *(Callipepla californica)*	

Fowl-like Birds (Phasianidae)

☐ Chukar *(Alectoris chukar)*

☐ Ring-necked Pheasant *(Phasianus colchicus)*

☐ Greater Sage-Grouse *(Centrocercus urophasianus)*

☐ White-tailed Ptarmigan *(Lagopus leucura)*

☐ Sooty Grouse *(Dendragapus fuliginosus)*

☐ Wild Turkey *(Meleagris gallopavo)*

Loons (Gaviidae)

☐ Red-throated Loon *(Gavia stellata)*, CASUAL

☐ Pacific Loon *(Gavia pacifica)*

☐ Common Loon *(Gavia immer)*

☐ Yellow-billed Loon *(Gavia adamsii)*, ACCIDENTAL

Grebes (Podicipedidae)

☐ Pied-billed Grebe *(Podilymbus podiceps)*

☐ Horned Grebe *(Podiceps auritus)*

☐ Red-necked Grebe *(Podiceps grisegena)*, CASUAL

☐ Eared Grebe *(Podiceps nigricollis)*

☐ Western Grebe *(Aechmophorus occidentalis)*

☐ Clark's Grebe *(Aechmophorus clarkii)*

Storm-Petrels (Hydrobatidae)

☐ Black Storm-Petrel *(Oceanodroma melania)*, ACCIDENTAL

Frigatebirds (Fregatidae)

☐ Magnificent Frigatebird *(Fregata magnificens)*, ACCIDENTAL

Boobies (Sulidae)

☐ Blue-footed Booby *(Sula nebouxii)*, ACCIDENTAL

Cormorants (Phalacrocoracidae)

☐ Double-crested Cormorant *(Phalacrocorax auritus)*

Name	Notes

Pelicans (Pelecanidae)

☐ American White Pelican *(Pelecanus erythrorhynchos)*

☐ Brown Pelican *(Pelecanus occidentalis)*, CASUAL

Herons and Relatives (Ardeidae)

☐ American Bittern *(Botaurus lentiginosus)*

☐ Least Bittern *(Ixobrychus exilis)*, CASUAL

☐ Great Blue Heron *(Ardea herodias)*

☐ Great Egret *(Ardea alba)*

☐ Snowy Egret *(Egretta thula)*

☐ Little Blue Heron *(Egretta caerulea)*, ACCIDENTAL

☐ Tricolored Heron *(Egretta tricolor)*, ACCIDENTAL

☐ Reddish Egret *(Egretta rufescens)*, ACCIDENTAL

☐ Cattle Egret *(Bubulcus ibis)*, CASUAL

☐ Green Heron *(Butorides virescens)*

☐ Black-crowned Night-Heron *(Nycticorax nycticorax)*

Ibis (Threskiornithidae)

☐ Glossy Ibis *(Plegadis falcinellus)*, ACCIDENTAL

☐ White-faced Ibis *(Plegadis chihi)*

New World Vultures (Cathartidae)

☐ Black Vulture *(Coragyps atratus)*, ACCIDENTAL

☐ Turkey Vulture *(Cathartes aura)*

☐ California Condor *(Gymnogyps californianus)*

Osprey (Pandionidae)

☐ Osprey *(Pandion haliaetus)*

Hawks and Relatives (Accipitridae)

☐ White-tailed Kite *(Elanus leucurus)*

☐ Mississippi Kite *(Ictinia mississippiensis)*, ACCIDENTAL

☐ Bald Eagle *(Haliaeetus leucocephalus)*

Name	Notes

☐ Northern Harrier *(Circus cyaneus)*

☐ Sharp-shinned Hawk *(Accipiter striatus)*

☐ Cooper's Hawk *(Accipiter cooperii)*

☐ Northern Goshawk *(Accipiter gentilis)*

☐ Red-shouldered Hawk *(Buteo lineatus)*

☐ Broad-winged Hawk *(Buteo platypterus)*, CASUAL

☐ Swainson's Hawk *(Buteo swainsoni)*

☐ Zone-tailed Hawk *(Buteo albonotatus)*, ACCIDENTAL

☐ Red-tailed Hawk *(Buteo jamaicensis)*

☐ Ferruginous Hawk *(Buteo regalis)*

☐ Rough-legged Hawk *(Buteo lagopus)*

☐ Golden Eagle *(Aquila chrysaetos)*

Falcons (Falconidae)

☐ Crested Caracara *(Caracara cheriway)*, ACCIDENTAL

☐ American Kestrel *(Falco sparverius)*

☐ Merlin *(Falco columbarius)*

☐ Peregrine Falcon *(Falco peregrinus)*

☐ Prairie Falcon *(Falco mexicanus)*

Rails and Relatives (Rallidae)

☐ Yellow Rail *(Coturnicops noveboracensis)*, CASUAL

☐ Black Rail *(Laterallus jamaicensis)*

☐ Virginia Rail *(Rallus limicola)*

☐ Sora *(Porzana carolina)*

☐ Common Gallinule *(Gallinula galeata)*

☐ American Coot *(Fulica americana)*

Cranes (Gruidae)

☐ Sandhill Crane *(Grus canadensis)*

Name	Notes

Plovers (Charadriidae)

☐ Black-bellied Plover *(Pluvialis squatarola)*

☐ American Golden-Plover *(Pluvialis dominica)*, CASUAL

☐ Pacific Golden-Plover *(Pluvialis fulva)*, CASUAL

☐ Snowy Plover *(Charadrius nivosus)*

☐ Semipalmated Plover *(Charadrius semipalmatus)*

☐ Killdeer *(Charadrius vociferus)*

☐ Mountain Plover *(Charadrius montanus)*, CASUAL

Stilts and Avocets (Recurvirostridae)

☐ Black-necked Stilt *(Himantopus mexicanus)*

☐ American Avocet *(Recurvirostra americana)*

Sandpipers and Relatives (Scolopacidae)

☐ Spotted Sandpiper *(Actitis macularius)*

☐ Solitary Sandpiper *(Tringa solitaria)*, RARE

☐ Wandering Tattler *(Tringa incana)*, CASUAL

☐ Greater Yellowlegs *(Tringa melanoleuca)*

☐ Willet *(Tringa semipalmata)*

☐ Lesser Yellowlegs *(Tringa flavipes)*, RARE

☐ Upland Sandpiper *(Bartramia longicauda)*, ACCIDENTAL

☐ Whimbrel *(Numenius phaeopus)*, CASUAL

☐ Long-billed Curlew *(Numenius americanus)*

☐ Marbled Godwit *(Limosa fedoa)*, RARE

☐ Ruddy Turnstone *(Arenaria interpres)*, CASUAL

☐ Black Turnstone *(Arenaria melanocephala)*, ACCIDENTAL

☐ Red Knot *(Calidris canutus)*, CASUAL

☐ Sanderling *(Calidris alba)*, RARE

☐ Semipalmated Sandpiper *(Calidris pusilla)*, CASUAL

☐ Western Sandpiper *(Calidris mauri)*

☐ Little Stint *(Calidris minuta)*, ACCIDENTAL

Name	Notes

☐ Least Sandpiper *(Calidris minutilla)*

☐ White-rumped Sandpiper *(Calidris fuscicollis)*, ACCIDENTAL

☐ Baird's Sandpiper *(Calidris bairdii)*, RARE

☐ Pectoral Sandpiper *(Calidris melanotos)*, RARE

☐ Dunlin *(Calidris alpina)*

☐ Stilt Sandpiper *(Calidris himantopus)*, CASUAL

☐ Ruff *(Philomachus pugnax)*, ACCIDENTAL

☐ Short-billed Dowitcher *(Limnodromus griseus)*, RARE

☐ Long-billed Dowitcher *(Limnodromus scolopaceus)*

☐ Wilson's Snipe *(Gallinago delicata)*

☐ Wilson's Phalarope *(Phalaropus tricolor)*

☐ Red-necked Phalarope *(Phalaropus lobatus)*

☐ Red Phalarope *(Phalaropus fulicarius)*, CASUAL

Gulls and Terns (Laridae)

☐ Black-legged Kittiwake *(Rissa tridactyla)*, ACCIDENTAL

☐ Sabine's Gull *(Xema sabini)*, RARE

☐ Bonaparte's Gull *(Chroicocephalus philadelphia)*

☐ Little Gull *(Hydrocoloeus minutus)*, ACCIDENTAL

☐ Laughing Gull *(Leucophaeus atricilla)*, ACCIDENTAL

☐ Franklin's Gull *(Leucophaeus pipixcan)*, RARE

☐ Heermann's Gull *(Larus heermanni)*, CASUAL

☐ Mew Gull *(Larus canus)*, RARE

☐ Ring-billed Gull *(Larus delawarensis)*

☐ Western Gull *(Larus occidentalis)*, CASUAL

☐ Yellow-footed Gull *(Larus livens)*, ACCIDENTAL

☐ California Gull *(Larus californicus)*

☐ Herring Gull *(Larus argentatus)*

☐ Thayer's Gull *(Larus thayeri)*, CASUAL

☐ Lesser Black-backed Gull *(Larus fuscus)*, CASUAL

Name	Notes

☐ Glaucous-winged Gull *(Larus glaucescens)*, CASUAL

☐ Glaucous Gull *(Larus hyperboreus)*, CASUAL

☐ Least Tern *(Sternula antillarum)*, CASUAL

☐ Caspian Tern *(Hydroprogne caspia)*

☐ Black Tern *(Chlidonias niger)*

☐ Common Tern *(Sterna hirundo)*, RARE

☐ Arctic Tern *(Sterna paradisaea)*, CASUAL

☐ Forster's Tern *(Sterna forsteri)*

Jaegers and Relatives (Stercorariidae)

☐ Pomarine Jaeger *(Stercorarius pomarinus)*, ACCIDENTAL

☐ Parasitic Jaeger *(Stercorarius parasiticus)*, RARE

☐ Long-tailed Jaeger *(Stercorarius longicaudus)*, RARE

Murrelets and Relatives (Alcidae)

☐ Long-billed Murrelet *(Brachyramphus perdix)*, ACCIDENTAL

☐ Ancient Murrelet *(Synthliboramphus antiquus)*, ACCIDENTAL

Pigeons and Doves (Columbidae)

☐ Rock Pigeon *(Columba livia)*

☐ Band-tailed Pigeon *(Patagioenas fasciata)*

☐ Eurasian Collared-Dove *(Streptopelia decaocto)*

☐ White-winged Dove *(Zenaida asiatica)*, CASUAL

☐ Mourning Dove *(Zenaida macroura)*

☐ Common Ground-Dove *(Columbina passerina)*, ACCIDENTAL

☐ Ruddy Ground-Dove *(Columbina talpacoti)*, ACCIDENTAL

Cuckoos and Roadrunners (Cuculidae)

☐ Yellow-billed Cuckoo *(Coccyzus americanus)*

☐ Black-billed Cuckoo *(Coccyzus erythropthalmus)*, ACCIDENTAL

☐ Greater Roadrunner *(Geococcyx californianus)*

Name	Notes

Chickadees and Titmice (Paridae)

☐ Mountain Chickadee *(Poecile gambeli)*

☐ Chestnut-backed Chickadee *(Poecile rufescens)*

☐ Oak Titmouse *(Baeolophus inornatus)*

☐ Juniper Titmouse *(Baeolophus ridgwayi)*

Verdin (Remizidae)

☐ Verdin *(Auriparus flaviceps)*

Bushtit (Aegithalidae)

☐ Bushtit *(Psaltriparus minimus)*

Nuthatches (Sittidae)

☐ Red-breasted Nuthatch *(Sitta canadensis)*

☐ White-breasted Nuthatch *(Sitta carolinensis)*

☐ Pygmy Nuthatch *(Sitta pygmaea)*

Creepers (Certhiidae)

☐ Brown Creeper *(Certhia americana)*

Wrens (Troglodytidae)

☐ Cactus Wren *(Campylorhynchus brunneicapillus)*

☐ Rock Wren *(Salpinctes obsoletus)*

☐ Canyon Wren *(Catherpes mexicanus)*

☐ Bewick's Wren *(Thryomanes bewickii)*

☐ House Wren *(Troglodytes aedon)*

☐ Pacific Wren *(Troglodytes pacificus)*

☐ Winter Wren *(Troglodytes hiemalis)*, ACCIDENTAL

☐ Marsh Wren *(Cistothorus palustris)*

Gnatcatchers (Polioptilidae)

☐ Blue-gray Gnatcatcher *(Polioptila caerulea)*

Dippers (Cinclidae)

☐ American Dipper *(Cinclus mexicanus)*

Name	Notes

Kinglets (Regulidae)

☐ Golden-crowned Kinglet *(Regulus satrapa)*

☐ Ruby-crowned Kinglet *(Regulus calendula)*

Wrentit (Sylviidae)

☐ Wrentit *(Chamaea fasciata)*

Thrushes and Relatives (Turdidae)

☐ Western Bluebird *(Sialia mexicana)*

☐ Mountain Bluebird *(Sialia currucoides)*

☐ Townsend's Solitaire *(Myadestes townsendi)*

☐ Veery *(Catharus fuscescens)*, ACCIDENTAL

☐ Swainson's Thrush *(Catharus ustulatus)*

☐ Hermit Thrush *(Catharus guttatus)*

☐ Wood Thrush *(Hylocichla mustelina)*, ACCIDENTAL

☐ American Robin *(Turdus migratorius)*

☐ Varied Thrush *(Ixoreus naevius)*

Mockingbirds and Thrashers (Mimidae)

☐ Gray Catbird *(Dumetella carolinensis)*, CASUAL

☐ Northern Mockingbird *(Mimus polyglottos)*

☐ Sage Thrasher *(Oreoscoptes montanus)*

☐ Brown Thrasher *(Toxostoma rufum)*, CASUAL

☐ Bendire's Thrasher *(Toxostoma bendirei)*, RARE

☐ California Thrasher *(Toxostoma redivivum)*

☐ Le Conte's Thrasher *(Toxostoma lecontei)*

Starlings (Sturnidae)

☐ European Starling *(Sturnus vulgaris)*

Pipits (Motacillidae)

☐ Red-throated Pipit *(Anthus cervinus)*, ACCIDENTAL

☐ American Pipit *(Anthus rubescens)*

Name	Notes

Waxwings (Bombycillidae)

☐ Bohemian Waxwing *(Bombycilla garrulus)*, RARE

☐ Cedar Waxwing *(Bombycilla cedrorum)*

Silky Flycatchers (Ptilogonatidae)

☐ Phainopepla *(Phainopepla nitens)*

Longspurs (Calcariidae)

☐ Lapland Longspur *(Calcarius lapponicus)*, RARE

☐ Chestnut-collared Longspur *(Calcarius ornatus)*, RARE

☐ McCown's Longspur *(Rhynchophanes mccownii)*, CASUAL

☐ Snow Bunting *(Plectrophenax nivalis)*, ACCIDENTAL

Wood-Warblers (Parulidae)

☐ Ovenbird *(Seiurus aurocapilla)*, CASUAL

☐ Worm-eating Warbler *(Helmitheros vermivorum)*, CASUAL

☐ Northern Waterthrush *(Parkesia noveboracensis)*, RARE

☐ Golden-winged Warbler *(Vermivora chrysoptera)*, CASUAL

☐ Blue-winged Warbler *(Vermivora cyanoptera)*, CASUAL

☐ Black-and-white Warbler *(Mniotilta varia)*, RARE

☐ Prothonotary Warbler *(Protonotaria citrea)*, CASUAL

☐ Tennessee Warbler *(Oreothlypis peregrina)*, CASUAL

☐ Orange-crowned Warbler *(Oreothlypis celata)*

☐ Lucy's Warbler *(Oreothlypis luciae)*, CASUAL

☐ Nashville Warbler *(Oreothlypis ruficapilla)*

☐ Virginia's Warbler *(Oreothlypis virginiae)*

☐ Connecticut Warbler *(Oporornis agilis)*, ACCIDENTAL

☐ MacGillivray's Warbler *(Geothlypis tolmiei)*

☐ Mourning Warbler *(Geothlypis philadelphia)*, ACCIDENTAL

☐ Kentucky Warbler *(Geothlypis formosa)*, CASUAL

☐ Common Yellowthroat *(Geothlypis trichas)*

Name	Notes

☐ Hooded Warbler *(Setophaga citrina)*, CASUAL

☐ American Redstart *(Setophaga ruticilla)*, RARE

☐ Cerulean Warbler *(Setophaga cerulea)*, ACCIDENTAL

☐ Northern Parula *(Setophaga americana)*, RARE

☐ Magnolia Warbler *(Setophaga magnolia)*, CASUAL

☐ Bay-breasted Warbler *(Setophaga castanea)*, ACCIDENTAL

☐ Blackburnian Warbler *(Setophaga fusca)*, CASUAL

☐ Yellow Warbler *(Setophaga petechia)*

☐ Chestnut-sided Warbler *(Setophaga pensylvanica)*, RARE

☐ Blackpoll Warbler *(Setophaga striata)*, CASUAL

☐ Black-throated Blue Warbler *(Setophaga caerulescens)*, CASUAL

☐ Palm Warbler *(Setophaga palmarum)*, CASUAL

☐ Pine Warbler *(Setophaga pinus)*, ACCIDENTAL

☐ Yellow-rumped Warbler *(Setophaga coronata)*

☐ Yellow-throated Warbler *(Setophaga dominica)*, CASUAL

☐ Prairie Warbler *(Setophaga discolor)*, CASUAL

☐ Grace's Warbler *(Setophaga graciae)*, ACCIDENTAL

☐ Black-throated Gray Warbler *(Setophaga nigrescens)*

☐ Townsend's Warbler *(Setophaga townsendi)*

☐ Hermit Warbler *(Setophaga occidentalis)*

☐ Black-throated Green Warbler *(Setophaga virens)*, ACCIDENTAL

☐ Canada Warbler *(Cardellina canadensis)*, CASUAL

☐ Wilson's Warbler *(Cardellina pusilla)*

☐ Red-faced Warbler *(Cardellina rubrifrons)*, ACCIDENTAL

☐ Painted Redstart *(Myioborus pictus)*, CASUAL

☐ Yellow-breasted Chat *(Icteria virens)*

Sparrows and Relatives (Emberizidae)

☐ Green-tailed Towhee *(Pipilo chlorurus)*

☐ Spotted Towhee *(Pipilo maculatus)*

Name	Notes

☐ Rufous-crowned Sparrow *(Aimophila ruficeps)*

☐ California Towhee *(Melozone crissalis)*

☐ Cassin's Sparrow *(Peucaea cassinii)*, ACCIDENTAL

☐ American Tree Sparrow *(Spizella arborea)*, RARE

☐ Chipping Sparrow *(Spizella passerina)*

☐ Clay-colored Sparrow *(Spizella pallida)*, CASUAL

☐ Brewer's Sparrow *(Spizella breweri)*

☐ Black-chinned Sparrow *(Spizella atrogularis)*

☐ Vesper Sparrow *(Pooecetes gramineus)*

☐ Lark Sparrow *(Chondestes grammacus)*

☐ Black-throated Sparrow *(Amphispiza bilineata)*

☐ Sage Sparrow *(Amphispiza belli)*

☐ Lark Bunting *(Calamospiza melanocorys)*, RARE

☐ Savannah Sparrow *(Passerculus sandwichensis)*

☐ Grasshopper Sparrow *(Ammodramus savannarum)*

☐ Fox Sparrow *(Passerella iliaca)*

☐ Song Sparrow *(Melospiza melodia)*

☐ Lincoln's Sparrow *(Melospiza lincolnii)*

☐ Swamp Sparrow *(Melospiza georgiana)*, RARE

☐ White-throated Sparrow *(Zonotrichia albicollis)*

☐ Harris's Sparrow *(Zonotrichia querula)*, rare

☐ White-crowned Sparrow *(Zonotrichia leucophrys)*

☐ Golden-crowned Sparrow *(Zonotrichia atricapilla)*

☐ Dark-eyed Junco *(Junco hyemalis)*

Grosbeaks and Relatives (Cardinalidae)

☐ Summer Tanager *(Piranga rubra)*

☐ Western Tanager *(Piranga ludoviciana)*

☐ Rose-breasted Grosbeak *(Pheucticus ludovicianus)*

☐ Black-headed Grosbeak *(Pheucticus melanocephalus)*

Name	Notes

☐ Blue Grosbeak *(Passerina caerulea)*

☐ Lazuli Bunting *(Passerina amoena)*

☐ Indigo Bunting *(Passerina cyanea)*

☐ Painted Bunting *(Passerina ciris),* CASUAL

☐ Dickcissel *(Spiza americana),* CASUAL

Blackbirds and Relatives (Icteridae)

☐ Bobolink *(Dolichonyx oryzivorus),* CASUAL

☐ Red-winged Blackbird *(Agelaius phoeniceus)*

☐ Tricolored Blackbird *(Agelaius tricolor)*

☐ Western Meadowlark *(Sturnella neglecta)*

☐ Yellow-headed Blackbird *(Xanthocephalus xanthocephalus)*

☐ Rusty Blackbird *(Euphagus carolinus),* ACCIDENTAL

☐ Brewer's Blackbird *(Euphagus cyanocephalus)*

☐ Common Grackle *(Quiscalus quiscula),* CASUAL

☐ Great-tailed Grackle *(Quiscalus mexicanus)*

☐ Brown-headed Cowbird *(Molothrus ater)*

☐ Orchard Oriole *(Icterus spurius),* ACCIDENTAL

☐ Hooded Oriole *(Icterus cucullatus)*

☐ Bullock's Oriole *(Icterus bullockii)*

☐ Baltimore Oriole *(Icterus galbula),* CASUAL

☐ Scott's Oriole *(Icterus parisorum)*

Finches and Relatives (Fringillidae)

☐ Gray-crowned Rosy-Finch *(Leucosticte tephrocotis)*

☐ Black Rosy-Finch *(Leucosticte atrata),* CASUAL

☐ Pine Grosbeak *(Pinicola enucleator)*

☐ Purple Finch *(Carpodacus purpureus)*

☐ Cassin's Finch *(Carpodacus cassinii)*

☐ House Finch *(Carpodacus mexicanus)*

☐ Red Crossbill *(Loxia curvirostra)*

Name	Notes

☐ Common Redpoll *(Acanthis flammea),* ACCIDENTAL

☐ Pine Siskin *(Spinus pinus)*

☐ Lesser Goldfinch *(Spinus psaltria)*

☐ Lawrence's Goldfinch *(Spinus lawrencei)*

☐ American Goldfinch *(Spinus tristis)*

☐ Evening Grosbeak *(Coccothraustes vespertinus)*

Old World Sparrows (Passeridae)

☐ House Sparrow *(Passer domesticus)*

APPENDIX 2

Rare, Casual, and Accidental Birds of the Sierra Nevada

This list summarizes the general occurrence of unusual bird species in the Sierra region, defined as:

RARE. Rarely encountered and often highly localized, never in large numbers, but one or a few individuals occur in the region in all or most years.

CASUAL. Not encountered in the region in most years, but a pattern of occurrence may exist over many years or decades.

ACCIDENTAL. Encountered in the region on one or a few occasions, and the species is well out of its normal range.

The list is based on Sierra records through December 2011. We provide comments on the status of each rare or casual species, along with notes on where and when they are most often encountered; occurrences of accidental species are also noted briefly. Species considered unusual enough in California or Nevada to require review by the California or Nevada Bird Records committees (CBRC or NBRC) are included only if there is at least one accepted record within the region. This is not intended to be an exhaustive list of all reported observations. We relied heavily on records published in *North American Birds* and *Rare Birds of California* (Hamilton, Patten, and Erickson 2007), the knowledge of many local experts, and online updates of CBRC and NBRC actions (http://www.californiabirds.org/cbrc_book/update.pdf and http://www.gbbo.org/nbrc/FullReport.htm).

The order of families and taxonomy of individual species follow the American Ornithologists' Union *Check-list of North American Birds* (52nd supplement, 2011). All locations are in California unless otherwise indicated. The Sierra "bird seasons" listed here do not follow strict calendar dates but are intended to capture seasonal changes from a bird's perspective and are consistent with the dates used in *North American Birds*: Winter (December–February), Spring (March–May), Summer (June–July), and Fall (August–November).

RARE SPECIES

Brant *(Branta bernicla)* Mainly spring and fall. While found on occasion on West Side reservoirs, most occurrences are from Mono County, where the species is nearly annual.

Blue-winged Teal *(Anas discors)* Spring through fall. Regular visitors, especially in spring, to East Side marshes, slow-moving streams, and shallow waters of large reservoirs; usually seen in association with much more common Cinnamon Teal. Other than Lake Isabella and Lake Almanor, relatively few records from the West Side; casual on the East Side in winter.

Greater Scaup *(Aythya marila)* Late fall through early spring. Mostly found on large lakes and reservoirs, somewhat more regular on the East Side than the West Side. Difficulty in differentiating from the more common Lesser Scaup may lead to overreporting of this species.

Surf Scoter *(Melanitta perspicillata)* Fall through early spring. Mainly on large lakes, reservoirs, and sewage ponds; most West Side records from spring or fall; East Side records mainly October through

December, but a few birds linger until April before migrating.

White-winged Scoter *(Melanitta fusca)* Mainly fall or spring. On large lakes and reservoirs; more frequently encountered than Surf Scoter in most East Side locations, especially fall through winter.

Solitary Sandpiper *(Tringa solitaria)* Spring and late summer/fall migrants. Marshes and ponds with some vegetative cover; mostly East Side records with the species recently nearly annual in late July in the Owens Valley; also recorded in Mono County and Tahoe Basin from July to September.

Lesser Yellowlegs *(Tringa flavipes)* Mainly late summer through early fall migrants. Wetlands, sewage ponds, and pond edges; usually with the more common Greater Yellowlegs.

Marbled Godwit *(Limosa fedoa)* Late spring and late summer/early fall migrants. Usually found in small flocks at lake edges and muddy river deltas; more regularly observed on the East Side (especially at south Lake Tahoe) than the West Side, with most records from August.

Sanderling *(Calidris alba)* Spring and fall migrants. Like Marbled Godwit, often seen in small groups frequenting muddy lake or delta edges and also more regular on the East Side. The transitional plumage seen in migrant Sanderlings often causes confusion among birders only familiar with basic-plumaged birds.

Baird's Sandpiper *(Calidris bairdii)* Late summer/ early fall migrants. Usually seen with other shorebirds on muddy edges of lakes and reservoirs with the great majority of records on the East Side from south Lake Tahoe to the Owens Valley.

Pectoral Sandpiper *(Calidris melanotos)* Mainly fall migrants. Generally prefers more vegetative cover than other Calidris sandpipers; like Baird's Sandpiper, most records are on the East Side, but generally more rare than Baird's; peak numbers are found in September.

Short-billed Dowitcher *(Limnodromus griseus)* Spring and late August/early September migrants. Found at muddy edges of lakes, reservoirs, and river deltas on the East Side, occasionally at West Side sewage ponds. Most reports are of juveniles in fall, which are more easily distinguished from Long-billed Dowitchers.

Sabine's Gull *(Xema sabini)* Fall migrants. Most often found on East Side lakes and reservoirs; most records from September; accidental in May and June.

Franklin's Gull *(Leucophaeus pipixcan)* Spring and summer/fall migrants. Great majority of records from East Side lakes and reservoirs; very nearly annual at Honey Lake (Lassen County), in the Owens Valley, and at Crowley Lake; West Side records from South Fork Kern River Valley and Salt Springs Reservoir (Calaveras County).

Mew Gull *(Larus canus)* Winter visitors. Has occurred widely on both East and West Side lakes and reservoirs, mostly in winter.

Common Tern *(Sterna hirundo)* Mainly fall migrants. All but a few records from East Side lakes and reservoirs; mostly in September.

Parasitic Jaeger *(Stercorarius parasiticus)* Late spring/early summer and fall migrants. Nearly all records are from East Side lakes and reservoirs, with the species most frequently found at Mono Lake, Crowley Lake, and Lake Tahoe; mostly first-year birds.

Long-tailed Jaeger *(Stercorarius longicaudus)* Mainly fall migrants. Fall observations from Lake Tahoe, Mono and Crowley Lakes, and the Owens Valley account for most records. An August record from Tuolumne County in 1998 may be the only documented West Side occurrence. It appears that although Long-tailed is the jaeger more regularly found well inland (excluding coastal locations) in most of California, Parasitic may occur with nearly the same frequency as Long-tailed in the Sierra. Difficulties differentiating between juveniles of these species make their status uncertain.

Barred Owl *(Strix varia)* Year-round residents. While still rare, this species has recently expanded its range into the Sierra, with records as far south as Kings Canyon National Park (Tulare County). Barred × Spotted Owl hybrids have been found as far south as southern Placer County on the West Side and south Lake Tahoe (El Dorado County) on the East Side.

Broad-tailed Hummingbird *(Selasphorus platycercus).* Occasional breeder? Although this species is a regular Great Basin breeder up to the eastern edge of the Sierra, other than historical breeding records and occasional breeding season reports from western Mono County, this species is most likely casual in the Sierra on the East Side from late July through August and accidental on the West Side. Difficulty in separating female/immatures from other Selasphorus hummers is underappreciated by many birders.

Cordilleran Flycatcher *(Empidonax occidentalis)* Spring/summer breeders. Since the Western

Flycatcher was split into Pacific-slope and Cordilleran Flycatchers, identification challenges abound where their ranges meet. Most experts agree that the birds found in the canyons on the western edge of Mono County south into northern Inyo County give vocalizations more like those of Cordilleran than Pacific-slope. The taxonomy of this recently described species continues to be under study.

Cassin's Kingbird *(Tyrannus vociferans)* Occasional breeders and spring visitors. Has bred in recent years in Onyx (Kern County) near the southern edge of the region; other records in Mono County and the Owens Valley are mainly from spring.

Gray Jay *(Perisoreus canadensis)* Winter visitors. This species' range barely crosses into the Sierra from the Cascades, but birds are occasionally found in the vicinity of Lake Almanor on Christmas Bird Counts and in fall and winter in Butte County above 5,800 feet.

Bendire's Thrasher *(Toxostoma bendirei)* Occasional breeders. Has bred near the southern edge of the Sierra south of Weldon (Kern County) and has been seen on several occasions in Butterbredt Canyon and the Kelso Valley, though breeding has not been confirmed in those locations.

Bohemian Waxwing *(Bombycilla garrulus)* Winter visitors. Flocks, sometimes numbering into the hundreds, occasionally winter as far south as Tulare County; most records are from the East Side.

Lapland Longspur *(Calcarius lapponicus)* Fall through early spring. Usually found in small numbers mixed with Horned Larks or American Pipits in open fields with short vegetation. All Sierra records from the East Side, with individuals present most winters in the Honey Lake area.

Chestnut-collared Longspur *(Calcarius ornatus)* Fall through early spring. Found in the same habitats and species associations as Lapland Longspur (see above), but more rarely and generally in fewer numbers. Nearly all Sierra records are from the East Side.

Northern Waterthrush *(Parkesia noveboracensis)* Spring and fall migrants. Frequents wooded streams and swampy areas; more regular on the East Side with most records from Mono County south to Kern County.

Black-and-white Warbler *(Mniotilta varia)* Spring and fall migrants. Areas with large deciduous trees are the best places to look for this bird; much more regular on the East Side than the West.

American Redstart *(Setophaga ruticilla)* Late spring/early summer and fall migrants. Lower elevation riparian areas in Mono and Lassen Counties and Butterbredt Spring produce records of this species nearly every year.

Northern Parula *(Setophaga americana)* Late spring/early summer migrants. A male singing in a riparian area with a mature cottonwood/willow canopy is a nearly annual event in Mono, Inyo, or Kern Counties, where the vast majority of reports originate.

Chestnut-sided Warbler *(Setophaga pensylvanica)* Late spring/early summer and fall migrants. Most records are from the East Side riparian areas below 8,000 feet.

American Tree Sparrow *(Spizella arborea)* Winter visitors. Usually found in areas with dense, shrubby vegetation. Nearly annual on the East Side in Lassen County, rare or casual as one moves south of Mono County.

Lark Bunting *(Calamospiza melanocorys)* Spring and fall migrant/winter visitors. Usually found in the company of White- and Golden-crowned Sparrow flocks in winter, this species prefers open areas with patches of shrubby vegetation. Most records are from the low foothills of the West Side and the Owens Valley on the East Side.

Swamp Sparrow *(Melospiza georgiana)* Late fall-winter visitors. Found in or near wetland areas with willow thickets or other dense vegetation. West Side records in low foothills and East Side records mostly from basins around Honey Lake, Mono Lake, and the Owens Valley; a few spring records from the East Side.

Harris's Sparrow *(Zonotrichia querula)* Winter visitors. Nearly always in flocks of White- and Golden-crowned sparrows, East Side records of this species outnumber West Side records more than 10 to 1. The Honey Lake and Mono Basin areas host this species most winters.

CASUAL SPECIES

Trumpeter Swan *(Cygnus buccinator)* All but one accepted record (western Nevada County, December 2008) from the East Side in winter through early spring.

Long-tailed Duck *(Clangula hyemalis)* Most Sierra records from late fall to winter in the Owens Valley and other East Side lakes and reservoirs.

Red-throated Loon *(Gavia stellata)* Nearly all records from November and December in large lakes and reservoirs on both slopes, with Owens Valley producing the most records.

Red-necked Grebe *(Podiceps grisegena)* Mostly fall and winter, but with a few spring records in large lakes and reservoirs. East Side records outnumber West Side records more than 2 to 1.

Brown Pelican *(Pelecanus occidentalis)* Most records are of juveniles in late summer/early fall, with Mono Lake accounting for three-fourths of Sierra occurrences.

Least Bittern *(Ixobrychus exilis)* Historically bred in marshes near Lake Tahoe, Mono Lake, and probably Honey Lake. With the exception of a July record at Lake Tahoe, recent records are all from May and June in the Mono Basin and Owens Valley and possibly represent breeding attempts. Several breeding season records on the West Side from Kern River Preserve (Kern County).

Cattle Egret *(Bubulcus ibis)* All Sierra records are from the East Side in spring and fall except for occasional records from the Kern River Valley.

Broad-winged Hawk *(Buteo platypterus)* Spring and summer records from Mono Basin and Owens Valley, fall records from Alpine County, Kern River Valley, Mono Basin, and Lake Tahoe (Placer County).

Yellow Rail *(Coturnicops noveboracensis)* Nested in Mono County near Mono Lake and Bridgeport into the 1950s; no recent records in the Sierra but calling birds at Willow Lake (Plumas County) in 2010 and 2011 and at three locations in Lassen County in 2011 were just beyond the northern border of the region. Like Least Bittern above, the secretive nature of the species and the lack of coverage of many potential breeding sites make status uncertain.

Pacific *(Pluvialis fulva)* and **American** *(Pluvialis dominica)* **Golden-Plovers** Exact status for each difficult to assess because most records prior to the split of these species were not identified to subspecies and difficulty of differentiation requires excellent documentation. Mostly fall records, all from the East Side, and nearly all from Mono County or the Owens Valley; well-documented fall records of American Golden-Plovers from Crowley Lake, Owens Valley (3), Lake Tahoe, and an alternate-plumaged male at Honey Lake suggest that American Golden-Plovers may be the more expected species in the Sierra.

Mountain Plover *(Charadrius montanus)* Records almost exclusively from fall or winter on the East

Side with the Owens Valley accounting for nearly half the total observations.

Wandering Tattler *(Tringa incana)* Mostly fall records with a few from spring and all from the East Side and all but two from Mono County.

Whimbrel *(Numenius phaeopus)* Mainly spring records with a fewer from July into fall; most East Side observations from Lake Tahoe, Mono Lake, and the Owens Valley; despite this species being a common spring migrant in the Central Valley, we are aware of West Side records only from the Kern River Valley.

Ruddy Turnstone *(Arenaria interpres)* May, July, and August account for nearly all Sierra records, all from the East Side except for one record at Lake Isabella.

Red Knot *(Calidris canutus)* Most found in late summer/early fall but with a few spring records; all from the East Side, mainly the Owens Valley and Mono County.

Semipalmated Sandpiper *(Calidris pusilla)* Nearly all observations are from August in the Owens Valley and Mono County.

Stilt Sandpiper *(Calidris himantopus)* Four of five Owens Valley records from July or August with one from May; two September records from Honey Lake; several Mono County records all from August to early September.

Red Phalarope *(Phalaropus fulicarius)* Occurrences almost exclusively in fall from the East Side; summer records from Honey Lake and Mono Lake; West Side records from Lake Success and Yosemite Valley.

Heermann's Gull *(Larus heermanni)* Mostly in spring; all records from East Side lakes and reservoirs.

Western Gull *(Larus occidentalis)* A smattering of records from spring, fall, and winter from both slopes with one well-described adult in late June at Lake Almanor.

Thayer's Gull *(Larus thayeri)* Like other "coastal" gulls, this species seems to be increasing in late fall and winter inland with Sierra records from most of the major lakes and reservoirs; nearly annual in the past decade.

Lesser Black-backed Gull *(Larus fuscus)* All from fall-winter and all but one (Kaweah Lake, Tulare County) from the East Side.

Glaucous-winged Gull *(Larus glaucescens)* As with Thayer's Gull above, the Glaucous-winged Gull appears to be increasing in winter in the region.

Glaucous Gull *(Larus hyperboreus)* All late fall through winter occurrences and all from the East Side, with Virginia Lake in Reno accounting for more than half.

Least Tern *(Sternula antillarum)* Spring and summer occurrences all from the Owens Valley and Mono County.

Arctic Tern *(Sterna paradisaea)* Summer and fall records from East Side lakes and reservoirs; one spring record from the Owens Valley.

White-winged Dove *(Zenaida asiatica)* Records, nearly always of single birds, evenly distributed between spring and fall from Alpine County south into Kern County.

Chimney Swift *(Chaetura pelagica)* Mainly spring and early summer and nearly all from either Mono or Inyo Counties; fall records from Lake Tahoe; usually observed with Vaux's Swifts.

Allen's Hummingbird *(Selasphorus sasin)* Actual status uncertain due to difficulty of separation from Rufous Hummingbird; specimen records and birds identified in the hand from West Side locations in Mariposa and Tulare Counties support some occurrence in spring and late summer/early fall. Banding data from the Kern River Preserve in July and August suggest that the ratio of Allen's to Rufous may be 1 to 20 or even higher at that location!

Yellow-bellied Sapsucker *(Sphyrapicus varius)* Nearly all observations are from winter and mostly from the southern half of the region.

Least Flycatcher *(Empidonax minimus)* Almost all Sierra records are of singing birds in late spring and summer in riparian areas.

Eastern Phoebe *(Sayornis phoebe)* Mainly winter records from both slopes, but a couple of May occurrences in Mono County.

Eastern Kingbird *(Tyrannus tyrannus)* Great majority found in summer and fall on the East Side with the Owens Valley and Honey Lake accounting for more than half the records; one confirmed breeding record from Honey Lake.

Scissor-tailed Flycatcher *(Tyrannus forficatus)* Records from May through September, other than a fall record from the Kern River Valley, all from the East Side from Lake Tahoe south to Butterbredt Spring.

White-eyed Vireo *(Vireo griseus)* All but two Sierra records fall into the May-June time frame typical of this species and all are from the East Side except for one record from the Kern River Preserve; Mono Basin records from August and September account for two of only five state records for these months.

Yellow-throated Vireo *(Vireo flavifrons)* Nearly all from May or June and all East Side except for one Kern River Valley July record.

Red-eyed Vireo *(Vireo olivaceus)* All records from late spring/early summer and nearly all from the Kern River Valley (has nested there) or the East Side; Butterbredt Spring accounts for a large portion of the East Side records.

Gray Catbird *(Dumetella carolinensis)* Slightly more records from fall than spring and all but one from the East Side.

Brown Thrasher *(Toxostoma rufum)* Mostly spring records from the East Side, with a few from fall; West Side records include two from Springville (Tulare County) and one from the Kern River Valley.

McCown's Longspur *(Rhynchophanes mccownii)* Mainly fall records and all from the East Side; one April occurrence in Mono County.

Ovenbird *(Seiurus aurocapilla)* Nearly every occurrence from May or June with half from the Mono Lake basin; West Side records from Kings Canyon National Park and Mariposa County.

Worm-eating Warbler *(Helmitheros vermivorum)* All records from April through June except for one occurrence in Bishop (Inyo County) in November; five of the seven Sierra records from Butterbredt Spring.

Golden-winged Warbler *(Vermivora chrysoptera)* Spring records from Butterbredt Spring and late summer records from the Kern River Preserve and locations in Mono County; fall record from the Owens Valley.

Blue-winged Warbler *(Vermivora cyanoptera)* All records from late May into June and all but one found at Butterbredt Spring; single exception was in Bridgeport in June 1984. A Blue-winged × Golden-winged hybrid specimen was obtained near Westwood (Lassen County) in July 1984.

Prothonotary Warbler *(Protonotaria citrea)* Mostly fall records from the East Side with the Owens Valley accounting for nearly half; a spring record from Canebrake Ecological Reserve (Kern County) may be the only West Side observation.

Tennessee Warbler *(Oreothlypis peregrina)* Nearly all observations from East Side locations throughout the region from spring and fall; West Side record from Ackerson Meadow (Tuolumne County) is also the only July record for the region.

Lucy's Warbler *(Oreothlypis luciae)* Most records from spring in the Owens Valley; one August Owens Valley occurrence; one observation of adult with fledglings in the Kern River Valley suggests possible breeding nearby.

Kentucky Warbler *(Geothlypis formosa)* Nearly all records from May or June on the East Side; Butterbredt Spring accounts for more than half the Sierra records.

Hooded Warbler *(Setophaga citrina)* Most occurrences are of singing males in spring and most are from the East Side, with a large proportion from the Owens Valley; one nesting attempt documented in the Kern River Valley.

Magnolia Warbler *(Setophaga magnolia)* Slightly more spring records than fall; nearly all from the East Side in Mono and Inyo Counties and from Butterbredt Spring.

Blackburnian Warbler *(Setophaga fusca)* Mainly spring records with nearly all from East Side locations, from Plumas to Kern Counties.

Blackpoll Warbler *(Setophaga striata)* Most observations are from September and nearly all from East Side from Lassen County to Kern; only West Side records from Yosemite Valley (Mariposa County) in April 1980 and Kern River Valley in May 2007.

Black-throated Blue Warbler *(Setophaga caerulescens)* A mix of early summer and mid-fall records scattered throughout the region from both slopes; winter records include one on the West Side in Auburn (Placer County) and another in Rovana (Inyo County).

Palm Warbler *(Setophaga palmarum)* Spring and fall/winter occurrences approximately equal with nearly all from the East Side in either the Owens Valley or Kern County.

Yellow-throated Warbler *(Setophaga dominica)* All but one record from spring/early summer either at Butterbredt Spring or the Owens Valley; a fall record from El Dorado County.

Prairie Warbler *(Setophaga discolor)* Mix of spring and fall observations with most from Mono Basin. Only one West Side record from Grass Lake (Plumas County).

Canada Warbler *(Cardellina canadensis)* Mostly spring records from Butterbredt Spring and other East Side locations; one West Side occurrence near Springville (Tulare County) in July 1969.

Painted Redstart *(Myioborus pictus)* Mostly spring records from Butterbredt Spring but has occurred on the West Side in Tulare County in summer 1969 and Placer County in March 2010.

Clay-colored Sparrow *(Spizella pallida)* About half the records are from fall with a couple from June and a couple from winter; occurrences scattered throughout the region.

Painted Bunting *(Passerina ciris)* Occurrences of escaped cage birds makes exact status uncertain for many California records, but all CBRC-accepted Sierra records (species reviewed through 2004) are from the Owens Valley in August and September.

Dickcissel *(Spiza americana)* Mainly spring and fall records from the Owens Valley.

Bobolink *(Dolichonyx oryzivorus)* About half the records from September, with the other half from May through June and nearly all from the East Side, mostly from the Owens Valley.

Common Grackle *(Quiscalus quiscula)* Mostly fall and winter and all but two records from the Owens Valley; single West Side record from Sierra County in June.

Baltimore Oriole *(Icterus galbula)* A mix of spring and fall records, mostly from the East Side except for a few occurrences in the Kern River Valley.

Black Rosy-Finch *(Leucosticte atrata)* Most records are from Aspendell (Inyo County) November to April; December records from Bridgeport and Carson City, Nevada.

ACCIDENTAL SPECIES

Whooper Swan *(Cygnus cygnus)* Lake Almanor in winter 2001.

Falcated Duck *(Anas falcata)* Honey Lake in spring 2002 and winter-spring in 2003.

Baikal Teal *(Anas formosa)* Honey Lake in winter 1974.

Tufted Duck *(Aythya fuligula)* Winter records from Alpine, Plumas, and Kern Counties.

Black Scoter *(Melanitta americana)* November records from Mono County (2), the Owens Valley (2), and Lake Tahoe (Placer County).

Smew (*Mergellus albellus*) Male in Tuolumne County foothills two successive winters, in January 2007 and December 2007–February 2008.

Yellow-billed Loon (*Gavia adamsii*) Grant Lake (Mono County) in winter 1976 and Mud Lake, Nevada, in winter 1996.

Black Storm-Petrel (*Oceanodroma melania*) Kings Canyon National Park (Tulare County) in fall 1994; exhausted bird rehabilitated and released.

Magnificent Frigatebird (*Fregata magnificens*) Remains found at Mono Lake in 1985 and birds documented at Lake Isabella in fall 1997; Diaz Lake (Inyo County) and Mono Lake, both summer 1998; Crowley Lake in fall 2002.

Blue-footed Booby (*Sula nebouxii*) New Hogan Reservoir (Calaveras County) in fall 1976.

Little Blue Heron (*Egretta caerulea*) Spring and fall records from the Owens Valley, a spring record in Sierra Valley (Plumas County), and a West Side spring record from Mariposa County.

Tricolored Heron (*Egretta tricolor*) Honey Lake in fall 1971.

Reddish Egret (*Egretta rufescens*) Little Lake (Inyo County) in fall 2001.

Glossy Ibis (*Plegadis falcinellus*) The Owens Valley in April 2009 and the Sierra Valley (Plumas County) in June 2010.

Black Vulture (*Coragyps atratus*) The Owens Valley in August 2011.

Mississippi Kite (*Ictinia mississippiensis*) Records all late May–early June; two in the Owens Valley, one each at Mono Basin and the Kern River Preserve.

Zone-tailed Hawk (*Buteo albonotatus*) Fall record from the Owens Valley and fall and spring records from the Kern River Valley.

Crested Caracara (*Caracara cheriway*) Fall record from Mono Basin and spring records from the Owens Valley and Kern River Valley.

Upland Sandpiper (*Bartramia longicauda*) Owens Valley records from June 1993 and October 2006.

Black Turnstone (*Arenaria melanocephala*) The Owens Valley in August 2007 and Mono Lake in July 1999 and September 2011.

Little Stint (*Calidris minuta*) The Owens Valley in August 2009 and 2010.

White-rumped Sandpiper (*Calidris fuscicollis*) Both records from Mono County, one in June 1981 and one in May 2003.

Ruff (*Philomachus pugnax*) Three Owens Valley records and one from Bridgeport Reservoir, all from September; a male in breeding plumage was south of Honey Lake in July 1999.

Black-legged Kittiwake (*Rissa tridactyla*) Mono Lake in December 2006.

Little Gull (*Hydrocoloeus minutus*) Two Crowley Lake records from August through September and a June record from Lake Tahoe (El Dorado County).

Laughing Gull (*Leucophaeus atricilla*) Late summer records from Mono Lake and Lake Isabella.

Yellow-footed Gull (*Larus livens*) Crowley Lake in July 1991.

Pomarine Jaeger (*Stercorarius pomarinus*) Three September records from Mono Lake cited in Gaines (1992), but we are unaware of any with specimen or photographic documentation.

Long-billed Murrelet (*Brachyramphus perdix*) Dead birds found at Mono Lake in August 1981, 1983, and 1986; one in September 1983.

Ancient Murrelet (*Synthliboramphus antiquus*) As above, a dead bird found at Mono Lake in December 1985.

Common Ground-Dove (*Columbina passerina*) All records from Kern County except for an August 1987 bird in Mono County.

Ruddy Ground-Dove (*Columbina talpacoti*) Two records from the Owens Valley in November 1998 and November 2003 through the following January.

Black-billed Cuckoo (*Coccyzus erythropthalmus*) Near Mono Lake in August 1986.

Mexican Whip-poor-will (*Caprimulgus arizonae*) One from Tulare County north of Springville in June 1983; another near Forbestown (Butte County) in June and July 2010.

Broad-billed Hummingbird (*Cynanthus latirostris*) Three fall records from the Owens Valley and Kern River Valley; one spring bird in Short Canyon (Kern County).

Violet-crowned Hummingbird (*Amazilia violiceps*) Grass Valley (Nevada County) in February 2006.

Blue-throated Hummingbird (*Lampornis clemenciae*) Only California record near Three Rivers (Tulare County) from December 1977 well into May 1978.

Ruby-throated Hummingbird *(Archilochus colubris)* Both Sierra records from Nevada County in May 1975 and August 2008.

Eastern Wood-Pewee *(Contopus virens)* Summer 1998 in South Fork Wildlife Area (Kern County) and near Mono Lake and fall 2011 in Birchim Canyon (Inyo County).

Alder Flycatcher *(Empidonax alnorum)* South Fork Wildlife Area in July 1991 and Butterbredt Spring in May 1992.

Dusky-capped Flycatcher *(Myiarchus tuberculifer)* Bishop in November 1977 and November 1997.

Tropical Kingbird *(Tyrannus melancholicus)* South Fork Wildlife Area in September 1996.

Thick-billed Kingbird *(Tyrannus crassirostris)* Lone Pine (Inyo County) in December 1991 through April 1992.

Gray Vireo *(Vireo vicinior)* Specimen collected in 1920s near Walker Pass (Kern County).

Blue-headed Vireo *(Vireo solitarius)* Butterbredt Spring in April 2004.

Philadelphia Vireo *(Vireo philadelphicus)* Two summer records from Mono Lake basin; fall record from Owens Valley.

Blue Jay *(Cyanocitta cristata)* Lake Tahoe (El Dorado County) in November 1983 through March 1984.

Winter Wren *(Troglodytes hiemalis)* Lundy Canyon (Mono County) in December 2004.

Veery *(Catharus fuscescens)* Near Mono Lake in June 2004 and Chester (Plumas County) in June 2010.

Wood Thrush *(Hylocichla mustelina)* June records from Mono Basin and Sagehen Creek (Nevada County) and a fall (specimen) record from Lee Vining (Mono County).

Red-throated Pipit *(Anthus cervinus)* Crowley Lake in May 2007.

Snow Bunting *(Plectrophenax nivalis)* Winter records from Honey Lake, Kelso Valley (Kern County), and Sierra Valley (Plumas County).

Connecticut Warbler *(Oporornis agilis)* Butterbredt Spring in May 2006.

Mourning Warbler *(Geothlypis philadelphia)* Bishop in August 2002.

Cerulean Warbler *(Setophaga cerulea)* Yosemite National Park (Mariposa County) in October 1981 and Birchim Canyon (Inyo County) in May 1997.

Bay-breasted Warbler *(Setophaga castanea)* Summer and fall records from Lassen, Mono, and Inyo Counties, and one from western Nevada County in August 2009.

Pine Warbler *(Setophaga pinus)* Ackerson Meadow (Tuolumne County) in November 1987.

Grace's Warbler *(Setophaga graciae)* Chimney Creek Campground (Tulare County), May–June 2004.

Black-throated Green Warbler *(Setophaga virens)* October records from the Owens Valley (2) and near Mono Lake; May record from Butterbredt Spring and June record from the Owens Valley.

Red-faced Warbler *(Cardellina rubrifrons)* Bishop, May 1998.

Cassin's Sparrow *(Peucaea cassinii)* June records from Mono Lake in 1984 and Kern River Valley in 2001.

Rusty Blackbird *(Euphagus carolinus)* Fall records from Alpine County and Mono Basin, a spring record from Honey Lake, and one December record from the Owens Valley.

Orchard Oriole *(Icterus spurius)* Mono Basin in August 1974.

Common Redpoll *(Acanthis flammea)* January records from Sierra Valley (Plumas County) in 2004 and Honey Lake in 2005 and an early December 2011 record at June Lake. A February 2012 record near Taylorsville (Plumas County) is the only West Side occurrence.

NEAR MISSES

The list above includes only records of species that fall strictly within the boundaries of the Sierra as we defined them (see our Introduction). Below are two species that occurred very close to the edge of the region:

Iceland Gull *(Larus glaucoides)* One record in Reno less than one-quarter mile east of Highway 395.

White Wagtail *(Motacilla alba)* A bird near Woodlake (Tulare County) was just below 500 feet.

APPENDIX 3

Methods Used to Determine Population Trends

FROM BREEDING BIRD SURVEY (BBS) DATA

We used data from 44 BBS routes that are within the boundaries we have used to define the Sierra (see Figure 13 in the chapter "Recent Trends in Sierra Bird Populations and Ranges") and that have been run between 1971 and 2010. Of the species detected on those routes, we selected 107 for analyses based on their relatively high abundances and widespread ranges. For each species we totaled all observations from all routes and performed linear regression analyses using number per routes versus year. Trends were determined to be significant only if they met the following criteria:

- trend from linear regression using data from all routes showed $p < 0.0005$ (based on applying Bonferroni adjustment), and
- trend from linear regression using data from only those routes that had been run at least eight times from 1971 to 1990 *and* at least eight times from 1991 to 2010 (24 of the 44 routes) showed $p < 0.0005$.

FROM CHRISTMAS BIRD COUNT (CBC) DATA

We used data from 25 CBC circles that are within the boundaries we have used to define the Sierra (see Figure 13 in the chapter "Recent Trends in Sierra Bird Populations and Ranges") and that have been run between 1971 and 2010. Of the approximately 200 species detected on those circles, we selected 101 species for analyses based on their relatively high abundances and widespread ranges. For each species we totaled all observations from

all circles and performed linear regression analyses using number per party hour versus year. Trends were determined to be significant only if they met the following criteria:

- trend from linear regression using data from all circles showed $p < 0.0005$ (based on applying Bonferroni adjustment), and
- trend from linear regression using data from only those circles that had been run at least eight times from 1971 to 1990 *and* at least eight times from 1991 to 2010 (12 of the 25 circles) showed $p < 0.0005$.

We applied the second criterion in each case because CBCs and BBS routes are not run consistently every year, and we wanted to be assured that apparent trends were not merely artifacts of when certain routes or circles were run. For example, if a route or circle with high abundance of a given species was only run in recent years, that could produce a false appearance of a positive trend over time when data from the full set of routes or circles were analyzed. Note that because our definition of the Sierra boundaries does not match exactly with the "Sierra Nevada Bird Conservation Region (BCR)" used by U.S. Geological Survey, our trend analyses cannot be compared directly with those of Sauer et al. 2011. Compared to the area we defined as the Sierra, BCR borders go further north, do not extend as far south, and do not include much of the lower elevation areas on the West and East sides. Sauer et al. 2011 used a Bayesian hierarchical statistical method to analyze trends and also used a slightly different time frame (1968–2009) than ours.

APPENDIX 4

Common and Scientific Names
of Plant Species

CONIFEROUS TREES

white fir *(Abies concolor)*

red fir *(A. magnifica)*

incense cedar *(Calocedrus decurrens)*

western juniper *(Juniperus occidentalis)*

Utah juniper *(J. osteosperma)*

whitebark pine *(Pinus albicaulis)*

foxtail pine *(P. balfouriana)*

lodgepole pine *(P. contorta* var. *murrayana)*

Jeffrey pine *(P. jeffreyi)*

sugar pine *(P. lambertiana)*

singleleaf pinyon pine *(P. monophylla)*

western white pine *(P. monticola)*

pondcrosa (yellow) pine *(P. ponderosa)*

gray (foothill) pine *(P. sabiniana)*

Douglas-fir *(Pseudotsuga menziesii)*

giant sequoia *(Sequoiadendron giganteum)*

mountain hemlock *(Tsuga mertensiana)*

BROAD-LEAVED TREES

big-leaf maple *(Acer macrophyllum)*

California buckeye *(Aesculus californica)*

mountain alder *(Alnus incana* var. *tenuifolia)*

white alder *(A. rhombifolia)*

Pacific madrone *(Arbutus menziesii)*

Pacific dogwood *(Cornus nuttallii)*

western sycamore *(Plantanus racemosa)*

Fremont cottonwood *(Populus fremontii)*

quaking aspen *(P. tremuloides)*

black cottonwood *(P. balsamifera* ssp. *trichocarpa)*

western chokecherry *(Prunus virginiana* var. *demissa)*

canyon live oak *(Quercus chrysolepis)*

blue oak *(Q. douglasii)*

California black oak *(Q. kelloggii)*

valley oak *(Q. lobata)*

interior live oak *(Q. wizlizenii)*

California bay laurel *(Umbellularia californica)*

WOODY SHRUBS

chamise *(Adenostoma fasciculatum)*

greenleaf manzanita *(Arctostaphylos patula)*

whiteleaf manzanita *(A. viscida)*

Great Basin sagebrush *(Artemisia tridentata)*

golden chinquapin *(Castanopsis chrysophylla)*

bush chinquapin *(C. sempervirens)*

snowbrush *(Ceanothus cordulatus)*

buckbrush *(C. cuneatus)*

redbud *(Cercis occidentalis)*

curl-leaf mountain mahogany *(Cercocarpus ledifolius)*

kit-kit-dizze (mountain misery) *(Chamabatia foliolosa)*

common rabbitbrush *(Chrysothamnus viscidiflorus)*

American dogwood *(Cornus sericea)*

Oregon ash *(Fraxinus latifolia)*

toyon *(Heteromeles arbutifolia)*
desert bitterbrush *(Purshia glandulosa)*
huckleberry oak *(Quercus vaccinifolia)*
California coffeeberry *(Rhamnus californica)*
Sierra currant *(Ribes nevadense)*
mountain gooseberry *(R.monitigenum)*
Himalayan blackberry *(Rubus armeniacus)*
California blackberry *(R. ursinus)*
willow *(Salix* spp.)
poison oak *(Toxicodendron diversilobum)*
California wild grape *(Vitis californica)*

DESERT TREES AND SHRUBS
catclaw acacia *(Acacia greggii)*
desert agave *(Agave deserti)*
blue paloverde *(Cercidium floridum)*
smoketree *(Dalea spinosa)*
burrowbrush *(Hymenoclea salsola)*

desert ironwood *(Olneya tesota)*
teddybear cholla *(Opuntia biglovii)*
beavertail pricklypear *(Opuntia basilaris)*
mesquite *(Prosopis glandulosa)*
tamarisk *(Tamarix* sp.)
Joshua tree *(Yucca brevifolia)*
Mohave yucca *(Yucca shidigera)*

NONWOODY PLANTS
wild onion *(Allium* spp.)
sedge *(Carex* spp.)
paintbrush *(Castilleja* spp.)
shooting star *(Dodecantheon* spp.)
rush *(Juncus* spp.)
penstemon *(Penstemon* spp.)
bracken fern *(Pteridium aquilinum)*
tule (bulrush) *(Schoenoplectus* spp.)
cattail *(Typha* spp.)

GLOSSARY

ALTRICIAL Species in which the young are hatched featherless, helpless, and in need significant parental care such as feeding and brooding (see *precocial*).

CALLS Vocalizations used by birds for locating each other in a flock, sending alarm signals, and for other purposes; calls are distinguished from songs by being less complex and not used for mate-attraction or territorial defense (see *song*).

COLOR MORPH A term applied to adult birds that have distinctly different plumage than the standard colors for their species; these do not represent a distinct lineage (e.g., subspecies), and such birds maintain color differences throughout different seasonal molts; examples include dark Red-tailed Hawks and "blue" Snow and Ross's Geese.

ENDANGERED SPECIES Those species officially designated by state or federal resource agencies as being at risk of extinction. They receive the highest level of conservation concern and protection (see *threatened*).

ENDEMIC SPECIES Those species (or subspecies) found exclusively within a given geographic area; Sierra examples include breeding "Thick-billed" Fox Sparrows and "Kern" Red-winged Blackbirds.

EXTINCT SPECIES Those species believed to no longer exist anywhere (e.g., Passenger Pigeon).

EXTIRPATED SPECIES Those species believed to no longer exist in a given geographic area; a Sierra example is the California Condor.

FULLY PROTECTED SPECIES Those designated by the California Department of Fish and Game in the 1960s to protect declining populations of birds; the state Endangered Species Act now provides safeguards for fully protected bird species.

FLEDGLING BIRDS Those that have left the nest but are inexperienced fliers and foragers and still require parental care such as feeding and protection from predators.

HYBRID INDIVIDUALS These have parents of two different species (see *intergrade*).

IMMATURE BIRDS These have molted out of juvenile plumage but have not attained fully adult plumage.

IRRUPTIVE SPECIES These sometimes display mass seasonal movements involving intermittent shifts into areas beyond their usual range, generally in the nonbreeding season; Sierra examples include Lewis's Woodpeckers and Pinyon Jays.

INTERGRADE INDIVIDUALS These have parents of two different subspecies (see *hybrid*).

INTRODUCED SPECIES These were released by humans into an area outside their native range.

JUVENILE BIRDS Birds in juvenile plumage (see "plumage, juvenile").

NESTLING BIRDS Birds that have hatched but cannot fly and are still in the nest.

NONPASSERINE SPECIES These are not in the order *Passeriformes* (perching birds); for the purposes of this book, they include all species from waterfowl through woodpeckers.

OSCINE SPECIES A subcategory of the order *Passeriformes* (perching birds) characterized by a highly specialized vocal apparatus for singing complex songs (see *suboscine*); for the purposes of this book, they include all species from shrikes through House Sparrow.

PASSERINE SPECIES In the order *Passeriformes* (perching birds); for the purposes of this book, they include all species from flycatchers through House Sparrow.

PLUMAGE, BREEDING The distinct set of feathers, often colorful and vivid, characterizing many species in the breeding season (typically spring and summer in the Sierra); often, but not always, equivalent to "alternate" plumage-a term not used in this book.

PLUMAGE, NONBREEDING A distinct, often less colorful set of feathers characterizing many species in the nonbreeding season (generally fall through winter for Sierra birds); often, but not always, equivalent to "basic" plumage-a term not used in this book.

PLUMAGE, JUVENILE A bird's first full plumage after leaving the nest; in some bird families (e.g., most sparrows) this plumage is retained only briefly; in others (e.g., many raptors) it may be retained for a full year.

POLYANDROUS SPECIES Those that use a breeding strategy where females mate with more than one male; Sierra examples include Spotted Sandpipers and Wilson's Phalaropes.

POLYGYNOUS SPECIES Those that use a breeding strategy where males mate with more than one female; Sierra examples include Wild Turkeys and Red-winged Blackbirds.

PRECOCIAL SPECIES Fully feathered and able to feed themselves and regulate body temperature just after hatching (see *altricial*).

SONGS The typically complex and often melodious vocalizations used by many bird species for mate-attraction and territorial defense (see *calls*).

SUBOSCINE SPECIES In the order *Passeriformes*, perching birds that lack the vocal apparatus capable of producing complex songs (see *oscine species*); their songs are usually innate, rather than learned; for the purposes of this book, they include only the flycatchers.

SUBSPECIES Distinct populations showing differences in structural, plumage, behavioral, or genetic characteristics but are not considered full species.

SYRINX A unique vocal organ of birds located at the end of the windpipe or trachea.

TAXONOMY The theory and practice of grouping individual taxa (subspecies, species, genera, and so on) and organizing them into larger groups based on shared characteristics, genetics, and evolutionary history.

THREATENED SPECIES Officially designated by state or federal resource agencies as being at risk of becoming endangered, and have the second-highest level of conservation concern and protection (see *endangered*).

TYPE SPECIMEN Defines the original, collected individual used to classify a species.

BIBLIOGRAPHY

INTRODUCTION AND FREQUENTLY CONSULTED REFERENCES

American Ornithologists' Union. 1998. Checklist of North American birds. 7th edition. Allen Press, Inc., Lawrence, KS.

Baicich, P. J., and C. J. O. Harrison.1997. A guide to the nests, eggs, and nestlings of North American birds. 2nd edition. Academic Press, New York, NY.

Beedy, E. C., and S. L. Granholm. 1985. Discovering Sierra birds. Yosemite Natural History Association and Sequoia Natural History Association.

Bent, A. C. 1961–1968. Life histories of North American birds. 26 volumes. Dover Publications, Inc., New York, NY.

Choate, E. A. 1973. The dictionary of American bird names. Gambit, Boston, MA.

Cogswell, H. L. 1977. Water birds of California. Univ. of Calif. Press, Berkeley, CA.

Dawson, W. L. 1923. The birds of California. 4 volumes. South Moulton Co., San Francisco, CA.

DeSante, D. F. 1999. Species accounts for the Sierra Nevada bird conservation plan. Available at http://www.prbo.org/calpif/htmldocs/sierra/spec accts.html. [accessed March 2012].

eBird. 2011. eBird. An online database of bird distribution and abundance [web application]. Version 2. eBird, Ithaca, NY. Available at http://www.ebird.org October. [accessed October–November 2011].

Ehrlich, P. R., D. S. Dobkin, and D. Wheye. 1988. The birder's handbook: A field guide to the natural history of North American birds. Simon & Schuster, New York, NY.

Evens, J., and I. Tait. 2005. Introduction to California birdlife. Univ. of Calif. Press, Berkeley, CA.

Floyd, T., C. S. Elphick, G. Chisholm, K. Mack, R. G. Elston, E. M. Ammon, and J. D. Boone. 2007. Atlas of the breeding birds of Nevada. Univ. of Nevada Press, Reno, NV.

Gabrielson, I. A., and S. G. Jewett. 1940. Birds of Oregon. Oregon State College, Corvallis, OR.

Gaines, D. 1992. Birds of Yosemite and the East Slope. 2nd printing. Artemisia Press, Lee Vining, CA.

Garrett, K., and J. Dunn. 1981. Birds of southern California: Status and distribution. Los Angeles Audubon Society, Los Angeles, CA.

Glover, S. 2009. Breeding bird atlas of Contra Costa County. Mount Diablo Audubon Society, Walnut Creek, CA.

Grinnell, J., and A. H. Miller 1944. The distribution of the birds of California. Pacific Coast Avifauna, No. 27. Cooper Ornithological Club, Berkeley, CA.

Grinnell, J., and T. I. Storer. 1924. Animal life in the Yosemite. Univ. of Calif. Press, Berkeley, CA.

Grinnell, J., J. Dixon, and J. M. Linsdale. 1930. Vertebrate natural history of a section of northern California through the Lassen Peak Region. No. 35. Univ. of Calif. Press, Berkeley, CA.

Gruson, E. S. 1972. Words for birds: A lexicon of North American birds with biographical notes. Quadrangle Books, New York, NY.

Hoffmann, R. 1927. Birds of the Pacific states. Riverside Press, Cambridge, MA.

Howell, S. N. G. 2010. Molt in North American birds. Houghton Mifflin, New York, NY.

Howell, S. N. G., M. O'Brien, B. L. Sullivan, C. L. Wood, I. Lewington, and R. Crossley. 2009. The purpose of field guides: Taxonomy vs. utility? Birding 41: 44–49.

Kaufman, K. 1996. Lives of North American birds. Houghton Mifflin, New York, NY.

———. 2011. Field guide to advanced birding. Houghton Mifflin, Boston, MA.

Kemper, J. 1999. Birding northern California. Falcon, Helena, MT.

Martin, A. C., H. S. Zim, and A. I. Nelson. 1951. American wildlife and plants: A guide to wildlife food habits. McGraw-Hill Book Co., New York, NY.

Moritz, C. 2007. Final report: A re-survey of the historic Grinnell-Storer vertebrate transect in Yosemite National Park, California. Available at http://mvz.berkeley.edu/Grinnell/pdf/2007 _Yosemite_report.pdf. [accessed June 2011]

National Audubon Society. 2010. The Christmas Bird Count historical results. Available at http:// www.christmasbirdcount.org. [accessed June 2011].

Orr, R. T., and J. Moffitt. 1971. Birds of the Lake Tahoe region. Calif. Acad. of Sciences, San Francisco, CA.

Paulson, R. W., E. B. Chase, R. S. Roberts, and D. W. Moody. 1991. National water summary 1988–89: Hydrological events and floods and droughts. U.S. Geological Survey, Water Supply Paper 2375. U.S.G.P.O.; Books and Open-file Reports Section (Distributor).

Peterson, R. T. 1990. Western birds. Houghton Mifflin, Boston, MA.

Pyle, P. 1997. Identification guide to North American birds, Part I. Slate Creek Press, Bolinas, CA.

———. 2008. Identification guide to North American birds, Part II. Slate Creek Press, Point Reyes Station, CA.

Remsen Jr., J. V. 1978. Bird species of special concern in California. California Dept. Fish and Game, Sacramento. Nongame Wildl. Manage. Branch Admin. Rep. No. 78-1.

Ryser Jr., F. A. 1985. Birds of the Great Basin. Univ. of Nevada Press, Reno, NV.

Sauer, J. R., J. E. Hines, J. E. Fallon, K. L. Pardieck, D. J. Ziolkowski Jr., and W. A. Link. 2011. The North American Breeding Bird Survey, results and analysis 1966–2009. Version 3.23.2011 USGS Patuxent Wildlife Research Center, Laurel, MD. [accessed March 2012].

Schram, B. 2007. A birder's guide to Southern California. Am. Birding Assoc., Asheville, NC.

Shuford, W. D. 1993. The Marin County breeding bird atlas. Bushtit Books, Bolinas, CA.

Shuford, W. D., and T. Gardali, eds. 2008. California bird species of special concern: A ranked assessment of species, subspecies, and distinct populations of birds of immediate conservation concern in California. Studies of Western Birds 1. Western Field Ornithologists, Camarillo, California, and Calif. Dept. of Fish and Game, Sacramento, CA. (See citations for individual authors below).

Sibley, D. A. 2001. The Sibley guide to bird life and behavior. Chanticleer Press, New York, NY.

Siegel, R. B., R. L. Wilkerson, J. F. Saracco, and Z. L. Steel. 2011. Elevation ranges of birds on the Sierra Nevada West Slope. West. Birds 42: 2–26.

Sullivan, B. L., C. L. Wood, M. J. Iliff, R. E. Bonney, D. Fink, and S. Kelling. 2009. eBird: A citizen-based bird observation network in the biological sciences. Biological Conservation 142: 2282–92.

Sumner, L., and J. S. Dixon. 1953. Birds and mammals of the Sierra Nevada. Univ. of Calif. Press, Berkeley, CA.

Terres, J. K. 1980. The Audubon Society encyclopedia of North American birds. Alfred A. Knopf, New York, NY.

Verner, J., E. C. Beedy, S. L. Granholm, L. V. Ritter, and E. F. Toth. Birds. 1980. In J. Verner and A. S. Boss, tech. coords. California wildlife and their habitats: Western Sierra Nevada. U.S. Dep. Agric. For. Serv. Gen. Tech. Rep. PSW-37, Berkeley, CA.

Walton, R. K., and R. W. Lawson. 1990. Birding by ear, western [audio CD set]. Houghton Mifflin, Boston, MA.

Zeiner, D. C., W. F. Laudenslayer Jr., K. E. Mayer, and M. White, eds. 1990. California's wildlife. Volume II: Birds. California Statewide Wildlife Habitat Relationships System. California Dept. of Fish and Game, Sacramento, CA.

SIERRA ECOLOGICAL ZONES AND BIRD HABITATS

Beardsley, D., C. Bolsinger, and R. Warbington. 1999. Old-growth forests in the Sierra Nevada. U.S. Dept. of Agric. For. Ser. Res. Paper PNW-RP-516. Portland, OR.

Johnston, V. R. 1998. Sierra Nevada: The naturalist's companion. Univ. of Calif. Press, Berkeley, CA.

McCreary, D. D., and M. R. George. 2005. Managed grazing and seedling shelters enhance oak regeneration on rangelands. Calif. Agric. 59: 217–22.

Muick P. C., and J. W. Bartolome. 1986. Oak regeneration on California's hardwood rangelands.

Transactions West Section of the Wildlife Society 22: 121–25.

Pavlik, B. M., P. C Muick, S. G. Johnson, and M. Popper. 2006. Oaks of California. California Oak Foundation, Oakland, CA.

Quinn, R. D., and S. C. Keeley. 2006. Introduction to California chaparral. California Natural History Guides Series No. 90. Univ. of Calif. Press, Berkeley, CA.

Schoenherr, A. A. 1992. A natural history of California. Univ. of Calif. Press, Berkeley, CA.

Smith, G. 2003. Sierra east: Edge of the Great Basin. Univ. of Calif. Press, Berkeley, CA.

Stromberg, M. R., J. D. Corbin, and C. M. D'Antonio. 2007. California grasslands ecology and management. Univ. of Calif. Press, Berkeley, CA.

Sugihara, N. G., J. W. van Wagtendonk, K. E. Shaffer, J. Fites-Kaufman, and A. E. Thode. 2006. Fire in California's ecosystems. Univ. of Calif. Press, Berkeley, CA.

Swiecki, T. J., and E. A. Bernhardt. 1998. Understanding Blue Oak regeneration. Fremontia 26: 19–26.

Zack, S. 2002. The Oak Woodland Bird Conservation Plan. California Oak Foundation, Oakland, CA.

RECENT TRENDS IN SIERRA NEVADA BIRD POPULATIONS AND RANGES

Airola, D. A., and R. H. Barrett. 1985. Foraging and habitat relationships of insect-gleaning birds in a Sierra Nevada mixed-conifer forest. Condor 87: 205–16.

Beedy, E. C. 1981. Bird communities and forest structure in the Sierra Nevada of California. Condor 83: 97–105.

———. 1982. Bird community structure in coniferous forests of Yosemite National Park, California. PhD dissertation, Univ. of California, Davis.

Brennan, L., and M. L. Morrison. 1991. Long-term trends of chickadee populations in Western North America. Condor 93: 130–37.

California Department of Fish and Game. 2004. Strategic plan for wild turkey management. Available at http://www.dfg.ca.gov/wildlife/hunting/uplandgame/docs/turkplan_04.pdf. [accessed June 2011].

Crase, F. T. 1976. Occurrence of the Chestnut-backed Chickadee in the Sierra Nevada mountains, California. Am. Birds 30: 673–75.

DeSante, D. F. 1990. The role of recruitment in the dynamics of a Sierran subalpine bird community. Am. Nat. 136: 429–55.

DeSante, D. F., and D. R. Kaschube. 2007. The Monitoring Avian Productivity and Survivorship (MAPS) program 2002 and 2003 report. Bird Populations 8: 46–115.

DeSante, D. F., and T. L. George. 1994. Population trends in the landbirds of Western North America. In J. R. Jehl and N. K. Johnson, eds. A century of avifaunal change in Western North America. Stud. Avian Bio. 15: 173–90.

Dettinger, M. D., and D. R. Cayan. 1995. Large-scale atmospheric forcing of recent trends toward early snowmelt runoff in California. J. Climate 8: 606–23.

Farmer, C. J., and Smith, J. P. 2009. Migration monitoring indicates widespread declines of American Kestrels (Falco sparverius) in North America. J. Raptor Res. 43: 263–73.

Granholm, S. L. 1982. Effects of surface fires on birds and their habitat associations in coniferous forests of the Sierra Nevada, California. PhD dissertation, Univ. of California, Davis.

———. 1983. Bias in density estimates due to movement of birds. Condor 85: 243–48.

Grinnell, J. 1904. The origin and distribution of the Chestnut-backed Chickadee. Auk 21: 364–382.

———. 1911. Early summer birds in Yosemite Valley. Sierra Club Bull. 8: 118–124.

Grinnell Resurvey Project. 2011. Data available at http://mvz.berkeley.edu/Grinnell/index.html. [accessed December 2011].

Hampton, S. 2006. The expansion of the Eurasian Collared-Dove into the Central Valley of California. Central Valley Bird Club Bull. 9: 7–14.

Hayhoe, K., D. Cayan, C. B. Field, P. C. Frumhoff, E. P. Maurer, N. L. Miller, S. C. Moser, S. H. Schneider, K. N. Cahill, E. E. Cleland, L. Dale, R. Drapek, R. M. Hanemann, L. S. Kalkstein, J. Lenihan, C. K. Lunch, R. P. Neilson, S. C. Sheridan, and J. H. Verville. 2004. Emissions pathways, climate change, and impacts on California. Proc. Nat. Acad. Sci. 101: 12422–27.

Hejl, S. J. 1987. Bird assemblages of true fir forests of the western Sierra Nevada. PhD dissertation, Northern Arizona Univ., Flagstaff, AZ.

Hejl, S. J., and E. C. Beedy. 1986. A problem for wildlife habitat relationships programs: Temporal variations in bird abundance. Pp. 241–44 in Wildlife 2000: Modeling habitat relationships of terrestrial vertebrates. Univ. of Wisconsin Press, Madison, WI.

Kelly, E. G., E. D. Forsman, and R. G. Anthony. 2003. Are Barred Owls displacing Spotted Owls? Condor 105: 45–53.

Kotliar, N. B. 2007. Olive-sided Flycatcher (Contopus cooperi): A technical conservation assessment. USDA For. Ser., Rocky Mountain Region. Available at http://www.fs.fed.us/r2/projects/scp/assessments/olivesidedflycatcher.pdf. [accessed June 2011].

North American Bird Conservation Initiative, U.S. Committee. 2009. The State of the Birds, United States of America, 2009. U.S. Dept. of Interior, Washington, DC.

Patten, M. A., G. McCaskie, and P. Unitt. 2003. Birds of the Salton Sea. Univ. of California Press, Berkeley, CA.

Pandolfino, E. R., B. E. Deuel, and L. Young. 2009. Colonization of the California's Central Valley by the Great-tailed Grackle. Central Valley Bird Club Bull. 12: 77–95.

Pandolfino, E. R., J. Kwolek, and K. Kreitinger. 2006. Expansion of the breeding range of the Hooded Merganser within California. West. Birds 37: 228–36.

Pandolfino, E. R., and K. Suedkamp Wells. 2009. Changes in the winter distribution of the Rough-legged Hawk in North America. West. Birds 40: 210–24.

Pettyjohn, B. G., and J. R. Sauer 1999. Population status of North American grassland birds from the North American Breeding Bird Survey, 1966–1996. Stud. in Avian Biol. 19: 27–44.

Richardson, T. W. 2007. Avian use, nest site selection, and nesting success in Sierra Nevada aspen. PhD dissertation, Univ. of Nevada, Reno, NV.

———. 2004. Expansion of the Breeding Range of the Bufflehead in California. West. Birds 35: 168–72.

Robertson, B. A., and R. L. Hutto. 2007. Is selectively-harvested forest an ecological trap for Olive-sided Flycatchers? Condor 109: 109–21.

Romagosa, C. M., and T. McEneaney. 1999. Eurasian Collared-Dove in North America and the Caribbean. North Am. Birds 53: 348–53.

Rothstein, S. I., J. Verner, and E. Stevens. 1980. Range expansion and diurnal changes in dispersion of the Brown-headed Cowbird in the Sierra Nevada. Auk 97: 253–67.

Rottenborn, S. C. 2000. Nest site selection and reproductive success of urban Red-shouldered Hawks in central California. J. Raptor Res. 34: 18–25.

Saracco, J. F., R. B. Siegel, and R. L. Wilkerson. 2011. Occupancy modeling of Black-backed Woodpeckers on burned Sierra Nevada forests. Ecosphere 2: 1–17.

Siegel, R. B., and D. F. DeSante. 2003. Bird communities in thinned versus unthinned stands of Sierran mixed conifer forest. Wilson Bull. 115: 155–65.

Siegel, R. B., R. L. Wilkerson, J. F. Saracco, and Z. L. Steel. 2011. Elevation ranges of birds on the Sierra Nevada's West Slope. West. Birds 42: 2–26.

Smallwood, J. A., M. F. Causey, D. H. Mossop, J. R. Klucsarits, B. Robertson, S. Robertson, J. Mason, M. J. Maurer, R. Melvin, R. D. Dawson, G. R. Bortolotti, J. W. Parrish Jr., J. Breen, and K. Boyd. 2009. Why are American kestrel (Falco sparverius) populations declining in North America? Evidence from nest-box programs. J. Raptor Res. 43: 274–82.

Stralberg, D., and B. Williams. 2002. Effects of residential development and landscape composition on the breeding birds of Placer County's foothill oak woodlands. In R. B. Standiford, D. McCreary, and K. L. Purcell, eds. Proceedings of the 5th Oak Symposium: Oaks in California's changing landscape. USDA For. Ser. Gen. Tech. Rep. PSW-GTR-184.

Stralberg D., D. Jongsomjit, C. A. Howell, M.A. Snyder, J. D. Alexander, J. A. Wiens, and T. Root. 2009. Re-shuffling of species with climate disruption: A no-analog future for California Birds? PLoS ONE 4(9) Available at www.plosone.org/home.action. [accessed June 2011].

Tingley, M. W., W. B. Monahan, S. R. Beissinger, and C. Moritz. 2009. Colloquium papers: Birds track their Grinnellian niche through a century of climate change. Proc. Nat. Acad. Sci. 106 (Supplement 2): 19637–43.

Wehtje, W. 2001. Range expansion of the Great-tailed Grackle in western North America. West. Birds 32:141–43.

BIRD CONSERVATION IN THE SIERRA

Beesley, D. 2004. Crow's range: An environmental history of the Sierra Nevada. Univ. of Nevada Press, Reno, NV.

California Resources Agencies. 1996. Sierra Nevada ecosystem project. Regents of the Univ. of California, Berkeley, CA. Available at http://ceres.ca.gov/snep/pubs/es.html. [accessed May 2011].

Cayan, D. R., E. P. Maurer, M. D. Dettinger, M. Tyree, and K. Hayhoe. Climate change scenarios for the California region. 2008. Climate Change 87 (Supplement 1): S21–S42.

Das, T., M. D. Dettinger, D. R. Cayan, and H. G. Hidalgo. 2011. Potential increase in floods in California's Sierra Nevada under future climate projections. Climatic Change 109: 71–94.

Estes, J. A., J. Terborgh, J. S. Brashares, M. E. Power, J. Berger, W. J. Bond, S. R. Carpenter, T. E. Essington, R. D. Holt, J. B. C. Jackson, R. J. Marquis, L. Oksanen, T. Oksanen, R. T. Paine, E. K.

Pikitch, W. J. Ripple, S. A. Sandin, M. Scheffer, T. W. Schoener, J. B. Shurin, A. R. E. Sinclair, M. E. Soulé, R. Virtanen, and D. A. Wardle. 2011. Trophic downgrading of planet Earth. Science 333: 301–6.

Farquhar, F. P. 2007. History of the Sierra Nevada. Univ. of Calif. Press, Berkeley, CA.

Hutto, R. L., and S. M. Gallo. 2006. The effects of post-fire salvage logging on cavity-nesting birds. Condor 108: 817–31.

Jackson, L. A. 2010. The Sierra Nevada before history. Mountain Press Publishing Co., Missoula, MT.

Laymon, S. A. 1987. Brown-headed Cowbirds in California: Historical perspectives and management opportunities in riparian habitats. West. Birds 18: 63–70.

Maurer, E. P., I. T. Stewart, C. Bonfils, P. B. Duffy, and D. Cayan. 2007. Detection, attribution, and sensitivity of trends toward earlier streamflow in the Sierra Nevada. Available at: www.agu.org/pubs/crossref/2007/2006JD008088.shtml. [accessed May 2011]

Monahan, W., A. Jones, K. Velas, and G. Langham. 2009. Conservation science policy brief: Curbing greenhouse gas emissions will reduce future california bird loss. Audubon California, Sacramento, CA.

PRBO Conservation Science. 2011. Projected effects of climate change in California: Ecoregional summaries emphasizing consequences for wildlife. Version 1.0. Available at http://data.prbo.org/apps/bssc/climatechange. [accessed January 2012].

Raphael, M. G., and M. White. 1984. Use of snags by cavity-nesting birds in the Sierra Nevada. Wildl. Monogr. 86: 3–66.

Rothstein, S. I. 1994. The cowbird's invasion of the Far West: History, causes, and consequences experienced by host species. In J. R. Jehl and N. K. Johnson, eds. A century of avifaunal change in western North America. Studies in Avian Biology No. 15. Pp. 301–15.

Saab, V., W. Block, R. Russell, J. Lehmkuhl. L. Bate, and R. White. 2007. Birds and burns of the interior west: Descriptions, habitats, and management in western forests. U.S.D.A. For. Ser. Gen. Tech Rep. PNW-GTR-712 (July 2007).

Saab, V. A., R. E. Russell, and J. G. Dudley. 2007. Nest densities of cavity-nesting birds in relation to postfire salvage logging and time since wildfire. Condor 109: 97–108.

———. 2009. Nest-site selection by cavity-nesting birds in relation to postfire salvage logging. Forest Ecology and Management 257: 151–59.

U.S. Forest Service. 2001. Record of decision: Sierra Nevada Forest plan amendment environmental impact statement. U.S.D.A., Washington, DC Available at http://www.fs.fed.us/r5/snfpa/library/archives/rod/rod.pdf. [accessed June 2011].

Winkler, D. W., C. P. Weigen, B. Engstrom, and E. Burch. 1977. An ecological study of Mono Lake, California. Univ. of Calif. Davis, Inst. of Ecol. Publ. No. 12 .

WATERFOWL

Austin, J. E., and M. R. Miller. 1995. Northern Pintail *(Anas acuta)*. In A. Poole, ed. The birds of North America online, No. 163. Cornell Lab of Ornith. [accessed September 2011].

Austin, J. E., C. M. Custer, and A. D. Afton. 1998. Lesser Scaup *(Aythya affinis)*. In A. Poole, ed. The birds of North America online, No. 338. Cornell Lab of Ornith. [accessed September 2011].

Beedy, E. C. 2008. Harlequin Duck *(Histrionicus histrionicus)*. In W. D. Shuford and T. Gardali, eds. California bird species of special concern. Studies of West. Birds 1: 91–95.

Beedy, E. C., and B. E. Deuel. 2008. Redhead *(Aythya Americana)*. In W. D. Shuford and T. Gardali, eds. California bird species of special concern. Studies of West. Birds 1: 85–90.

Belding, L. 1891. Notices of some California birds. Zoo 2: 97–100.

Bellrose, F. C. 1980. Ducks, geese, and swans of North America. Stackpole Books, Harrisburg, PA.

Brua, R. B. 2002. Ruddy Duck *(Oxyura jamaicensis)*. In A. Poole, ed. The birds of North America online, No. 696. Cornell Lab of Ornith. [accessed September 2011].

Drilling, N., R. Titman, and F. McKinney. 2002. Mallard *(Anas platyrhynchos)*. In A. Poole and F. Gill, ed. The birds of North America, No. 658. The Birds of North America, Inc. Philadelphia, PA.

Dubowy, P. J. 1996. Northern Shoveler *(Anas clypeata)*. In A. Poole, ed. The birds of North America online, No. 217. Cornell Lab of Ornith. [accessed September 2011].

Dugger, B. D., K. M. Dugger, and L. H. Fredrickson. 2009. Hooded Merganser *(Lophodytes cucullatus)*. In A. Poole, ed. The birds of North America online, No. 98. Cornell Lab of Ornith. [accessed August 2011].

Eadie, J. M., M. L. Mallory, and H. G. Lumsden. 1995. Common Goldeneye *(Bucephala clangula)*. In A. Poole, ed. The birds of North America online, No. 170. Cornell Lab of Ornith. [accessed September 2011].

Eadie, J. M., J-P. L. Savard, and M. L. Mallory. 2000. Barrow's Goldeneye *(Bucephala islandica)*. In A. Poole and F. Gill, eds. The birds of North America, No. 548. The Birds of North America, Inc. Philadelphia, PA.

Ely, C. R. and A. X. Dzubin. 1994. Greater White-fronted Goose *(Anser albifrons)*. In A. Poole and F. Gill, eds. The birds of North America, No. 131. Academy of Natural Sciences, Philadelphia, and American Ornithologists' Union, Washington, DC.

Gammonley, J. H. 1996. Cinnamon Teal *(Anas cyanoptera)*. In A. Poole and F. Gill, eds. The birds of North America, No. 209. Academy of Natural Sciences, Philadelphia, and American Ornithologists' Union, Washington, DC.

Gauthier, G. 1993. Bufflehead *(Bucephala albeola)*. In A. Poole, ed. The birds of North America online, No. 67. Cornell Lab of Ornith. [accessed September 2011].

Hepp, G. R., and F. C. Bellrose. 1995. Wood Duck *(Aix sponsa)*. In A. Poole, ed. The birds of North America online, No. 169. Cornell Lab of Ornith. [accessed September 2011].

Hohman, W. L., and R. T. Eberhardt. 1998. Ring-necked Duck *(Aythya collaris)*. In A. Poole, ed. The birds of North America online, No. 329. Cornell Lab of Ornith. [accessed September 2011].

Johnsgard, P. A. 1978. Ducks, geese, and swans of the world. Univ. of Nebraska Press, Lincoln, NB.

Johnson, K. 1995. Green-winged Teal *(Anas crecca)*. In A. Poole and F. Gill, eds. The birds of North America, No. 193. Academy of Natural Sciences, Philadelphia, and American Ornithologists' Union, Washington, DC.

Leschack, C. R., S. K. Mckinght, and G. R. Hepp. 1997. Gadwall *(Anas strepera)*. In A. Poole, ed. The birds of North America online, No. 283. Cornell Lab of Ornith. [accessed September 2011].

Limpert, R. J., and S. L. Earnst. 1994. Tundra Swan *(Cygnus columbianus)*. In A. Poole, ed. The birds of North America online, No. 89. Cornell Lab of Ornith. [accessed September 2011].

Mallory, M., and K. Metz. Common Merganser *(Mergus merganser)*. In A. Poole and F. Gill, eds. The birds of North America, No. 442. The Birds of North America, Inc. Philadelphia, PA.

Mowbray, T. B. 1999. American Wigeon *(Anas americana)*. In A. Poole, ed. The birds of North America online, No. 401. Cornell Lab of Ornith. [accessed September 2011].

———. 2000. Snow Goose *(Anser caerulescens)*. In A. Poole and F. Gill, eds. The birds of North America, No. 514. The Birds of North America, Inc. Philadelphia, PA.

———. 2002. Canvasback *(Aythya valisineria)*. In A. Poole, ed. The Birds of North America online, No. 659. Cornell Lab of Ornith. [accessed September 2011].

Mowbray, T. B., C. R. Ely, J. D. Sedinger, and R. E. Trost. 2002. Canada Goose *(Branta canadensis)*. In A. Poole and F. Gill, eds. The birds of North America, No. 682. The Birds of North America, Inc. Philadelphia, PA.

Pandolfino, E. R., J. Kwolek, and K. Kreitinger. 2006. Expansion of the breeding range of the Hooded Merganser within California. West. Birds 37: 228–36.

Robertson, G. J., and I. Goudie. 1999. Harlequin Duck *(Histrionicus histrionicus)*. In A. Poole and F. Gill, eds. The birds of North America, No. 466. The Birds of North America, Inc. Philadelphia, PA.

Ryder, J. P., and R. T. Alisaukas. 1995. Ross's Goose *(Anser rossii)*. In A. Poole and F. Gill, eds. The birds of North America, No. 162. Academy of Natural Sciences, Philadelphia, and American Ornithologists' Union, Washington, DC.

Stallcup, R. 2002. Hooded Merganser *(Lophodytes cucullatus)*: Recent breeding range expansion in California and potential for breeding in Nevada. Great Basin Birds 5: 45–46.

Titman, R. D. 1999. Red-breasted Merganser *(Mergus serrator)*. In A. Poole and F. Gill, eds. The birds of North America, No. 443. The Birds of North America, Inc. Philadelphia, PA.

Woodin, M. C., and T. C. Michot. 2002. Redhead *(Aythya americana)*. In A. Poole, ed. The birds of North America online, No. 695. Cornell Lab of Ornith. [accessed September 2011].

QUAIL

Calkins, J. D., J. C. Hagelin, and D. F. Lott. 1999. California Quail *(Callipepla californica)*. In A. Poole, ed. The birds of North America online, No. 473. Cornell Lab of Ornith. [accessed September 2011].

Gutiérrez, R. J., and D. J. Delehanty. 1999. Mountain Quail *(Oreortyx pictus)*. In A. Poole, ed. The birds of North America online, No. 457. Cornell Lab of Ornith. [accessed September 2011].

FOWL-LIKE BIRDS

Bland, J. D. 2008. Mount Pinos Sooty Grouse *(Dendragapus fuliginosus howardi)*. In W. D. Shuford and T. Gardali, eds. California bird species of special concern. Studies of West. Birds 1: 102–6.

Braun, C. E., and L. A. Robb. 1993. White-tailed

Ptarmigan *(Lagopus leucurus)*. In A. Poole and F. Gill, eds. The birds of North America, No. 68. Academy of Natural Sciences, Philadelphia, and American Ornithologists' Union, Washington, DC.

Christensen, G. C. 1996. Chukar *(Alectoris chukar)*. In A. Poole and F. Gill, eds. The birds of North America, No. 258. Academy of Natural Sciences, Philadelphia, and American Ornithologists' Union, Washington, DC.

Eaton, S. W. 1992. Wild Turkey *(Meleagris gallopavo)*. In A. Poole, ed. The birds of North America online, No. 22. Cornell Lab of Ornith. [accessed September 2011].

Frederick, G. P., and R. J. Gutiérrez. 1992. Habitat use and population characteristics of the White-tailed Ptarmigan in the Sierra Nevada, California. Condor 94: 889–902.

Grinnell, J., H. C. Bryant, and T. I. Storer. 1918. The game birds of California. Univ. of Calif. Press, Berkeley, CA.

Guidice, J. H., and J. T. Ratti. 2001. Ring-necked Pheasant *(Phasianus colchicus)*. In A. Poole and F. Gill, eds. The Birds of North America, No. 443. The Birds of North America, Inc. Philadelphia, PA.

Hall, F. A., S. C. Gardner, and D. S. Blankenship. 2008. Greater Sage-Grouse *(Centrocercus urophasianus)*. In W. D. Shuford and T. Gardali, eds. California bird species of special concern. Studies of West. Birds 1: 96–101.

Muir, J. 1901. Our national parks. Houghton-Mifflin Co., Boston, MA.

Schroeder, M. A., J. R. Young, and C. E. Braun. 1999. Greater Sage-Grouse *(Centrocercus urophasianus)*. In A. Poole, ed. The birds of North America online, No. 425. Cornell Lab of Ornith. [accessed September 2011].

Zwickel, F. C., and J. F. Bendell. 2005. Blue Grouse *(Dendragapus obscurus)*. In A. Poole, ed. The birds of North America online, No. 15. Cornell Lab of Ornith. [accessed September 2011].

LOONS

McIntyre, J. W., and J. F. Barr. 1997. Common Loon *(Gavia immer)*. In A. Poole and F. Gill, eds. The birds of North America, No. 313. Academy of Natural Sciences, Philadelphia, and American Ornithologists' Union, Washington, DC.

Remsen Jr., J. V. 1978. Bird species of special concern in California. Calif. Dep. Fish and Game, Sacramento. Nongame Wildl. Manage. Branch Admin. Rep. No. 78-1.

Russell, R. W. 2002. Pacific Loon *(Gavia pacifica)*

and Arctic Loon *(Gavia arctica)*. In A. Poole and F. Gill, eds. The birds of North America, No. 657. The Birds of North America, Inc. Philadelphia, PA.

Townsend, C. H. 1887. Field-notes on the mammals, birds, and reptiles of northern California. Proc. U.S. Nat. Mus. 10: 159–241.

GREBES

Cullen, S. A., J. R. Jehl Jr., and G. L. Nuechterlein. 1999. Eared Grebe *(Podiceps nigricollis)*. In A. Poole, ed. The birds of North America online, No. 433. Cornell Lab of Ornith. [accessed September 2011].

Fischer, D. L. 2001. Early southwest ornithologists, 1528–1900. Univ. of Arizona Press, Tucson, AZ.

Jehl Jr., J. R. 1996. Mass mortality events of Eared Grebes in North America. J. Field Ornithol. 67: 471–76.

Muller, M. J., and R. W. Storer. 1999. Pied-billed Grebe *(Podilymbus podiceps)*. In A. Poole and F. Gill, eds. The birds of North America, No. 410. The Birds of North America, Inc. Philadelphia, PA.

Stedman, S. J. 2000. Horned Grebe *(Podiceps auritus)*. In A. Poole, ed. The birds of North America online, No. 505. Cornell Lab of Ornith. [accessed September 2011].

Storer, R. W., and G. L. Nuechterlein. 1992. Western Grebe *(Aechmophorus occidentalis)* and Clark's Grebe *(Aechmophorus clarkii)*. In A. Poole, ed. The birds of North America online, No. 26. Cornell Lab of Ornith. [accessed September 2011].

Winkler, D. W., C. P. Weigen, B. Engstrom, and E. Burch. 1977. An ecological study of Mono Lake, California. Univ. of Calif. Davis, Inst. of Ecol. Publ. No. 12.

CORMORANTS

Hatch, J. J., and D. V. Weseloh. 1999. Double-crested Cormorant *(Phalacrocorax auritus)*. In A. Poole, ed. The birds of North America online, No. 441. Cornell Lab of Ornith. [accessed September 2011].

PELICANS

Knopf, F. L., and R. M. Evans. 2004. American White Pelican *(Pelecanus erythrorhynchos)*. In A. Poole, ed. The birds of North America online, No. 57. Cornell Lab of Ornith. [accessed September 2011].

Shuford, W. D. 2008. American White Pelican

(*Pelecanus erythrorhynchos*). In W. D. Shuford and T. Gardali, eds. California bird species of special concern. Studies of West. Birds 1: 130–35.

Yates, M. 1999. Satellite and conventional telemetry study of American White Pelicans in northern Nevada. Great Basin Birds 2: 4–9.

HERONS AND RELATIVES

Davis, W. E. 1993. Black-crowned Night-Heron *(Nycticorax nycticorax)*. In A. Poole and F. Gill, eds. The birds of North America, No. 74. Academy of Natural Sciences, Philadelphia, and American Ornithologists' Union, Washington, DC.

Davis, W. E., and J. A. Kushlan. 1994. Green Heron *(Butorides virescens)*. In A. Poole and F. Gill, eds. The birds of North America, No. 74. Academy of Natural Sciences, Philadelphia, and American Ornithologists' Union, Washington, DC.

Lowther, P., A. F. Poole, J. P. Gibbs, S. Melvin, and F. A. Reid. 2009. American Bittern *(Botaurus lentiginosus)*. In A. Poole, ed. The birds of North America online, No. 18. Cornell Lab of Ornith. [accessed September 2011].

McCrimmon Jr., D. A., J. C. Ogden, and G. T. Bancroft. 2011. Great Egret *(Ardea alba)*. In A. Poole, ed. The birds of North America online, No. 570. Cornell Lab of Ornith. [accessed September 2011].

Parsons, K. C., and T. L. Master. Snowy Egret *(Egretta thula)*. In A. Poole and F. Gill, eds. The birds of North America, No. 489. The Birds of North America, Inc. Philadelphia, PA.

Vennesland, R. G., and R. W. Butler. 2011. Great Blue Heron *(Ardea herodias)*. In A. Poole, ed. The birds of North America online, No. 25. Cornell Lab of Ornith. [accessed September 2011].

IBIS

Pandolfino, E. R. 2006. Christmas Bird Counts reveal wintering bird status and trends in California's Central Valley. Central Valley Bird Club Bull. 9: 21–36.

———. 2010. Review of the 110th Christmas Bird Count in the Central Valley of California: December 2009–January 2010. Central Valley Bird Club Bull. 13: 25–34.

Ryder, R. A., and D. E. Manry. 1994. White-faced Ibis *(Plegadis chihi)*. In A. Poole, ed. The birds of North America online, No. 130. Cornell Lab of Ornith. [accessed September 2011].

Shuford, W. D., C. M. Hickey, R. I. Safran, and G. W. Page. 1996. A review of the status of the White-faced Ibis in winter in California. West. Birds 27: 169–96.

NEW WORLD VULTURES

Kirk, D. A., and M. J. Mossman. 1998. Turkey Vulture *(Cathartes aura)*. In A. Poole, ed. The birds of North America online, No. 339. Cornell Lab of Ornith. [accessed September 2011].

Lynch, K. 2000. The hard season. Poetry Magazine. January 2000, p. 175.

Snyder, N. F., and N. J. Schmitt. 2002. California Condor *(Gymnogyps californianus)*. In A. Poole, ed. The birds of North America online, No. 610. Cornell Lab of Ornith. [accessed September 2011].

OSPREY

Poole, A. F., R. O. Bierregaard, and M. S. Martell. 2002. Osprey *(Pandion haliaetus)*. In A. Poole, ed. The birds of North America online, No. 683. Cornell Lab of Ornith. [accessed October 2011].

HAWKS AND RELATIVES

Bechard, M. J., and J. K. Schmutz. 1995. Ferruginous Hawk *(Buteo regalis)*. In A. Poole and F. Gill, eds. The birds of North America, No. 172. Academy of Natural Sciences, Philadelphia, and American Ornithologists' Union, Washington, DC.

Bechard, M. J., and T. R. Swem. 2002. Rough-legged Hawk *(Buteo lagopus)*. In A. Poole and F. Gill, eds. The birds of North America, No. 641. The Birds of North America, Inc. Philadelphia, PA.

Bent, A. C. 1937. Life histories of familiar North American birds. U.S. Govt. Printing Office, Washington, DC.

Bildstein, K. L., and K. Meyer. 2000. Sharp-shinned Hawk *(Accipiter striatus)*. In A. Poole and F. Gill, eds. The birds of North America, No. 482. The Birds of North America, Inc. Philadelphia, PA.

Bradbury, M., J. A. Estep, and D. Anderson. In preparation. Migratory patterns and wintering range of the Central Valley Swainson's Hawk.

Buehler, D. A. 2000. Bald Eagle *(Haliaetus leucocephalus)*. In A. Poole and F. Gill, eds. The birds of North America, No. 506. The Birds of North America, Inc. Philadelphia, PA.

Crocoll, S. T. 1994. Red-shouldered Hawk *(Buteo lineatus)*. In A. Poole and F. Gill, eds. The birds of North America, No. 107. Academy of Natural Sciences, Philadelphia, and American Ornithologists' Union, Washington, DC.

Davis, J. N., and C. A. Niemla. 2008. Northern Harrier *(Circus cyaneus)*. In W. D. Shuford and T.

Gardali, eds. California bird species of special concern. Studies of West. Birds 1: 149–55.

Dunk, J. R. 1995. White-tailed Kite *(Elanus leucurus)*. In A. Poole, ed. The birds of North America online, No. 178. Cornell Lab of Ornith. [accessed October 2011].

Hoffman, S. W., J. P. Smith, and T. D. Meehan. 2002. Breeding grounds, winter ranges, and migratory routes of raptors in the Mountain West. J. Raptor Res. 36: 97–110.

Keane, J. J. 2008. Northern Goshawk *(Accipiter gentilis)*. In W. D. Shuford and T. Gardali, eds. California bird species of special concern. Studies of West. Birds 1: 156–72.

Kochert, M. N., K. Steenhof, C. L. McIntyre, and E. H. Craig. 2002. Golden Eagle *(Aquila chrysaetos)*. In A. Poole and F. Gill, eds. The birds of North America, No. 684. The Birds of North America, Inc. Philadelphia, PA.

MacWhirter, R. B., and K. L. Bildstein. 1996. Northern Harrier *(Circus cyaneus)*. In A. Poole and F. Gill, eds. The birds of North America, No. 210. Academy of Natural Sciences, Philadelphia, and American Ornithologists' Union, Washington, DC.

Pandolfino, E. R. 2008. Review of the 108th Christmas Bird Count in the Central Valley of California: December 2007–January 2008. Central Valley Bird Club Bull. 11: 53–61.

Pandolfino, E. R., M. P. Herzog, S. L. Hooper, and Z. Smith. 2011. Winter habitat associations of diurnal raptors in California's Central Valley. West. Birds 42: 62–84.

Preston, C. R., and R. D. Beane. 1993. Red-tailed Hawk *(Buteo jamaicensis)*. In A. Poole and F. Gill, eds. The birds of North America, No. 52. Academy of Natural Sciences, Philadelphia, and American Ornithologists' Union, Washington, DC.

Pruett-Jones, S. G., M. A. Pruett-Jones, and R. L. Knight. 1980. The White-tailed Kite in North and Middle America: Current status and recent population changes. Am. Birds 34: 682–88.

Rosenfield, R. N., and J. Bielefeldt. 1993. Cooper's Hawk *(Accipiter cooperii)*. In A. Poole and F. Gill, eds. The birds of North America, No. 75. Academy of Natural Sciences, Philadelphia, and American Ornithologists' Union, Washington, DC.

Squires, J. R., and R. T. Reynolds. 1997. Northern Goshawk *(Accipiter gentilis)*. In A. Poole and F. Gill, eds. The birds of North America, No. 298. Academy of Natural Sciences, Philadelphia, and American Ornithologists' Union, Washington, DC.

Wheeler, B. K. 2003. Raptors of western North America. Princeton Univ. Press, Princeton, NJ.

FALCONS

Hackett, S. J., R. T. Kimball, S. Reddy, R. C. K. Bowie, E. L. Braun, M. J. Braun, J. L. Chojnowski, W. A. Cox, K. L. Han, J. Harshman, C. J. Huddleston, B. D. Marks, K. J. Miglia, W. S. Moore, F. H. Sheldon, D. W. Steadman, C. C. Witt, and T. Yuri. 2008. A phylogenomic study of birds reveals their evolutionary history. Science 320: 1763–68.

Pandolfino, E. R., M. P. Herzog, and Z. Smith. 2011. Sex-related differences in habitat associations of wintering American Kestrels in California's Central Valley. J. Raptor Res. 45: 38–45.

Smallwood, J. A., and D. M. Bird. 2002. American Kestrel *(Falco sparverius)*. In A. Poole, ed. The birds of North America online, No. 602. Cornell Lab of Ornith. [accessed November 2011].

Sodhi, N. S., L. W. Oliphant, F. C. James, and I. G. Warkentin. 1993. Merlin *(Falco columbarius)*. In A. Poole and F. Gill, eds. The birds of North America, No. 44. Academy of Natural Sciences, Philadelphia, and American Ornithologists' Union, Washington, DC.

Steenhof, K. 1998. Prairie Falcon *(Falco mexicanus)*. In A. Poole and F. Gill, eds. The birds of North America, No. 346. The Birds of North America, Inc. Philadelphia, PA.

White, C. M., N. J. Clum, T. J. Cade, and W. G. Hunt. 2002. Peregrine Falcon *(Falco peregrinus)*. In A. Poole and F. Gill, eds. The birds of North America, No. 660. The birds of North America, Inc. Philadelphia, PA.

RAILS AND RELATIVES

Aigner, P., J. Tecklin, and C. Koehler. 1995. Probable breeding population of the black rail in Yuba County, California. West. Birds 26: 157–60.

Bannor, B. K., and E. Kiviat. 2002. Common Gallinule *(Gallinula chloropus)*. In A. Poole, ed. The birds of North America online, No. 685. Cornell Lab of Ornith. [accessed November 2011].

Brisbin Jr., I. L., and T. B. Mobray. 2002. American Coot *(Fulica Americana)*. In A. Poole and F. Gill, eds. The birds of North America, No. 697. The Birds of North America, Inc. Philadelphia, PA.

Conway, C. J. 1995. Virginia Rail *(Rallus limicola)*. *In* A. Poole and F. Gill, eds. The birds of North America, No. 173. Academy of Natural Sciences, Philadelphia, and American Ornithologists' Union, Washington, DC.

Eddleman, W., R. Flores, and M. Legare. 1994. Black rail *(Laterallus jamaicensis)*. In A. Poole, eds. The birds of North America online, No. 123. Cornell Lab of Ornith. [accessed November 2011].

Evens, J. G., G. W. Page, S. A. Laymon, and R. W. Stallcup. 1991. Distribution, relative abundance, and status of the California black rail in western North America. Condor 93: 952–66.

Girard, P., J. Y. Takekawa, and S. R. Beissinger. 2010. Uncloaking a cryptic, threatened rail with molecular markers: Origins, connectivity, and demography of a recently-discovered population. Conservation Genetics 11: 2409–18.

Manolis, T. 1978. Status of the black rail in central California. West. Birds 9: 151–58.

Melvin, S. M., and J. P. Gibbs. 1996. Sora *(Porzana carolina)*. In A. Poole and F. Gill, eds. The birds of North America, No. 250. Academy of Natural Sciences, Philadelphia, and American Ornithologists' Union, Washington, DC.

Richmond, O. M., J. Tecklin, and S. R. Beissinger. 2008. Distribution of California Black Rails in the Sierra Nevada foothills. J. Field Ornithol. 79: 381–90.

Tecklin, J. 1999. Distribution and abundance of the California black rail *(Laterallus jamaicensis coturniculus)* in the Sacramento Valley region with accounts of ecology and call behavior of the subspecies. Draft report for the California Dept. of Fish and Game, Contract Nos. FG6154WM and FG6154-1WM.

CRANES

Littlefield, C. D. 2008. Lesser Sandill Crane *(Grus canadensis canadensis)*. In W. D. Shuford and T. Gardali, eds. California bird species of special concern. Studies of Western Birds 1: 167–72.

Tacha, T. C., S. A. Nesbitt, and P. A. Vohs. 1992. Sandhill Crane *(Grus canadensis)*. In A. Poole, eds. The birds of North America online, No. 31. Cornell Lab of Ornith. [accessed November 2011].

PLOVERS

Jackson, B. J. S., and J. A. Jackson. 2000. Killdeer *(Charadrius vociferous)*. In A. Poole and F. Gill, eds. The birds of North America, No. 517. The Birds of North America, Inc. Philadelphia, PA.

Nol, E., and M. S. Blanken. 1999. Semipalmated Plover *(Charadrius semipalmatus)*. In A. Poole, ed. The birds of North America online, No. 444. Cornell Lab of Ornith. [accessed December 2011].

Page, G. W., L. E. Stenzel, J. S. Warriner, J. C. Warriner, and P. W. Paton. 2009. Snowy Plover *(Charadrius nivosus)*. In A. Poole, ed. The birds of North America online, No. 154. Cornell Lab of Ornith. [accessed December 2011].

Page, G. W., M. A. Stern, and P. W. C. Paton. 1995. Differences in wintering areas of Snowy Plovers from inland breeding sites in western North America. Condor 97: 258–62.

Paulson, D. R. 1995. Black-bellied Plover *(Pluvialis squatarola)*. In A. Poole, ed. The birds of North America online, No. 186. Cornell Lab of Ornith. [accessed December 2011].

Shuford, W. D., S. Abbott, and T. D. Rhulen. 2008. Snowy Plover *(Charadruis alexandria)* (interior population). In W. D. Shuford and T. Gardali, eds. California bird species of special concern. Studies of West. Birds 1: 173–79.

Warriner, J. S., J. C. Warriner, G. W. Page, and L. E. Stenzel. 1986. Mating system and reproductive success of a small population of polygamous Snowy Plovers. Wilson Bull. 98: 15–37.

STILTS AND AVOCETS

Robinson, J. A., L. W. Oring, J. P. Skorupa, and R. Boettcher. 1997. American Avocet *(Recurvirostra americana)*. In A. Poole, ed. The birds of North America online, No. 275. Cornell Lab of Ornith. [accessed December 2011].

Robinson, J. A., J. M. Recd, J. P. Skorupa, and L. W. Oring. 1999. Black-necked Stilt *(Himantopus mexicanus)*. In A. Poole and F. Gill, eds. The birds of North America, No. 449. The Birds of North America, Inc. Philadelphia, PA.

SANDPIPERS AND RELATIVES

Brown, S., C. Hickey, B. Harrington, and R. Gill, eds. 2001. The U.S. Shorebird Conservation Plan, 2nd edition. Manomet Center for Conservation Sciences, Manomet, MA.

Colwell, M. A., and J. R. Jehl Jr. 1994. Wilson's Phalarope *(Phalaropus tricolor)*. In A. Poole, ed. The birds of North America online, No. 83. Cornell Lab of Ornith. [accessed December 2011].

Cooper, J. M. 1994. Least Sandpiper *(Calidris minutella)*. In A. Poole and F. Gill, eds. The birds of North America, No. 115. Academy of Natural Sciences, Philadelphia, and American Ornithologists' Union, Washington, DC.

Dugger, B. D., and K. M. Dugger. 2002. Long-billed Curlew *(Numenius americanus)*. In A. Poole, ed. The birds of North America online, No. 628. Cornell Lab of Ornith. [accessed December 2011].

Elphick, C. S., and T. L. Tibbits. 1998. Greater Yellowlegs *(Tringa melanoleuca)*. In A. Poole and

F. Gill, eds. The birds of North America, No. 355. The Birds of North America, Inc. Philadelphia, PA.

Hayman, P., J. Marchant, and T. Prater. 1986. Shorebirds: An identification guide. Houghton Mifflin Company, Boston, MA.

Jehl Jr., J. R. 1987. Moult and migration in a transequatorially migrating shorebird: Wilson's Phalarope. Ornis Scand. 18: 173–78.

Lowther, P. E., H. D. Douglas III, and C. L. Gratto-Trevor. 2001. Willet *(Tringa semipalmata)*. In A. Poole, ed. The birds of North America online, No. 579. Cornell Lab of Ornith. [accessed December 2011].

Mueller, H. 1999. Wilson's Snipe *(Gallinago delicata)*. In A. Poole, ed. The birds of North America online, No. 417. Cornell Lab of Ornith. [accessed December 2011].

Oring, L. W., E. M. Gray, and J. M. Reed. 1997. Spotted Sandpiper *(Actitis macularius)*. In A. Poole, ed. The birds of North America online, No. 289. Cornell Lab of Ornith. [accessed December 2011].

Rubega, M. 1992. Feeding limitations and ecology of Red-necked Phalaropes at Mono Lake with incidental observations of other species. Unpublished report submitted to Jones & Stokes Associates, Sacramento, CA.

Rubega, M. A., D. Schamel, and D. M. Tracy. 2000. Red-necked Phalarope *(Phalaropus lobatus)*. In A. Poole and F. Gill, eds. The birds of North America, No. 538. The Birds of North America, Inc. Philadelphia, PA.

Takekawa, J. Y., and N. Warnock. 2000. Long-billed Dowitcher *(Limnodromus scolopaceus)*. In A. Poole, ed. The birds of North America online, No. 493. Cornell Lab of Ornith. [accessed December 2011].

Warnock, N. D., and R. E. Gill. 1996. Dunlin *(Calidris alpina)*. In A. Poole, ed. The birds of North America online, No. 203. Cornell Lab of Ornith. [accessed December 2011].

Wilson, W. H. 1994. Western Sandpiper *(Calidris mauri)*. In A. Poole, ed. The birds of North America online, No. 90. Cornell Lab of Ornith. [accessed December 2011].

GULLS AND TERNS

Burger, J., and M. Gochfeld. 2002. Bonaparte's Gull *(Chroicocephalus philadelphia)*. In A. Poole, ed. The birds of North America online, No. 634. Cornell Lab of Ornith. [accessed December 2011].

Cuthbert, F. J., and L. R. Wires. 1999. Caspian Tern *(Hydroprogne caspia)*. In A. Poole, ed. The birds

of North America online, No. 403. Cornell Lab of Ornith. [accessed September 2012].

Dunn, E. H., and D. J. Agro. 1995. Black Tern *(Chlidonias niger)*. In A. Poole and F. Gill, eds. The birds of North America, No. 147. Academy of Natural Sciences, Philadelphia, and American Ornithologists' Union, Washington, DC.

Howell, S. N. G., and J. Dunn. 2007. Gulls of the Americas. Houghton Mifflin Company, Boston, MA.

McNicholl, M. K., P. E. Lowther, and J. A. Hall. 2001. Forster's Tern *(Sterna forsteri)*. In A. Poole, ed. The birds of North America online, No. 595. Cornell Lab of Ornith. [accessed December 2011].

Pierotti, R. J., and T. P. Good. 1994. Herring Gull *(Larus argentatus)*. In A. Poole and F. Gill, eds. The birds of North America, No. 124. Academy of Natural Sciences, Philadelphia, and American Ornithologists' Union, Washington, DC.

Ryder, J. P. 1993. Ring-billed Gull *(Larus delawarensis)*. In A. Poole, ed. The birds of North America online, No. 33. Cornell Lab of Ornith. [accessed December 2011].

Shuford, W. D. 2008. Black Tern *(Chlidonisa niger)*. In W. D. Shuford and T. Gardali, eds. California bird species of special concern. Studies of West. Birds 1: 193–98.

Winkler, D. W. 1996. California Gull *(Larus californicus)*. In A. Poole, ed. The birds of North America online, No. 259. Cornell Lab of Ornith. [accessed December 2011].

PIGEONS AND DOVES

Dolton, D. D., and R. D. Rau. 2006. Mourning Dove population status, 2006. U.S. Fish and Wildlife Service, Laurel, MD.

Hampton, S. 2006. The expansion of the Eurasian Collared-Dove into the Central Valley of California. Central Valley Bird Club Bull. 9: 7–14.

Johnston, R. F. 1992. Rock Dove *(Columba livia)*. In A. Poole and F. Gill, eds. The birds of North America, No. 13. Academy of Natural Sciences, Philadelphia, and American Ornithologists' Union, Washington, DC.

Keppie, D. M., and C. E. Braun. 2000. Band-tailed Pigeon *(Patagioenas fasciata)*. In A. Poole, ed. The birds of North America online, No. 530. Cornell Lab of Ornith. [accessed September 2011].

Otis, D. L., J. H. Schulz, D. Miller, R. E. Mirarchi, and T. S. Baskett. 2008. Mourning Dove *(Zenaida macroura)*. In A. Poole, ed. The birds of North America online, No. 117. Cornell Lab of Ornith. [accessed September 2011].

Pandolfino, E. R. 2011. Review of the 111th Christmas Bird Count in the Central Valley of California: December 2010–January 2011. Central Valley Bird Club Bull. 14: 9–19.

Romagosa, C. M. 2002. Eurasian Collared-Dove (*Streptopelia decaocto*). In A. Poole, ed. The birds of North America online, No. 630. Cornell Lab of Ornith. [accessed June 2010].

Romagosa, C. M., and T. McEneaney. 1999. Eurasian Collared-Dove in North America and the Caribbean. North Am. Birds 53: 348–53.

Root, T. 1988. Atlas of wintering North American birds: An analysis of Christmas Bird Count data. Univ. of Chicago Press, Chicago, IL.

CUCKOOS AND ROADRUNNERS

Hughes, J. M. 1999. Yellow-billed Cuckoo *(Coccyzus americanus)*. In A. Poole, ed. The birds of North America online, No. 418. Cornell Lab of Ornith. [accessed February 2012].

———. 2011. Greater Roadrunner *(Geococcyx californianus)*. In A. Poole, ed. The birds of North America online, No. 244. Cornell Lab of Ornith. [accessed January 2012).

Thompson, K. 1961. Riparian forests of the Sacramento Valley, California. American Assoc. of Am. Geog. 51: 294–15.

BARN OWL

Marti, C. D. 1992. Barn Owl (*Tyto alba*). In A. Poole, ed. The birds of North America online, No. 1. Cornell Lab of Ornith. [accessed June 2012].

TYPICAL OWLS

Blakesley, J. A., M. E. Seamans, M. M. Conner, A. B. Franklin, G. C. White, R. J. Gutierrez, J. E. Hines, J. D. Nichols, T. E. Munton, D. W. H. Shaw, J. J. Keane, G. N. Steger, and T. L. McDonald. 2010. Population dynamics of Spotted Owls in the Sierra Nevada, California. Wildl. Monogr. 174: 1–36.

Bond, M. L., D. E. Lee, and R. B. Siegel. 2010. Winter movements by California Spotted Owls in a burned landscape. West. Birds 41: 174–80.

Bond, M. L., D. E. Lee, R. B. Siegel, and J. P. Ward Jr. 2009. Habitat use by California Spotted Owls in a postfire landscape. J. Wildl. Manage. 73: 1116–24.

Bull, E. L., and J. R. Duncan. 1993. Great Gray Owl *(Strix nebulosa)*. In A. Poole, ed. The birds of North America online, No. 41. Cornell Lab of Ornith. [accessed November 2011].

Cannings, R. J., and T. Angell. 2001. Western

Screech-Owl *(Otus kennicottii)*. In A. Poole and F. Gill, eds. The birds of North America, No. 597. The Birds of North America, Inc. Philadelphia, PA.

Crozier, M. L., M. E. Seamans, and R. J. Gutierrez. 2003. Forest owls detected in the central Sierra Nevada. West. Birds 34: 149–56.

Davis, J. N., and G. L. Gould Jr. 2008. California Spotted Owl *(Strix occidentalis occidentalis)*. In W. D. Shuford and T. Gardali, eds. California bird species of special concernStudies of Western Birds 1: 227–33.

Gervais, J. A., D. K. Rosenberg, and L. A. Comrack. 2008. Burrowing Owl *(Athene cunicularia)*. In W. D. Shuford and T. Gardali, eds. California bird species of special concernStudies of West. Birds 1: 218–26.

Groce, J. E., and M. L. Morrison. 2010. Habitat use by Saw-whet Owls in the Sierra Nevada. J. Wildl. Manage. 74: 1523–32.

Gutierrez, R. J., A. B. Franklin, and W. S. Lahaye. 1995. Spotted Owl *(Strix occidentalis)*. In A. Poole and F. Gill, eds. The birds of North America, No. 179. Academy of Natural Sciences, Philadelphia, and American Ornithologists' Union, Washington, DC.

Haug, E. A., B. A. Millsap, and M. S. Martell. 1993. Burrowing Owl *(Speotyto cunicularia)*. In A. Poole and F. Gill, eds. The birds of North America, No. 61. Academy of Natural Sciences, Philadelphia, and American Ornithologists' Union, Washington, DC.

Hayward, G. D., and J. Verner, eds. 1994. Flammulated, Boreal, and Great Gray Owls in the United States: A technical conservation assessment. USDA For. Serv. General Technical Report RM-253. Fort Collins, CO.

Henjum, M. G. and E. L. Bull. 1990. Ecology of the Great Gray Owl. Pacific Northwest Research Station, U.S. For. Ser., Portland, OR. Gen. Tech. Rep. PNW-GTR-265.

Holt, D. W., and J. L. Petersen. 2000. Northern Pygmy-Owl *(Glaucidium gnoma)*. In A. Poole, ed. The birds of North America online, No. 494. Cornell Lab of Ornith. [accessed November 2011].

Houston, C. S., D. G. Smith, and C. Rohner. 1998. Great Horned Owl *(Bubo virginianus)*. In A. Poole and F. Gill, eds. The birds of North America, No. 372. The Birds of North America, Inc. Philadelphia, PA.

Hull, J. M., J. J. Keane, W. K. Savage, S. A. Godwin, J. A. Shafer, E. P. Jepsen, R. Gerhardt, C. Stermer, and H. B. Ernest. 2010. Range-wide genetic differentiation among North American Great Gray Owls *(Strix nebulosa)* reveals a distinct lineage restricted to the Sierra Nevada Mountains. Molecular phylogenetics and evolution 56: 212–21.

Hunting, K. 2008. Long-eared Owl *(Asio otus)*. In W. D. Shuford and T. Gardali, eds. California bird species of special concern. Studies of West. Birds 1: 234–41.

Jepson, E. P., J. J. Keane, and H. B. Ernest. 2011. Winter distribution and conservation status of the Sierra Nevada Great Gray Owl. J. Wildl. Manage. 75: 1678–87.

Johnsgard, P. A. 1988. North American owls. Smithsonian Inst. Press, Washington, DC.

Marks, J. S., D. L. Evans, and D. W. Holt. 1994. Long-eared Owl *(Asio otus)*. In A. Poole and F. Gill, eds. The Birds of North America, No. 133. Academy of Natural Sciences, Philadelphia, and American Ornithologists' Union, Washington, DC.

McCallum, D. A. 1994. Flammulated Owl *(Otus flammeolus)*. In A. Poole, ed. The Birds of North America online, No. 93. Cornell Lab of Ornith. [accessed November 2011].

Peeters, H. 2007. Field guide to owls of California and the West. Univ. of Calif. Press, Berkeley, CA.

Powers, B., M. D. Johnson, J. A. Lamanna, and J. Rich. 2011. The influence of pocket gophers in the central Sierra mountains, California: Potential implications for Great Gray Owls. Northwestern Nat. 92: 13–18.

Rasmussen, J. Lee, S. G. Sealy, and R. J. Cannings. 2008. Northern Saw-whet Owl *(Aegolius acadicus)*. In A. Poole, ed. The Birds of North America online, No. 42. Cornell Lab of Ornith. [accessed November 2011].

Roberson, D. 2008. Short-eared Owl *(Asio falmmeus)*. In W. D. Shuford and T. Gardali, eds. California bird species of special concern. Studies of West. Birds 1: 242–48.

Sears, C. L. 2006. Assessing distribution, habitat suitability, and site occupancy of Great Gray Owls *(Strix nebulosa)* in California. MS thesis, Univ. of California, Davis.

Shuford, W. D., and S. D. Fitton. 1998. Status of owls in the Glass Mountain region, Mono County, California. West. Birds 29: 1–20.

van Riper III, C., and J. van Wagtendonk. 2006. Home range characteristics of Great Gray Owls in Yosemite National Park, California. J. Raptor Res. 40: 130–41.

Wiggins, D. A., D. W. Holt, and S. M. Leasure. 2006. Short-eared Owl *(Asio flammeus)*. In A. Poole, ed. The Birds of North America online, No. 62. Cornell Lab of Ornith. [accessed November 2011].

Wildman, A. M. 1992. The effect of human activity on Great Gray Owl hunting behavior in Yosem-ite National Park, California. MS thesis, Univ. of Calif., Davis.

Wilkerson, R. L., and R. B. Siegel. 2010. Assessing changes in the distribution and abundance of Burrowing Owls in California, 1993–2007. Bird Populations 10: 1–36.

Williams, P. J., R. J. Gutierrez, and S. A. Whitmore. 2011. Home range and habitat selection of Spotted Owls in the central Sierra Nevada. J. Wildl. Manage. 75: 333–43.

Winter, J. 1974. The distribution of the Flammulated Owl in California. West. Birds 5: 25–44.

NIGHTHAWKS AND RELATIVES

Brigham, R. M., J. Ng, R. G. Poulin, and S. D. Grindal. 2011. Common Nighthawk *(Chordeiles minor)*. In A. Poole, ed. The Birds of North America online, No. 213. Cornell Lab of Ornith. [accessed December 2011].

Latta, S. C., and M. E. Baltz. 1997. Lesser Nighthawk *(Chordeiles acutipennis)*. In A. Poole and F. Gill, eds. The Birds of North America, No. 314. Academy of Natural Sciences, Philadelphia, and American Ornithologists' Union, Washington, DC.

Woods, C. P., R. D. Csada, and R. M. Brigham. 2005. Common Poorwill *(Phalaenoptilus nuttallii)*. In A. Poole, ed. The birds of North America online, No. 32. Cornell Lab of Ornith. [accessed December 2011].

SWIFTS

Beason, J. P., C. Gunn, K. M. Potter, R. A. Sparks, and J. W. Fox. 2012. The northern Black Swift: Migration path and wintering area revealed. Wilson J. of Ornith. 124: 1–8.

Bull, E. L., and C. T. Collins. 2007. Vaux's Swift *(Chaetura vauxi)*. In A. Poole, ed. The birds of North America online, No. 77. Cornell Lab of Ornith. [accessed December 2011].

Hunter, J. E. 2008. Vaux's Swift *(Chaetura vauxi)*. In W. D. Shuford and T. Gardali, eds. California bird species of special concern. Studies of West. Birds 1: 254–59.

Levad, R. G., K. M. Potter, C. W. Shultz, C. Gunn, and J. G. Doerr. 2008. Distribution, abundance, and nest-site characteristics of Black Swifts in the southern Rocky Mountains of Colorado and New Mexico. Wilson J. of Ornith. 120: 331–38.

Lowther, P. E., and C. T. Collins. 2002. Black Swift *(Cypseloides niger)*. In A. Poole, ed. The birds of North America online, No. 676. Cornell Lab of Ornith. [accessed December 2011].

Roberson, D., and C. T. Collins. 2008. Black Swift

(*Cypseloides niger*). In W. D. Shuford and T. Gardali, eds. California bird species of special concern. Studies of West. Birds 1: 249–53.

Ryan, T. P., and C. T. Collins. 2000. White-throated Swift (*Aeronautes saxatalis*). In A. Poole, ed. The Birds of North America online, No. 526. Cornell Lab of Ornith. [accessed December 2011].

HUMMINGBIRDS

Baltosser, W. H., and P. E. Scott. 1996. Costa's Hummingbird (*Calypte costae*). In A. Poole and F. Gill, eds. The birds of North America, No. 251. Academy of Natural Sciences, Philadelphia, and American Ornithologists' Union, Washington, DC.

Baltosser, W. H., and S. M. Russell. 2000. Black-chinned Hummingbird (*Archilochus alexandri*). In A. Poole and F. Gill, eds. The birds of North America, No. 495. The Birds of North America, Inc. Philadelphia, PA.

Calder, W. A. 1993. Rufous Hummingbird (*Selasphorus rufus*). In A. Poole and F. Gill, eds. The birds of North America, No. 53. Academy of Natural Sciences, Philadelphia, and American Ornithologists' Union, Washington, DC.

Calder, W. A., and L. L. Calder. 1994. Calliope Hummingbird (*Stellula calliope*). In A. Poole, ed. The birds of North America online, No. 135. Cornell Lab of Ornith. [accessed October 2011].

Des Granges, J. L. 1979. Organization of a tropical nectar feeding bird guild in a variable environment. Living Bird 17: 199–236.

Howell, S. N. G. 2002. Hummingbirds of North America. Princeton Univ. Press, Princeton, NJ.

Johnsgard, P. A. 1997. The hummingbirds of North America. Smithsonian Inst. Press, Washington, DC.

Russell, S. M. 1996. Anna's Hummingbird (*Calypte anna*). In A. Poole, ed. The birds of North America online, No. 226. Cornell Lab of Ornith. [accessed October 2011].

KINGFISHERS

Kelly, J. F., E. S. Bridge, and M. J. Hamas. 2009. Belted Kingfisher (*Megaceryle alcyon*). In A. Poole, ed. The birds of North America online, No. 84. Cornell Lab of Ornith. [accessed February 2012].

WOODPECKERS

Bull, E. L., and J. A. Jackson. 2011. Pileated Woodpecker (*Dryocopus pileatus*). In A. Poole, ed. The birds of North America online, No. 148. Cornell Lab of Ornith. [accessed June 2012].

Devillers, P. 1970. Identification and distribution in California of the *Sphyrapicus varius* group of sapsuckers. Calif. Birds 1: 47–76.

Dixon, R. D., and V. A. Saab. 2000. Black-backed Woodpecker (*Picoides arcticus*). In A. Poole, ed. The birds of North America online, No. 509. Cornell Lab of Ornith. [accessed June 2012].

Dobbs, R. C., T. E. Martin, and C. J. Conway. 1997. Williamson's Sapsucker (*Sphyrapicus thyroideus*). In A. Poole, ed. The birds of North America online, No. 285. Cornell Lab of Ornith. [accessed October 2011].

Garrett, K. L., M. G. Raphael, and R. A. Dixon. 1996. White-headed Woodpecker (*Picoides albolarvatus*). In A. Poole, ed. The birds of North America online, No. 252. Cornell Lab of Ornith. [accessed June 2012].

Hanson, C. T., and M. T. North. 2008. Postfire woodpecker foraging in salvage-logged and unlogged forests of the Sierra Nevada. Condor 110(4): 777–82.

Hutto, R. L. 2006. Toward meaningful snag-management guidelines for postfire salvage logging in North American conifer forests. Conserv. Biol. 20: 984–93.

———. 2008. The ecological importance of severe wildfires: Some like it hot. Ecological Applications 18(8): 1827–34.

Jackson, J. A., and H. R. Ouellet. 2002. Downy Woodpecker (*Picoides pubescens*). In A. Poole, ed. The birds of North America online, No. 613. Cornell Lab of Ornith. [accessed June 2012].

Jackson, J. A., H. R. Ouellet, and B. J. S. Jackson. 2002. Hairy Woodpecker (*Picoides villosus*). In A. Poole, ed. The Birds of North America online, No. 702. Cornell Lab of Ornith. [accessed June 2012].

Kirk, A., and L. Kirk. 2004. Expansion of the breeding range of the Acorn Woodpecker east of the Sierra Nevada, California. West. Birds 35: 221–23.

Koenig, W. D., P. B. Stacey, M. T. Stanback, and R. L. Mumme. 1995. Acorn Woodpecker (*Melanerpes formicivorus*). In A. Poole, ed. The birds of North America online, No. 194. Cornell Lab of Ornith. [accessed October 2011].

Ligon, J. D., and P. B. Stacey. 1996. Land use, lag times, and the detection of demographic change: An example provided by the Acorn Woodpecker. Conserv. Biol. 10(3): 840–46.

Lowther, P. E. 2001. Ladder-backed Woodpecker (*Picoides scalaris*). In A. Poole, ed. The birds of North America online, No. 565. Cornell Lab of Ornith. [accessed June 2012].

Lowther, P. E. 2000. Nuttall's Woodpecker (*Picoides nuttallii*). In A. Poole, ed. The birds of North

America online, No. 555. Cornell Lab of Ornith. [accessed June 2012].

McKeever, S., and L. Adams. 1960. Acorn Woodpecker resident east of the Sierra Nevada in California. Condor 62: 297.

Moore, W. S. 1995. Northern Flicker *(Colaptes auratus)*. In A. Poole, ed. The birds of North America online, No. 166. Cornell Lab of Ornith. [accessed June 2012].

Raitt, R. J. 1959. Rocky Mountain race of the Williamson Sapsucker wintering in California. Condor 62: 142.

Tobalske, B. W. 1997. Lewis's Woodpecker *(Melanerpes lewis)*. In A. Poole, ed. The Birds of North America online, No. 284. Cornell Lab of Ornith. [accessed June 2012].

Walters, E. L., E. H. Miller, and P. E. Lowther. 2002. Red-breasted Sapsucker *(Sphyrapicus ruber)* and Red-naped Sapsucker *(Sphyrapicus nuchalis)*. In A. Poole, ed. The birds of North America online, No. 663. Cornell Lab of Ornith. [accessed June 2012].

TYRANT FLYCATCHERS

Altman, B., and R. Sallabanks. 2000. Olive-sided Flycatcher *(Contopus cooperi)*. In A. Poole, ed. The birds of North America online, No. 502. Cornell Lab of Ornith. [accessed December 2011].

Bemis, C., and J. D. Rising. 1999. Western Wood-Pewee *(Contopus sordidulus)*. In A. Poole, ed. The birds of North America online, No. 451. Cornell Lab of Ornith. [accessed December 2011].

Cardiff, S. W., and D. L. Dittmann. 2000. Brown-crested Flycatcher *(Myiarchus tyrannulus)*. In A. Poole, ed. The birds of North America online, No. 496. Cornell Lab of Ornith. [accessed December 2011].

Cardiff, S. W., and D. L. Dittmann. 2002. Ash-throated Flycatcher *(Myiarchus cinerascens)*. In A. Poole, ed. The birds of North America online, No. 664. Cornell Lab of Ornith. [accessed December 2011].

Flett, M. A., and S. D. Sanders. 1987. Ecology of a Sierra Nevada population of Willow Flycatchers. West. Birds. 18: 37–42.

Floyd, T., C. S. Elphick, G. Chisholm, K. Mack, R. G. Elston, E. M. Ammon, and J. D. Boone. 2007. Atlas of the breeding birds of Nevada. Univ. of Nevada Press, Reno, NV.

Fowler, C., B. Valentine, S. D. Sanders, and M. Stafford. 1991. Habitat Suitability Index Model: Willow Flycatcher *(Empidonax traillii)*. USDA For. Ser., Tahoe Nat. For., Nevada City, CA.

Gamble, L. R., and T. M. Bergin. 1996. Western Kingbird *(Tyrannus verticalis)*. In A. Poole, ed. The birds of North America online, No. 227. Cornell Lab of Ornith. [accessed December 2011].

Harris, J. D., S. D. Sanders, and M. A. Flett. 1987. Willow Flycatcher surveys in the Sierra Nevada. West. Birds 18: 27–36.

Kotliar, N. B. 2007. Olive-sided Flycatcher *(Contopus cooperi)*: A technical conservation assessment. USDA For. Ser., Rocky Mountain Region. Available at http://www.fs.fed.us/r2/projects/scp/assessments/olivesidedflycatcher.pdf [accessed November 2011].

Lowther, P. E. 2000. Pacific-slope Flycatcher *(Empidonax difficilis)*. In A. Poole, ed. The birds of North America online, No. 556. Cornell Lab of Ornith. [accessed December 2011].

Myers, S. J. 2008. Vermilion Flycatcher *(Pyrocephalus rubinus)*. In W. D. Shuford and T. Gardali, eds. California bird species of special concern. Studies of West. Birds 1: 266–70.

Pereyra, M. E. 2011. Effects of snow-related environmental variation on breeding schedules and productivity of a high-altitude population of Dusky Flycatchers *(Empidonax oberholseri)*. Auk 128: 746–58.

Phillips, A. R. 1944. Some differences between the Wright's and Gray Flycatchers. Auk 61: 293–294.

Robertson, B. A., and R. L. Hutto. 2007. Is selectively harvested forest an ecological trap for Olive-sided Flycatchers? Condor 109: 109–21.

Roosevelt, T. 1913. Theodore Roosevelt: An autobiography. MacMillan, New York, NY.

Sanders, S. D., and M. A. Flett. 1989. The ecology of a Sierra Nevada population of Willow Flycatchers *(Empidonax traillii)*, 1986 and 1987. California Management Branch Administrative Report No. 89-3, Calif. Dept. of Fish and Game, Sacramento, CA.

Schukman, J. M., and B. O. Wolf. 1998. Say's Phoebe *(Sayornis saya)*. In A. Poole, ed. The birds of North America online, No. 374. Cornell Lab of Ornith. [accessed December 2011].

Sedgwick, J. A. 1993. Dusky Flycatcher *(Empidonax oberholseri)*. In A. Poole, ed. The birds of North America online, No. 78. Cornell Lab of Ornith. [accessed December 2011].

———. 1994. Hammond's Flycatcher *(Empidonax hammondii)*. In A. Poole, ed. The birds of North America online, No. 109. Cornell Lab of Ornith. [accessed December 2011].

———. 2000. Willow Flycatcher *(Empidonax traillii)*. In A. Poole, ed. The birds of North America online, No. 533. Cornell Lab of Ornith. [accessed December 2011].

Siegel, R. B., R. L. Wilkerson, and D. F. DeSante. 2008. Extirpation of the Willow Flycatcher from Yosemite National Park. West. Birds 39: 8–21.

Sterling, J. C. 1999. Gray Flycatcher *(Empidonax wrightii)*. In A. Poole, ed. The birds of North America online, No. 458. Cornell Lab of Ornith. [accessed December 2011].

Unitt, P. 1987. *Empidonax trailii extimus*: An endangered subspecies. West. Birds 18: 137–62.

Whitfield, M. J., K. M. Enos, and S. P. Rowe. 1999. Is Brown-headed Cowbird trapping effective for managing populations of the endangered Southwestern Willow Flycatcher? Stud. Avian Bio. 18: 260–66.

Widdowson, W. P. 2008. Olive-sided Flycatcher *(Contopus cooperi)*. In W. D. Shuford and T. Gardali, eds. California bird species of special concern. Studies of West. Birds 1: 260–65.

Wolf, B. O. 1997. Black Phoebe *(Sayornis nigricans)*. In A. Poole, ed. The birds of North America online, No. 268. Cornell Lab of Ornith. [accessed December 2011].

SHRIKES

Cade, T. J., and E. C. Atkinson. 2002. Northern Shrike *(Lanius excubitor)*. In A. Poole and F. Gill, eds. The birds of North America, No. 671. The Birds of North America, Inc. Philadelphia, PA.

Humple, D. 2008. Loggerhead Shrike *(Lanius ludovicianus)* (mainland populations). In W. D. Shuford and T. Gardali, eds. California bird species of special concern. Studies of West. Birds 1: 271–77.

Pandolfino, E. R. 2008. Population trends of the Loggerhead Shrike in California: Possible impact of the West Nile virus in the Central Valley. Central Valley Bird Club Bull. 11: 37–44.

Peterjohn, B. G., and J. R. Sauer. 1995. Population trends of the Loggerhead Shrike from the North American breeding bird survey. In R. Yosef and F. E. Lohrer, eds. Shrikes *(Laniidae)* of the world: Biology and conservation. Proc. of the West. Found. of Vert. Zool. 6: 117–21.

Yosef, R. 1996. Loggerhead Shrike *(Lanius ludovicianus)*. In A. Poole, ed. The birds of North America online, No. 231. Cornell Lab of Ornith. [accessed December 2011].

VIREOS

Bent, A. C. 1950. Life histories of North American wagtails, shrikes, vireos, and their allies. U.S. Natl. Mus. Bull. No. 197.

Curson, D. R., and C. B. Goguen. 1998. Plumbeous Vireo *(Vireo plumbeus)*. In A. Poole, ed. The birds of North America online, No. 366. Cornell Lab of Ornith. [accessed December 2011].

Davis, J. N. 1995. Hutton's Vireo *(Vireo huttoni)*. In A. Poole, ed. The birds of North America online, No. 189. Cornell Lab of Ornith. [accessed December 2011].

Gardali, T., and G. Ballard. 2000. Warbling Vireo *(Vireo gilvus)*. In A. Poole, ed. The birds of North America online, No. 551. Cornell Lab of Ornith. [accessed December 2011].

Goguen, C. B., and D. R. Curson. 2002. Cassin's Vireo *(Vireo cassinii)*. In A. Poole, ed. The birds of North America online, No. 615. Cornell Lab of Ornith. [accessed December 2011].

Kus, B., S. L. Hopp, R. R. Johnson, and B. T. Brown. 2010. Bell's Vireo *(Vireo bellii)*. In A. Poole, ed. The birds of North America online, No. 35. Cornell Lab of Ornith. [accessed December 2011].

Kus, B. E., and M. J. Whitfield. 2005. Parasitism, productivity, and population growth: Response of least Bell's vireos *(Vireo bellii pusillus)* and southwestern willow flycatchers *(Empidonax trailii extimus)* to cowbird *(Molothrus* spp.) control. Ornith. Monogr. 57: 16–27.

JAYS AND RELATIVES

Balda, R. P. 2002. Pinyon Jay *(Gymnorhinus cyanocephalus)*. In A. Poole and F. Gill, eds. The birds of North America, No. 605. The Birds of North America, Inc. Philadelphia, PA.

Birkhead, T. R. 1991. The magpies: Ecology and behavior of Black-billed and Yellow-billed magpies. T. & A. D. Poyser, London.

Boarman, W. I., and B. Heinrich. 1999. Common Raven *(Corvus corax)*. In A. Poole, ed. The birds of North America online, No. 476. Cornell Lab of Ornith. [accessed October 2011].

Crosbie, S. P., W. D. Koenig, W. Reisen, V. L. Kramer, L. Marcus, R. Carney, E. R. Pandolfino, G. M. Bolen, L. R. Crosbie, D. A. Bell, and H. B. Ernest. 2008. A preliminary assessment: Impact of West Nile Virus on the Yellow-billed Magpie, a California endemic. Auk 125: 542–50.

Curry, R. L., A. T. Peterson, and T. A. Langen. 2002. Western Scrub-Jay *(Aphelocoma californica)*. In A. Poole and F. Gill, eds. The birds of North America, No. 712. The Birds of North America, Inc. Philadelphia, PA.

Delaney, K. S., S. Zafar, and R. K. Wayne. 2008. Genetic divergence and differentiation within the Western Scrub-Jay *(Aphelocoma californica)*. Auk 125: 839–49.

Greene, E., W. Davison, and V. R. Muehter. 1998. Steller's Jay *(Cyanocitta stelleri)*. In A. Poole, ed.

The birds of North America online, No. 343. Cornell Lab of Ornith. [accessed October 2011].

Heinrich, B. 1999. Mind of the raven. Harper Collins, New York, NY.

Hutchins, H. E., and R. M. Lanner. 1982. The central role of Clark's Nutcracker in the dispersal and establishment of whitebark pine. Oecologia 55: 192–201.

Kilham, L. 1989. The American Crow and the Common Raven. Texas A&M Univ. Press, College Station, TX.

Marzluff, J. M., and T. Angell. 2005. In the company of crows and ravens. Yale Univ. Press, New Haven, CT.

Reynolds, M. D. 1995. Yellow-billed Magpie *(Pica nuttalli)*. In A. Poole and F. Gill, eds. The birds of North America, No. 180. Academy of Natural Sciences, Philadelphia, and American Ornithologists' Union, Washington, DC.

Tomback, D. F. 1978. Pre-roosting flight of the Clark's Nutcracker. Auk 95: 554–62.

———. 1998. Clark's Nutcracker *(Nucifraga columbiana)*. In A. Poole, ed. The birds of North America online, No. 331. Cornell Lab of Ornith. [accessed October 2011].

Tomback, D. F., and K. A. Kramer. 1980. Limber pine seed harvest by Clark's Nutcracker in the Sierra Nevada: Timing and foraging behavior. Condor 82: 467–68.

Trost, C. H. 1999. Black-billed Magpie *(Pica pica)*. In A. Poole and F. Gill, eds. The birds of North America, No. 389. The Birds of North America, Inc. Philadelphia, PA.

Verbeek, N. A. M., and C. Caffrey. 2002. American Crow *(Corvus brachyrhynchos)*. In A. Poole and F. Gill, eds. The birds of North America, No. 647. The Birds of North America, Inc. Philadelphia, PA.

LARKS

Beason, R. C. 1995. Horned Lark *(Eremophila alpestris)*. In A. Poole, ed. The birds of North America online, No. 195. Cornell Lab of Ornith. [accessed January 2012].

SWALLOWS

Airola, D. A., and J. Grantham. 2003. Purple Martin population status, nesting habitat characteristics, and management in Sacramento. West. Birds 34: 235–51.

Airola, D. A., and D. Kopp. 2009. Recent Purple Martin declines in the Sacramento region of California: Recovery implications. West. Birds 40: 254–59.

Airola, D. A., and B. D. C. Williams. 2008. Purple Martin *(Progne subis)*. In W. D. Shuford and T. Gardali, eds. California bird species of special concern. Studies of West. Birds 1: 293–99.

Brown, C. R. 1997. Purple Martin *(Progne subis)*. In A. Poole, ed. The birds of North America online, No. 287. Cornell Lab of Ornith. [accessed January 2012].

Brown, C. R., and M. B. Brown. 1995. Cliff Swallow *(Petrochelidon pyrrhonota)*. In A. Poole, ed. The birds of North America online, No. 149. Cornell Lab of Ornith. [accessed January 2012].

———. 1999. Barn Swallow *(Hirundo rustica)*. In A. Poole, ed. The birds of North America online, No. 452. Cornell Lab of Ornith. [accessed January 2012].

Brown, C. R., A. M. Knott, and E. J. Damrose. 2011. Violet-green Swallow *(Tachycineta thalassina)*. In A. Poole, ed. The birds of North America online, No. 14. Cornell Lab of Ornith. [accessed January 2012].

Cousens, B., J. Rowoth, and D. A. Airola. 2010. Large Tree Swallow roost verified in California's Central Valley. Central Valley Bird Club Bull. 13: 1–17.

De Jong, M. J. 1996. Northern Rough-winged Swallow *(Stelgidopteryx serripennis)*. In A. Poole, ed. The birds of North America online, No. 234. Cornell Lab of Ornith. [accessed January 2012].

Eltzroth, K. E., and S. R. Robinson. 1984. Violet-green Swallows help Western Bluebirds at the nest. J. Field Ornithol. 55: 259–61.

Franzreb, K. E. 1976. Nest site competition between Mountain Chickadees and Violet-green Swallows. Auk 93: 836–37.

Garrison, B. A. 1999. Bank Swallow *(Riparia riparia)*. In A. Poole, ed. The birds of North America online, No. 414. Cornell Lab of Ornith. [accessed January 2012].

Garrison, B. A., J. Humphrey, and S. A. Laymon. 1987. Bank Swallow distribution and nesting ecology on the Sacramento River, California. West. Birds 18: 71–76.

Schlorff, R. W. 1992. Recovery Plan: Bank Swallow *(Riparia riparia)*. Calif. Dept. of Fish and Game, Sacramento, CA.

Sylvester, V., and D. A. Airola. 2010. Purple Martins nesting in low elevation transmission towers in the San Joaquin Valley, California. Central Valley Bird Club Bull. 13: 69–75.

Williams, B. D. C. 1998. Distribution, habitat associations, and conservation of Purple Martins in California. MS thesis, Calif. State Univ., Sacramento.

Winkler, D. W., K. K. Hallinger, D. R. Ardia, R. J.

Robertson, B. J. Stutchbury, and R. R. Cohen. 2011. Tree Swallow *(Tachycineta bicolor)*. In A. Poole, ed. The birds of North America online, No. 11. Cornell Lab of Ornith. [accessed January 2012].

CHICKADEES AND TITMICE

Brennan, L. A., and M. L. Morrison. 1991. Long-term trends of chickadee populations in western North America. Condor 93: 130–37.

Chaplin, S. B., 1976. The physiology of hypothermia in the Black-capped Chickadee, *Parus atricapillus*. Jour. of Comp. Physio. 112: 335–44.

Cicero, C. 1996. Sibling species of titmice in the *Parus inornatus* complex. Zoology vol. 128. Univ. of Calif. Press, Berkeley, CA.

———. 2000. Oak Titmouse *(Baeolophus inornatus)* and Juniper Titmouse *(Parus ridgwayi)*. In A. Poole, ed. The birds of North America online, No. 485. Cornell Lab of Ornith. [accessed January 2012].

Crase, F. T. 1976. Occurrence of the Chestnut-backed Chickadee in the Sierra Nevada mountains, California. Am. Birds 30: 673–75.

Dahlsten, D. L., L. A. Brennan, D. A. Mccallum and S. I. Gaunt. 2002. Chestnut-backed Chickadee *(Poecile rufescens)*. In A. Poole, ed. The birds of North America online, No. 689. Cornell Lab of Ornith. [accessed January 2012].

Dahlsten, D. L., and S. G. Herman. 1965. Birds as predators of destructive forest insects. Calif. Agric. September 1965: 8–10.

Dahlsten, D. L., W. A. Copper, D. L. Rowney, and P. L. Kleintjes. 1990. Quantifying bird predation of arthropods in forests. Stud. Avian Bio. 13: 44–52.

Hutto, R. L., and J. S. Young. 1999. Habitat relationships of landbirds in the Northern Region, USDA For. Ser. Gen. Tech. Rep. RMRS-GTR-32. Rocky Mtn. Res. Stn., Ogden, UT.

McCallum, D. A., R. Grundel, and D. L. Dahlsten. 1999. Mountain Chickadee *(Poecile gambeli)*. In A. Poole, ed. The birds of North America online, No. 453. Cornell Lab of Ornith. [accessed January 2012].

Morrison, M. L., I. C. Timosi, K. A. With, and P. N. Manley. 1985. Use of tree species by forest birds in winter and summer. J. Wildl. Manage. 49: 1098–102.

VERDIN

Webster, M. D. 1999. Verdin *(Auriparus flaviceps)*. In A. Poole, ed. The birds of North America online, No. 470. Cornell Lab of Ornith. [accessed Mar 2012].

BUSHTIT

Sloane, S. A. 2001. Bushtit *(Psaltriparus minimus)*. In A. Poole, ed. The birds of North America online, No. 598. Cornell Lab of Ornith. [accessed January 2012].

NUTHATCHES

Dunn, J. L., and K. L. Garrett. In preparation. Ranges of White-breasted Nuthatch subspecies in California.

Ghalambor, C. K., and T. E. Martin. 1999. Red-breasted Nuthatch *(Sitta canadensis)*. In A. Poole, ed. The birds of North America online, No. 459. Cornell Lab of Ornith. [accessed January 2012].

Grubb Jr., T. C., and V. V. Pravosudov. 2008. White-breasted Nuthatch *(Sitta carolinensis)*. In A. Poole, ed. The birds of North America online, No. 54. Cornell Lab of Ornith. [accessed January 2012].

Kingery, H. E., and C. K. Ghalambor. 2001. Pygmy Nuthatch *(Sitta pygmaea)*. In A. Poole, ed. The birds of North America online, No. 567. Cornell Lab of Ornith. [accessed January 2012].

Spellman, G. M., and J. Klicka. 2007. Phylogeography of the White-breasted Nuthatch *(Sitta carolinensis)*: Diversification in North American pine and oak woodlands. Molecular Ecology 16: 1729–40.

Walstrom, V. W., J. Klicka, and G. M. Spellman. 2012. Speciation in the White-breasted Nuthatch *(Sitta carolinensis)*: A multilocus perspective. Molecular Ecology 21: 907–20.

CREEPERS

Adams, E. M., and M. L. Morrison. 1993. Effects of forest stand structure and composition on Red-breasted Nuthatches and Brown Creepers. J. Wildl. Manage. 57: 616–29.

Hejl, S. J., K. R. Newlon, M. E. Mcfadzen, J. S. Young, and C. K. Ghalambor. 2002. Brown Creeper *(Certhia americana)*. In A. Poole, ed. The birds of North America online, No. 669. Cornell Lab of Ornith. [accessed January 2012].

WRENS

Hamilton, R. A., G. A. Proudfoot, D. A. Sherry, and S. Johnson. 2011. Cactus Wren *(Campylorhynchus brunneicapillus)*. In A. Poole, ed. The birds of North America online, No. 558. Cornell Lab of Ornith. [accessed January 2012].

Hejl, S. J., J. A. Holmes, and D. E. Kroodsma. 2002.

Winter Wren *(Troglodytes hiemalis)*. In A. Poole, ed. The birds of North America online, No. 623. Cornell Lab of Ornith. [accessed January 2012].

Johnson, L. S. 1998. House Wren *(Troglodytes aedon)*. In A. Poole, ed. The birds of North America online, No. 380. Cornell Lab of Ornith. [accessed January 2012].

Jones, S. L., and J. S. Dieni. 1995. Canyon Wren *(Catherpes mexicanus)*. In A. Poole, ed. The birds of North America online, No. 197. Cornell Lab of Ornith. [accessed January 2012].

Kennedy, E. D., and D. W. White. 1997. Bewick's Wren *(Thryomanes bewickii)*. In A. Poole, ed. The birds of North America online, No. 315. Cornell Lab of Ornith. [accessed January 2012].

Kroodsma, D. E. 1989. Two North American song populations of the Marsh Wren reach distributional limits in the central Great Plains. Condor 91: 332–40.

Kroodsma, D. E., and J. Verner. 1997. Marsh Wren *(Cistothorus palustris)*. In A. Poole, ed. The birds of North America online, No. 308. Cornell Lab of Ornith. [accessed January 2012].

Lowther, P. E., D. E. Kroodsma, and G. H. Farley. 2000. Rock Wren *(Salpinctes obsoletus)*. In A. Poole, ed. The birds of North America online, No. 486. Cornell Lab of Ornith. [accessed January 2012].

Phillips, A. R. 1986. The known birds of North and Middle America, Part I. R. Phillips, Denver, CO.

Verner, J., and K. L. Purcell. 1999. Fluctuating populations of House Wrens and Bewick's Wrens in the foothills of the western Sierra Nevada of California. Condor 101: 219–29.

GNATCATCHERS

Ellison, W. G. 1992. Blue-gray Gnatcatcher *(Polioptila caerulea)*. In A. Poole, ed. The birds of North America online, No. 23. Cornell Lab of Ornith. [accessed January 2012].

DIPPERS

Epanchin, P. N., R. A. Knapp, and S. P. Lawler. 2010. Nonnative trout impact an alpine-nesting bird by altering aquatic-insect subsidies. Ecology 91: 2406–15.

Muir, J. 1878. The humming-bird of California water-falls. Scribner's Monthly 15: 545–54. Reprinted in J. Muir. 1894. The mountains of California. The Century Company, New York, NY.

Price, F. E., and C. E. Bock. 1983. Population ecology of the American Dipper *(Cinclus mexicanus)*

in the Front Range of Colorado. Stud. Avian Bio. No. 7, publication of the Cooper Ornithological Society.

Willson, M. F., and H. E. Kingery. 2011. American Dipper *(Cinclus mexicanus)*. In A. Poole, ed. The birds of North America online, No. 229. Cornell Lab of Ornith. [accessed January 2012].

KINGLETS

Heinrich, B. 2003. Overnighting of Golden-crowned Kinglets during winter. Wilson Bull. 115: 113–14.

Ingold, J. L., and R. Galati. 1997. Golden-crowned Kinglet *(Regulus satrapa)*. In A. Poole, ed. The birds of North America online, No. 301. Cornell Lab of Ornith. [accessed January 2012].

Swanson, D. L., J. L. Ingold, and G. E. Wallace. 2008. Ruby-crowned Kinglet *(Regulus calendula)*. In A. Poole, ed. The birds of North America online, No. 119. Cornell Lab of Ornith. [accessed January 2012].

WRENTIT

Geupel, G. R., and G. Ballard. 2002. Wrentit *(Chamaea fasciata)*. In A. Poole, ed. The birds of North America online, No. 654. Cornell Lab of Ornith. [accessed January 2012].

THRUSHES AND RELATIVES

Bowen, R. V. 1997. Townsend's Solitaire *(Myadestes townsendi)*. In A. Poole and F. Gill, eds. The birds of North America, No. 269. The Birds of North America, Inc. Philadelphia, PA.

Feihler, C. M., W. D. Teitje, and W. R. Fields. 2006. Nesting success of Western Bluebirds *(Sialia mexicana)* using nest boxes in vineyards and oak-savannah habitats of California. Wilson Jour. of Ornithol. 118: 553–57.

George, T. L. 2000. Varied Thrush *(Ixoreus naevius)*. In A. Poole, ed. The birds of North America online, No. 541. Cornell Lab of Ornith. [accessed December 2011].

Germaine, H. L., and S. S. Germaine. 2006. Forest restoration treatment effects on the nesting success of Western Bluebirds *(Sialia mexicana)*. Restor. Ecology 10: 362–67.

Guinan, J. A., P. A. Gowaty, and E. K. Eltzroth. 2008. Western Bluebird *(Sialia mexicana)*. In A. Poole and F. Gill, eds. The birds of North America. No. 510. Academy of Natural Sciences, Philadelphia, PA.

Johnson, M. D., and G. R. Geupel. 1998. The

importance of productivity to the dynamics of a Swainson's Thrush population. Condor 98: 134–42.

Jones, P. W., and T. M. Donovan. 1996. Hermit Thrush *(Catharus guttatus)*. In A. Poole, ed. The birds of North America online, No. 261. Cornell Lab of Ornith. [accessed December 2011].

Mack, D. E., and W. Yong. 2000. Swainson's Thrush *(Catharus ustulatus)*. In A. Poole, ed. The birds of North America online, No. 540. Cornell Lab of Ornith. [accessed December 2011].

Marshall, J. 1988. Birds lost from a giant Sequoia forest during fifty years. Condor 90: 359–72.

Power, H. W., and M. P. Lombardo. 1996. Mountain Bluebird *(Sialia currucoides)*. In A. Poole and F. Gill, eds. The birds of North America. No. 222. Academy of Natural Sciences, Philadelphia, PA.

Sallabanks, R., and F. C. James. 1999. American Robin *(Turdus migratorius)*. In A. Poole and F. Gill, eds. The birds of North America, No. 462. The Birds of North America, Inc. Philadelphia, PA.

MOCKINGBIRDS AND THRASHERS

Cody, M. L. 1998. California Thrasher *(Toxostoma redivivum)*. In A. Poole, ed. The birds of North America online, No. 323. Cornell Lab of Ornith. [accessed February 2012].

Farnsworth, G., G. A. Londono, J. Ungvari Martin, K. C. Derrickson, and R. Breitwisch. 2011. Northern Mockingbird *(Mimus polyglottos)*. In A. Poole, ed. The birds of North America online, No. 7. Cornell Lab of Ornith. [accessed February 2012].

Fitton, S. D. 2008. Le Conte's Thrasher *(Toxostoma lecontei)*. In W. D. Shuford and T. Gardali, eds. California bird species of special concern. Studies of West. Birds 1: 321–26.

Reynolds, T. D., T. D. Rich, and D. A. Stephens. 1999. Sage Thrasher *(Oreoscoptes montanus)*. In A. Poole, ed. The birds of North America online, No. 463. Cornell Lab of Ornith. [accessed February 2012].

Sheppard, J. M. 1996. Le Conte's Thrasher *(Toxostoma lecontei)*. In A. Poole, ed. The birds of North America online, No. 230. Cornell Lab of Ornith. [accessed February 2012].

STARLINGS

Cabe, P. R. 1993. European Starling *(Sturnus vulgaris)*. In A. Poole, ed. The birds of North America online, No. 48. Cornell Lab of Ornith. [accessed January 2012].

PIPITS

Gaines, D. 1992. Birds of Yosemite and the East Slope, 2nd printing. Artemisia Press, Lee Vining, CA.

Knorr, O. A. 2000. Breeding of the American Pipit *(Anthus rubescens)* in Nevada. Great Basin Birds 3: 7–9.

Miller, J. H., and M. T. Greene. 1987. Distribution, status, and origin of Water Pipits breeding in California. Condor 89: 788–97.

Verbeek, N. A., and P. Hendricks. 1994. American Pipit *(Anthus rubescens)*. In A. Poole, ed. The birds of North America online, No. 95. Cornell Lab of Ornith. [accessed February 2012].

WAXWINGS

Witmer, M. C., D. J. Mountjoy, and L. Elliot. 1997. Cedar Waxwing *(Bombycilla cedrorum)*. In A. Poole, ed. The birds of North America online, No. 309. Cornell Lab of Ornith. [accessed December 2011].

SILKY FLYCATCHERS

Chu, M., and G. Walsberg. 1999. Phainopepla *(Phainopepla nitens)*. In A. Poole, ed. The birds of North America online, No. 415. Cornell Lab of Ornith. [accessed December 2011].

Walsberg, G. E. 1977. Ecology and energetics of contrasting social systems in *Phainopepla nitens (Aves: Ptilogonatidae)*. Univ. of Calif. Pub. in Zool. 108: 1–63.

WOOD-WARBLERS

Ammon, E. M., and W. M. Gilbert. 1999. Wilson's Warbler *(Cardellina pusilla)*. In A. Poole, ed. The birds of North America online, No. 478. Cornell Lab of Ornith. [accessed December 2011].

Chase, M. K., N. Nur, and G. R. Geupel. 1997. Survival, productivity, and abundance in a Wilson's Warbler population. Auk 114: 354–66.

Comrack, L. A. 2008. Yellow-breasted Chat *(Icteria virens)*. In W. D. Shuford and T. Gardali, eds. California bird species of special concern. Studies of West. Birds 1: 351–58.

Dunn, J., and K. L. Garrett. 1997. Warblers. Houghton Mifflin, Boston, MA.

Eckerle, K. P., and C. F. Thompson. 2001. Yellow-breasted Chat *(Icteria virens)*. In A. Poole, ed. The birds of North America online, No. 575. Cornell Lab of Ornith. [accessed June 2012].

Gilbert, W. M., M. K. Sogge, and C. Van Riper III. 2010. Orange-crowned Warbler *(Oreoth-*

lypis celata). In A. Poole, ed. The birds of North America online, No. 101. Cornell Lab of Ornith. [accessed December 2011].

Guzy, M. J., and P. E. Lowther. 1997. Black-throated Gray Warbler *(Setophaga nigrescens)*. In A. Poole, ed. The birds of North America online, No. 319. Cornell Lab of Ornith. [accessed December 2011].

Guzy, M. J., and G. Ritchison. 1999. Common Yellowthroat *(Geothlypis trichas)*. In A. Poole and F. Gill, eds. The birds of North America, No. 448. The Birds of North America, Inc. Philadelphia, PA.

Heath, S. K. 2008. Yellow Warbler *(Dendroica petechia)*. In W. D. Shuford and T. Gardali, eds. California bird species of special concern. Studies of West. Birds 1: 332–39.

Humple, D. L., and R. D. Burnett. 2010. Nesting ecology of Yellow Warblers *(Dendroica petechia)* in montane chaparral habitat in the northern Sierra Nevada. West. N. Am. Naturalist 70: 355–63.

Hunt, P. D., and D. J. Flaspohler. 1998. Yellowrumped Warbler *(Setophaga coronata)*. In A. Poole and F. Gill, eds. The birds of North America, No. 376. The Birds of North America, Inc. Philadelphia, PA.

Lowther, P. E., and J. McI. Williams. 2011. Nashville Warbler *(Oreothlypis ruficapilla)*. In A. Poole, ed. The birds of North America online, No. 205. Cornell Lab of Ornith. [accessed December 2011].

Lowther, P. E., C. Celada, N. K. Klein, C. C. Rimmer, and D. A. Spector. 1999. Yellow Warbler *(Setophaga petechia)*. In A. Poole and F. Gill, eds. The birds of North America, No. 454. The Birds of North America, Inc. Philadelphia, PA.

Olson, C. R., and T. E. Martin. 1999. Virginia's Warbler *(Oreothlypis virginiae)*. In A. Poole, ed. The birds of North America online, No. 477. Cornell Lab of Ornith. [accessed December 2011].

Pearson, S. F. 1997. Hermit Warbler *(Setophaga occidentalis)*. In A. Poole, ed. The birds of North America online, No. 303. Cornell Lab of Ornith. [accessed December 2011].

Pitocchelli, J. 1995. MacGillivray's Warbler *(Geothlypis tolmiei)*. In A. Poole, ed. The birds of North America online, No. 159. Cornell Lab of Ornith. [accessed December 2011].

Stewart, R. M. 1973. Breeding behavior and life history of the Wilson's Warbler. Wilson Bull. 85: 21–30.

Stewart, R. M., R. P. Henderson, and K. Darling. 1977. Breeding ecology of the Wilson's Warbler in the high Sierra Nevada, California. Living Bird 16: 83–192.

Wright, A. L., G. D. Hayward, S. M. Matsuoka, and P. H. Hayward. 1998. Townsend's Warbler *(Setophaga townsendi)*. In A. Poole and F. Gill, eds. The birds of North America, No. 333. The Birds of North America, Inc. Philadelphia, PA.

SPARROWS AND RELATIVES

Ammon, E. M. 1995. Lincoln's Sparrow *(Melospiza lincolnii)*. In A. Poole, ed. The birds of North America online, No. 191. Cornell Lab of Ornith. [accessed November 2011].

Arcese, P., M. K. Sogge, A. B. Marr, and M. A. Patten. 2002. Song Sparrow *(Melospiza melodia)*. In A. Poole, ed. The birds of North America online, No. 704. Cornell Lab of Ornith. [accessed November 2011].

Beadle, D., and J. Rising. 2003. Sparrows of the United States and Canada. Princeton Univ. Press, Princeton, NJ.

Benedict, L., M. R. Kunzmann, K. Ellison, K. L. Purcell, R. R. Johnson, and L. T. Haight. 2011. California Towhee *(Melozone crissalis)*. In A. Poole, ed. The birds of North America online, No. 632. Cornell Lab of Ornith. [accessed November 2011].

Berger, A. J. 1968. Vesper Sparrow. In O. L. Austin, ed. Life histories of North American cardinals, grosbeaks, buntings, towhees, finches, sparrows, and their allies. Pt. 2. U.S. Natl. Mus. Bull. 237: 868–86.

Chilton, G., M. C. Baker, C. D. Barrentine, and M. A. Cunningham. 1995. White-crowned Sparrow *(Zonotrichia leucophrys)*. In A. Poole, ed. The birds of North America online, No. 183. Cornell Lab of Ornith. [accessed November 2011].

Cicero, C. 2010. The significance of subspecies: A case study of Sage Sparrows *(Emberizidae, Amphispiza belli)*. Chapter 9 in K. Winker, and S. M. Haig, eds. Avian Subspecies. Ornith. Monogr. no. 67, 103–113.

Cicero, C., and N. K. Johnson. 2007. Narrow contact of desert Sage Sparrows *(Amphispiza belli nevadensis and A. b. canescens)* in Owens Valley, eastern California: Evidence from mitochondrial DNA, morphology, and GIS-based niche models. Ornith. Monogr. 63: 78–95.

Collins, P. W. 1999. Rufous-crowned Sparrow *(Aimophila ruficeps)*. In A. Poole and F. Gill, eds. The birds of North America, No. 472. The Birds of North America, Inc. Philadelphia, PA.

Cord, B., and J. R. Jehl Jr. 1979. Distribution, biology, and status of a relict population of Brown Towhee *(Pipilo fuscus eremophilus)*. West. Birds 10: 131–56.

Dobbs, R. C., P. R. Martin, and T. E. Martin. 1998.

Green-tailed Towhee *(Pipilo chlorurus)*. In A. Poole, ed. The birds of North America online, No. 368. Cornell Lab of Ornith. [accessed October 2011].

Erickson, R. A. 2008. Oregon Vesper Sparrow *(Pooecetes gramineus affinis)*. In W. D. Shuford and T. Gardali, eds. California bird species of special concern. Studies of West. Birds 1: 377–81.

Falls, J. B., and J. G. Kopachena. 1994. White-throated Sparrow *(Zonotrichia albicollis)*. In A. Poole and F. Gill, eds. The birds of North America, No. 128. The Birds of North America, Inc. Philadelphia, PA.

Gennet, S., M. Hammond, and J. Bartolome. In preparation. Association of vegetation composition and canopy structure with songbirds in California valley grasslands.

Greenlaw, J. S. 1996. Spotted Towhee *(Pipilo maculatus)*. In A. Poole, ed. The birds of North America online, No. 263. Cornell Lab of Ornith. [accessed October 2011].

Hendricks, C. J. N., and S Santonocito. 1998. Golden-crowned Sparrow *(Zonotrichia atricapilla)*. In A. Poole and F. Gill, eds. The birds of North America, No. 352. The Birds of North America, Inc. Philadelphia, PA.

Johnson, M. J., C. van Riper III, and K. M. Pearson. 2002. Black-throated Sparrow *(Amphispiza bilineata)*. In A. Poole, ed. The birds of North America online, No. 637. Cornell Lab of Ornith. [accessed November 2011].

Jones, S. L., and J. E. Cornely. 2002. Vesper Sparrow *(Pooecetes gramineus)*. In A. Poole, ed. The birds of North America online, No. 624. Cornell Lab of Ornith. [accessed November 2011].

Kaspari, M., and H. O'Leary. 1988. Nonparental attendants in a north-temperate migrant. Auk 105: 792–93.

Klicka, J., and R. C. Banks. 2011. A generic name for some sparrows *(Aves: Emberizidae)*. Zootaxa 2793: 67–68.

Martin, J. W., and B. A. Carlson. 1998. Sage Sparrow *(Amphispiza belli)*. In A. Poole, ed. The birds of North America online, No. 326. Cornell Lab of Ornith. [accessed November 2011].

Martin, J. W., and J. R. Parrish. 2000. Lark Sparrow *(Chondestes grammacus)*. In A. Poole, ed. The birds of North America online, No. 488. Cornell Lab of Ornith. [accessed November 2011].

Middleton, A. L. 1998. Chipping Sparrow *(Spizella passerina)*. In A. Poole, ed. The birds of North America online, No. 334. Cornell Lab of Ornith. [accessed November 2011].

Nolan Jr., V., E. D. Ketterson, D. A. Cristol, C. M. Rogers, E. D. Clotfelter, R. C. Titus, S. J. Schoech, and E. Snajdr. 2002. Dark-eyed Junco *(Junco hyemalis)*. In A. Poole, ed. The birds of North America online, No. 716. Cornell Lab of Ornith. [accessed November 2011].

Orejuela, J. E., and M. L. Morton. 1975. Song dialects in several populations of Mountain White-crowned Sparrows *(Zonotrichia leucophrys oriantha)* in the Sierra Nevada. Condor 77: 145–53.

Patten, M. A., and M. Fugate. 1998. Systematic relationships among Emberizid sparrows. Auk 111: 412–24.

Patten, M. A., and P. Unitt. 2002. Diagnosability versus mean differences of Sage Sparrow subspecies. Auk 119: 26–35.

Petersen, K. L., and L. B. Best. 1985. Brewer's Sparrow nest-site characteristics in a sagebrush community. J. Field Ornithol. 56: 23–27.

Rotenberry, J. T., M. A. Patten, and K. L. Preston. 1999. Brewer's Sparrow *(Spizella breweri)*. In A. Poole, ed. The birds of North America online, No. 390. Cornell Lab of Ornith. [accessed November 2011].

Sedgwick, J. A. 1987. Avian habitat relationships in pinyon-juniper woodlands. Wilson Bull. 99: 413–31.

Sibley, C. G., and D. A. West. 1959. Hybridization in the Rufous-sided Towhees of the Great Plains. Auk 76: 326–38.

Smyth, M., and G. A. Bartholomew. 1966. The water economy of the Black-throated Sparrow and the Rock Wren. Condor 68: 447–58.

Tenney, C. R. 1997. Black-chinned Sparrow *(Spizella atrogularis)*. In A. Poole, ed. The birds of North America online, No. 270. Cornell Lab of Ornith. [accessed November 2011].

Unitt, P. 2004. San Diego County bird atlas. San Diego Natural History Museum, San Diego, CA.

———. 2008. Grasshopper Sparrow *(Ammodramus savannarum)*. In W. D. Shuford and T. Gardali, eds. California bird species of special concern. Studies of West. Birds 1: 393–99.

Vickery, P. D. 1996. Grasshopper Sparrow *(Ammodramus savannarum)*. In A. Poole, ed. The birds of North America online, No. 239. Cornell Lab of Ornith. [accessed 2011].

Weckstein, J. D., D. E. Kroodsma, and R. C. Faucett. 2002. Fox Sparrow *(Passerella iliaca)*. In A. Poole, ed. The birds of North America online, No. 715. Cornell Lab of Ornith. [accessed November 2011].

Wheelwright, N. T., and J. D. Rising. 2008. Savannah Sparrow *(Passerculus sandwichensis)*. In A. Poole, ed. The birds of North America online, No. 45. Cornell Lab of Ornith. [accessed November 2011].

Wolf, L. L. 1977. Species relationships in the avian genus *Aimophila*. Ornithol. Monogr. 23, Allen Press, Lawrence, KS.

GROSBEAKS AND RELATIVES

Greene, E., V. R. Muehter, and W. Davison. 1996. Lazuli Bunting *(Passerina amoena)*. In A. Poole, ed. The birds of North America online, No. 232. Cornell Lab of Ornith. [accessed December 2011].

Hudon, J. 1999. Western Tanager *(Piranga ludoviciana)*. In A. Poole, ed. The birds of North America online, No. 432. Cornell Lab of Ornith. [accessed December 2011].

Lowther, P. E., and J. L. Ingold. 2011. Blue Grosbeak *(Passerina caerulea)*. In A. Poole, ed. The birds of North America online, No. 79. Cornell Lab of Ornith. [accessed December 2011].

Ortega, C., and G. E. Hill. 2010. Black-headed Grosbeak *(Pheucticus melanocephalus)*. In A. Poole, ed. The birds of North America online, No. 143. Cornell Lab of Ornith. [accessed December 2011].

Pandolfino, E. R., C. Risser, and L. Risser. 2010. First evidence suggesting hybridization between Summer and Western Tanagers. West. Birds 41: 181–83.

Payne, R. B. 2006. Indigo Bunting *(Passerina cyanea)*. In A. Poole, ed. The birds of North America online, No. 4. Cornell Lab of Ornith. [accessed December 2011].

Unitt, P. 2008. Summer Tanager *(Piranga rubra)*. In W. D. Shuford and T. Gardali, eds. California bird species of special concern. Studies of West. Birds 1: 359–64.

Wyatt, V. E., and C. M. Francis. 2002. Rose-breasted Grosbeak *(Pheucticus ludovicianus)*. In A. Poole, ed. The birds of North America online, No. 692. Cornell Lab of Ornith. [accessed March 2012].

BLACKBIRDS AND RELATIVES

Ammon, E. M., and J. Woods. 2008. Status of Tricolored Blackbirds in Nevada. Great Basin Birds 10: 63–66.

Beedy, E. C. 2008. Tricolored Blackbird *(Agelaius tricolor)*. In W. D. Shuford and T. Gardali, eds. California bird species of special concern. Studies of West. Birds 1: 437–43.

Beedy, E. C., and W. J. Hamilton III. 1999. Tricolored Blackbird *(Agelaius tricolor)*. In A. Poole, ed. The birds of North America online, No. 423. Cornell Lab of Ornith. [accessed February 2012].

Davis, S. K., and W. E. Lanyon. 2008. Western Meadowlark *(Sturnella neglecta)*. In A. Poole, ed. The birds of North America online, No. 104. Cornell Lab of Ornith. [accessed February 2012].

Flood, N. J. 2002. Scott's Oriole *(Icterus parisorum)*. In A. Poole, ed. The birds of North America online, No. 608. Cornell Lab of Ornith. [accessed February 2012].

Gallion, T. 2008. Kern Red-winged Blackbird *(Agelaius phoeniceus aciculatus)*. In W. D. Shuford and T. Gardali, eds. California bird species of special concern. Studies of West. Birds 1: 432–36.

Hamilton III, W. J. 1998. Tricolored Blackbird itinerant breeding in California. Condor 100: 218–26.

Heath, S. K., L. A. Culp, and C. A. Howell. 2010. Brood parasitism and nest survival of Brown-headed Cowbird hosts at high-elevation sites in the eastern Sierra Nevada, California. West. N. Amer. Nat. 70: 364–76.

Jaramillo, A. 2008. Yellow-headed Blackbird *(Xanthocephalus xanthocephalus)*. In W. D. Shuford and T. Gardali, eds. California bird species of special concern. Studies of West. Birds 1: 444–50.

Jaramillo, A., and P. Burke. 1999. New World blackbirds: The Icterids. Princeton Univ. Press, Princeton, NJ.

Johnson, K., and B. D. Peer. 2001. Great-tailed Grackle *(Quiscalus mexicanus)*. In A. Poole, ed. The birds of North America online, No. 576. Cornell Lab of Ornith. [accessed February 2012].

Kyle, K. 2011. Results of the 2011 Tricolored Blackbird statewide survey. Prepared by Audubon California. Available on the Tricolored Blackbird Portal at http://tricolor.ice.ucdavis.edu/files/trbl/Results of the 2011 Statewide Survey.pdf. [accessed February 2011].

Lowther, P. E. 1993. Brown-headed Cowbird *(Molothrus ater)*. In A. Poole, ed. The birds of North America online, No. 47. Cornell Lab of Ornith. [accessed February 2012].

Martin, S. G. 2002. Brewer's Blackbird *(Euphagus cyanocephalus)*. In A. Poole, ed. The birds of North America online, No. 616. Cornell Lab of Ornith. [accessed February 2012].

Meese, R. J. 2012. Cattle egret predation causing reproductive failures of nesting Tricolored Blackbirds. Calif. Fish and Game 98: 47–50.

Neff, J. A. 1937. Nesting distribution of the Tri-colored Red-wing. Condor 39: 61–81.

Orians, G. H. 1960. Autumnal breeding in the Tricolored Blackbird. Auk 77: 379–98.

———. 1985. Blackbirds of the Americas. Univ. of Washington Press, Seattle, WA.

Orians, G. H., and G. M. Christman. 1968. A comparative study of the behavior of Red-winged, Tricolored, and Yellow-headed blackbirds. Univ. of Calif. Press, Los Angeles, CA.

Pleasants, B. Y., and D. J. Albano. 2001. Hooded Oriole *(Icterus cucullatus)*. In A. Poole, ed. The birds of North America online, No. 568. Cornell Lab of Ornith. [accessed February 2012].

Rising, J. D., and P. L. Williams. 1999. Bullock's Oriole *(Icterus bullockii)*. In A. Poole, ed.The birds of North America online, No. 416. Cornell Lab of Ornith. [accessed February 2012].

Rothstein, S. I., J. Verner, and E. Stevens. 1984. Radio-tracking reveals a unique diurnal pattern of spatial occurrence in the parasitic Brown-Headed Cowbird. Ecology 65: 77–88.

Stallcup, R. 2004. Late nesting Tricolored Blackbirds in western Marin County. Central Valley Bird Club Bull. 7: 51–52.

Twedt, D. J., and R. D. Crawford. 1995. Yellow-headed Blackbird *(Xanthocephalus xanthocephalus)*. In A. Poole, ed. The birds of North America online, No. 192. Cornell Lab of Ornith. [accessed February 2012].

Yasukawa, K., and W. A. Searcy. 1995. Red-winged Blackbird *(Agelaius phoeniceus)*. In A. Poole, ed. The birds of North America online, No. 184. Cornell Lab of Ornith. [accessed February 2012].

FINCHES AND RELATIVES

Adkisson, C. S. 1996. Red Crossbill *(Loxia curvirostra)*. In A. Poole, ed. The birds of North America online, No. 256. Cornell Lab of Ornith. [accessed February 2012].

Adkisson, C. S. 1999. Pine Grosbeak *(Pinicola enucleator)*. In A. Poole, ed. The birds of North America online, No. 456. Cornell Lab of Ornith. [accessed February 2012].

Airola D. A. 1981. Recent colonization of Lassen Peak, California, by the Gray-Crowned Rosy Finch *(Leucosticte tephrocotis)*. West. Birds. 12: 117–24.

Benkman, C. W. 1999. The selection mosaic and diversifying coevolution between crossbills and lodgepole pine. Am. Nat. 153: 75–31.

———. 2007. Red Crossbill types in Colorado: Their ecology, evolution, and distribution. Colo. Birds 41: 153–63.

Davis, J. N. 1999. Lawrence's Goldfinch *(Spinus lawrencei)*. In A. Poole, ed. The birds of North America online, No. 480. Cornell Lab of Ornith. [accessed February 2012].

Davis, J. N., E. Pandolfino, S. C. Rottenborn, and M. M. Rogers. 2011. Northern California: Fall migration. North American Birds 65: 156–61.

Dawson, W. R. 1997. Pine Siskin *(Spinus pinus)*. In A. Poole, ed. The birds of North America online, No. 280. Cornell Lab of Ornith. [accessed February 2012].

Gillihan, S. W., and B. Byers. 2001. Evening Grosbeak *(Coccothraustes vespertinus)*. In A. Poole, ed. The birds of North America online, No. 599. Cornell Lab of Ornith. [accessed February 2012].

Hahn, T. P. 1996. Cassin's Finch *(Carpodacus cassinii)*. In A. Poole, ed. The birds of North America online, No. 240. Cornell Lab of Ornith. [accessed February 2012].

Hill, G. E. 1993. House Finch *(Carpodacus mexicanus)*. In A. Poole, ed. The birds of North America online, No. 46. Cornell Lab of Ornith. [accessed February 2012}.

Irwin, K. 2010. A new and cryptic call type of the Red Crossbill. West. Birds 41: 10–25.

Kelsey, T. R. 2008. Biogeography, foraging ecology, and population dynamics of Red Crossbills in North America. PhD dissertation, Univ. of Calif., Davis.

Macdougall-Shackleton, S. A., R. E. Johnson, and T. P. Hahn. 2000. Gray-crowned Rosy-Finch *(Leucosticte tephrocotis)*. In A. Poole, ed. The birds of North America online, No. 559. Cornell Lab of Ornith. [accessed February 2012].

McGraw, K. J., and A. L. Middleton. 2009. American Goldfinch *(Spinus tristis)*. In A. Poole, ed. The birds of North America online, No. 80. Cornell Lab of Ornith. [accessed February 2012].

Salt, G. W. 1952. The relation of metabolism to climate and distribution in three finches in the genus *Carpodacus*. Ecol. Monogr. 22: 121–52.

Sewall, K., R. Kelsey, and T. P. Hahn. 2004. Discrete variants of Evening Grosbeak flight calls. Condor 106: 161–65.

Watt, D. J., and E. J. Willoughby. 1999. Lesser Goldfinch *(Spinus psaltria)*. In A. Poole, ed. The birds of North America online, No. 392. Cornell Lab of Ornith. [accessed February 2012].

Wootton, J. T. 1996. Purple Finch *(Carpodacus purpureus)*. In A. Poole, ed. The birds of North America online, No. 208. Cornell Lab of Ornith. [accessed February 2012].

Young, M. 2008. Introduction to differences in crossbill vocalizations. Available at http://ebird.org/content/ebird/news/introduction-to%20 crossbill-vocalizations. [accessed February 2012].

OLD WORLD SPARROWS

Lowther, P. E., and C. L. Cink. 2006. House Sparrow *(Passer domesticus)*. In A. Poole, ed. The birds of North America online, No. 12. Cornell Lab of Ornith. [accessed February 2012].

RARE, CASUAL, AND ACCIDENTAL BIRDS OF THE SIERRA

Dark, S. J., R. J. Gutierrez, and G. I. Gould Jr. 1998. The Barred Owl *(Strix varia)* invasion in California. Auk 115: 50–56.

Hamilton, R. A., M. A. Patten, and R.A. Erickson. 2007. Rare birds of California.West. Field Ornithologists, Camarillo, CA.

Heindel, M. 2000. Birds of East Kern County. Available at http://fog.ccsf.org/~jmorlan/eastkern .pdf. [accessed July 2011].

McCaskie, G., P. DeBenedictis, R. Erickson, and J. Morlan. 1979. Birds of northern California. 2nd edition. Golden Gate Audubon Society, San Francisco, CA.

Seamans, M. E., J. Corcoran, and A. Rex. 2004. Southernmost record of a Spotted Owl × Barred Owl hybrid in the Sierra Nevada. West. Birds 35: 173–74.

Snowden, J., and T. Manolis. 2001. Annotated bird list for Butte County, California. Altacal Audubon Society, Chico, CA.

Steger, G. N., L. R. Werner, and T. E. Munton. 2006. First documented record of the Barred Owl in the southern Sierra Nevada. West. Birds 37: 106–9.

OTHER NATURAL HISTORY INFORMATION

Barbour, M., and J. Major, eds. 1988. Terrestrial vegetation of California. 2nd edition. Special Publication No. 9. Calif. Native Plant Society, Sacramento, CA.

Block, W. M., M. L. Morrison, and J. Verner. 1990. Wildlife and oak-woodland interdependency. Fremontia 18(3): 72–76.

Griffin, J. R., and W. B. Critchfield. 1976. The distribution of forest trees in California. Research paper PSW-82. USDA, For. Ser., Pacific Southwest Research Station, Berkeley, CA.

Hickman, J. C. 1993. The Jepson Manual: Higher plants of California. Univ. of Calif. Press, Berkeley, CA.

Hunter, J. C., E. C. Beedy, V. Machek, and T. Sinnot. 2011. Sierra Cascade Foothill Area Conservation Report. Prepared for the Sierra Cascade Land Trust Council. Available at http://www .sierracascadelandtrustcouncil.org/foothills-area -conservation-report/. [accessed July 2011].

Laws, J. M. 2007. The Laws field guide to the Sierra Nevada. Heyday Books, Berkeley, CA.

Mayer, K. E., and W. F. Laudenslayer Jr., eds. 1988. A guide to wildlife habitats of California. USDA For. Ser. Pacific Southwest Forest and Range Experiment Station, Calif. Dept. of Fish and Game, Pacific Gas and Electric Company, and USDA For. Ser. Region 5, Sacramento, CA.

Muir, J. 1894. The mountains of California. The Century Company, New York, NY.

Munz, P. A. 1959. A California flora. Univ. of Calif. Press, Berkeley, CA.

———. 1968. Supplement to "A California Flora." Univ. of Calif. Press, Berkeley, CA.Sawyer, J. O., T. Keeler-Wolf, and J. M. Evens. 2009. A manual of California vegetation. 2nd edition. Native Plant Society, Sacramento, CA.

Storer, T., R. Usinger, and D. Lukas. 2004. Sierra Nevada natural history. Univ. of Calif. Press, Berkeley, CA.

Whitney, S. 1979. A Sierra Club naturalist's guide to the Sierra Nevada. Sierra Club Books, San Francisco, CA.

INDEX OF COMMON NAMES

Page numbers in **boldface type** *indicate main family and species accounts.*

INDEX OF SCIENTIFIC NAMES

ABOUT THE AUTHORS AND ARTIST

EDWARD C. (TED) BEEDY Ted first encountered the Sierra Nevada at the age of three months, when his parents brought him to their family cabin near Donner Summit in 1950. They reported that he has always been a "problem birdwatcher," as he required physical restraint as a three-year-old (i.e., a harness and rope) to keep him from following birds, chipmunks, or ant swarms off into the deep woods. He later earned his bird-watching merit badge in Cub Scouts and was hooked for life. Ted received his BSc (zoology and English), MSc (zoology), and PhD (zoology) from the University of California–Davis. His master's thesis concerned birds of the Donner Summit region, and his dissertation research focused on conifer forest bird communities of Yosemite. Ted taught ornithology and other biology classes at UCD and California State University–Sacramento before starting his career in 1985 as an environmental consultant at Jones & Stokes in Sacramento. During the more than 20 years he worked at J&S, Ted conducted long-term studies of Tricolored Blackbirds as well as sensitive bird species at Mono Lake, Kern River Valley, and elsewhere in the Sierra. In 2006 he started his own firm, Beedy Environmental Consulting, at his home in Nevada City, where he lives with his wife, two children, and a pond full of Mallards and Wood Ducks.

EDWARD R. (ED) PANDOLFINO Ed spent most of his early life on the move, living in many different states and countries and attending 13 different schools between first grade and high school. His first exposure to the Sierra came as a teenager— hunting, fishing, and backpacking but somehow remaining essentially unaware of the birds. After a checkered and inconsistent college experience that included dropping-out and touring Europe as a drummer for a rock and roll band, Ed finally settled down and earned a PhD in biochemistry at Washington State University. He spent over 20 years working in various management positions in the medical device industry. After an eyeball-to-eyeball encounter with a Spotted Towhee (then known as the Rufous-sided Towhee), Ed's relationship with birds transformed instantaneously from oblivious to obsessed. Since retiring in 1999, he has devoted his life to birds, working on habitat conservation and avian research. Ed is on the board of Sierra Foothills Audubon Society, president of Western Field Ornithologists, and one of the regional editors for *North American Birds* for Northern California. Over the past several years he has rediscovered his inner scientist, publishing more than two dozen papers on status and distribution of western birds. Ed lives in Sacramento with his wife and close to both their children.

KEITH HANSEN Keith has always felt at home in nature's embrace. His earliest memories are from Hawaii. Keith's Navy pilot father taught his sons observational skills, pointing out field marks on Navy planes. Keith's three older brothers kept lists of the aircraft, battleships, and other military hardware they identified. The Hansens, now with five boys, left the tropics for the lush woods of Maryland, where Keith entered first grade. Exploring the wonders of his naturally diverse neighborhood, Keith saw a Cedar Waxwing—a life-changing vision that would forever focus his attention on birds. Transfixed, he wanted more! The Hansen family, now five boys and one girl, moved to Fresno, California, where Keith's father opened a hobby shop. Art is a great part of Hansen life, with Keith's mother, Jan, teaching her kids the fundamentals of illustration. Keith began drawing in twelfth grade and has since illustrated for nature organizations, conservation groups, and Audubon Societies. He has prepared the art for a dozen books, including *Discovering Sierra Birds,* *Distributional Checklist of North American Birds,* and *More Tales of a Low-Rent Birder.* Keith has undertaken countless private art commissions as well as a 128-foot mural that the Hansen family created for the Fresno Art Museum. He and his artistically gifted wife, Patricia, have a tour company for birding and leading cultural trips to Mexico and Central America. Keith creates from his gallery in Bolinas, California, and is especially thankful for the Big Bang and the inspiration that nature provides.